Other Kaplan Books for College-Bound Students

College Admissions and Financial Aid

Straight Talk on Paying for College
Parent's Guide to College Admissions
The Unofficial, Unbiased Guide to the 328 Most Interesting Colleges

Test Preparation

ACT
AP Biology
AP Calculus AB: An Apex Learning Guide
AP English Literature and Composition
AP English Language and Composition
AP Macroeconomics/Microeconomics: An Apex Learning Guide
AP Physics B: An Apex Learning Guide
AP Statistics: An Apex Learning Guide
AP U.S. Government & Politics: An Apex Learning Guide
AP U.S. History: An Apex Learning Guide
SAT & PSAT
SAT Vocabulary Words Flip-O-Matic
SAT Math Workbook
SAT Verbal Workbook
SAT II: Biology E/M
SAT II: Chemistry
SAT II: Literature
SAT II: Mathematics
SAT II: Physics
SAT II: Spanish
SAT II: U.S. History
SAT II: Writing
SAT II: World History

AP* CHEMISTRY

2004 EDITION

**By Albert S. Tarendash
with Frederick J. Rowe
and the Staff of Kaplan**

Simon & Schuster

SYDNEY · LONDON · NEW YORK · TORONTO

Kaplan Publishing
Published by Simon & Schuster
1230 Avenue of the Americas
New York, New York 10020

For bulk sales to schools, colleges, and universities, please contact: Order Department, Simon and Schuster, 100 Front Street, Riverside, NJ 08075. Phone: (800) 223-2336. Fax: (800) 943-9831.

Contributing Editors: Seppy Basili and Jon Zeitlin
Project Editor: Charli Engelhorn
Chemistry Editor: Robert C. Atkins
Interior Design and Page Layout: Dave Chipps
Cover Design: Cheung Tai
Production Manager: Michael Shevlin
Managing Editor: Déa Alessandro
Executive Editor: Del Franz

The section on managing stress was adapted from *The Kaplan Advantage Stress Management System* by Dr. Ed Newman and Bob Verini, © 1996 by Kaplan, Inc.

Special thanks to Mark Anestis, Stephen Foley, Harvey W. Gendreau (of Framingham High School), and Megan Duffy.

Manufactured in the United States of America
Published simultaneously in Canada

January 2004
10 9 8 7 6 5 4 3 2

ISBN: 0-7432-4161-4

Table of Contents

ABOUT THE AUTHORS

Albert S. Tarendash is in his 37th year as a chemistry and physics instructor. In addition to teaching courses in high school and AP chemistry and physics, he has authored review texts in these subjects. The author has retired as Assistant Principal of Department of Chemistry and Physics at Stuyvesant High School in New York City, and is currently at The Frisch School in Paramus, New Jersey.

Frederick J. Rowe, who authored the practice questions, has a B.S. (City College of New York) and an M.S. (New Jersey Institute of Technology) degree in Chemical Engineering. In addition to teaching Regents Chemistry and AP Chemistry at Northport High School in Long Island, N.Y., Mr. Rowe has chaired the American Chemical Society Examinations Institute Advanced High School Chemistry Test Committee, producing tests widely used by teachers as an AP Chemistry Final Exam. Mr. Rowe participated several times as a reader/grader of the AP Chemistry test. He is a recipient of the Catalyst Award (Chemical Manufacturer's Association) and the Advanced Placement Recognition Award for his teaching. He enjoys spending time on his J/24 racing sailboat on Long Island Sound and speed walking with his Pomeranian pal, Peaches.

PREFACE

To the Student:

Someone once asked a real estate mogul what the three most important factors for selling real estate were. His famous answer: "LOCATION! LOCATION! LOCATION!"

What are the three most important factors for doing well (getting a 4 or 5) on the AP Chemistry exam? "PREPARATION! PREPARATION! PREPARATION!"

This book was written with that thought in mind. If you want to do really well on the AP exam, there is no substitute for being well prepared. Preparation begins when the school year starts and continues throughout the year. If you work steadily and consistently, the AP exam will pose no problem. I'm not trying to stand on a soapbox and preach to you, I'm just trying to make you aware of the facts: AP Chemistry is not an easy subject; there will be many concepts and skills to master by the time your AP course comes to a conclusion in May.

Why use this review book rather than those published by the competition? A number of other review books take the "quick fix" approach to the AP exam: They are short on conceptual material and long on strategies for beating The College Board at its own game. In contrast, this book develops each area of chemistry in considerable depth, but without the wordiness (I hope!) of traditional textbooks. Some of the material even goes *beyond* the items that will be tested in the exam. I chose to include this extra material because I felt that it would help you achieve a greater understanding of chemistry, a necessity if you expect to get that 4 or 5 on the exam.

The best time to buy this book is near the beginning of the school year, so you can use it along with your textbook and class notes. The sample problems within each chapter, the end-of-chapter exercises, and the full-length practice tests will also aid you in preparing for the real one in May. A word of advice, however: The College Board spends gazillions of dollars in test development, and the quality of their questions on actual exams reflects their efforts. Part of your preparation MUST include answering as many *authentic* AP exam questions as you can beg, borrow, or steal from your friends, your teachers, or the Internet. There is absolutely no substitute for this crucial activity.

If you have any questions or comments, you can email me at MrT@tarendash.net, or you can write to me in care of the publisher.

Best wishes for success on the AP Chemistry exam.

Albert S. Tarendash

This book is dedicated to:

My wife Bea, who supported my efforts in spite of her misgivings about the writing of this book,
my children, Franci and Jeff, David and Janet, and my grandchildren Brittany,
Alexa, Danielle, and Raina, who are just too much for words.

Albert S. Tarendash

THE BASICS

Section I

Introduction to AP Chemistry

Before you plunge into studying for the AP Chemistry exam, let's take a step back and look at the big picture. What is the AP Chemistry exam all about? How can you prepare for it? How is it scored? This chapter and the next will answer these questions and more.

THE AP CHEMISTRY COURSE

The AP Chemistry course is the equivalent of a first-year general chemistry course given at the college level. It has both lecture and laboratory components that are typical of college chemistry courses. It is highly recommended that a high school chemistry course be taken prior to the AP course. Studies conducted by The College Board indicate that the likelihood of achieving a good grade on the AP examination (3 or better) is considerably greater for those students who have completed a high school chemistry course.

The topics covered in the AP Chemistry examination, which we will review in this book, include the structure of matter, states of matter, chemical reactions, and descriptive chemistry. The test also includes some questions based on experiences and skills students acquire in the laboratory.

THE AP CHEMISTRY EXAMINATION

The AP Chemistry exam is divided into two sections.

Section I consists of 75 multiple-choice questions and constitutes 45 percent of the final grade. Each question is followed by five choices labeled A through E. The time allotted to this section is 90 minutes. Calculators can *not* be used on this section and only a "bare bones" Periodic Table (atomic number, symbol, and atomic mass) is supplied to the student.

Section II of the examination is the free response section, which consists of eight questions. All of the questions have multiple parts. This section constitutes 55 percent of the examination and the time allotted to it is 90 minutes. It is divided into the following parts with a total weighting of 100 points:

- **Part A.** The total time allotment for this part is 40 minutes. Students can use any type of calculator *except* ones having a QWERTY (typewriter) keyboard. In addition to the Periodic Table, students receive a table of Standard Reduction Potentials and a list of AP Chemistry Equations and Constants. This part contains three questions:

  Question 1 must be answered by *all* students and focuses principally on some type of equilibrium problem, although other related aspects may appear within the question. (20 points)

  Questions 2 and 3 are quantitative questions which draw from a variety of topics such as stoichiometry, kinetics, electrochemistry, and thermodynamics. Students must answer *only one* of the two questions. (20 points)

- **Part B.** The total time allotment for this part is 50 minutes. Students can *not* use a calculator, but they can continue to use the tables and lists used for Part A. This part contains the remaining five (5) questions:

  Question 4 consists of eight word-descriptions of reactions. The student must answer any five by writing the *net reaction* for the description. Reactions need not be balanced and phases need not be included. Students must *not* include any molecules or ions that are *unchanged* by the reaction. [See Appendix A for more information about Question 4.] (15 points)

  Questions 5 and 6 must *both* be answered by all students. One of the questions will be a laboratory-related topic and the other will focus on some aspect of chemistry deemed especially important to the examiners. The weighting for *each* question is 15 points. (30 points)

  Questions 7 and 8 will focus on other aspects of chemistry. Students must answer *only one* of the two questions. (15 points)

SUMMARY OF PREVIOUS AP CHEMISTRY EXAMS

As the saying goes, "to be forewarned is to be forearmed." One of the ingredients for success on the AP test is to be able to anticipate which topics are likely to be selected for the examination. In order to aid you in this area, the following summaries have been compiled for previous examinations.

KAPLAN

Summary of Part A of Previous AP Chemistry Exams

YEAR	QUESTION (1)	QUESTIONS (2) AND (3)
2003	Acids/Bases	(2) Gas laws (3) Kinetics; Electrochemistry
2002	Acid-base equilibria; Titration	(2) Electrochemistry buffers; Structure and acid strength (3) Stoichiometry; Thermochemistry; Organic isomers
2001	Solution equilibrium, K_{sp}	(2) Thermodynamics; Bond energies (3) Stoichiometry; Acid-base titration; pH
1997	Acid/Base	(2) Electochemistry (3) Kinetics
1996	Acid/Base	(2) Thermodynamics; Bond energies (3) Solution concentrations
1995	Gas equilibrium	(2) Gas stoichiometry; Heat of reaction (3) Decomposition of solid carbonates (stoichiometry)
1994	Precipitation	(2) Rate law, mechanisms (3) Gas laws
1993	Acid/Base	(2) Empirical formula; Colligative properties (3) Oxidation-reduction titration
1992	Gas equilibrium	(2) Electrochemistry, Thermodynamics (3) Thermodynamics, Enthalpy, and Entropy
1991	Acid/Base; Buffer	(2) Empirical formula; Freezing point lowering (3) Rate law, mechanisms
1990	Precipitation	(2) Gas laws; Stoichiometry (3) Thermodynamics
1989	Acid/Base	(2) Electrolysis (3) Thermodynamics
1988	Gas equilibrium	(2) Thermodynamics (3) Electrolysis
1987	Acid/Base/Precipitation	(2) Rate laws (3) Acid-base titration
1986	Acid/Base	(2) Redox/Electrolysis (3) Empirical formula
1985	Precipitation	(2) Redox/Electrolysis (3) Freezing point lowering; Empirical formula
1984	Acid/Base	(2) Rates (3) Thermodynamics

Summary of Part B of Previous AP Chemistry Exams

YEAR	QUESTIONS 5–8* †
2003	(5) Lab: Spectrophotometry (6) General Chemical Principles (7) Thermodynamics (8) Organic Chemistry
2002	(5) Lab: Heat of neutralization (6) Atomic Structure—Periodic properties (7) Kinetics—Reaction mechanisms (8) Thermodynamics and equilibrium
2001	(5) Lab: Solution properties (6) Reaction types; Kinetics—Reaction rate and order (7) Electrochemical cells (8) Intermolecular forces
2000	(5) Lab: Molar mass by freezing point depression (6) Kinetics—Reaction rate and order (7) Isotopes; Electron configuration; ionization energy; molecular structure (8) Acid-base titration, indicators
1999	(5) Lab: Gas collection (6) Thermodynamics (7) Properties of solutions (8) Bonding and molecular structure
1998	(5) Acid-base titration curves (6) Reaction kinetics; reaction orders (7) LeChâtelier's principle (8) Electrochemical cells (9) Altitude and boiling point; Copper-ammonia complex; molecular polarity; redox agents
1997	(5) Molecular geometry and Lewis structures, polarity, Group V fluorides (6) Atomic/molecular structure related to ionization energies and radii (7) Thermodynamics, $\Delta S°$, $\Delta G°$, LeChâtelier's principle (8) Nuclear decay process, mass defect, particle properties (9) Lab process in determination of mass percent of sulfate in an unknown

1996	(5) Gases, kinetic molecular theory, effusion (6) Titration of acid, effects of lab errors (7) K_{eq}, Cell potentials and changes (8) Kinetics, rate law, rate constant (9) Electronic structure and bonding for differences in boiling point, polarity, bond length, and Group 16 fluorides
1995	(5) Conductivity explanation based on chemical bonding and/or atomic or molecular structure (6) Phase diagram explanation (7) Explanation in terms of electronic structure and bonding (8) Solubility, thermodynamics explanation (9) Chemical reaction potential energy diagram explanation
1994	(5) Provide explanations for various physical and chemical phenomena (6) Thermodynamics—$\Delta S°$, $\Delta G°$, $\Delta H°$, and spontaneity (7) Acid-base titration curve (8) Various chemical principles (9) Atomic structure and bonding explanations
1993	(5) Explanation of various reactions of H_2SO_4 (6) Principles of atomic structure (7) Galvanic cells (8) Thermodynamics—$\Delta S°$, $\Delta G°$, $\Delta H°$, and spontaneity (9) Kinetic-molecular theory
1992	(5) Rate law, LeChâtelier's principle, enthalpy profiles, kinetic-molecular theory (6) Buffer systems (7) Identification of substances (8) Explanations of physical properties of substances (9) Lewis structures, bond angles, and hybridization
1991	(5) Thermodynamics (6) Lab: Molar mass of liquid by vapor density method (7) Electrolysis (8) Explanations of physical properties of substances (9) Nuclear chemistry
1990	(5) Molecular structure and models of chemical bonding (6) Ionization energy differences explained by atomic structure (7) Factors that affect reaction rates (8) Explaining the strength of acids (9) Lab: Empirical formula

1989	(5) Lewis structures and the VSEPR model (6) Melting point differences as explained by bonding principles (7) Descriptive chemistry—Identification of three metals by chemical tests (8) Factors affecting reaction rates (9) Nuclear chemistry
1988	(5) Explaining physical properties based on bonding and intermolecular forces (6) LeChâtelier's principle (7) Acid/base titration (8) Phase diagram (9) Lab: Determining enthalpy of neutralization
1987	(5) Explanation of periodic properties based on atomic theory (6) Electrolysis (7) Explanation of ionization of salts in water (8) Thermodynamics (9) Heisenberg's uncertainty principle; Bohr's theory of the hydrogen atom
1986	(5) Factors affecting the heat of formation (6) Rate law and reaction mechanisms (7) Strength of oxoacids (8) Various chemical principles (9) Activity series of metals
1985	(5) Periodic properties explained by atomic structure (6) Thermodynamics—Prediction of enthalpy, entropy, and free energy changes (7) Lab: Preparation of salts (8) Reaction rates (9) Melting point trends explained by bonding and intermolecular forces
1984	(5) Acid-base titrations, indicators (6) Van der Waals equation and van der Waals constants (7) Physical differences between metals and nonmetals

* Between 1985 and 1998, Part B consisted of 9 questions.

† Descriptions of Question 4 have been omitted; see Appendix A.

Summary of Multiple-Choice Questions on Prior Examinations

1999 Multiple Choice Section (75 Questions)

TOPIC	# OF QUESTIONS	% OF EXAM
Stoichiometry	5	6.7
Atomic Theory	4	5.3
Bonding/Intermolecular Forces	9	12.0
Periodic Properties	5	6.7
Phases	9	12.0
Solutions/Solubility	10	13.3
Kinetics/Equilibrium	7	9.3
Acid/Base	8	10.7
Thermodynamics	4	5.3
Redox/Electrochemistry	5	6.7
Organic	1	1.3
Nuclear	1	1.3
Laboratory	7	9.3

1994

TOPIC	# OF QUESTIONS	% OF EXAM
Stoichiometry	6	8.0
Gas Laws/Kinetic Theory	5	6.7
Atomic Theory	6	8.0
Periodic Properties	9	12.0
Solutions/Phases	9	12.0
Rates and Equilibrium	6	8.0
Precipitation	1	1.3
Acid/Base/Buffers	8	10.7
Redox/Electrochemistry	5	6.7
General	4	5.3
Thermodynamics	4	5.3
Qualitative/Descriptive	4	5.3
Nuclear	2	2.7
Organic	1	1.3
Laboratory	5	6.7

1989

TOPIC	# OF QUESTIONS	% OF EXAM
Stoichiometry	7	9.3
Gas Laws/Kinetic Theory	4	5.3
Atomic Theory	5	5.3
Bonding/Intermolecular Forces	8	10.7
Periodic Properties	3	4.0
Solutions/Phase Diagrams	10	13.3
Rates and Equilibrium	4	5.3
Precipitation	2	2.7
Acid/Base/Buffers	10	13.3
Redox/Electrochemistry	5	6.7
General	5	6.7
Thermodynamics	3	4.0
Qualitative	3	4.0
Reactions	1	1.3
Nuclear	3	4.0
Laboratory	2	2.7

SCORING THE AP CHEMISTRY EXAM

After Sections I and II have been scored, the scores are converted to a final grade that is based on a five-point system:

GRADE	DESCRIPTION	APPROXIMATE GRADE DISTRIBUTION	
		2002 EXAM	**2003 EXAM**
5	Extremely Well Qualified	15%	15%
4	Well Qualified	18%	18%
3	Qualified	24%	24%
2	Possibly Qualified	18%	19%
1	No Recommendation	24%	24%
	Mean Grade	**2.8**	**2.8**

While the distribution of grades and the mean grade will change slightly from year to year, they will be close to the figures given above. Therefore, approximately 60 percent of the students taking the examination will achieve a grade of 3 or better.

The maximum possible score on the May 2002 examination was 160 points. The range of points needed for each AP grade was as follows:

GRADE	RANGE OF POINTS	APPROXIMATE % OF MAXIMUM
5	107–160	67–100
4	106–85	53–66
3	84–61	38–52
2	60–42	26–38
1	41–0	0–26

As you can see from the table, achieving a grade of 3 or better, while not easy, is not as difficult as one would expect. This is the main reason why adequate preparation is so important.

How Do I Get My Grade?

AP Grade Reports are sent in July to each student's home, high school, and any colleges designated by the student. Students may designate the colleges they would like to receive their grade on the answer sheet at the time of the test. Students may also contact AP Services to forward their grade to other colleges after the exam, or to cancel or withhold a grade.

AP Grades by Phone

AP Grades by phone are available for $15 a call beginning in early July. A touch-tone phone is needed. The toll-free number is (888) 308-0013.

REGISTRATION

To register for the AP Chemistry exam, contact your school guidance counselor or AP Coordinator. If your school does not administer the exam, contact AP Services for a listing of schools in your area that do.

FEES

At press time, the fee for each AP exam is $82. The College Board offers a $22 credit to qualified students with acute financial need. A portion of the exam fee may be refunded if a student does not take the test. There is a $20 late fee for late exam orders. Check with AP Services for application deadlines.

ADDITIONAL RESOURCES

The College Board offers a number of publications about the Advanced Placement Program, including: *AP Course Description for Chemistry*, *AP Program Guide*, and *AP Bulletin for Students and Parents*. You can order these and other publications online at store.collegeboard.com and at www.collegeboard.com/ap/library/; or you can call AP Order Services at (609) 771-7243.

FOR MORE INFORMATION

For more information about the AP Program and/or the AP Chemistry examination, contact your school's AP Coordinator or guidance counselor, or contact AP Services at:

AP Services
P.O. Box 6671
Princeton, NJ 08541-6671
Phone: (609) 771-7300; (888) CALL-4-AP
TTY: (609) 882-4118
Fax: (609) 530-0482
Email: apexams@ets.org
Website: www.collegeboard.com/ap/students/

Strategies for Succeeding on the AP Chemistry Exam

The AP Chemistry exam is a comprehensive examination that covers one year of college-level work. No matter what all of the other review books say, there is no "quick fix" for achieving a high grade. (Sorry about that!) While an intensive end-of-year review is very helpful, last minute "cramming" is counterproductive because it will increase your level of anxiety before the examination. The best solution is continuous, yearlong study. Nevertheless, here are some pointers for improving your performance on the examination.

Before the Examination

- Use this book in conjunction with your notes and textbook.

- Answer as many **authentic** AP Chemistry test questions as you possibly can. Your teacher and The College Board website (www.collegeboard.com/ap/students/chemistry/) are the best sources for obtaining these exams. The questions contained in *all* review books, including this one, have not undergone professional test development and, consequently, they can only approximate what you will see on the real exam.

- Get a good night's rest before the examination and be certain that you eat a decent meal.

- Arrive at school well before the examination begins so that you will not rush to complete all of the pre-examination paper work.

- Be certain that you remember to bring all of the necessary materials to the examination: pens, pencils (with erasers), a calculator with fresh batteries, and a watch. If necessary bring a photo ID, your secondary school code, and your social security number. If your school allows it, bring some light mid-examination snacks.

- Do NOT bring any cell phones or pagers into the examination room.

Section I Strategies

Pace Yourself

Time management is crucial! On Section I (multiple-choice) you have 90 minutes to answer 75 questions. This translates to slightly more than one minute per question. If you are clueless about a particular question, skip it and move on—you will be able to return to it later.

Guess Intelligently and with Caution

Should you guess on Section I questions? The College Board levies a one-quarter point penalty for each incorrect answer. If can eliminate one or two obviously incorrect choices, there is a statistical advantage to guessing. When you come across such questions, ask yourself: "Which choices are obviously incorrect?"

Section II Strategies

Keep Track of Time

As with Section I, the key to tackling Section II is to pace yourself. The time you spend on a question should be related to the time available for the entire part (Part A, 40 minutes; Part B, 50 minutes) and the point value of that question. For example, you should plan to spend approximately:

- 20 minutes each on Questions 1 and 2 or 3
- 12 minutes each on Questions 4–6, and 7 or 8

If you follow these guidelines you will not run out of time!

Be Familiar with the Information Contained on the Reference Tables

This does not mean that you should try and memorize them, rather, you should know where you can locate a given equation or the value of a particular constant.

Answer Each Question as Directly and as Succinctly as You Possibly Can

The AP Readers score *thousands* of examination papers and they can spot "padding" immediately. As your paper is read, it will be scanned by the Reader for obviously correct phrases, diagrams, or graphs. Readers are human, and if a Reader has to wade through a morass of extraneous junk, he or she might fail to notice that you have actually answered the question correctly.

Read the Question Twice Before You Begin to Organize Your Answer

Be sure you're clear on what you're being asked to do. All free-response questions will contain one or more words that will tell you *how* the question is to be answered: *arrange, calculate, compare or contrast, explain, predict,* . . . In your exam book, list the relevant facts and principles you need to answer the question, and then write your response in clear, coherent

English. Add any labeled diagrams or equations that you deem necessary. Again, be as brief as you possibly can without neglecting important details.

Use Scratch Work to Your Advantage

In performing calculations, do some scratch work on the green insert before you transfer it to your examination book. If you use an equation, be certain to show substitution *with units* and be certain to "box" your answers. *Be certain to include units in your answers.* If you use your calculator and obtain an incorrect answer, the substituted equation will show the Reader that you knew what you were doing.

Pay Attention to Significant Figures

Your answer can deviate by ±1 significant figure, after which you will incur a one-point penalty. Generally, answers contain 2 or 3 significant figures; if you express *all* your answers to 3 significant figures, you should have no problem in this area.

THE BASICS OF STRESS MANAGEMENT

The countdown has begun. Your date with the test is looming on the horizon. Anxiety is on the rise. The butterflies in your stomach have gone ballistic. Your thinking is getting cloudy. Maybe you think you won't be ready. Maybe you already know your stuff, but you're going into panic mode anyway. Don't freak! It's possible to tame that anxiety and stress—before and during the test.

Remember, a little stress is good. Anxiety is a motivation to study. The adrenaline that gets pumped into your bloodstream when you're stressed helps you stay alert and think more clearly. But if you feel that the tension is so great that it's preventing you from using your study time effectively, here are some things you can do to get it under control.

Take Control

Lack of control is a prime cause of stress. Research shows that if you don't have a sense of control over what's happening in your life, you can easily end up feeling helpless and hopeless. Try to identify the sources of the stress you feel. Which ones can you do something about? Can you find ways to reduce the stress you're feeling about any of these sources?

Focus on Your Strengths

Make a list of areas of strength you have that will help you do well on the test. We all have strengths, and recognizing your own is like having reserves of solid gold at Fort Knox. You'll be able to draw on your reserves as you need them, helping you solve difficult questions, maintain confidence, and keep test stress and anxiety at a distance. And every time you recognize a new area of strength, solve a challenging problem, or score well on a practice test, you'll increase your reserves.

Imagine Yourself Succeeding

Close your eyes and imagine yourself in a relaxing situation. Breathe easily and naturally. Now, think of a real-life situation in which you scored well on a test or did well on an assignment. Focus on this success. Now turn your thoughts to the test, and keep your thoughts and feelings in line with that successful experience. Don't make comparisons between them; just imagine yourself taking the upcoming test with the same feelings of confidence and relaxed control.

Set Realistic Goals

Facing your problem areas gives you some distinct advantages. What do you want to accomplish in the time remaining? Make a list of realistic goals. You can't help feeling more confident when you know you're actively improving your chances of earning a higher test score.

Exercise Your Frustrations Away

Whether it's jogging, biking, pushups, or a pickup basketball game, physical exercise will stimulate your mind and body, and improve your ability to think and concentrate. A surprising number of students fall out of the habit of regular exercise, ironically because they're spending so much time prepping for exams. A little physical exertion will help to keep your mind and body in sync and sleep better at night.

Avoid Drugs

Using drugs (prescription or recreational) specifically to prepare for and take a big test is definitely self-defeating. (If they are illegal drugs, you may end up with a bigger problem on your hands than the AP Chemistry examination!) Mild stimulants, such as coffee or cola can sometimes help as you study, since they keep you alert. On the down side, too much of these can also lead to agitation, restlessness, and insomnia. It all depends on your tolerance for caffeine.

Eat Well

Good nutrition will help you focus and think clearly. Eat plenty of fruits and vegetables, low-fat protein such as fish, skinless poultry, beans, and legumes, and whole grains such as brown rice, whole wheat bread, and pastas. Don't eat a lot of sugar and high-fat snacks, or salty foods.

Work at Your Own Pace

Don't be thrown if other test takers seem to be working more furiously than you during the exam. Continue to spend your time patiently thinking through your answers; it will lead to better results. Don't mistake the other people's sheer activity as signs of progress and higher scores.

Keep Breathing

Conscious attention to breathing is an excellent way to manage stress while you're taking the test. Most of the people who get into trouble during tests take shallow breaths: They breathe using only their upper chests and shoulder muscles, and may even hold their breath for long periods of time. Conversely, those test takers who breathe deeply in a slow, relaxed manner are likely to be in better control during the session.

Stretch

If you find yourself getting spaced out or burned out as you're taking the test, stop for a brief moment and stretch. Even though you'll be pausing on the test for a moment, it's a moment well spent. Stretching will help to refresh you and refocus your thoughts.

CHEMISTRY REVIEW

Matter and Measurement

All sciences study matter and energy in some way. In particular, the science of **chemistry** focuses on:

- The structure and composition of matter
- The interactions among samples of matter
- The role played by energy in accomplishing these interactions

CLASSIFICATION OF MATTER

Substances

Matter is anything that has mass and volume. It is classified according to its properties and composition. A **substance** is a sample of matter whose composition cannot be varied. All substances are necessarily uniform in composition; that is, each sample of a given substance has a composition that is *identical* to every other sample of that substance. If a substance is composed of one type of atom, it cannot be decomposed by ordinary chemical methods. We call this type of matter an **element**. Iron, oxygen, and gold are three of the more familiar elements. The Alphabetical List of the Elements, found in Appendix B, lists the names and symbols of most of the known elements. If a substance is composed of more than one type of atom, it can be decomposed by ordinary chemical methods. For example, the substance known as hydrogen peroxide decomposes into water and oxygen in the presence of light. We call this type of substance a **compound**. All substances can be represented by **chemical formulas**. For example, the element iron is represented by the formula Fe; the element oxygen, by the formula O_2; and the compound hydrogen peroxide, by the formula H_2O_2. Chemical formulas will be studied in chapter 4.

Properties of Matter

We describe and distinguish substances by the properties they possess: Certain properties, such as *density* and *temperature*, do not depend on the amount of substance present: these are known as **intensive properties**. Other properties, such as *mass* and *volume*, do depend on the amount of substance present: these are known as **extensive properties**.

If a property can be observed or measured without changing the nature of the substance, it is known as a **physical property**. Examples of physical properties include *color*, *odor*, and *physical state*. If a property can only be observed or measured when accompanied by a change in the nature of the substance, it is called a **chemical property**. An example of a chemical property is *combustibility* (the ability of a compound to burn in air).

Changes within Matter

Changes that accompany the observation of physical properties are known as **physical changes**. The melting of ice is categorized as a physical change because the compound water has changed from its solid form to its liquid form. The essential nature of the water (i.e., its formula) has *not* changed. Changes that accompany the observation of chemical properties are known as **chemical changes**. When methane burns in air, it forms two new substances: water and carbon dioxide. Chemical changes can always be represented by **chemical reactions**, in which the nature of the change is specified explicitly. We will have many opportunities to explore chemical reactions throughout this book.

Mixtures

Samples of matter that are not substances are known as **mixtures**. Mixtures can be classified as **homogeneous** (uniform) or **heterogeneous** (nonuniform). A solution of sugar in water is an example of a homogeneous mixture; vegetable soup is an example of a heterogeneous mixture. Mixtures are collections of substances whose components can be separated by a variety of "physical" methods, including evaporation, filtration, distillation, and chromatography. These techniques are discussed briefly in the following paragraphs.

Evaporation

Evaporation is employed when one desires to recover a solid from a solid-liquid solution. For example, a salt-water solution can be separated by heating the solution in an **evaporating dish** until the water evaporates. This procedure allows the solid salt to be recovered from the original mixture. The diagram below illustrates how this technique is performed in the laboratory.

Evaporation

Filtration

Filtration is employed when one of the components of a mixture can be dissolved in a liquid while the other component(s) cannot. For example, if a mixture of sand and sugar is mixed with water, the sugar will be dissolved while the sand will remain undissolved. When the sand-sugar-water mixture is poured into a funnel containing **filter paper**, the sugar-water solution passes through the paper and is collected in a tube and is known as the **filtrate**. The sand residue remains on the filter paper. The diagram below illustrates this technique.

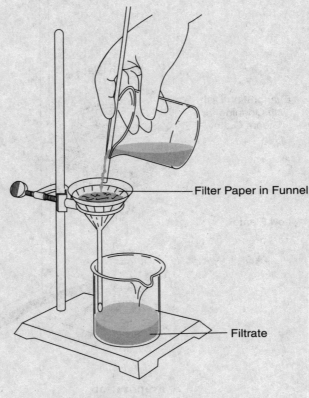

Filter Paper in Funnel

Filtrate

Filtering

Distillation

There are a number of types of distillation, but in the simplest, a *volatile* liquid such as water can be separated from a less volatile component such as salt. The mixture is heated to boiling and the water vapor is passed through a **condenser** that is cooled with water. As a result, the vapor is transformed back into a liquid and is collected as the **distillate**. The salt remains as a residue in the boiling flask. The diagram below illustrates simple distillation.

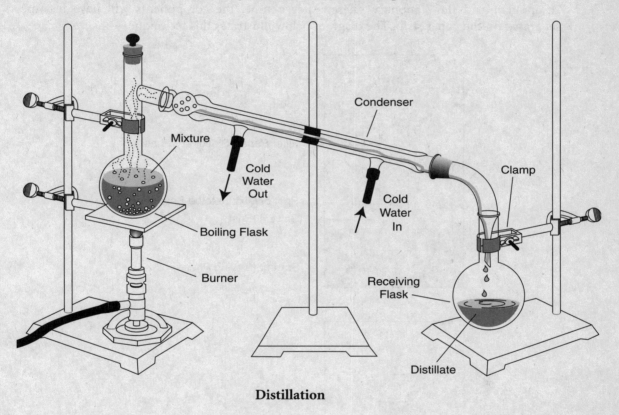

Distillation

Chromatography

There are a number of types of chromatography, but each has the following in common: A mixture is placed on some material such as a paper strip, which is known as the **stationary phase**. The strip is then immersed in a liquid, which is known as the **moving phase**. As the liquid moves along the paper strip, it carries the original mixture with it. Each component of the mixture will travel at a speed that depends on the strength of attraction between the component and the stationary phase. After a time, the components will have become separated on the paper strip. The diagram below illustrates this technique.

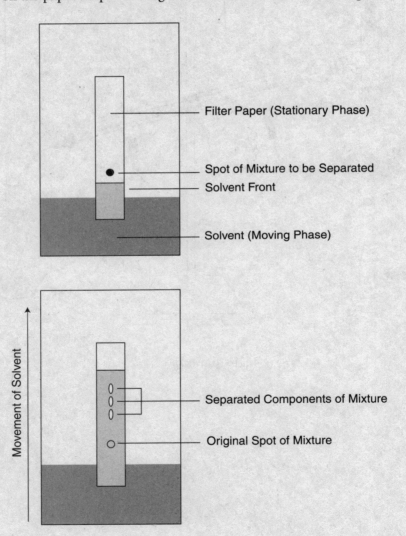

In summary, the diagram below is a flow chart for distinguishing among samples of matter.

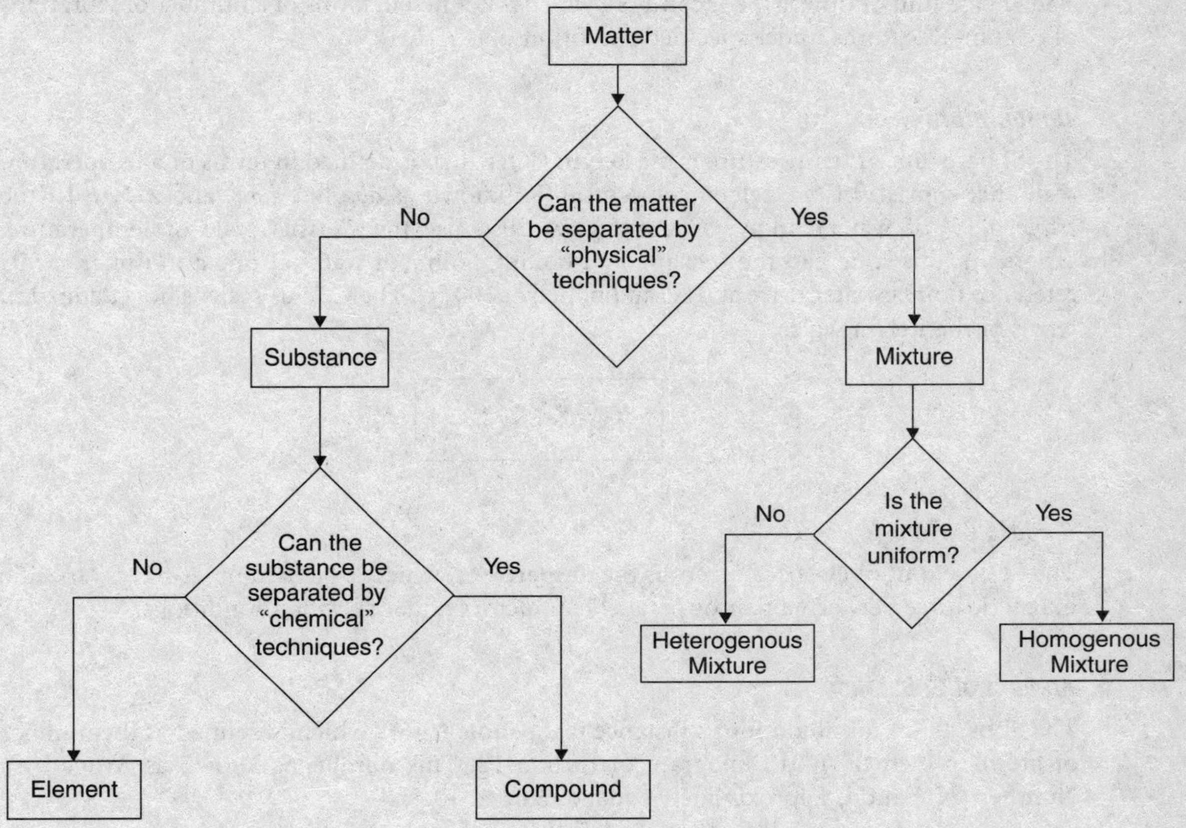

```
                          ┌──────────┐
                          │  Matter  │
                          └──────────┘
                               │
                         ◇ Can the matter ◇
              No        be separated by        Yes
         ┌──────────── "physical" ────────────┐
         │               techniques?           │
         ▼                                      ▼
   ┌───────────┐                          ┌───────────┐
   │ Substance │                          │  Mixture  │
   └───────────┘                          └───────────┘
         │                                      │
    ◇ Can the ◇                            ◇ Is the ◇
  No substance be  Yes                  No  mixture  Yes
 ┌── separated by ──┐                 ┌── uniform? ──┐
 │   "chemical"     │                 │              │
 │   techniques?    │                 ▼              ▼
 ▼                  ▼          ┌─────────────┐ ┌─────────────┐
┌─────────┐  ┌──────────┐     │ Heterogenous│ │ Homogenous  │
│ Element │  │ Compound │     │   Mixture   │ │   Mixture   │
└─────────┘  └──────────┘     └─────────────┘ └─────────────┘
```

MEASUREMENT AND THE METRIC SYSTEM

All sciences rely on the ability of experimenters to perform measurements. The basis for modern measurements is the *Systéme International* (SI). Seven *fundamental* quantities have been established by the SI. The SI *base units* for six of the seven quantities are described below:

Length

The SI base unit of length is the **meter** (m), which is defined as the distance traveled by light in space in 1/299,792,458 second. One meter is approximately 39 inches.

Mass

The SI base unit of mass is the **kilogram** (kg), which is defined as the mass of a platinum-iridium cylinder that is housed in France. On earth, one kilogram weighs approximately 2.2 pounds. (Note that the *mass* of an object is quite distinct from its *weight*.) In AP Chemistry, we generally use the **gram** (g), which is one-thousandth the mass of a kilogram.

Time

The SI base unit of time is the **second** (s), which is defined in terms of a number of vibrations of cesium-133 atoms under specified conditions.

Temperature

The SI base unit of temperature is the **kelvin** (K), which is defined in terms of a temperature scale that consists of two reference points: 0 K (known as *absolute zero*) and 273.16 K (the *triple point* of water). In AP Chemistry, we also use the **Celsius** scale of temperature. Originally, this scale had the freezing and boiling points of water at one atmosphere as its reference points: water froze at 0 °C and boiled at 100 °C. The Celsius scale is now defined in terms of the Kelvin scale:

$$°C = K - 273.15$$

Electric Current

The SI base unit of electric current is the **ampere** (A), which is defined in terms of a specific magnetic force between a pair of parallel conductors under specified conditions.

Amount of Substance

The SI base unit for amount of substance is the **mole** (mol), which is defined as the number of atoms present in 0.012 kilogram of carbon-12. This number is known as **Avogadro's Number** (N_A) and is approximately equal to 6.02×10^{23}.

The seventh SI base unit, the *candela* (cd), is a measure of a quantity known as *luminous intensity*. We will not have occasion to use it in this book.

Metric Prefixes

An SI base unit is not always convenient for expressing measurements. For example, the meter is not useful for describing either the length of a bacterium or interstellar distances. There are two ways to deal with this problem: Scientific (exponential) notation can be used with a base unit, or a set of **metric prefixes** can be used to create multiples and subdivisions of the base unit. The accompanying table lists some metric prefixes along with their symbols and values.

Metric Prefixes

Factor	Prefix	Symbol	Factor	Prefix	Symbol
10^{15}	peta	P	10^{-1}	deci	d
10^{12}	tera	T	10^{-2}	centi	c
10^{9}	giga	G	10^{-3}	milli	m
10^{6}	mega	M	10^{-6}	micro	μ
10^{3}	kilo	k	10^{-9}	nano	n
10^{2}	hecto	h	10^{-12}	pico	p
10^{1}	decka	da	10^{-15}	femto	f

For example, the term *nanometer* (nm) is equivalent to 10^{-9} meter, and the term *megasecond* (Ms) is equivalent to 10^6 seconds.

Derived Metric Units

A *derived metric unit* is a combination of metric base units. For example, we are all familiar with the concept of *speed*. The unit *meters per second* (m/s or m s^{-1}) is the combination of the units for length and time.

An important derived unit is **volume**, which is the amount of three-dimensional space occupied by an object. The SI unit of volume is the *cubic meter* (m^3). In AP Chemistry, the volume units *cubic centimeter* (cm^3) and *cubic decimeter* (dm^3) are used more often. Since one centimeter is equal to 10^{-2} meter, one cubic centimeter is equal to *one millionth* of a cubic meter [$(10^{-2})^3$]. Similarly, one cubic decimeter is equal to *one thousandth* of a cubic meter. It follows that one cubic decimeter is equal to 1,000 cubic centimeters (Why?). A non-SI unit of volume that is frequently used is the *liter* (L), which is equivalent to a cubic decimeter. In the older English system of measurement, one liter is slightly more than one quart. A subdivision of the liter, the *milliliter* (mL) is therefore equivalent to a cubic centimeter. In summary:

$$1 \text{ m}^3 = 10^3 \text{ dm}^3 = 10^6 \text{ cm}^3$$

$$1 \text{ dm}^3 = 10^3 \text{ cm}^3 = 1 \text{ L}$$

$$1 \text{ cm}^3 = 1 \text{ mL}$$

Another important derived unit is **density** (*d*), which is the ratio of the mass of a substance (*m*) to its volume (*V*):

$$d = \frac{m}{V}$$

Density is an intensive property that depends on the nature of the substance rather than on its size; it measures the compactness of a substance. The SI unit of density is the *kilogram per cubic meter* (kg/m^3 or kg m^{-3}). In AP Chemistry, we use a variety of units to report the density of a substance. The relationships among these units are summarized below:

$$1 \text{ kg/m}^3 = 1 \text{ g/dm}^3 = 1 \text{ g/L} = 10^3 \text{ g/cm}^3 = 10^3 \text{ g/mL}$$

A number of substances and their densities are listed in the table below:

Density of Selected Elements, Compounds, and Mixtures *

Element	Density /(g/dm^3)	Compound	Density /(g/dm^3)	Mixture	Density /(g/dm^3)
hydrogen(g)	0.0899	ammonia(g)	0.760	corn oil(ℓ)	922$^{20°}$
oxygen(g)	1.429	hydrogen chloride(g)	1.63	steel(s)	7,890$^{20°}$
mercury (ℓ)	13,550$^{20°}$	ethanol(ℓ)	789$^{20°}$	brass(s)	8,470$^{20°}$
calcium(s)	1,550$^{20°}$	water(ℓ)	1,000$^{4°}$		
phosphorus(s)	1,823$^{20°}$	ice (s)	917$^{0°}$		
carbon(s)	2,267$^{20°}$	sucrose(s)	1,580$^{17°}$		
gold(s)	19,300$^{20°}$	sodium chloride(s)	2,170$^{20°}$		

* (s) = solid; (ℓ) = liquid; (g) = gas

Gas densities were measured at one atmosphere and 273.15 K (0 °C).

The temperature (in °C) at which a liquid or solid density was measured, appears as a *superscript* to the right of the value.

UNCERTAINTY IN MEASUREMENT

All measurements contain some degree of uncertainty. Even if we discount *mistakes* made by an experimenter, *errors* invariably are part of any experiment. A **systematic error** skews a series of measurements in the same direction: this type error will lead to a measurement that is always too large or too small. Systematic errors arise from the use of faulty measuring devices and they affect the **accuracy** of a measurement: the agreement between the measured value of a quantity and its true value.

A **random error** arises when a measuring device (even an accurate one) is used a number of times to make the same measurement. These errors are distributed in a way that causes the measurements to cluster on both sides of the true value of the quantity. Random errors are closely linked to **precision**: the closeness of a set of measurements. Note that precision need not imply accuracy. It is entirely possible to produce a set of very precise set of measurements using an inaccurate measuring device.

Precision is also related to the concept of *significant figures*: the number of digits that are properly part of any measurement. Consider the liquid level in the graduated cylinder shown below:

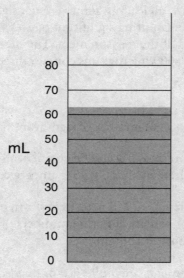

The measured volume is reported as 62 milliliters. Let us analyze this measurement. The "6" is called a "certain" digit because there is no doubt that the volume lies between 60 and 70 milliliters. The "2" is known as a "doubtful" or "uncertain" digit because it represents an educated guess as to how far between 60 and 70 milliliters the liquid level lies.

Correctly reported measurements can contain only *one* doubtful digit. Reporting the measurement as 62.37 milliliters is clearly incorrect. Since the "2" is known to be a guess, the "3" and the "7" can only be figments of an overworked experimenter's imagination!

The collection of all of the certain digits and the one uncertain digit in a measurement is known as the number of **significant figures** contained in the measurement. The measurement 62 milliliters contains two significant figures. A measurement such as 148.935 grams contains six significant figures: the trailing "5" is the doubtful digit.

There are occasions when not every digit in a measurement is significant. For example, the measurement 93,000,000 miles (the average Earth-sun distance) contains only *two* significant figures. The zeroes are not part of the measurement: they indicate the *size* of the number. A number of rules have been established for determining the number of significant figures in a measurement:

- All nonzero digits are significant. The measurement 123 seconds contains three significant figures.

- Zeroes that are "sandwiched" between nonzero numbers are significant. The measurement 1,002.69 kilograms contains six significant figures.

- If a number is greater than 1 and ends in one or more zeroes, the zeroes are significant *only* if a decimal point is expressed. The measurements 200, 200., and 200.00 meters contain one, three, and five significant figures, respectively.

- If a number is less than 1, all leading zeroes are *not* significant: they indicate the size of the number. We begin counting significant figures with the first nonzero number and continue to the end of the measurement. The measurements 0.000478, 0.0004780, and 0.000478000 micrograms contain three, four, and six significant figures, respectively.

- *Counted* numbers, such as 12 bagels, and *defined* constants, such as 273.16 K, are considered to have an infinite number of significant figures.

- If a measurement is expressed in *scientific notation*, all of the digits are significant. For example, the measurement 1.230×10^{-5} ampere contains four significant figures.

In solving AP Chemistry problems, we will need to perform mathematical manipulations on a variety of measurements that contain differing numbers of significant figures. How many significant figures should the answer contain?

- If two or more measurements are *multiplied* or *divided*, the answer contains the same number of significant figures as the measurement with the *smallest* number of significant figures. For example:

$$6.32 \text{ cm} \times 2.282 \text{ cm} = 14.42224 \text{ cm}^2 = \textbf{14.4 cm}^2$$

The answer contains three significant figures because the measurement 6.32 cm contains three significant figures, while the measurement 2.282 cm contains four.

Note that the final digit in the answer is "4" because the *next* digit in the provisional answer ("2") is *less* than 5. This is known as **rounding down**.

- If two or more measurements are *added* or *subtracted*, the answer contains the same number of digits *to the right of the decimal point* as does the measurement with the *smallest* number of digits to the right of the decimal point. For example:

$$34.232 \text{ g} + 161.92 \text{ g} + 4.7 \text{ g} = 200.852 \text{ g} = \textbf{200.9 g}$$

The answer contains only one digit to the right of the decimal point because the measurement 4.7 g contains one digit to the right of the decimal point, while the other two measurements contain three and two digits, respectively.

Note that the final digit in the answer is "9" because the *next* digit in the provisional answer ("5") is *greater than or equal to 5*. This is known as **rounding up**.

Suppose we are confronted with a "mixed" operation such as:

$$\frac{(14.651 - 3.22) \times (13.4 + 160.)}{65.99 \times 1420}$$

There is an elaborate set of rules for performing this calculation and reporting it to the appropriate number of significant figures. While this technique (which is not discussed here) is perfectly appropriate for a laboratory report, it takes up valuable time, given the pressures of examination conditions.

It is recommended (by this author, at least) that you report *all* answers to such mixed operations to **three significant figures**. The "significant-figure penalty" established by the AP Chemistry examiners is at most one point per question. Moreover, since there is some leeway, it is unlikely that an answer containing three significant figures will incur such a penalty.

DIMENSIONAL ANALYSIS AND PROBLEM SOLVING

Dimensional Analysis is a tool that manipulates units in order to solve problems. Consider the identity: 1 meter $\equiv$ 100 centimeters. Although the numbers (1 and 100) are very different, both measurements represent the same length. Consequently, as far as length is concerned, the fractions

$$\frac{1 \text{ m}}{100 \text{ cm}} \text{ and } \frac{100 \text{ cm}}{1 \text{ m}}$$

are essentially equal to 1. The only function served by these fractions is to allow a conversion between meters and centimeters. In using dimensional analysis, units are treated as numbers for the purposes of addition, subtraction, multiplication, and division.

As an example, consider the problem: How many meters are equivalent to 78 centimeters? We solve this problem by multiplying the "given" (78 centimeters) by the first fraction. This allows the replacement of centimeters with meters as shown below:

$$78 \text{ cm} \cdot \left(\frac{1 \text{m}}{100 \text{ cm}}\right) = 0.78 \text{ m}$$

When using dimensional analysis, we need not limit ourselves to converting between the units of a single quantity (such as length). For example, consider the quantity we call *density*. The density of gold is 19.2 grams per cubic centimeter (19.2 g/cm^3) at 20°C. We can use the density to form two fractions:

$$\frac{19.2 \text{ g}}{1 \text{ cm}^3} \text{ and } \frac{1 \text{ cm}^3}{19.2 \text{ g}}$$

The value of each of these fractions is also considered to be 1 because each fraction expresses the same fixed relationship for *all* gold at 20°C. We can use these fractions to convert between the mass of a sample of gold and its corresponding volume (at 20°C), as shown in the following example:

Sample Problem

What is the volume of 152 grams of gold?

Solution

$$152\,\cancel{g} \cdot \left(\frac{1\ cm^3}{19.2\ \cancel{g}}\right) = 7.92\ cm^3$$

If we need to convert more than one unit in a measurement, we treat each unit separately.

Sample Problem

Convert 60. miles per hour to feet per second.

Solution

We will utilize the relationships 1 mile = 5280 feet and 1 hour = 3600 seconds:

$$60.\,\frac{\cancel{mi}}{\cancel{h}} \cdot \left(\frac{5280\ ft}{1\ \cancel{mi}}\right) \cdot \left(\frac{1\ \cancel{h}}{3600\ s}\right) = 88\,\frac{ft}{s}$$

Finally, let us convert a unit involving exponents.

Sample Problem

How many cubic centimeters are there in 1 cubic decimeter?

Solution

To solve this problem, we use the *linear relationships* to convert from decimeters to meters, and then from meters to centimeters. (We will utilize the relationships 10 decimeters = 1 meter, and 1 meter =100 centimeters.) Finally, we insert the exponents as shown below:

$$1\,\cancel{dm}^3 \cdot \left(\frac{1\ \cancel{m}}{10\ \cancel{dm}}\right)^3 \cdot \left(\frac{100\ cm}{1\ \cancel{m}}\right)^3 = 1{,}000\ cm^3$$

The Language of Chemistry

CHAPTER FOUR

At the beginning of the nineteenth century, the English schoolteacher John Dalton proposed the first scientific model of atomic structure, known as the **Dalton Model**, which is summarized below:

- All matter is composed of indivisible particles called atoms.

- An element is a sample of matter, in which all of the atoms are identical to one another; different elements contain different types of atoms.

- In a chemical reaction, atoms combine, but they are not created, destroyed, or changed into other types of atoms.

- A sample of matter that contains a combination of different types of atoms is known as a compound.

As a result of this model, it was recognized that atoms combined with one another in whole-number ratios and a formula could be written for the compound. For example, the formula CO_2 (carbon dioxide) means that each unit of this compound contains one atom of carbon and two atoms of oxygen. Furthermore, the model gave rise to two important principles:

- **The Law of Definite Composition:** *Each sample of matter has a fixed composition by mass.* For example, the compound carbon dioxide is composed of 27.3% carbon and 72.7% oxygen by mass.

- **The Law of Multiple Proportions:** *When two elements, A and B, form more than one compound, the masses of B combining with a fixed mass of A form small whole-number ratios.* For example, hydrogen and oxygen form two compounds: water and hydrogen peroxide. In water,

8 grams of oxygen combine with 1 gram of hydrogen. In hydrogen peroxide, 16 grams of oxygen combine with 1 gram of hydrogen. The combining masses of oxygen in both compounds (8 and 16 grams) form a small whole-number ratio (1:2).

The experimental verification of these two principles lent powerful support to Dalton's Atomic Model.

THE ELECTRON

By 1850, it was already suspected that atoms were composed of smaller subatomic particles. The production of "cathode rays," negatively charged particles in evacuated glass tubes, supported this idea. In 1897, the British physicist J. J. Thomson examined the charge-to-mass ratios of these cathode rays and found that all had the same value. As a result of this work, Thomson was credited with the discovery of the *electron*. In 1909, the American physicist Robert Millikan succeeded in measuring the charge on an electron by observing the motion of electrically charged oil droplets. The mass of an electron is 9.1094×10^{-31} kilogram, and its charge is 1.6022×10^{-19} coulomb.

RUTHERFORD'S NUCLEAR MODEL

In 1910, the New Zealand physicist Ernest Rutherford and his assistants (Hans Geiger and Ernest Marsden) bombarded thin metallic foils with positively charged alpha particles (the nuclei of helium-4 atoms). In examining the paths of the *alpha particles* as they emerged from the foils, Rutherford concluded that while most of the *volume* of an atom was empty space, most of the *mass* of the atom was concentrated in a small, dense, positively charged core known as the **nucleus**. The electrons present in an atom were located in the space outside the nucleus. The Rutherford model was a *planetary* model, analogous to that of our solar system: the nucleus was located at the center, and the electrons were in orbit around the nucleus.

Between 1910 and 1932, two more fundamental particles were discovered: the positively charged **proton** and the neutral **neutron**. The characteristics of the three fundamental particles are compared in the table below:

Particle	Location	Charge /C	Mass /kg
proton	nucleus	$+1.6022 \times 10^{-19}$	1.6726×10^{-27}
neutron	nucleus	0	1.6749×10^{-27}
electron	outside nucleus	-1.6022×10^{-19}	9.1094×10^{-31}

In examining the table, we note that the proton and the electron have equal, but opposite electric charges. We also note that the mass of a proton is nearly equal to the mass of a neutron, but an electron is nearly 2,000 times less massive than either of these particles.

KAPLAN

An alternative way of expressing charge is to assign a value of +1 to a proton and –1 to an electron. In this book, we write such relative charges in the following way:

- The magnitude of the charge precedes its sign: "2+" instead of "+2"

- A "1" is never written out explicitly: "+" instead of "1+"

A SIMPLIFIED VIEW OF ATOMIC STRUCTURE

An atom consists of a nucleus containing protons and neutrons; the electrons are located outside the nucleus. The atom of each element is identified by three unique characteristics: its *name*, its *symbol*, and its *atomic number*. If we examine the Alphabetical List of the Elements found in Appendix B, we can inspect all three characteristics.

Elements are named in a variety of ways: they may represent places (francium–France), famous persons (einsteinium–Einstein), or properties of the element itself (iodine–from the Greek *iodes* [violet]).

Each element is associated with a unique one- or two-letter **symbol**. For example hydrogen has the symbol H; helium, He; and iron, Fe (from its old Latin name, *ferrum*). In writing a symbol, only the *first* letter of the symbol is capitalized. (Unnamed elements are assigned three-letter symbols that reflect the atomic number of the element.)

Each element is also associated with a number known as the **atomic number** (denoted by the symbol Z): the number of protons in the nucleus of each atom of the element. The element chromium (Cr) has an atomic number of 24, meaning that the nucleus of each atom of chromium contains 24 protons. If an atom is electrically neutral, it follows that it must contain equal numbers of protons and electrons. Therefore, a neutral atom of chromium also contains 24 electrons surrounding the nucleus.

Neutrons add mass to an atom, but do not change its identity: Consider the element hydrogen, whose atomic number is 1. A hydrogen nucleus containing 0, 1, or 2 neutrons is still considered to be hydrogen because each of the nuclei contains exactly 1 proton. We call these different varieties of hydrogen atoms **isotopes**. An isotope of an element is distinguished by its **mass number** (denoted by the symbol A): the *sum of the number of protons and neutrons* in the nucleus of the atom in question. For example, the mass numbers of hydrogen isotopes introduced at the beginning of this paragraph are 1 [1 + 0], 2 [1 + 1], and 3 [1 + 2]. We signify isotopes in a variety of ways:

- We can write the name of the element followed by its mass number, as in: hydrogen-1, hydrogen-2, and hydrogen-3.

- We can write the symbol of an element followed by its mass number: H-1, H-2, and H-3.

- We can write the symbol, placing the atomic number as a *subscript* to the left of the symbol, and placing the mass number as a *superscript* to the left of the symbol, as in: ^1_1H, ^2_1H, ^3_1H. Occasionally, the atomic number is omitted (for example, ^2H), since it is readily obtainable from the symbol of the element.

Given a particular isotope of an element, we can readily find the number of neutrons (N) in the nucleus by *subtracting* the atomic number (Z) of the element from the *mass number* (A) of the isotope. In symbolic form:

$$N = A - Z$$

For example, the number of neutrons in the nucleus of an atom of potassium-40 (atomic number = 19, mass number = 40) is: $(40 - 19) = 21$ neutrons.

INTRODUCTION TO THE PERIODIC TABLE OF THE ELEMENTS

The physical and chemical properties of elements possess a considerable degree of regularity. For this reason, the elements have been arranged in order of increasing atomic number into an approximately rectangular array known as the **Periodic Table** (see Appendix B). The word "periodic" implies that the properties of the elements recur at regular intervals.

The (vertical) columns of the Table are known as **groups** (or **chemical families**), and are numbered from 1 to 18. Elements within a single group have similar physical and chemical properties. For example, the elements comprising Group 17 are all characteristically colored and they tend to gain a single electron in simple chemical reactions. Some of the groups have special names: Group 1 elements are named **alkali metals**; Group 2 metals are named **alkaline earth elements**; Group 17 elements are named **halogens**; Group 18 elements are named **noble gases**.

The (horizontal) rows of the Table are known as **periods**, which are numbered from 1 to 7. Across a period, the properties of the elements change from **metallic** to **nonmetallic**. For example, in Period 3, the elements Na, Mg, and Al are metallic; the elements P, Cl, and Ar are nonmetallic; and the element Si is known as a **metalloid** (or **semimetal**) because it possess properties common to both metallic and nonmetallic elements.

The elements of Groups 1, 2, and 13–18 are known as the **representative elements**: their physical and chemical properties are most easy to predict. The elements of Groups 3–11 are known as the **transition elements**, and display certain special properties. Another collection of elements is found as two rows placed at the bottom of the main body of the Table. These rows are named the **lanthanide** and **actinide** series, respectively. We will examine the Periodic Table in detail in chapter 8.

SIMPLE IONS

An atom that acquires a positive or negative electric charge by virtue of having lost or gained electrons is known as an **ion**. Consider the following examples:

$$Na - 1e^- \rightarrow Na^+$$

$$S + 2e^- \rightarrow S^{2-}$$

A positive ion is also known as a **cation**, and a negative ion is also known as an **anion**. Note that ionic charges are written as relative charges.

The elements located in Groups 1, 2, and 13 of the Periodic Table most commonly form simple ions with respective relative charges of $(1)+$, $2+$, and $3+$. The elements located in Groups 17, 16, and 15 most commonly form ions with respective relative charges of $(1)-$, $2-$, and $3-$. Transition metals tend to form ions with more than one charge: For example, iron can lose either two or three electrons to form the ions Fe^{2+} and Fe^{3+}.

CHEMICAL FORMULAS

A **chemical formula** is basically a shorthand notation that describes the composition of a substance or an ion. A formula takes on different meanings depending on the context in which it is used. At this point we need to define two terms that will be used in the succeeding paragraphs. We can think of an element or compound as a collection of repeating units. Each single unit is known as a **formula unit**, and a chemical formula represents this unit. If the formula unit is able to exist *independently* of the other formula units, it is known as a **molecule**.

Elements

The formula unit of the element helium is a single helium atom and is represented by the formula "He." Since each helium atom can exist independently of other helium atoms, it is also called a **monatomic molecule**. In general, all of the Group 18 elements exist as monatomic molecules.

At room temperature and pressure, the formula unit of oxygen consists of two oxygen atoms that are joined by the *sharing* of electrons. This **diatomic molecule** is represented by the formula "O_2." (Note that the number "2" is placed as a subscript to the right of the elemental symbol.) Other elements that form diatomic molecules include hydrogen (H_2), nitrogen (N_2), and the Group 17 elements. A number of elements form **polyatomic molecules** such as phosphorus (P_4), sulfur (S_8), and a type of oxygen known as *ozone* (O_3).

At room temperature and pressure, the element sodium is composed of a complex array of sodium ions (Na^+) and its separated electrons. Molecules of sodium do *not* exist independently and we can think of the formula unit as a single sodium ion and its separated electron. This formula unit is represented by the formula "Na."

Compounds

A compound can be formed by the sharing of electrons among atoms, a situation known as **covalent bonding**. This type of compound is known as a **molecular compound** because it consists of *independent molecules*. The formula of the compound represents an individual molecule. Common examples include water (H_2O), carbon dioxide (CO_2), and glucose ($C_6H_{12}O_6$).

A compound can also be formed by the association of oppositely charged metallic and nonmetallic simple ions, such as potassium fluoride (KF) and calcium chloride ($CaCl_2$). This type of compound is known as an **ionic compound**. Individual molecules do *not* exist and the formulas represent the numerical ratios of the ions. For example, the formula KF means that potassium fluoride consists of one K^+ ion for each F^- ion present. Similarly, the formula $CaCl_2$ means the ionic ratio $Ca^{2+}:Cl^-$ is 1:2.

Certain compounds contain ions that have internal covalent bonding. Examples are the compounds $KMnO_4$ and NH_4Cl. The $[MnO_4]^-$ ion consists of one manganese atom covalently bonded to four oxygen atoms. An *additional electron* gives this ion its 1– charge. Similarly, the $[NH_4]^+$ ion consists of one nitrogen atom bonded to four hydrogen atoms. Since the assembly is *deficient by one electron*, the ion has a charge of 1+. These ions, which contain more than one atom, are known as **polyatomic ions**.

If an ionic or molecular compound contains exactly two elements, we call it a **binary compound**. Examples of binary compounds include NaCl, CO_2, Al_2O_3, and P_4O_{10}.

WRITING FORMULAS AND NAMING COMPOUNDS

Ionic Compounds

In writing the formula of an ionic compound, the positive ion precedes the negative ion and the formula is constructed so that the sum of all charges totals zero. Refer to the listing of common positive and negative ions given in Appendix B. When Li^+ and Cl^- ions combine to form a compound, an ionic ratio of 1:1 will cause the positive and negative charges to add to zero: the formula is LiCl. When Al^{3+} and F^- ions combine, we require an ionic ratio of 1:3, and the formula is AlF_3. When Ba^{2+} and NO_3^- ions combine, we enclose the NO_3^- ion in parentheses; the formula is $Ba(NO_3)_2$.

Binary compounds are named so that the positive ion has the name of the element, while the negative ion adds the suffix "–ide." The ionic compounds KI, AlF_3, and CaO are named potassium iodide, aluminum fluoride, and calcium oxide, respectively. If the compound contains a polyatomic ion, the full name of the ion is used in the name. The compounds $Mg(NO_3)_2$, $KMnO_4$, and $(NH_4)_2SO_4$ are named magnesium nitrate, potassium permanganate, and ammonium sulfate, respectively.

If an element can form a positive ion with more than one charge, more than one compound can exist. For example, iron forms 2+ and 3+ ions, and the ionic compound iron nitrate comes in two forms: $Fe(NO_3)_2$ and $Fe(NO_3)_3$. We name these compounds according to the charge on the *positive* ion (Fe) by inserting a Roman numeral in parentheses in the name of the formula: $Fe(NO_3)_2$ is named iron (II) nitrate and $Fe(NO_3)_3$ is named iron (III) nitrate. (An older system of naming uses the suffixes "–ous" and "–ic" as part of the name of the positive ion. We will not have occasion to use this system.)

Negative polyatomic ions that contain oxygen are known as **oxyanions** and they have their own special naming system. Consider the group of oxyanions: ClO^-, ClO_2^-, ClO_3^-, and ClO_4^-. The most common ion (in this case, ClO_3^-) takes on the ending "–ate," hence its name, *chlorate*. An ion that contains one *less* oxygen atom (ClO_2^-) takes on the ending "–ite," hence its name, *chlorite*. An ion with *two* less oxygen atoms (ClO^-) adds the prefix "hypo–" to the "–ite" ending, hence its name, *hypochlorite*. If an ion has one *more* oxygen atom than the most common ion (ClO_4^-), it adds the prefix "per–" to the "–ate" ending, hence its name, *perchlorate*.

Binary Molecular Compounds

Interpreting formulas for binary molecular compounds poses a special problem because the atoms in these compounds are not electrically charged. Nevertheless, we can overcome this problem if we assign apparent charges—known as **oxidation numbers**—to the atoms. A simplified list of rules follows:

- A *free element* (such as He, Al, N_2, or P_4) is assigned an oxidation number of 0.

- Oxygen in molecular compounds nearly always has an oxidation number of 2–. (A notable exception is the class of compounds known as *peroxides*, such as H_2O_2, in which the oxidation number is 1–.)

- Hydrogen in molecular compounds nearly always has an oxidation number of 1+. (A notable exception is the class of compounds known as *hydrides*, such as BH_3, in which the oxidation number is 1–.)

- The sum of all the apparent charges in a compound must total 0.

For example, in the compound P_2O_5, The total apparent charges of the five oxygen atoms is 10–. Therefore, the phosphorus atoms must have a total apparent charge of 10+. Since there are two phosphorus atoms, each atom must have an oxidation number of 5+.

When a formula for a binary molecular compound is written:

- The element with the *lower* periodic Group number is written first. For example, nitrogen is placed first in the formula NO_2 since it is in Group 15, while oxygen is in Group 16.

- If two elements are in the same Group, the *lower* one is written first, as in ICl.

Naming binary molecular compounds is similar to naming binary ionic compounds except that prefixes such as *mono, di, tri, tetra,* . . . are used to indicate the number of atoms present in the compound. The prefix *mono* is never used for the first element. For example, CO_2 is named carbon *di*oxide, and N_2O_4 is named *di*nitrogen *tetr*oxide. (Note that the *a* in *tetra* has been omitted because "oxide" begins with a vowel.)

Empirical Formulas

An empirical formula is one in which the elements appear in *smallest whole-number* ratios. For example, the empirical formulas of N_2O_4, H_2O_2, and Na_2SO_4 are, respectively, NO_2, HO, and Na_2SO_4.

Naming Acids

For the present an **acid** can be thought of as an anion with enough H^+ ions added to neutralize it. For example, HNO_3, H_2SO_4, and H_3PO_4 are acids. The acid is named according to the name of its anion:

- If the anion ends in *–ide*, the acid name takes on the prefix *hydro–* and the suffix *–ic*. For example, HCN (from CN^- [cyanide ion]) is named *hydro*cyan*ic* acid.

- If the anion ends in *–ate*, the acid name takes on the suffix *–ic*. For example, H_2SO_4 (from SO_4^{2-} [sulfate ion]) is named sulfur*ic* acid.

- If the anion ends in *–ite*, the acid name takes on the suffix *–ous*. For example, HClO (from ClO^- [hypochlorite ion]) is named hypochlor*ous* acid.

CHEMICAL REACTIONS AND EQUATIONS

A chemical equation is a shorthand notation for a chemical reaction. Consider the reaction ion which solid calcium hydroxide is reacted with an *aqueous* solution of hydrogen chloride (that is, hydrochloric acid). As a result of the reaction, liquid water and an aqueous solution of calcium chloride are produced. The equation for this reaction is:

$$Ca(OH)_2(s) + HCl(aq) \rightarrow H_2O(\ell) \overset{\Delta}{} CaCl_2(aq)$$

The substances to the *left* of the arrow are known as **reactants**, and those to the *right* of the arrow are known as **products**. The phases of a substance is written in parentheses to the right of its formula. (Had a gas been present, the notation (g) would have been used.) The arrow represents the word "yields." Occasionally, some of the conditions needed for the reaction may be written over the arrow. For example, the symbol $\longrightarrow$ means that heat has been added to the reactants.

Balancing Chemical Equations

The equation shown on the previous page is not entirely correct as written because the number of atoms on each side of the arrow do not match. This equation is said to be *unbalanced*. In order for an equation to be written properly, it must obey three **conservation** laws: mass, electric charge, and energy. At present, we will rewrite the equation so that mass is conserved:

$$Ca(OH)_2(s) + 2HCl(aq) \rightarrow 2H_2O(\ell) + CaCl_2(aq)$$

Such an equation is said to be **balanced**. In order to balance this equation, we used **coefficients**, numbers placed *to the left* of the formula. We could not change the subscripts, because that would change the identity of the substance. Generally, an equation is balanced so that the smallest whole-number coefficients are used (as in the case above). However, the following equations are also considered balanced:

$$3Ca(OH)_2(s) + 6HCl(aq) \rightarrow 6H_2O(\ell) + 3CaCl_2(aq)$$

$$\frac{1}{2}Ca(OH)_2(s) + HCl(aq) \rightarrow H_2O(\ell) + \frac{1}{2}Ca\,Cl_2(aq)$$

Classifying Simple Chemical Reactions

Synthesis (Combination) Reactions

In a **synthesis reaction**, a single compound is the sole product of the reaction. Consider the following examples of synthesis reactions:

$$2Mg(s) + O_2(g) \rightarrow 2MgO(s)$$

$$H_2O(\ell) + CO_2(g) \rightarrow H_2CO_3(aq)$$

Decomposition Reactions

A **decomposition reaction** is the *opposite* of a synthesis reaction: a single reactant is reduced to simpler products, as shown in the following example:

$$CaCO_3(s) \rightarrow CO_2(g) + CaO(s)$$

Combustion Reactions

For our purposes, a **combustion reaction** is one in which a carbon-hydrogen or a carbon-hydrogen-oxygen compound is reacted rapidly with oxygen. If the oxygen is present in sufficient supply, the products in each case are carbon dioxide and water. The following are examples of combustion reactions:

$$C_3H_8(g) + 5O_2(g) \rightarrow 3CO_2(g) + 4H_2O(g)$$

$$C_2H_4O_2(\ell) + 2O_2(g) \rightarrow 2CO_2(g) + 2H_2O(g)$$

If the supply of oxygen is somewhat limited, the poisonous gas carbon monoxide (CO) will be produced instead of carbon dioxide. If the supply of oxygen is severely limited, particles of carbon, known as soot, will be produced.

CHEMICAL REACTIONS IN AQUEOUS SOLUTIONS

When an ionic compound is dissolved in water, the ions are surrounded by water molecules and are separated during the solution process:

$$KBr(s) \xrightarrow{H_2O} K^+(aq) + Br^-(aq)$$

When molecular compounds are dissolved in water, there are three possibilities:

· The molecules separate in the water, but each molecule remains intact:

$$CH_3OH\ (\ell) \xrightarrow{H_2O} CH_3OH(aq)$$

· The molecules *ionize completely* in water:

$$HCl(g) \xrightarrow{H_2O} H^+(aq) + Cl^-(aq)\ [{\sim}100\%]$$

· The molecules *ionize partially* in water:

$$HC_2H_3O_2(\ell) \xrightarrow{H_2O} H^+(aq) + C_2H_3O_2^-(aq) + HC_2H_3O_2(aq)$$

$$[{\sim}1\%{-}2\%] \qquad\qquad\qquad [{\sim}98\%{-}99\%]$$

When a substance dissolves in water and produces ions, the substance is known as an **electrolyte** because the solution will conduct an electric current.

Net Ionic Equations

Consider the reaction:

$$HCl(aq) + NaOH(aq) \rightarrow NaCl(aq) + H_2O(\ell)$$

In solution, the HCl, NaOH, and NaCl exist entirely in ionic form and we can rewrite the equation as:

$$H^+(aq) + Cl^-(aq) + Na^+(aq) + OH^-(aq) \rightarrow Na^+(aq) + Cl^-(aq) + H_2O(\ell)$$

The Na^+ and Cl^- ions exist on both sides of the equation and are known as **spectator ions**. A **net ionic equation** excludes all spectator ions; the net ionic equation for this reaction is given below:

$$H^+(aq) + OH^-(aq) \rightarrow H_2O(\ell)$$

Metathesis (Double Replacement) Reactions

A **metathesis** or **double replacement** reaction is one that occurs by the exchange of ions in solution. In order for the reaction to occur, at least one of the products must:

- separate from the solution as a solid **precipitate**, or
- be a molecular compound, or
- escape as a gas.

Precipitation Reactions

Consider the following reaction between the ionic compounds barium nitrate and sodium sulfate:

$$Ba(NO_3)_2(aq) + Na_2SO_4(aq) \rightarrow BaSO_4(s) + 2NaNO_3(aq)$$

The Na^+ and NO_3^- remain in solution and are spectator ions. Therefore, the net ionic equation for this reaction is:

$$Ba^{2+}(aq) + SO_4^{2-}(aq) \rightarrow BaSO_4(s)$$

How do we know when a precipitate will form? The following set of solubility rules is a valuable guide to the solubility of ionic compounds in water:

SOLUBILITY RULES FOR IONIC COMPOUNDS IN WATER

Compounds that are mostly soluble (i.e., dissolve appreciably in water):

- Compounds containing NH_4^+ or Group 1 ions.
- Compounds containing NO_3^- or $C_2H_3O_2^-$
- Compounds containing Cl^-, Br^-, or I^- (*exceptions*: Pb^{2+}, Ag^+, and Hg_2^{2+})
- Compounds containing SO_4^{2-} (*exceptions*: Sr^{2+}, Ba^{2+}, Pb^{2+}, and Hg_2^{2+})

Compounds that are mostly *insoluble* (i.e., do not dissolve appreciably in water):

- Compounds containing S^{2-} (*exceptions*: NH_4^+, ions of Groups 1 and 2)
- Compounds containing OH^- (*exceptions*: Group 1 ions)
  (Ca^{2+}, Sr^{2+}, and Ba^{2+} are *slightly soluble*)
- Compounds containing CO_3^{2-} or PO_4^{3-} (*exceptions*: NH_4^+ and Group 1 ions)

By inspecting the rules, we can see that barium sulfate is an insoluble ionic compound. It is essential that this set of rules be learned thoroughly.

Reactions Leading to the Formation of a Molecular Compound

Consider the following reaction:

$$HCl(aq) + NaOH(aq) \rightarrow NaCl\ (aq) + H_2O(\ell)$$

As we saw above, the Na^+ and Cl^- ions remain in solution while the H^+ and OH^- ions combine to form the molecular compound H_2O. A similar reaction occurs when solid magnesium hydroxide is dissolved by aqueous hydrochloric acid:

$$2HCl(aq) + Mg(OH)_2(s) \rightarrow MgCl_2(aq) + 2H_2O(\ell)$$

The net ionic equation for this reaction is:

$$2H^+(aq) + Mg(OH)_2(s) \rightarrow Mg^{2+}(aq) + 2H_2O(\ell)$$

Reactions in Which a Gas is Formed

When a (molecular) gas is formed by the exchange of ions, the gas escapes from the solution, as shown below:

$$2HCl(aq) + Na_2S(aq) \rightarrow 2NaCl(aq) + H_2S(g)$$

The Na^+ and Cl^- ions remain in solution; the net ionic equation for this reaction is:

$$2H^+(aq) + S^{2-}(s) \rightarrow H_2S(g)$$

Single Replacement Reactions

Consider the reaction between aqueous copper(II) nitrate and metallic zinc:

$$Cu(NO_3)_2(aq) + Zn(s) \rightarrow Cu(s) + Zn(NO_3)_2(aq)$$

The net ionic equation for this reaction is:

$$Cu^{2+}(aq) + Zn(s) \rightarrow Cu(s) + Zn^{2+}(aq)$$

In this **single replacement** reaction, aqueous copper ions were replaced by ions formed from metallic zinc, while the copper ions were changed into metallic copper. Single replacement reactions are but one example of a reaction type known as **oxidation–reduction**. Such reactions occur by the *transfer of electrons*. In the reaction described above, it follows that $Zn(s)$ must lose electrons more easily than $Cu(s)$, otherwise the reaction would proceed in the *reverse* direction.

How can we predict the reaction that will occur between metals and metallic ions? We need to construct a table, known as an **activity series**, that places metals in the order of their ability to lose electrons and form positive ions.

Activity Series**

Most

Metals
Li
Rb
K
Cs
Ba
Sr
Ca
Na
Mg
Al
Ti
Mn
Zn
Cr
Fe
Co
Ni
Sn
Pb
**H_2
Cu
Ag
Au

Least

**Activity Series Based
on Hydrogen Standard

For example, we note that aluminum is higher on the table than iron. Therefore, metallic aluminum will form a positive ion more easily than metallic iron. If Al(s) is reacted with Fe^{2+}(aq), the following **half-reactions** will occur simultaneously:

$$Al(s) \rightarrow Al^{3+}(aq) + 3e^-$$

$$Fe^{2+}(aq) + 2e^- \rightarrow Fe(s)$$

Since all chemical equations must be balanced for *electric charge* as well as mass, we must "multiply" the first half-reaction by 2 and the second one by 3 in order to insure that equal numbers of electrons (6 in this case) are transferred between the two half-reactions. The net ionic equation for the reaction is:

$$2Al(s) + 3Fe^{2+}(aq) \rightarrow 2Al^{3+}(aq) + 3Fe(s)$$

We can also construct an activity series table that ranks the ability of nonmetals to form negative ions. A simplified table listing the Group 17 elements is shown below:

Activity Series

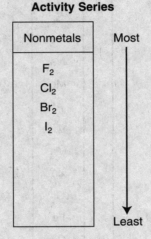

For example, consider the reaction between aqueous chlorine and aqueous sodium bromide:

$$Cl_2(aq) + 2NaBr(aq) \rightarrow Br_2(aq) + 2NaCl(aq)$$

The net ionic equation for this reaction is:

$$Cl_2(aq) + 2Br^-(aq) \rightarrow Br_2(aq) + 2Cl^-(aq)$$

As we can see, the element chlorine forms a negative ion more readily than the element bromine. We will have occasion to examine oxidation–reduction reactions in greater detail in chapter 15.

Chapter 4: Practice Questions

Multiple-Choice Questions

1. Which of these is a strong electrolyte?

   A. $CaCO_3(s)$

   B. $CHCl_3(\ell)$

   C. $CH_3CO_2H(aq)$

   D. $KCl(s)$

   E. $KCl(aq)$

2. Which will produce an acidic solution when dissolved in water?

   A. NaF

   B. MgO

   C. NH_4Cl

   D. $Ca(NO_3)_2$

   E. $LiC_2H_3O_2$

3. Which is a displacement reaction?

   A. $N_2(g) + 3\,H_2(g) \rightarrow 2\,NH_3(g)$

   B. $HCl(g) + NH_3(g) \rightarrow NH_4Cl$

   C. $C_3H_8(g) + 5\,O_2(g) \rightarrow 3\,CO_2(g) + 4\,H_2O(\ell)$

   D. $Cu(NO_3)_2(aq) + Zn(s) \rightarrow Zn(NO_3)_2(aq) + Cu(s)$

   E. $(NH_4)_2CO_3(s) + heat \rightarrow 2\,NH_3(g) + H_2O(g) + CO_2(g)$

4. What is produced by the reaction between sulfur trioxide and water?

   $$SO_3(g) + H_2O(\ell) \rightarrow ?$$

   A. $SO_2(g) + 2\,OH^-(aq)$

   B. $SO_4(g) + 2\,H^+(aq)$

   C. $HSO_3^-(aq) + OH^-(aq)$

   D. $H_2S(g) + 2\,O_2(g)$

   E. $H_2SO_4(aq)$

5. Solid silver sulfide and aqueous nitric acid result from bubbling hydrogen sulfide gas through aqueous silver nitrate. Which are the spectator ions?

   $$2\,AgNO_3(aq) + H_2S(g) \rightarrow Ag_2S(s) + 2\,HNO_3(aq)$$

   A. H^+ only

   B. NO_3^- only

   C. Ag^+ and S^{2-}

   D. Ag^+ and H^+

   E. NO_3^- and S^{2-}

Free-Response Question

1. (A) Write a balanced net ionic equation for the reaction occurring when solutions of sodium hydroxide and magnesium chloride are mixed.

    (B) Write a balanced net ionic equation for the reaction occurring when solutions of potassium chloride and copper(II) sulfate are mixed.

    (C) An unknown aqueous solution is combined with aqueous sodium sulfate forming a solid precipitate and another aqueous solution. State the formula of a compound which could be in the unknown aqueous solution, and state a reason supporting this choice?

    (D) Iron(III) ions, calcium ions, and lead(II) ions are all in a solution. You want to separate the ions so that you have each one in solid form without the other ions. Solutions of Na_2CO_3, NaBr, NaCl, NaOH, and NaI are available. What precipitates should you form and in what order?

Multiple-Choice Answers

1. E   2. C   3. D   4. E   5. B

1. E

An electrolyte is a substance that breaks up into ions when dissolved in water or melted. Once ionized, the electrolyte conducts electricity. The more ions that are formed, the stronger the electrolyte is. In a problem like this it is best to try and find out which choices are covalent substances, which are acids or bases, and which are ionic. Answer choices (B), $CHCl_3$, and (C), CH_3CO_2H, are covalently bonded, with the acetic acid being a very weak acid. Answer choices (A), (D), and (E) are ionic compounds, with only (E) being dissolved in water.

2. C

When dissolved in water, ammonium chloride will dissociate to form chloride ions and ammonium ions. The ammonium ion is a weak acid and will dissociate partially to produce ammonia and H+ ions in the solution, making it acidic.

3. D

The Zn is being oxidized (0 to 2+) while the Cu ion is being reduced (2+ to 0) in a redox displacement reaction. (A) is a redox combination reaction, (B) and (D) are not redox, and (C) is a redox combustion reaction.

4. E

The correct answer is H_2SO_4(aq). Nonmetal oxides react with water to form strong acids. Sulfur trioxide gas will react with water to form sulfuric acid, which will then dissociate completely to produce hydrogen ions in solution.

5. B

When net ionic equations are written, some ions appear on both sides of the equation. These ions are known as "spectator ions" because they do not appear to participate in the reaction. They are thus left out of the final net ionic equation.

The ionic equation is $2 Ag^+(aq) + 2 NO_3^-(aq) + H_2S(g) \rightarrow Ag_2S(s) + 2 H^+(aq) + 2 NO_3^-(aq)$

Deleting the NO_3^- gives the net ionic equation is $2 Ag^+(aq) + H_2S(g) \rightarrow Ag_2S(s) + 2 H^+(aq)$

Free-Response Answer

1. (A) $Mg^{2+}(aq) + 2 Cl^-(aq) + 2 Na^+(aq) + 2 OH^-(aq) \rightarrow Mg(OH)_2(s) + 2 Na^+(aq) + 2 Cl^-(aq)$

$Mg^{2+}(aq) + 2 OH^-(aq) \rightarrow Mg(OH)_2(s)$

All Group 1 compounds are soluble. Hydroxide compounds are insoluble except for Ca^{2+}, Ba^{2+}, and Sr^{2+}.

(B) $2 KCl(aq) + CuSO_4(aq) \rightarrow K_2SO_4(aq) + CuCl_2(aq)$

$2 K^+(aq) + 2 Cl^-(aq) + Cu^{2+}(aq) + SO_4^{2-}(aq) \rightarrow 2 K^+(aq) + SO_4^{2-}(aq) + Cu^{2+}(aq) + 2 Cl^-(aq)$

There is no chemical reaction between the reactants, and there is no net ionic equation.

(C) $Na_2SO_4(aq) + ? \rightarrow ppt + solution$

Since sodium compounds are always soluble, the precipitate must be a cation which is insoluble with the sulfate ion. The insoluble sulfates are Ca^{2+}, Ba^{2+}, Sr^{2+}, Ag^+, Hg_2^{2+}, and Pb^{2+}. Thus the aqueous compound to combine with the $Na_2SO_4(aq)$ can include *any* of these six listed cations in combination with any anion that provides an aqueous solution. Nitrate compounds ($Ca(NO_3)_2$, $AgNO_3$, and so on) are always good as a source of soluble salts.

(D) To separate Fe^{3+}, Ca^{2+}, and Pb^{2+} in solution, one must look for the exceptions in the solubility rules. We need to precipitate and separate the ions in a consecutive manner. If NaCl, NaBr, or NaI are added first, the Pb^{2+} is precipitated, but not the Fe^{3+} or Ca^{2+}. Centrifuge off the precipitate and pour off the solution containing the Fe^{3+} and Ca^{2+}. Next add a source of OH^- (NaOH or NH_4OH). This precipitates the Fe^{3+} ion but not the Ca^{2+}. Centrifuge off the precipitate and pour off the solution containing the Ca^{2+}. Now a source of carbonate ion (Na_2CO_3) is added to completely precipitate the Ca^{2+}.

Chemistry Calculations

In order to measure any mass we need to establish a standard and a unit. In the SI system, the standard is a platinum-iridium cylinder and the unit is the kilogram. In the world of atoms, an **atomic mass scale** has been established: the standard is one atom of carbon-12 and the unit of measure is the **atomic mass unit** (u). By definition, one atom of carbon-12 is assigned a mass of *exactly* 12 atomic mass units (12 u). In the SI system, $1u = 1.661 \times 10^{-27}$ kg. The masses of other atoms, molecules, ions, and subatomic particles are measured by a device known as a *mass spectrometer* and the values are reported relative to the mass of carbon-12. The table below lists the masses of the familiar subatomic particles in the SI and atomic mass systems:

Particle	/kg	/u
proton	1.6726×10^{-27}	1.0073
neutron	1.6749×10^{-27}	1.0087
electron	9.1094×10^{-31}	0.00054858

As we can see from the table the proton and neutron each have masses nearly equal to 1 atomic mass unit, while the electron's mass is small enough (≈ 0.0005 u) to be neglected for most purposes.

ATOMIC MASSES OF ELEMENTS

Most elements are mixtures of isotopes, and in order to calculate the atomic mass of an element we need to know three things:

- Which isotopes are present naturally
- The masses of each of the isotopes
- The abundance of each isotope in the element

The following table lists this data for the element zinc:

Isotope	Isotopic mass /u	Abundance /%
^{64}Zn	63.9291	48.63
^{66}Zn	65.9260	27.90
^{67}Zn	66.9271	4.10
^{68}Zn	67.9248	18.75
^{70}Zn	69.9253	0.62

The atomic mass of an element is the **weighted average** of the masses of the naturally occurring isotopes forming the element. We calculate the weighted average by multiplying the mass of each isotope by the decimal equivalent of its abundance and then we add each of these products together:

$$(63.9291 \text{ u})\cdot(0.4863) + (65.9260 \text{ u})\cdot(0.2790) + (66.9271 \text{ u})\cdot(0.0410) +$$

$$(67.9248 \text{ u})\cdot(0.1875) + (69.9253 \text{ u})\cdot(0.0062) = \textbf{65.39 u}$$

The atomic masses of the elements are found in a table located in Appendix B.

FORMULA MASS

The **formula mass** is simply an extension of the concept of atomic mass. For example, the formula mass of the compound $C_6H_{12}O_6$ is calculated by multiplying the number of atoms of each element in the formula by the atomic mass of the element and adding the results together:

$$6\cdot(12.01 \text{ u}) + 12\cdot(1.008 \text{ u}) + 6\cdot(16.00 \text{ u}) = \textbf{180.2 u}$$

PERCENT COMPOSITION FROM FORMULA MASS

The **percent composition** of an element is the relative contribution of the mass of the element to the mass of the formula in which it appears. We can use the following equation in order to calculate percent composition:

$$\text{Percent Composition} = \frac{\text{mass of element in formula}}{\text{formula mass}} \times 100$$

Therefore, the percent of carbon in $C_6H_{12}O_6$ is:

$$\text{percent carbon} = \frac{72.06 \text{ u}}{180.2 \text{u}} \times 100 = \textbf{40.00\%}$$

When the mass of *each* element in a substance is specified, either as a percentage or in grams, it is known as the **mass composition** of the substance.

Sample Problem

What is the mass composition of the elements associated with the formula CH_2?

Solution

$$\%C = \frac{12.01\ u}{14.03\ u} \times 100 = \mathbf{85.60\%}$$

$$\%H = \frac{2.016\ u}{14.03\ u} \times 100 = \mathbf{14.40\%}$$

Sample Problem

CH_2 is the empirical formula (see page 42) of the compounds C_2H_4, C_3H_6, and C_4H_8. What are the mass compositions of each of these compounds?

Solution

The three compounds have the same mass composition as that of CH_2 because the masses of C and H in each compound are proportional to the masses of C and H in CH_2.

Sample Problem

Calculate the mass composition in grams of a 2.474-gram sample of C_3H_6.

Solution

We know from above that the mass composition in percent for C_3H_6 is 85.60% C and 14.40% H. Therefore, the mass composition in grams is:

$$C: (2.474\ g) \cdot (0.8560) = \mathbf{2.118\ g}$$

$$H: (2.474\ g) \cdot (0.1440) = \mathbf{0.356\ g}$$

THE MOLE

Atoms, ions, and simple molecules are extremely small entities, therefore, it is necessary to work with very large numbers of these entities in order to obtain quantities that are measurable in a laboratory situation. For this purpose, the amount of substance known as the **mole** (mol) has been established. One mole is the *number* of atoms contained in exactly 0.012 kilogram (12 grams) of carbon-12. This number is known as **Avogadro's number** (N_A) and is equal to approximately 6.022×10^{23}. For example, if we wanted to calculate the number of H_2 molecules in 2.20 moles of this substance, we could use dimensional analysis to solve the problem:

$$2.20\ \text{mol}\ H_2 \cdot \left(\frac{6.02 \times 10^{23}\ H_2\ \text{molecules}}{1.00\ \text{mol}\ H_2} \right) = \mathbf{1.32 \times 10^{24}\ H_2\ molecules}$$

Any chemical formula can also be interpreted as representing 1 mole of the substance. The formula $C_6H_{12}O_6$ is interpreted to mean 1 mole of molecules, consisting of 6 moles of carbon atoms, 12 moles of hydrogen atoms, and 6 moles of oxygen atoms. The formula $CaCl_2$ is interpreted to mean 1 mole of this compound, consisting of 1 mole of Ca^{2+} ions and 2 moles of Cl^- ions.

MOLAR MASS

The **molar mass** ($\mathcal{M}$) of a substance is the mass, in grams, associated with one mole of the substance. The units of molar mass are grams per mole (g/mol). Since the mole and the atomic mass scale are both based on carbon-12, it follows that there must be a connection between molar mass and atomic mass: *the molar mass of an element is its atomic mass in grams*. For example, the molar mass of Fe is 55.85 grams per mole. *The molar mass of a compound or ion is its formula mass in grams*. For example, the molar mass of $C_6H_{12}O_6$ is 180.2 grams per mole. If m is the mass (in grams) of n moles of a substance, and $\mathcal{M}$ is its molar mass (in grams per mole), then

$$n = \frac{m}{\mathcal{M}}$$

We can easily use the formula shown above or dimensional analysis to convert among mass, molar mass and number of moles.

Sample Problem

Calculate the mass of 1.223 mol of H_2O ($\mathcal{M}$ = 18.02 g/mol).

Solution

$$1.223 \text{ mol} \cdot \left(\frac{18.02 \text{ g}}{1 \text{ mol}}\right) = \textbf{22.04 g}$$

Sample Problem

Calculate the number of moles of Fe_2S_3 ($\mathcal{M}$ = 207.9 g/mol) present in 150.0 g of the compound.

Solution

$$150.0 \text{ g} \cdot \left(\frac{1 \text{ mol}}{207.9 \text{ g}}\right) = \textbf{0.7215 mol}$$

Alternate Solution

We can substitute the data directly into the formula given above:

$$n = \frac{m}{\mathcal{M}} = \frac{150.0 \text{ g}}{207.9 \frac{\text{g}}{\text{mol}}} = \textbf{0.7215 mol}$$

We can also use Avogadro's number in order to incorporate numbers of particles into our problems.

Sample Problem

Calculate the number of NO_2 molecules ($\mathcal{M}$ = 46.01 g/mol) present in 30.2 g of the compound.

Solution

$$30.2 \text{ g} \cdot \left(\frac{1 \text{ mol}}{46.01 \text{ g}}\right) \cdot \left(\frac{6.02 \times 10^{23} \text{ molecules}}{1 \text{ mol}}\right) = \textbf{3.95} \times \textbf{10}^{\textbf{23}} \textbf{ molecules}$$

CALCULATION OF EMPIRICAL FORMULAS

Calculation of Empirical Formulas from Mass Composition

Recall that a chemical formula can represent 1 mole of a substance. For example, the formula H_2SO_3 represents a 1-mole sample that contains 2 moles of H atoms, 1 mole of S atoms, and 3 moles of O atoms. If we know the mass composition of a substance, we can convert it to the number of moles of each element present in the sample. The following steps show how this is done:

- If the mass composition is expressed as a percentage, convert it to grams by assuming a 100.00-gram sample.

- Divide the mass of each element present by the atomic mass of that element in order to convert it to moles.

- Divide each number of moles by the smallest number of moles present in order to obtain the relative numbers of moles of each element in the formula.

- Express the relative numbers of moles in terms of smallest whole numbers.

Note that this technique *only* allows one to calculate an empirical formula of a substance.

Sample Problem

Calculate the empirical formula of a compound whose mass analysis is 20.00% Mg, 26.67% S, and 53.33% O.

Solution

$$Mg \frac{20.00 \text{ g}}{24.31 \frac{g}{mol}} \quad S \frac{26.67 \text{ g}}{32.07 \frac{g}{mol}} \quad O \frac{53.33 \text{ g}}{16.00 \frac{g}{mol}} =$$

$$Mg_{0.8227 \text{ mol}} \ S_{0.8316 \text{ mol}} \ O_{3.333 \text{ mol}}$$

$$Mg_{\frac{0.8227 \text{ mol}}{0.8227 \text{ mol}}} \ S_{\frac{0.8316 \text{ mol}}{0.8227 \text{ mol}}} \ O_{\frac{3.333 \text{ mol}}{0.8227 \text{ mol}}}$$

$$Mg_1 \ S_{1.011} \ O_{4.051} = MgSO_4$$

Sample Problem

A 6.463-gram sample of aluminum oxide is found to contain 3.420 grams of aluminum. Calculate the empirical formula of the compound.

Solution

Since the compound contains 3.420 grams of aluminum, it must contain 6.463g − 3.420g = 3.043g of oxygen.

$$Al_{\frac{3.420\,g}{26.98\,\frac{g}{mol}}}\ O_{\frac{3.043\,g}{16.00\,\frac{g}{mol}}}\ =$$

$$Al_{0.1268\,mol}\ O_{0.1902\,mol}$$

$$Al_{\frac{0.1268\,mol}{0.1268\,mol}}\ O_{\frac{0.1902\,mol}{0.1268\,mol}}$$

$$Al_1\,O_{1.500}\ =\ Al_2O_3$$

In the last step, we multiplied the entire formula by 2 in order to obtain smallest whole-number ratios.

Calculation of Empirical Formulas from Combustion Analysis

When carbon-hydrogen and carbon-hydrogen-oxygen compounds are combusted in an excess of O_2, the only products formed are CO_2 and H_2O. By measuring the mass of the original sample and the masses of the products, it is possible to calculate the empirical formula of the compound.

- Since all of the carbon and hydrogen appear as CO_2 and H_2O respectively, the masses of these elements can be determined.

- If the original compound contains oxygen, its mass is determined by subtracting the mass of the compound from the sum of the masses of carbon and hydrogen.

- The masses of the elements are converted to moles and the empirical formula is determined as shown above.

The following problem demonstrates the technique:

Sample Problem

Combustion of 11.5 grams of ethanol (a compound containing carbon, hydrogen, and oxygen) produces 22.0 grams of CO_2 and 13.5 grams of H_2O. Determine the empirical formula of ethanol.

Solution

$$22.0 \text{ g } CO_2 \cdot \left(\frac{1 \text{ mol } CO_2}{44.01 \text{ g } CO_2} \right) \cdot \left(\frac{1 \text{ mol C}}{1 \text{ mol } CO_2} \right) \cdot \left(\frac{12.01 \text{ g C}}{1 \text{ mol C}} \right) = \mathbf{6.00 \text{ g C}}$$

$$13.5 \text{ g } H_2O \cdot \left(\frac{1 \text{ mol } H_2O}{18.02 \text{ g } H_2O} \right) \cdot \left(\frac{2 \text{ mol H}}{1 \text{ mol } H_2O} \right) \cdot \left(\frac{1.008 \text{ g H}}{1 \text{ mol H}} \right) = \mathbf{1.51 \text{ g H}}$$

$$\text{mass of O} = 11.5 \text{ g} - (6.00 \text{ g} + 1.51 \text{ g}) = \mathbf{4.0 \text{ g O}}$$

$$C_{\frac{6.00 \text{ g}}{12.01 \frac{g}{mol}}} \; H_{\frac{1.5 \text{ g}}{1.00 \frac{g}{mol}}} \; O_{\frac{4.0 \text{ g}}{16.00 \frac{g}{mol}}} = C_{0.50 \text{ mol}} H_{1.5 \text{ mol}} O_{0.25 \text{ mol}} = \mathbf{C_2H_6O}$$

Determination of Molecular Formulas from Empirical Formulas

If the molar mass of a compound is known in addition to its empirical formula, its molecular formula can be determined by dividing the molar mass of the compound by the *empirical molar mass*. The empirical formula is then multiplied by the quotient obtained from the division.

Sample Problem

The empirical formula of a compound is CH_2O and its molar mass is 180.2 grams per mole. Determine the molecular formula of the compound.

Solution

The empirical formula mass is 30.03 grams per mole.

$$\frac{180.2 \frac{g}{mol}}{30.03 \frac{g}{mol}} = 6.001 \approx \mathbf{6}$$

$$6 \cdot (CH_2O) = \mathbf{C_6H_{12}O_6}$$

Miscellaneous Problems Involving Composition

Sample Problem

An 8.129-gram sample of hydrated magnesium sulfate, $MgSO_4 \cdot xH_2O$, is heated until all of the water of hydration is driven off. The mass of the remaining *anhydrous* magnesium sulfate, $MgSO_4$, is 3.967 grams. Determine the value of x in the hydrated compound.

Solution

The mass of the water of hydration is: 8.129 g – 3.967 g = 4.162 g. The fraction of water in the hydrated compound is: $\dfrac{4.162g}{8.129\ g} = 0.5120$.

The fraction of water can also be expressed in terms of the formula of the hydrate and the molar masses of $MgSO_4$ and H_2O:

$$\text{fraction of } H_2O = \frac{18.02 \cdot x}{120.4 + 18.02 \cdot x}$$

where the numerator represents the mass of the water present in one mole of $MgSO_4 \cdot xH_2O$, and the denominator represents the molar mass of $MgSO_4 \cdot xH_2O$. Equating the two expressions and solving for x, we obtain:

$$\frac{18.02 \cdot x}{120.4 + 18.02 \cdot x} = 0.5120; \ \boldsymbol{x = 7.01}$$

The formula of the hydrated compound is **$MgSO_4 \cdot 7H_2O$**.

CALCULATIONS INVOLVING CHEMICAL EQUATIONS

The mole concept can be extended to balanced chemical equations. For example, the coefficients in the balanced equation:

$$2C_2H_6 + 7O_2 \rightarrow 4CO_2 + 6H_2O$$

can be interpreted to mean that 2 moles of C_2H_6 combines with 7 moles of O_2 to produce 4 moles of CO_2 and 6 moles of H_2O. We can also focus our attention on a subset of the equation. For example, we can say that 2 moles of C_2H_6 combines with 7 moles of O_2, or we can say that (in this particular equation) 7 moles of O_2 will lead to the production of 6 moles of H_2O.

Since the number of moles of a substance is directly related to its mass in grams, we can solve a variety of problems with chemical equations.

Mole and Mass Problems

In solving the next three problems, we will use the equation: $2C_2H_6 + 7O_2 \rightarrow 4CO_2 + 6H_2O$. The solutions involve the use of dimensional analysis, the relevant equation coefficients, and (when necessary) the molar masses of the substances involved.

Sample Problem

If 1.5 moles of C_2H_6 reacts, how many moles of H_2O will be formed?

Solution

$$1.5 \text{ mol } C_2H_6 \left(\frac{6 \text{ mol } H_2O}{2 \text{ mol } C_2H_6} \right) = \textbf{4.5 mol } C_2H_6$$

Sample Problem

If 1.50 moles of C_2H_6 reacts, what mass of O_2 will also react?

Solution

$$1.50 \text{ mol } C_2H_6 \cdot \left(\frac{7 \text{ mol } O_2}{2 \text{ mol } C_2H_6} \right) \cdot \left(\frac{32.0 \text{ g } O_2}{1 \text{ mol } O_2} \right) = \textbf{168 g } O_2$$

Sample Problem

If 160. grams of O_2 reacts, how many grams of CO_2 will be formed?

Solution

$$160. \text{ g } O_2 \cdot \left(\frac{1 \text{ mol } O_2}{32.0 \text{ g } O_2} \right) \cdot \left(\frac{4 \text{ mol } CO_2}{7 \text{ mol } O_2} \right) \cdot \left(\frac{44.0 \text{ g } CO_2}{1 \text{ mol } CO_2} \right) = \textbf{126 g } CO_2$$

Sample Problem

Consider the reactions:

$$NaHCO_3 + HCl \rightarrow NaCl + CO_2 + H_2O$$

$$Na_2CO_3 + 2HCl \rightarrow 2NaCl + CO_2 + H_2O$$

(A) A 12.00-gram sample of solid $NaHCO_3$ is treated with an excess of HCl and heated to remove the water. What is the change in the mass of the solid?

(B) A 3.75-gram sample of a solid is either $NaHCO_3$ or Na_2CO_3. When treated with excess HCl and heated, the mass of the solid increases by 0.38 gram. Identify the original sample.

Solution

(A) After the reaction, the solid that is present is NaCl. From the equation, we know that 1 mole of $NaHCO_3$ produces 1 mole of NaCl.

$$12.00 \text{ g } NaHCO_3 \cdot \left(\frac{1 \text{ mol } NaHCO_3}{84.01 \text{ g } NaHCO_3} \right) \cdot \left(\frac{1 \text{ mol } NaCl}{1 \text{ mol } NaHCO_3} \right) \cdot \left(\frac{58.44 \text{ g } NaCl}{1 \text{ mol } NaCl} \right) = \textbf{8.348 g NaCl}$$

$$8.348 \text{ g (NaCl)} - 12.00 \text{ g (}NaHCO_3\text{)} = \textbf{--3.65 g}$$

The mass of the solid *decreases* by 3.65 grams.

(B) We calculate the quantities of NaCl produced by 3.75 grams of $NaHCO_3$ and Na_2CO_3 and we compare the differences in mass with that given in the problem statement.

$$3.75 \text{ g NaHCO}_3 \cdot \left(\frac{1 \text{ mol NaHCO}_3}{84.01 \text{ g NaHCO}_3}\right) \cdot \left(\frac{1 \text{ mol NaCl}}{1 \text{ mol NaHCO}_3}\right) \cdot \left(\frac{58.44 \text{ g NaCl}}{1 \text{ mol NaCl}}\right) = \textbf{2.61 g NaCl}$$

$$2.61 \text{ g (NaCl)} - 3.75 \text{ g (NaHCO}_3) = \textbf{–1.14 g}$$

$$3.75 \text{ g Na}_2\text{CO}_3 \cdot \left(\frac{1 \text{ mol Na}_2\text{CO}_3}{106.0 \text{ g Na}_2\text{CO}_3}\right) \cdot \left(\frac{2 \text{ mol NaCl}}{1 \text{ mol Na}_2\text{CO}_3}\right) \cdot \left(\frac{58.44 \text{ g NaCl}}{1 \text{ mol NaCl}}\right) = \textbf{4.13 g NaCl}$$

$$4.13 \text{ g (NaCl)} - 3.75 \text{ g (Na}_2\text{CO}_3) = \textbf{+0.38 g}$$

It is evident that the original sample is Na_2CO_3 since the mass *increases* by 0.38 gram.

Miscellaneous Problems Involving Equations

Sample Problem

Magnesium and calcium carbonates can be decomposed by heat according to the reactions:

$$MgCO_3(s) \rightarrow MgO(s) + CO_2(g)$$

$$CaCO_3(s) \rightarrow CaO(s) + CO_2(g)$$

A certain limestone is a mixture of $MgCO_3$ and $CaCO_3$. When a 1.000-gram sample of this limestone was decomposed by heat, the mass of the remaining solid product was 0.5027 g. Determine the composition of the limestone.

Solution

Compound	Molar Mass
MgO	40.30 g/mol
$MgCO_3$	84.31 g/mol
CaO	56.08 g/mol
$CaCO_3$	100.1 g/mol

In each reaction, the decomposition of one mole of the carbonate produces one mole of the oxide.

Let x = mass of $CaCO_3$; 1.000-x = mass of $MgCO_3$.

Let y = mass of CaO; 0.5027-y = mass of MgO.

$$n_{CaCO_3} = \frac{x \text{ g}}{100.1 \frac{\text{g}}{\text{mol}}}; \quad n_{CaO} = \frac{y \text{ g}}{56.08 \frac{\text{g}}{\text{mol}}}$$

$$n_{MgCO_3} = \frac{1.000 - x \text{ g}}{84.31 \frac{\text{g}}{\text{mol}}}; \quad n_{MgO} = \frac{0.5027 - y \text{ g}}{40.30 \frac{\text{g}}{\text{mol}}}$$

Since $n_{CaCO_3} = n_{CaO}$ and $n_{MgCO_3} = n_{MgO}$, we can solve the simultaneous equations for x and y:

$$x = 0.3000 \text{ g } CaCO_3; \quad y = 0.7000 \text{ g } MgCO_3.$$

The composition of the limestone is:
$$\textbf{CaCO}_3 = \textbf{30.00\%}; \quad \textbf{MgCO}_3 = \textbf{70.00\%}$$

Sample Problem

When 4.000 grams of the solid M_2S_3 is heated strongly in pure oxygen, the only solid that remains has the formula MO_2. If the oxide weighs 0.277 grams *less* than the sulfide, identify the element "M" by determining its molar atomic mass ($\mathcal{M}$).

Solution

When one mole of M_2S_3 reacts, *two* moles of MO_2 are formed. The atomic mass of M_2S_3 is $(2\mathcal{M} + 96.18)$ grams per mole; the atomic mass of MO_2 is $(\mathcal{M} + 32.00)$ grams per mole.

$$n_{M_2S_3} = \frac{4.000 \text{ g}}{(2\mathcal{M} + 96.18) \frac{\text{g}}{\text{mol}}}; \quad n_{MO_2} = \frac{4.000 - 0.277 \text{ g}}{(\mathcal{M} + 32.00) \frac{\text{g}}{\text{mol}}}$$

$2 \cdot n_{M_2S_3} = n_{MO_2}$ (2 moles of MO_2 are produced for every mole of M_2S_3 reacted.)

Solving for $\mathcal{M}$, we obtain: $\mathcal{M} = 183.9 \frac{\text{g}}{\text{mol}}$; **M is the element *tungsten* (symbol W).**

PERCENT YIELD

In one of the sample problems given on page 61, we calculated that the combustion of 160. grams of C_2H_6 would produce 126 grams of CO_2. This calculation is based on the assumption that the reaction goes to completion as written. The yield of CO_2 under these ideal conditions is known as the **theoretical yield**. Under laboratory conditions, the **actual yield** is usually less than the theoretical yield. We define the **percent yield** of a reaction as follows:

$$\% \text{ yield} = \frac{\text{actual yield}}{\text{theoretical yield}} \times 100$$

For example, if the combustion of 160. grams of C_2H_6 under laboratory conditions produced 72.0 grams of CO_2, the percent yield of the reaction would be:

$$\% \text{ yield} = \frac{72.0 \text{ g}}{126 \text{ g}} \times 100 = 57.1\%.$$

LIMITING REACTANTS

Consider the following reaction:

$$2AgNO_3(aq) + Na_2SO_4(aq) \rightarrow Ag_2SO_4(s) + 2NaNO_3(aq)$$

According to the reaction, 2 moles of $AgNO_3$ will react with 1 mole of Na_2SO_4 (under ideal conditions). Whenever the mole ratio of the $AgNO_3 : Na_2SO_4$ is 2 : 1, we say that the reactants are present in **stoichiometric quantities**. At the completion of the reaction, both reactants will have been entirely consumed.

What would happen if we mixed 3 moles of $AgNO_3$ with 1 mole of Na_2SO_4? Since 1 mole of Na_2SO_4 can react with only 2 moles of $AgNO_3$, it follows that 1 mole of $AgNO_3$ would remain unreacted. Under these conditions, we say that the $AgNO_3$ is present in **excess**. The masses of the products of the reaction are controlled entirely by the Na_2SO_4, which is known as the **limiting reactant.**

In order to determine the limiting reactant in a reaction, we employ the following steps:

- Use the equation to calculate the stoichiometric mole ratio of the reactants.

- Calculate the mole ratio of the reactants under the experimental conditions given.

- Compare the two mole ratios: If the experimental mole ratio is *larger* than the stoichiometric mole ratio, the reactant in the *denominator* is the limiting reactant; if it is *smaller*, the reactant in the *numerator* is the limiting reactant.

- In solving any problems involving calculations, always use the limiting reactant.

The following problem demonstrates the determination of a limiting reactant.

Sample Problem

How many grams of Fe_2O_3 can be produced by the reaction of 151.0 grams of Fe with 175.0 grams of O_2 according to the equation: $4Fe(s) + 3O_2(g) \rightarrow 2Fe_2O_3(s)$?

Solution

The stoichiometric mole ratio Fe : O_2 is 4 : 3 = **1.333**.

Under experimental conditions, the mole ratio is

$$\frac{151.0 \text{ g Fe}}{55.85 \frac{\text{g}}{\text{mol}}} : \frac{175.0 \text{ g } O_2}{32.00 \frac{\text{g}}{\text{mol}}} = \frac{2.704 \text{ mol Fe}}{5.469 \text{ mol } O_2} = \mathbf{0.4944.}$$

Since the experimental ratio is *smaller* than the stoichiometric ratio, **Fe is the limiting reactant**.

We now solve the problem in the usual way:

$$151.0 \text{ g Fe} \cdot \left(\frac{1 \text{ mol Fe}}{55.85 \text{ g Fe}}\right) \cdot \left(\frac{2 \text{ mol } Fe_2O_2}{4 \text{ mol Fe}}\right) \cdot \left(\frac{159.7 \text{ g } Fe_2O_3}{1 \text{ mol } Fe_2O_3}\right) = \mathbf{215.9 \text{ g } Fe_2O_3}$$

STOICHIOMETRY OF AQUEOUS SOLUTIONS

When reactions take place in aqueous solutions, the water generally acts as the "carrier" for the reactants and products of the reaction. The water is known as the **solvent** and any substance that is dissolved in it is known as a **solute**. In order to solve problems involving aqueous solutions, it is necessary to be able to calculate the **composition** or **concentration** of the solution. One method of expressing concentration is detailed below.

Molarity

Molarity is defined as the number of moles of solute dissolved in one liter of solution:

$$\boxed{M = \frac{n_{\text{solute}}}{V_{\text{solution}}}}$$

The unit of molarity is **moles per liter** (mol/L). An alternative—and equivalent—unit is **millimoles per milliliter** (mmol/mL). If a solution has 2.0 moles of solute dissolved in 5.0 liters of solution, its concentration is 0.40 mole per liter. The can be written in an abbreviated form: 0.40 M, and is referred to as a 0.40-molar solution.

Sample Problem

90.10 grams of $C_6H_{12}O_6$ ($\mathcal{M}$ = 180.2 g/mol) is dissolved in 250.0 milliliters of solution. Calculate the molarity of the solution.

Solution

In applying the definition of molarity, we need to convert the mass of the solute to moles and specify the volume of the solution in liters:

$$M = \frac{n_{\text{solute}}}{V_{\text{solution}}} = \frac{\dfrac{90.10 \text{ g}}{180.2 \dfrac{\text{g}}{\text{mol}}}}{250.0 \text{ mL} \cdot \left(\dfrac{1\text{L}}{1000 \text{ mL}}\right)} = \frac{0.5000 \text{ mol}}{0.2500 \text{ L}} = \mathbf{2.000 \text{ M}}$$

Dilution of Aqueous Solutions

When water is added to an aqueous solution, it is evident that the volume of the solution increases and the concentration of the solution decreases. What does *not* change is the number of moles (or millimoles) of solute present in the solutions before and after dilution. Since $n = M \cdot V$, we can write:

$$\boxed{(M \cdot V)_{\text{before dilution}} = (M \cdot V)_{\text{after dilution}}}$$

Sample Problem

500. mL of 2.00 M NaOH is diluted with water to 975 mL. Calculate the molarity of the resulting solution.

Solution

$$(2.00 \text{ M}) \cdot (500. \text{ mL}) = (x \text{ M}) \cdot (975 \text{ mL})$$

$$x = \mathbf{1.03 \text{ M}}$$

Sample Problem

It is necessary to prepare 250. mL of 1.50 M $C_6H_{12}O_6$ from a **stock solution** whose concentration is 5.60 M. How should this be done?

Solution

$$(5.60 \text{ M}) \cdot (x \text{ mL}) = (1.50 \text{ M}) \cdot (250. \text{ mL})$$

$$x = \mathbf{67.0 \text{ mL}}$$

Withdraw 67.0 mL of the stock solution and dilute it to 250. mL with water.

Molarity and Calculations Involving Equations

The following examples demonstrate how molarity can be used to solve problems involving reactions that take place in aqueous solution.

Sample Problem

How many grams of Zn(s) will react completely with 0.300 liter of 1.40 M HCl(aq) according to the equation:

$$Zn(s) + 2\ HCl(aq) \rightarrow H_2(g) + ZnCl_2(aq)$$

Solution

The volume of HCl solution is converted to moles, and the problem is then solved in the usual way.

$$0.300\ \text{L HCl} \cdot \left(\frac{1.40\ \text{mol HCl}}{1\ \text{L HCl}} \right) \cdot \left(\frac{1\ \text{mol Zn}}{2\ \text{mol HCl}} \right) \cdot \left(\frac{65.4\ \text{g Zn}}{1\ \text{mol Zn}} \right) = \textbf{13.7 g Zn}$$

Titration Problems

Titration is the technique in which a precise volume of a **standardized solution** (i.e., a solution of known concentration) is added to a measured volume of a solution whose concentration is *not* known. The technique employs the use of a long graduated tube, known as a **burette**, to deliver the standard solution. The delivery continues until some external sign, such as a color change, signifies the **endpoint** of the titration. Generally, a small amount of a substance known as an **indicator** is used to effect the color change. This technique is shown in the diagram below:

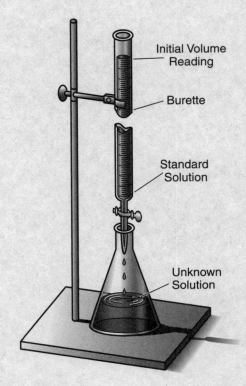

Initial Volume Reading

Burette

Standard Solution

Unknown Solution

Once the data is collected, the unknown concentration can be determined by using the balanced equation for the reaction. Titration is particularly useful in acid-base reactions. The following problem illustrates how a titration problem is solved.

Sample Problem

A burette is used to deliver standardized NaOH to a beaker containing 25.0 mL of H_2SO_4. The concentration of the standardized NaOH is 0.267 M and the endpoint is reached when 36.8 mL of the base has been delivered. The equation for the reaction is:

$$2NaOH(aq) + H_2SO_4(aq) \rightarrow Na_2SO_4(aq) + 2H_2O(\ell)$$

Calculate the molarity of the H_2SO_4.

Solution

We use dimensional analysis and we note that molarity and equation coefficients can be expressed in terms of millimoles and milliliters.

$$36.8 \text{ mL NaOH} \cdot \left(\frac{0.267 \text{ mmol NaOH}}{1 \text{ mL NaOH}} \right) \cdot \left(\frac{1 \text{ mmol } H_2SO_4}{2 \text{ mmol NaOH}} \right) = 4.91 \text{ mmol } H_2SO_4$$

Since the 4.91 mmol of H_2SO_4 is contained in 25.0 mL of solution, the molarity of the solution is:

$$\frac{4.91 \text{ mmol } H_2SO_4}{25.0 \text{ mL } H_2SO_4} = \textbf{0.197 M}$$

Chapter 5: Practice Questions

Multiple-Choice Questions

1. What is the empirical formula for the compound dichlorocyclohexane, $C_6H_{10}Cl_2$?

   A. CHCl

   B. $C_{1.5}H_{2.5}Cl$

   C. C_3H_5Cl

   D. $C_3H_5Cl_2$

   E. $C_6H_{10}Cl_2$

2. What is the empirical formula of a compound that is 43.64% phosphorus and 56.36% by weight oxygen?

   A. PO

   B. PO_3

   C. P_2O_3

   D. P_2O_5

   E. P_3O_2

3. 3.64 g of calcium hydroxide react with excess sodium sulfate in aqueous solution to produce solid calcium sulfate and aqueous sodium hydroxide. How many moles of calcium atoms are reacting here?

   $Ca(OH)_2(aq) + Na_2SO_4(aq) \rightarrow CaSO_4(s) + 2\ NaOH(aq)$

   A. 0.00982 mol
   B. 0.0246 mol
   C. 0.0266 mol
   D. 0.0491 mol
   E. 0.0909 mol

4. A 0.250 M solution of $AgNO_3$ is to be prepared. What mass of solid $AgNO_3$ do you need in order to prepare 50.0 mL of this solution?

   A. 2.12 g $AgNO_3$

   B. 4.98 g $AgNO_3$

   C. 6.66 g $AgNO_3$

   D. 9.87 g $AgNO_3$

   E. 12.5 g $AgNO_3$

5. Which compound has the highest percent by mass of nitrogen?

   A. $(CH_3)_3N(\ell)$

   B. $N_2O_4(\ell)$

   C. $HNO_3(g)$

   D. $NO_2(g)$

   E. $N_2(g)$

6. If the bacterial digestion reaction shown is allowed to go to completion, which is true regarding the relative mass of each of the products in the reaction?

$$5 CO_2(g) + 55 NH_4^+(aq) + 76 O_2(g) \rightarrow$$
$$C_5H_7O_2N(s) + 54 NO_2^-(aq) + 109 H^+(aq) +$$
$$52 H_2O(\ell)$$

A. $g\ C_5H_7O_2N > g\ H_2O > g\ NO_2^-$

B. $g\ H_2O > g\ NO_2^- > g\ C_5H_7O_2N$

C. $g\ H_2O > g\ C_5H_7O_2N > g\ NO_2^-$

D. $g\ NO_2^- > g\ C_5H_7O_2N > g\ H_2O$

E. $g\ NO_2^- > g\ H_2O > g\ C_5H_7O_2N$

7. If one mole of the rocket fuel ammonium perchlorate, NH_4ClO_4 (s) is allowed to react with excess Al so all of the NH_4ClO_4 is consumed, how many molecules of water will be produced?

$$3 NH_4ClO_4 (s) + 3 Al(s) \rightarrow Al_2O_3(s) +$$
$$AlCl_3(s) + 3 NO(g) + 6 H_2O(g)$$

A. 3.61×10^{23}

B. 1.00×10^{23}

C. 6.02×10^{23}

D. 1.20×10^{24}

E. 3.01×10^{24}

8. What volume of stock $AgNO_3$ solution with a concentration of 0.250 M is needed to prepare 150.0 mL of 0.015 M $AgNO_3$?

A. 6.00 mL
B. 9.00 mL
C. 16.7 mL
D. 48.8 mL
E. 78.0 mL

9. How many grams of potassium cyanide, PCl_3, is produced from 93.0 grams of $P_4(s)$ and 213 g of $Cl_2(g)$, assuming the reaction goes to completion? The balanced equation for the reaction is:

$$P_4(s) + 6 Cl_2(g) \rightarrow 4 PCl_3(g)$$

A. 277 g
B. 416 g
C. 213 g
D. 104 g
E. 69.3 g

10. A 58.90 g sample of a mystery compound containing carbon, hydrogen, and oxygen is subjected to combustion analysis. 86.35 g CO_2 and 35.35 g H_2O are produced in the combustion reaction. What percent by mass are carbon, hydrogen, and oxygen in the mystery compound?

A. 19.6% carbon, 3.3% hydrogen, and 77.1% oxygen

B. 40.0% carbon, 3.29% hydrogen, and 56.7% oxygen

C. 40.0% carbon, 1.6% hydrogen, and 58.4% oxygen

D. 78.5% carbon, 6.6% hydrogen, and 14.9% oxygen

E. 40% carbon, 6.7% hydrogen, and 53.4% oxygen

Free-Response Questions

1. The following four isotopes of magnesium are the only ones found in nature on earth:

Isotope	Percent Abundance	Atomic Mass
Magnesium-24	78.99%	23.9850
Magnesium-25	10.00%	24.9850
Magnesium-26	11.01%	25.9826

(A) Write the nuclide symbol for each isotope, showing both the mass number and the atomic number.

(B) Calculate the average atomic mass for the element magnesium from the data in the table.

2. Commercial brass, an alloy of Zn and Cu reacts with hydrochloric acid. Cu does not react with HCl. When a 5.000 gram sample of a certain brass alloy is reacted with 100.0 mL of 1.000 M HCl, 0.9722 g of $ZnCl_2(s)$ is eventually isolated.

$$Zn(s) + 2\ HCl(aq) \rightarrow ZnCl_2(s)$$

(A) How many grams of zinc are in the brass?

(B) What is the percent copper in the brass?

(C) During the isolation process the solution prepared in step (B) is transferred quantitatively to a 250.00 mL volumetric flask and diluted to the line. What is the molarity of the Zn^{2+} ion in this solution?

Multiple-Choice Answers

1. C   2. D   3. D   4. A   5. E   6. E   7. D   8. B   9. A   10. E

1. C

An empirical formula is analogous to a reduced fraction. It reduces the ratio of the elements in the compound to the lowest possible values—but all atoms must have whole number subscripts. $C_6H_{10}O_2$ is reduced to C_3H_5O when you divide each subscript by 2. It cannot be reduces further because the subscript for O is already down to 1. This makes answer choice (C) the correct answer.

2. D

Assume a 100. g sample of compound and calculate the number of moles of each element by dividing the amount of each element present by the molar mass of the element.

Moles of P: $43.64 \div 30.97 = 1.409$ moles of phosphorus

Moles of O: $56.36 \div 15.99 = 3.525$ moles of oxygen

From this information an informal molecular formula is written: $PO_{2.502}$

A formula is a simple, whole number ratio of atoms. At this point, this is not an acceptable formula because of the non-whole number in the oxygen subscript. If the formula is multiplied by 2, a correct whole-number ratio of P_2O_5 results.

3. D

Calcium hydroxide has a molar mass of 74.09 g/mol. There is 0.0491 mol in 3.64 g of $Ca(OH)_2$. In each mole of calcium hydroxide there is one mole of calcium atoms. So in 0.0491 mol $Ca(OH)_2$ there is 0.0491 mol Ca.

4. A

Molarity is expressed in units of moles per liter, so the volume must be converted to 0.0500 liters before multiplied by the molarity. The number of moles of $AgNO_3$ needed is 0.0125 mol. $AgNO_3$ has a molar mass of 169.9 g/mol, so 2.12 g of $AgNO_3$ is required to prepare the solution.

5. E

Computing the percent mass in a compound involves dividing the mass of the element present by the molar mass of the compound. $(CH_3)_3N$ (ℓ) is $14 \div 59 = 24\%$. Nitrogen gas, N_2, is made up entirely of nitrogen and gives the highest percent, 100%.

6. E

The relative mass of each product is based upon the stoichiometric relationship given in the balanced equation. There is 1 mole of bacterial tissue, $C_5H_7O_2N$, 54 moles of NO_2^-, and 52 moles of H_2O. The mass of each is found by multiplying the moles by their respective molar mass.

The total mass of $C_5H_7O_2N$ is $1 \times 113 = 113$ grams.

The total mass of NO_2^- is $54 \times 46 = 2484$ grams.

The total mass of H_2O is $52 \times 18 = 936$ grams.

So, $NO_2^- > H_2O > C_5H_7O_2N$.

7. D

This tests the understanding of the "limiting reagent" concept. Since there is an excess of Al and only one mole of NH_4ClO_4, the NH_4ClO_4 is the limiting reagent. The mole-ratio of NH_4ClO_4:H_2O is 3:6 or 1:2. From one mole of NH_4ClO_4, 2 moles of H_2O are formed. There are 6.02×10^{23} molecules per mole giving: 2 moles $\times 6.02 \times 10^{23}$ molecules/mole $= 1.20 \times 10^{24}$ molecules.

8. B

When a dilution is made, the number of moles in the sample remains the same; only the volume, and thus the molarity, changes. The final sample will be 0.150 L of 0.0150 M $AgNO_3$, which is 0.00225 mol. The starting sample is 0.250 M and must also contain 0.00225 mol $AgNO_3$. This 0.00225 mol divided by 0.250 mol/L gives 0.00900 L, equivalent to 9.00 mL of the original stock solution.

9. A

This problem uses the "limiting reagent" concept. This time it must be figured out which of the two reactants is the limiting reagent. How many moles of P_4 and Cl_2 at the start are calculated?

Moles of P_4 = 93.0 ÷ 124 = 0.750 moles; moles of Cl_2 = 213 g ÷ 71.0 = 3.00 moles

Based on the equation, the reactant molar ratio of Cl_2:P_4 is 6:1. Since 3.00 moles of Cl_2 react with only 0.500 mole of P_4 there is an *excess* of P_4 with Cl_2 being the limiting reagent.

The quantity of chlorine reacted (3.00 moles) determines that 2.00 moles of PCl_3 is produced. Multiplying this 2.00 moles by the molar mass of PCl_3 gives: 2.00 moles × 138.5 g/mol = 277 grams.

10. E

The masses of carbon dioxide and water are first converted to moles. The number of moles of carbon dioxide formed is equal to the number of moles of carbon in the original sample. The number of moles of water formed is one half the number of moles of hydrogen in the original sample. The number of moles of carbon and hydrogen is then converted to the mass of carbon and hydrogen using the molar masses of the elements. Any mass not accounted for by hydrogen or carbon is due to oxygen. The percent by mass of the elements can be determined by dividing the mass of the element by the total mass of the sample, and multiplying by 100.

Free-Response Answers

1. (A) The symbol for the element magnesium is Mg. The atomic number for magnesium is 26, as found in the Periodic Table. The mass number is the number following the name of the element in the table given.

$^{24}_{12}$Mg	$^{25}_{12}$Mg	$^{26}_{12}$Mg

(B) The average atomic mass is the average of the element's isotopes, weighted by natural abundance.

$$(0.7899 \times 23.9850) + (0.1000 \times 24.9850) + (0.1101 \times 25.9826) = 24.305 \text{ amu}$$

2. (A) Moles of $ZnCl_2$ is 0.9722 g ÷ 136.28 g/mol = 0.007134 moles

Just as a check, the HCl used is 0.1000 moles, which can react with 0.0500 moles of $ZnCl_2$. HCl is *not* the limiting reactant.

The mole ratio of Zn to $ZnCl_2$ is 1:1, so there are 0.007134 moles of zinc in the initial brass sample.

grams of Zn = 0.007134 mol × 65.38 g/mol = 0.4664 g

(B) grams of copper in the initial brass sample = 5.000 - 0.4664 = 4.534 g

percent copper = (4. 534 ÷ 5.000) × 100 = 90.67 %

(C) moles Zn^{2+} in solution = 0.007134 (from step A)

concentration of Zn^{2+} in solution = 0.007134 ÷ 0.250 L = 0.0285 M

Ideal Gas Behavior

CHAPTER SIX

Of the three phases of matter, the gaseous phase is in many respects the simplest one to study. A gas takes the shape and volume of its container and consists of independent particles in constant, random motion. In order to study gas behavior, only four variables need be specified: *pressure*, *volume*, *temperature*, and *number of particles*. If the conditions of temperature and pressure are not too stressful, the behavior of the gas will approximate the behavior of an abstract concept known as an *ideal gas*. Throughout this chapter we will assume that the gases under study approximate ideal behavior and we will explore this important concept in detail later in the chapter.

PRESSURE

Pressure is defined as the force exerted on or by a gas per unit of surface area:

$$P = \frac{F}{A}$$

The SI unit of pressure is the **newton per square meter** (N/m^2), which is also known as the pascal (Pa). The pascal is a small unit of pressure and, frequently, its multiple, the **kilopascal** (kPa), is used in calculations. **Standard pressure** is defined to be *exactly* 1×10^5 Pa (or 100 kPa). While the pressure of our atmosphere varies from day to day and from place to place, the **standard atmosphere** (atm) has been defined in terms of the SI: one standard atmosphere is *exactly* equal to 101.325 kPa. As you can see, standard pressure is nearly equal to the standard atmosphere.

Atmospheric pressure is measured with a device known as a **barometer**; it is illustrated below.

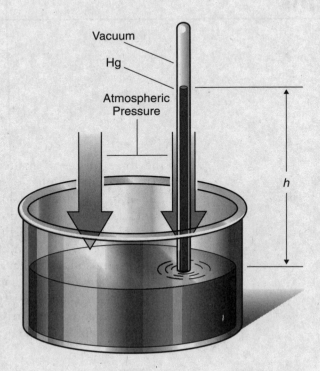

A barometer is essentially a long tube that is filled with liquid, inverted and immersed in an open vessel containing the same liquid. After inversion, the liquid in the tube falls to a specific height, which is directly proportional to the atmospheric pressure outside of the tube. If *mercury* is used as the liquid, a pressure of one atmosphere will support a column whose height is 760 millimeters. It is common to report pressure in terms of the height of the column, that is, in **millimeters of mercury** (mmHg). The millimeter of mercury is also known as the **torr**. In summary:

1 atm = 101.325 kPa = 760 mmHg = 760 torr

Gas pressure is also measured with an instrument known as a **manometer**, which consists of a u-tube that is partially filled with a liquid such as mercury. One end of the u-tube is connected to the container of gas that is being measured. If other end of the tube is sealed, and a vacuum exists above the liquid, the instrument is known as a **closed-tube manometer**; it is illustrated below:

Closed-Tube Maxometer

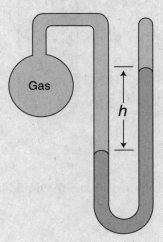

Closed-tube manometers are used to measure gas pressures that are considerably less than atmospheric pressure. Since the pressure of the gas causes the liquid levels to be different in height, it is this difference (h) that is the measure of the gas pressure in the container:

$$P_{gas} = P_h$$

If one end of the manometer is open to the atmosphere, it is known as an **open-tube manometer**, and is used to measure gas pressures that are at or near atmospheric pressure; several scenarios are illustrated on the next page:

Open-Tube Maxometer

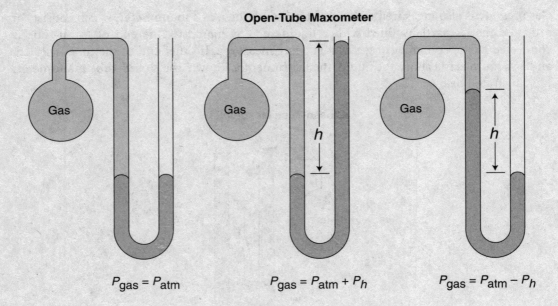

$P_{\text{gas}} = P_{\text{atm}}$ $\qquad$ $P_{\text{gas}} = P_{\text{atm}} + P_h$ $\qquad$ $P_{\text{gas}} = P_{\text{atm}} - P_h$

If the pressure of the gas in the container causes the liquid levels to be equal, then the gas is at atmospheric pressure:

$$P_{\text{gas}} = P_{\text{atm}}$$

If the liquid level in the tube connected to the gas is *lower* than the level in the open tube, then the gas pressure is *greater* than the atmospheric pressure by an amount that is equal to the difference in the heights of the liquid levels:

$$P_{\text{gas}} = P_{\text{atm}} + P_h$$

If the liquid level in the tube connected to the gas is *higher* than the level in the open tube, then the gas pressure is *less* than the atmospheric pressure by an amount that is equal to the difference in the heights of the liquid levels:

$$P_{\text{gas}} = P_{\text{atm}} - P_h$$

THE GAS LAWS

Boyle's Law

In the seventeenth century, the British chemist Robert Boyle investigated the relationship between the pressure of a gas and its volume. He observed that the pressure and the volume of a gas were inversely proportional if a fixed quantity of the gas (N) was maintained at constant temperature (T). The following relationships express **Boyle's law** in mathematical form:

$$P \cdot V = \text{constant}; \quad P_1 \cdot V_1 = P_2 \cdot V_2 \ (\text{constant N, T})$$

A graph of V versus P yields a hyperbola, while a graph of V versus $\frac{1}{P}$ yields a straight line as shown below:

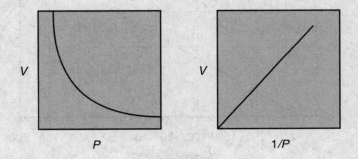

Sample Problem

A gas occupies a volume of 0.30 cubic decimeter at a pressure of 75 kilopascal. What volume does it occupy at 2.0 kilopascals?

Solution

In solving a Boyle's law problem, any units of pressure and volume can be used as long as there is consistency on both sides of the equation.

$$P_1 \cdot V_1 = P_2 \cdot V_2$$

$$75 \text{ kPa} \cdot 0.30 \text{ dm}^3 = 2.0 \text{ kPa} \cdot V_2$$

$$V_2 = 11 \text{ dm}^3$$

Charles's Law

In 1787, Jacques Charles observed that the volume of a fixed quantity of gas (N) varied linearly with the temperature if the pressure (P) was held constant. A graph of this relationship projects that the volume will fall to zero at −273°C, and is shown below:

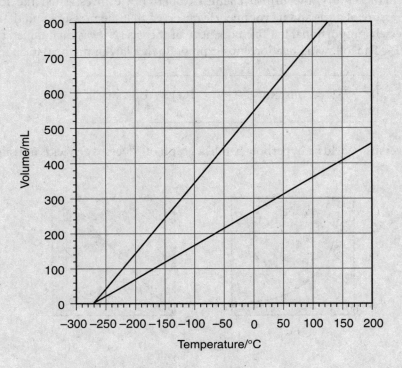

In 1848, William Thomson—known as Lord Kelvin—proposed an **absolute temperature scale** whose zero point coincided with the temperature at which the gas volume dropped to zero. The unit of this scale is the **kelvin** (K) and 0 K is exactly equal to −273.15°C. A temperature *change* of one kelvin is equal to a temperature *change* of one Celsius degree.

When the volume of a gas is graphed against the absolute temperature, the projected graph intersects the origin, as shown on the next page:

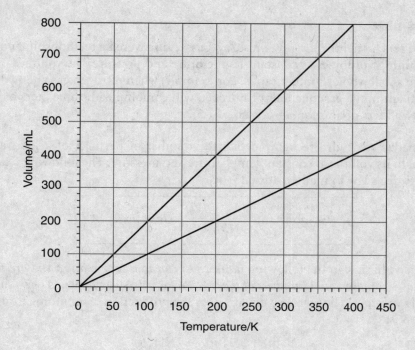

Under these conditions, the volume of the gas is *directly proportional* to the *absolute* temperature. The following relationships express **Charles's law** in mathematical form:

$$\frac{V}{T} = \text{constant}; \quad \frac{V_1}{T_1} = \frac{V_2}{T_2} \text{ (constant N, P)}$$

Sample Problem

A gas occupies 700.0 milliliters at 26.85°C. What volume will it occupy at 226.85°C?

Solution

We must convert both temperatures to their Kelvin equivalents:

$$T_1 = 26.85°C + 273.15 = 300.00 \text{ K}; \ T_2 = 226.85°C + 273.15 = 500.00 \text{ K}$$

$$\frac{V_1}{T_1} = \frac{V_2}{T_2}$$

$$\frac{700.0 \text{ mL}}{300.00 \text{ K}} = \frac{V_2}{500.00 \text{ K}}$$

$$V_2 = \textbf{1,166 mL}$$

Avogadro's Law

In 1808, the French scientist Joseph Louis Gay-Lussac discovered a relationship known as the **law of combining volumes**: At constant temperature and pressure, the volumes of reacting *gases* stand in small whole-number ratios. For example, when nitrogen gas and hydrogen gas react to form ammonia gas, one liter of nitrogen will combine with three liters of hydrogen to produce two liters of ammonia—a ratio of 1:3:2.

In 1811, Amadeo Avogadro recognized that the number of particles of a gas was directly proportional to its volume at constant temperature and pressure. The following relationships express **Avogadro's law** in mathematical form:

$$\frac{V}{N} = \text{constant}; \quad \frac{V_1}{N_1} = \frac{V_2}{N_2} \text{ (constant; } T, P)$$

Avogadro stated his law in the following fashion: At *constant temperature and pressure, equal volumes of gases contain equal numbers of particles*. Avogadro's law explains Gay-Luccac's law of combining volumes. The balanced equation for the production of ammonia from nitrogen and hydrogen is:

$$N_2(g) + 3H_2(g) \rightarrow 2NH_3(g)$$

Note that the coefficients—which represent the number of moles of particles reacting—stand in the ratio 1:3:2, the same ratio as the volumes of the gases that react.

The Ideal Gas Law

Boyle's law, Charles's law, and Avogadro's law can be combined into one relationship:

$$\frac{PV}{NT} = \text{constant}; \quad \frac{P_1V_1}{N_1T_1} = \frac{P_2V_2}{N_2T_2}$$

The constant is known as **Boltzmann's constant** and has the symbol k; its value is 1.381×10^{-23} joules per kelvin (J/K).

Sample Problem

A fixed quantity of gas occupies 300. mL at 200. kPa and 400. K. What volume will it occupy at 50.0 kPa and 200. K?

Solution

Since the quantity of gas is fixed, $N_1 = N_2$, and this term will drop out of the equation:

$$\frac{P_1V_1}{T_1} = \frac{P_2V_2}{T_2}$$

$$\frac{200. \text{ kPa} \cdot 300. \text{ mL}}{400. \text{ K}} = \frac{50.0 \text{ kPa} \cdot V_2}{200. \text{ K}}$$

$$V_2 = \textbf{600. mL}$$

Generally, the ideal gas law is written in the form:

$$PV = NkT$$

If the number of gas particles are expressed as *moles* (*n*), the ideal gas law takes the form:

$$PV = nRT$$

R is known as the **Gas constant**. Its value depends on the units in which pressure and volume are expressed. In the SI system, $R = 8.315$ joules per mole·kelvin (J/mol·K). If the pressure is measured in atmospheres and the volume in liters, $R = 0.08206$ liter·atmospheres per mole·kelvin (L·atm/mol·K).

If the two forms of the ideal gas law are compared, it is apparent that $Nk = nR$. Since $n/N = N_A$ (Avogadro's number), it follows that k and R are related by the expression: $k = R/N_A$.

Sample Problem

Standard Temperature and Pressure (STP) is defined as exactly 273.15 K and 1 atmosphere. What is the volume occupied by 1.000 mole of gas at STP?

Solution

$$PV = nRT$$

$$V = \frac{nRT}{P}$$

$$V = \frac{1.000 \text{ mol} \cdot 0.08206 \dfrac{\text{L} \cdot \text{atm}}{\text{mol} \cdot \text{K}} \cdot 273.15 \text{ K}}{1.000 \text{ atm}}$$

$$V = \textbf{22.41 L}$$

This volume is used so frequently in chemistry that is designated the **molar gas volume at STP** (V_m); its units are reported as liters (or cubic decimeters) per mole (L/mol or dm^3/mol).

Gas Density and Molar Mass

The density of a gas at a particular temperature and pressure can be related to its molar mass by using a modification of the ideal gas equation as shown below:

$$PV = nRT; \text{ Since } n = \frac{m}{\mathcal{M}}: PV = \frac{m}{\mathcal{M}} RT;$$

Solving for $\mathcal{M}$: $\mathcal{M} = \dfrac{mRT}{PV}$

Since $d = \dfrac{m}{V}$:

$$\mathcal{M} = \frac{dRT}{P}$$

Sample Problem

Calculate the molar mass of a gas if its density at 0.0800 atm and 400. K is 0.0391 g/L.

Solution

$$\mathcal{M} = \frac{dRT}{P} = \frac{0.0391 \, \frac{g}{L} \cdot 0.08206 \, \frac{L \cdot atm}{mol \cdot K} \cdot 400. \, K}{0.0800 \, atm}$$

$$\mathcal{M} = 16.0 \, \frac{g}{mol}$$

Sample Problem

Calculate the density of $N_2O_4(g)$ [$\mathcal{M}$ = 92.01 g/mol] at 700. torr and 350. K.

Solution

$$\mathcal{M} = \frac{dRT}{P}$$

$$d = \frac{\mathcal{M}P}{RT} = \frac{92.01 \, \frac{g}{mol} \cdot 700. \, torr \cdot \left(\frac{1.00 \, atm}{760. \, torr}\right)}{0.08206 \, \frac{L \cdot atm}{mol \cdot K} \cdot 350. \, K}$$

$$d = 2.95 \, \frac{g}{L}$$

Gas Volumes and Chemical Equations

Since the volume of a gas is related to the number of gas particles present, volume can be incorporated into problems involving chemical equations, as shown below:

Sample Problem

Given the balanced equation: $4NH_3(g) + 5O_2(g) \rightarrow 4NO(g) + 6H_2O(g)$

(A) How many liters of O_2, measured at STP, will react with 25.0 grams of NH_3?

(B) If 2.00 moles of NH_3 react, how many liters of NO will be formed at 700.°C and 6.00 atmospheres?

Solution

(A) $25.0 \, g \, NH_3 \cdot \left(\frac{1.00 \, mol \, NH_3}{17.0 \, g \, NH_3}\right) \cdot \left(\frac{5.00 \, mol \, O_2}{4.00 \, mol \, NH_3}\right) \cdot \left(\frac{22.4 \, L \, O_2}{1 \, mol \, O_2}\right)$

   $= 41.2 \, L \, O_2$ at STP

(B) $2.00 \text{ mol NH}_3 \cdot \left(\dfrac{4.00 \text{ mol NO}}{4.00 \text{ mol NH}_3}\right) = 2.00 \text{ mol NO}$

$PV = nRT$

$V = \dfrac{nRT}{P} = \dfrac{2.00 \text{ mol} \cdot 0.08206 \dfrac{\text{L} \cdot \text{atm}}{\text{mol} \cdot \text{K}} \cdot (700. \, ^{\circ}\text{C} + 273.15) \text{ K}}{6.00 \text{ atm}}$

$V = \mathbf{26.6 \text{ L NO}}$

Gas Mixtures: Dalton's Law of Partial Pressures

If a mixture of gases in a container behaves ideally, each gas acts as though it were the only gas in the container. It follows that:

- Each gas occupies the *entire* container

- Each gas exerts its own pressure—known as a **partial pressure**—on the walls of the container

- The total pressure is the *sum* of the individual partial pressures

Expressed mathematically, the last statement is known as **Dalton's law of partial pressures**:

$$P_t = P_1 + P_2 + P_3 + \ldots + P_i$$

where P_t equals the total pressure and $P_1, P_2, P_3, \ldots P_i$ equal the individual partial pressures. Since each component obeys the ideal gas law ($PV = nRT$) and has the same temperature and volume, it follows that the partial pressure of each gas in the container is *directly proportional* to the number of moles of gas present. In particular, we can write:

$$\frac{P_i}{P_t} = \frac{n_i}{n_t}$$

$$P_i = \frac{n_i}{n_t} \cdot P_t$$

$$P_i = X_i P_t$$

The quantity X_i is known as the **mole fraction** of gas component "i." Its use is shown in the following problem:

Sample Problem

A mixture of gases consists of 3.00 moles of helium, 4.00 moles of argon, and 1.00 mole of neon. The total pressure of the mixture is 1,200. torr.

(A) Calculate the mole fraction of each gas in the mixture.

(B) Calculate the partial pressure of each gas in the mixture.

Solution

(A) n_t = 3.00 mol + 4.00 mol + 1.00 mol = 8.00 mol

$$X_i = \frac{n_i}{n_t}$$

$$X_{He} = \frac{3.00 \text{ mol}}{8.00 \text{ mol}} = 0.375$$

$$X_{Ar} = \frac{4.00 \text{ mol}}{8.00 \text{ mol}} = 0.500$$

$$X_{Ne} = \frac{1.00 \text{ mol}}{8.00 \text{ mol}} = 0.125$$

(B) $P_i = X_i P_i$

P_{He} = 0.375 • 1,200. torr = **450. torr**

P_{Ar} = 0.500 • 1,200. torr = **600. torr**

P_{Ne} = 0.125 • 1,200. torr = **150. torr**

Sample Problem

Consider the experimental two-bulb arrangement shown in the diagram below:

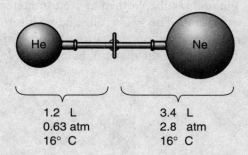

1.2 L   3.4 L
0.63 atm   2.8 atm
16° C   16° C

The valve between the bulbs is opened and the apparatus is kept at constant temperature. Calculate the pressure of the gas mixture when equilibrium has been attained.

Solution

We need to apply Boyle's law, followed by Dalton's law of partial pressures. Each gas expands to a total volume of 4.6 liters.

$$(0.63 \text{ atm}) \cdot (1.2 \text{ L}) = P_{He} \cdot (4.6 \text{ L})$$

$$P_{He} = 0.16 \text{ atm}$$

$$(2.8 \text{ atm}) \cdot (3.4 \text{ L}) = P_{Ne} \cdot (4.6 \text{ L})$$

$$P_{Ne} = 2.1 \text{ atm}$$

$$P_{total} = P_{He} + P_{Ne} = 0.16 \text{ atm} + 2.1 \text{ atm} = \textbf{2.2 atm}$$

Collecting Gases over Water

Gases that do not dissolve well in water can be collected by displacing the water in a collecting vessel. In practice, a long, *graduated* tube, known as a **eudiometer**, is filled with water and inverted over a pan filled with water. The gas is collected by inserting an inlet tube into the bottom of the eudiometer. The graduations allow the volume of gas collected to be read directly. This procedure is illustrated in the diagram below:

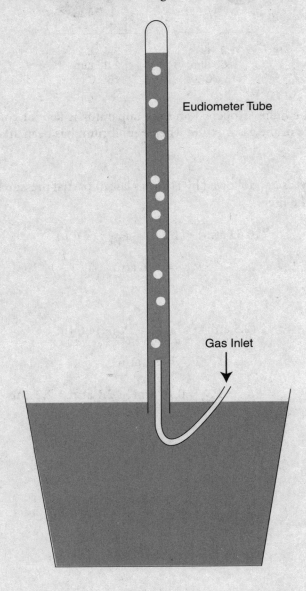

At the end of the collection, the height of the eudiometer is adjusted so that the level of water in the eudiometer is equal to the level of water in the pan. Under these conditions, the pressure of the gas inside the eudiometer is equal to the *atmospheric* pressure surrounding the tube.

The problem is that there are *two* gases in the eudiometer: the collected gas and the water vapor. The partial pressure of the water—known as the **equilibrium vapor pressure** of water—depends solely on the *temperature* of the system. A table of vapor pressures for water is given in Appendix B. Using Dalton's law of partial pressures, we can determine the partial pressure of the collected gas (P_g) by subtracting the vapor pressure of the water (P_w) from the atmospheric pressure (P_{atm}):

$$\boxed{P_g = P_{atm} - P_w}$$

Sample Problem

CH_4 gas is collected in a eudiometer at 26.0°C. When the water level is adjusted (as described above), the volume of gas inside the eudiometer is 78.0 mL. If the atmospheric pressure is 764 torr, how many grams of CH_4 were collected? (The vapor pressure of water at 26.0°C is 25 torr.)

Solution

The strategy is to determine the partial pressure of the CH_4, and then use the ideal gas equation to complete the solution.

$$P_{CH_4} = P_{atm} - P_w = 764 \text{ torr} - 25 \text{ torr} = 739 \text{ torr}$$

$$PV = nRT = \frac{m}{\mathcal{M}} RT$$

$$\text{Rearranging: } m = \frac{PV\mathcal{M}}{RT}$$

$$m = \frac{739 \text{ torr} \cdot \left(\dfrac{1 \text{ atm}}{760 \text{ torr}}\right) \cdot 78.0 \text{ mL} \left(\dfrac{1 \text{ L}}{1000 \text{ mL}}\right) \cdot 16.0 \dfrac{\text{g}}{\text{mol}}}{0.08206 \dfrac{\text{L} \cdot \text{atm}}{\text{mol} \cdot \text{K}} \cdot (26.0°C + 273.15) \text{ K}}$$

$$m = \textbf{0.0494 g } \mathbf{CH_4}$$

THE KINETIC-MOLECULAR THEORY OF GAS BEHAVIOR

The **kinetic-molecular theory** (KMT) is a mathematical model that was developed in the 18th and 19th centuries to *explain* gas behavior. The basis of the KMT is an abstraction known as an **ideal gas**, which has the following properties:

- An ideal gas consists of a large number of molecules that are in continuous, random motion.

- While the gas molecules have measurable mass, they are considered to have *negligible* volume (when compared to the volume of their container).

- Attractive (and repulsive) forces between molecules are negligible, except during molecular collisions.

- All molecular collisions are perfectly **elastic**, that is, kinetic energy (as well as momentum) is conserved during the collision.

The KMT begins by tracing the motion of a single molecule in a container, and then extrapolates the result to a container that contain a large number of molecules. As a result, the following are outcomes of the KMT:

- An equation of the form $PV = \frac{2}{3} N\bar{E}_k$ is derived, in which N represents the number of molecules and $\bar{E}_k$ represents the average kinetic energy per molecule.

- The volume of an ideal gas is the volume of its container.

- The pressure exerted by a gas is related to the force of the molecular collisions and the frequency with which they collide with the walls of the container.

- The *absolute* temperature of an ideal gas is proportional to the average kinetic energy of the molecules.

In calculating the average kinetic energy per molecule, a quantity known as **root-mean-square speed** (v_{rms})—rather than the ordinary average (mean) speed—is used. In a gas sample, the molecules possess a *distribution* of molecular speeds as is shown in the graph below:

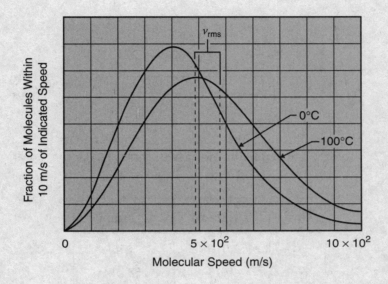

(Notice that at the higher temperature, the gas has a higher root-mean-square speed.)

The relationship between the absolute temperature and the root-mean-square speed is given by:

$$\bar{E}_k \text{ (per molecule)} = \frac{1}{2}\, mv^2_{rms} = \frac{3}{2}\, kT$$

where m is the mass of a molecule, k is Boltzmann's constant, and T is the absolute temperature.

In calculating the average kinetic energy *per mole* of gas molecules, both sides of the relationship are multiplied by Avogadro's number, yielding:

$$\bar{E}_k \text{ (per mole)} = \frac{1}{2}\, \mathcal{M}v^2_{rms} = \frac{3}{2}\, RT$$

where $\mathcal{M}$ is the molar mass and R is the molar gas constant. Solving for v_{rms}, we obtain:

$$v_{rms} = \sqrt{\frac{3RT}{\mathcal{M}}}$$

Sample Problem

Calculate the root-mean-square speed of HCl gas at 100.0°C, assuming it to be ideal.

Solution

$$v_{rms} = \sqrt{\frac{3RT}{\mathcal{M}}} = \sqrt{\frac{3 \cdot \left(8.315 \ \frac{J}{mol \cdot K}\right) \cdot (100.0° + 273.15) \ K}{\left(36.46 \ \frac{g}{mol}\right) \cdot \left(\frac{1 \ kg}{1,000 \ g}\right)}} = 505.3 \ \frac{m}{s}$$

A table of common gases illustrates the relationship of molar mass to root-mean-square speed at 25°C (298.15 K). Note that as the molar mass *increases*, the root-mean-square speed *decreases*.

Gas	$\mathcal{M}$/(g/mol)	v_{rms}/(m/s)
H_2	2.016	1856
Ne	20.18	584.1
O_2	32.00	482.1
CO_2	44.01	411.1
SO_2	64.06	340.7

The KMT has immediate application to the experimental gas laws:

- (Boyle's law) If the volume of a gas in a cylinder is *reduced* at constant temperature, the gas molecules will collide more frequently with the walls of the container even though they are not moving more rapidly on average. Consequently, the pressure exerted by the gas will increase.

- (Charles's law) If the temperature of a gas is *increased* they will move more rapidly, on average. If constant pressure is to be maintained, the rate of molecular collisions can not allowed to increase. It follows that the volume must increase in order to maintain the same collision rate.

- (Avogadro's law) If the number of molecules in a container is *increased* at constant temperature, there are more molecules moving at the same average speed. If the collision rate is to be maintained (constant pressure), the volume of the container will have to *increase*.

Effusion, Diffusion, and Graham's Law

Effusion is the process by which a gas escapes through a small opening into a vacuum. **Diffusion** is the process by which a gas spreads throughout a space, whether evacuated or not. Although the processes are different, both obey the same law. In 1846, Thomas Graham discovered that at the same temperature, the relative rate of effusion (r) of two gases was inversely proportional to the square root of their molar masses, a relationship that is now known as **Graham's law:**

$$\frac{r_1}{r_2} = \sqrt{\frac{\mathcal{M}_2}{\mathcal{M}_1}}$$

If we assume that the relative rate of effusion is equal to the ratio of the root-mean-square speeds of both gases, Graham's law follows immediately from the KMT and the relationship :

$$v_{rms} = \sqrt{\frac{3RT}{\mathcal{M}}} :$$

$$v_{rms} = \sqrt{\frac{3RT}{\mathcal{M}}}$$

$$\frac{v_{1rms}}{v_{2rms}} = \frac{\sqrt{\dfrac{3RT}{\mathcal{M}_1}}}{\sqrt{\dfrac{3RT}{\mathcal{M}_2}}} = \sqrt{\frac{\mathcal{M}_2}{\mathcal{M}_1}}$$

Sample Problem

Calculate the relative rate of effusion of NO_2 to SO_2.

Solution

$$\frac{r_{NO_2}}{r_{SO_2}} = \sqrt{\frac{\mathcal{M}_{SO_2}}{\mathcal{M}_{NO_2}}} = \sqrt{\frac{60.06 \dfrac{g}{mol}}{46.01 \dfrac{g}{mol}}} = \mathbf{1.143}$$

In other words, NO_2 effuses 1.143 times more rapidly than SO_2.

Chapter 6: Practice Questions

Multiple-Choice Questions

1. Gases are differentiated from solids and liquids because all gases have:

   A. color.
   B. life-giving properties.
   C. low chemical activity.
   D. molecules with a spherical shape.
   E. average densities around 10^{-3} g/mL.

2. The volume increases as a sample of gas is heated. Why?

   A. A gas softens when heated.

   B. The entire universe is expanding.

   C. A gas decomposes into more molecules.

   D. Increased molecular velocity increases collision forces.

   E. The lid on the container holding the gas loosens when heated.

3. A U-tube manometer has a sample of hydrogen in the closed end. The difference in the height of the columns of mercury in this manometer is 18.0 cm. What is the pressure of the contained hydrogen gas?

   A. 0.237 Pascals
   B. 1.80 mmHg
   C. 18.0 atm
   D. 1.82×10^6 Pascals
   E. 180 torr

4. A rescue balloon filled with 10.0 L of helium is placed in a chamber. The pressure in the chamber is reduced to 80% of atmospheric pressure, and the temperature reduced as well. The volume of the balloon does not change. If the original temperature is 27°C, to what must the temperature have been lowered to achieve this effect?

   A. −33°C
   B. 0°C
   C. 5.4°C
   D. 21.6°C
   E. 27 K

5. At 47.0°C and 16.0 atmospheres, the molar volume of NH3 (g) is approximately 10% less than that predicted by the ideal gas law. The actual volume of ammonia gas is smaller than the ideal gas volume at 47.0°C and 16.0 atmospheres because they:

   A. do not move randomly.
   B. decompose into the elements N_2 and H_2.
   C. occupy a significant fraction of the volume of a container.
   D. have attractive forces between molecules which are significant.
   E. move more slowly than predicted at 47.0°C and 16.0 atmospheres.

6. A mixture of three gases consists of a number of moles of fluorine and an equal number of moles of Xe and H_2, each of which is double the moles of F_2. The total pressure of the mixture is 1000.0 torr. What is the partial pressure of the fluorine gas?

   A. 100.0 torr
   B. 200.0 torr
   C. 250.0 torr
   D. 333.3 torr
   E. 400.0 torr

7. In a molar mass determination, the density of a sample of an unknown pure gas is 1.26 g/L at 25.0°C and 781 mmHg. What is the most likely identification for this gas?

   A. C_2H_6
   B. CH_4
   C. CO_2
   D. N_2O
   E. Ne

8. The rate of effusion of ammonia gas, NH_3 (g), is 3.32 times faster than that of an unknown gas. What is the molar mass of the unknown gas?

   A. 33.2 g/mol
   B. 45.5 g/mol
   C. 56.4 g/mol
   D. 112 g/mol
   E. 187 g/mol

9. Which is a postulate of the kinetic molecular theory?

   A. Gas molecules travel at or near the speed of light.

   B. The collisions between gas molecules are inelastic.

   C. The molecules of a sample of gas have very small volume.

   D. The energy of gas molecules is determined by quantum mechanics.

   E. The absolute temperature of a substance is proportional to the average kinetic energy of its molecules.

10. Which gas sample has the *lowest* average velocity at 298K?

    A. CH_4 at 0.20 atm
    B. CO_2 at 0.40 atm
    C. He at 0.60 atm
    D. Ne at 0.80 atm
    E. NO at 1.00 atm

11. Which statement explains how the kinetic molecular theory explains Charles's law?

    A. An increase in the number of molecules will result in an increase in the number of collisions.

    B. The increase in volume creates more space between molecules causing less frequent collisions.

    C. The increase in frequency and strength of molecular collisions results in an increased volume in order to keep temperature constant.

    D. The total pressure is dependent upon the total number of collisions inside the container, and is unrelated to the chemical nature of the gas molecules.

    E. The increase in temperature and the increase in average kinetic energy means the molecules collide with more force causing the volume to increase at a constant external pressure.

Free-Response Questions

1. At standard conditions, the ideal gas law is remarkably successful in relating the quantity of gas and its pressure, volume, and temperature.

   (A) Under what conditions of temperature will a gas deviate most from ideal behavior? Why?

   (B) At what relative pressure will a gas behave in a most ideal fashion? Why?

   (C) What is assumed about the size of an ideal gas molecule?

   (D) How do ideal gases interact when they collide?

2. A steel drum is filled with air and a mixture of several gases (including N_2, O_2, CO_2, and Ar), and sealed. The quantity of gas does not change because of the strength of the container seals.

   (A) What is the relationship between the total pressure in the drum and the partial pressures of its several component gases as postulated in Dalton's Law of Partial Pressures?

   (B) Write an equation representing this relationship.

   (C) How is Dalton's law explained by the kinetic molecular theory?

3. Laboratory procedure to study gases often involves collecting the gas by the displacement of water from an upside-down graduated cylinder. The volume displaced is recorded as an important piece of data. This procedure is often referred to as "collecting (the gas) over water" meaning that a pressure adjustment must be made.

   (A) Why is the vapor pressure of a substance a constant value at a given temperature?

   (B) Why does only the pressure and not the volume have to be adjusted when a gas is collected with water vapor mixed in?

4. What is the molar mass of a gas that has a density of 0.557 g/L at 0.800 atm and 77°C?

Multiple-Choice Answers

1. E   2. D   3. E   4. A   5. D   6. B   7. A   8. E   9. E   10. B   11. E

1. E

Gases are much less dense than liquids and solids. The molecules are much farther apart in gases, making for much lower mass in a given volume.

2. D

The increased temperature increases kinetic energy and causes more force and frequency of collisions, which expand a flexible container increasing its volume.

3. E

The column height difference of 18.0 cm is equal to 180 mm, with 1.0 mmHg is equal to 1.0 torr. The difference in the heights of the two columns measures the pressure of the gas using the described manometer.

4. A

The volume of the helium remains constant. The use of the ideal gas law equation requires the conversion of all temperatures into Kelvins. The original temperature is 27°C + 273 = 300 K. With the volume constant, $T_2/T_1 = P_2/P_1$ since temperature and pressure are directly proportional. Since $P_2 = 0.80 \times P_1$, then $T_2 = 0.80 \times T_1 = 0.80 \times 300 = 240$ K = –33°C.

5. D

Real gases differ from ideal gases because their molecules attract each other and take up space. The NH_3 is a strong hydrogen bonded dipole, and the attractions between the molecules can account for their taking up a smaller than anticipated space.

6. B

A little bit of algebra is helpful. If the moles of xenon and hydrogen are x, and the fluorine y, then Dalton predicts the total pressure is the sum of the partial pressures; or $x + x + y = 1000.0$. But $x = 2y$. So then $5y = 1000.0$, giving $y = 200.0$.

7. A

The ideal gas law can be rearranged to determine the molar mass of a gas from its density.

$\mu = d\, RT \div P$ where d is the density.

P (atm) = 781 mmHg ÷ 760. mmHg / atm = 1.0276 atm

(Rule: Hold insignificant digits in your calculator until the last answer. The last answer is then rounded out to the proper number of significant figures.)

= (1.26 g/L × 0.0821 L·atm/mol·K × 298 K) ÷ (0.895 atm) = 30.0 g/mol

The only gas in the list with a molar mass of 30.0 is C_2H_6.

8. E

Graham's Law predicts the ratio of the effusion rates is equal to the square root of the molar masses of two gases. The square of 3.32 is 11.0, which means the unknown gas has an molar mass 11 times greater than NH_3, or its $\mu = 11.0 \times 17.0 = 187$ g/mol.

9. E

The average kinetic energy of gas molecules is determined by the temperature of the gas. Ideal gas molecules do not travel at the speed of light, and have elastic rather than inelastic collisions. They are assumed to have negligible volume. Quantum mechanics determines the position of electrons with certain energies.

10. B

The velocities are found by Graham's Law where the ratio of the velocities is proportional to the square root of the inverse ratio of the molar mass of the gases at constant temperature. The partial pressure has nothing to do with the velocity, which is dependent only on temperature and molar mass. The heavier gas, CO_2 ($\mu = 32$), will have the lowest average velocity.

11. E

The pressure of the gas must increase if the volume is fixed; if the pressure is kept constant, the volume must grow larger. The relationship between temperature and volume at constant pressure is described by Charles's law.

Free-Response Answers

1. (A) At *low temperature* the molecules of a gas are moving slowly. They begin to approach the point where the attractive forces are now being able to take effect. Only when the molecules no longer have enough kinetic energy to overcome the attractive forces will the gas liquefy.

   (B) At *low pressure* the frequency of collisions are very low. This is usually due to large spaces between molecules. If the overall volume of a gas sample is very large, then the actual volume taken up by the gas molecules is relatively insignificant. The lower the pressure, the larger the relative volume, and the more the volume of the individual molecules can be ignored.

   (C) An ideal gas molecule is assumed to be a point mass. It has mass, it has momentum, it can cause pressure with the force of its collision, and its kinetic energy can be calculated. But theoretically it takes up no space, just like a mathematical point. This means that a sample of an ideal gas can be shrunk down to zero volume, which does not actually happen with real gases. Real gases at some point become noncompressible.

   (D) Ideal gases collide only with perfectly elastic collisions. There are no attractive or repulsive forces between gas molecules. In reality, most substances in the gaseous phase will have little interaction between molecules. They are at a high enough temperature so that they have overcome the attractive forces with their kinetic energy.

2. (A) The total pressure of a gas mixture is the sum of the partial pressures of its components.

   (B) $P_T = P_1 + P_2 + P_3 + P_4 + \ldots$

   (C) The total pressure depends on the total number of collisions inside the container, and is unrelated to the chemical nature of the gas particles themselves. So the pressure of a mixture can be thought of as the sum of the collisions which will be proportional to the number of moles of gas particles being considered at any one time.

3. (A) Vapor pressure is constant at any given temperature because an equilibrium is reached between the rate of evaporation and the rate of condensation. The two phases stay in the same proportion. The balancing of the rates is achieved at a point by the relationship between the strength of attraction between particles and the tendency to overcome these attractions due to the kinetic energy of the molecules. If the temperature stays the same, the rate of evaporation stays constant and eventually the rate of condensation matches it at a certain vapor pressure.

   (B) All gases completely fill their container, regardless if there is another gas mixed in with them. So, while the pressure is reduced when a partial pressure is subtracted from the total pressure, the volume will not change. This remaining pressure applies to the remaining gas that still fills the same space.

4. It is helpful to memorize the modifications of the ideal gas law involving density and molar mass, although if memorization is difficult, they can be derived fairly quickly.

   $$\text{Molar mass} = \frac{dRT}{P} = \frac{(0.557 \text{ g/L})(0.0821 \text{ L·atm/mol·K})(350 \text{ K})}{0.800 \text{ atm}} = 20.0 \text{ g/mol}$$

Thermodynamics I: Thermochemistry

Thermodynamics is the study of the interactions among work, energy, and heat (a specific type of energy). Before these terms are defined precisely, we need to define the nature of the universe in which these interactions occur. The object or objects under inspection is known as a **system**. The remainder of the universe is known as the **surroundings**. A familiar example of a system is a gas enclosed in a cylinder that is fitted with a piston, as shown in the diagram below:

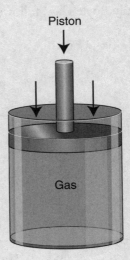

Piston

Gas

An **open system** is one that permits the transfer of mass, and/or work, and/or energy between the system and its surroundings. A **closed system** is one that permits the transfer of energy and/or work—but *not* mass—between the system and its surroundings. An **isolated system** is one that does not permit any exchanges between the system and its surroundings.

STATE FUNCTIONS

In thermodynamics, changes in the state of a system are studied. The state of a system is defined to the all of the values of the (relevant) macroscopic properties of the system such as energy, temperature, pressure, volume, and composition.

A **state function** is a property that is determined by the state of the system, rather than the way in which the state was achieved. As a result, any change in a state function is calculated by the simple process of subtracting the initial value of the property from its final value: no other considerations need be known. Consider an ideal gas that changes its volume from 10 liters to 4 liters. The change in volume is a decrease of 6 liters ($\Delta V = -6$ L), regardless of how the change was performed. In order to illustrate this concept, let us imagine that the change was brought about by changing the pressure from 0.2 atm to 1 atm, and changing the temperature from 300 K to 600 K. In the first scenario, the pressure is changed *before* the temperature is changed; in the second scenario, the pressure is changed *after* the temperature is changed. (NOTE: The subscripts i and f are used to denote initial and final values, respectively; the subscripts x and y are used to denote intermediate values.)

SCENARIO I

We apply Boyle's law, followed by Charles's law:

$$P_i V_i = P_f V_x$$
$$(0.2 \text{ atm}) \cdot (10 \text{ L}) = (1 \text{ atm}) \cdot (V_x)$$
$$V_x = 2 \text{ L}$$

$$\frac{V_x}{T_i} = \frac{V_f}{T_f}$$

$$\frac{2 \text{ L}}{300 \text{ K}} = \frac{V_f}{600 \text{ K}}$$

$$V_f = \mathbf{4 \text{ L}}$$

SCENARIO II

We apply Charles's law followed by Boyle's law:

$$\frac{V_i}{T_i} = \frac{V_y}{T_f}$$

$$\frac{10 \text{ L}}{300 \text{ K}} = \frac{V_y}{600 \text{ K}}$$

$$V_y = 20 \text{ L}$$

$$P_i V_y = P_f V_f$$
$$(0.2 \text{ atm}) \cdot (20 \text{ L}) = (1 \text{ atm}) \cdot (V_f)$$
$$V_f = 4 \text{ L}$$

Since the initial and final values of the volume are the same in both scenarios, it follows that the change in the volume is dependent *only* on these values, and *not* by the *paths* that were taken to produce the change.

In contrast, heat and work are *not* state functions: they are path dependent. For example, if the temperature of 1 mole of helium gas is raised by 1 kelvin at *constant pressure*, 20.8 joules of heat will be absorbed by the gas from its surroundings; if the change is performed at *constant volume*, only 12.5 joules of heat will be absorbed.

WORK

The physical concept of **work** involves the application of a force (F) that is associated with an object moving a specific distance (d). Assuming that the direction of the force and the displacement of the object are the same, work is defined as the product of force and distance:

$$w = F \cdot d$$

In the SI system, the unit of work is the newton-meter, which is also known as the **joule (J)**. If work is done on a system (for example, a gas in a cylinder is compressed by the piston), then it is taken to be a positive quantity; if work is done by the system (for example, a gas in a cylinder expands against the piston), then it is taken to be a negative quantity.

Of particular interest is the work done on or by a gas confined in a cylinder fitted with a piston. If pressure is applied to the piston, the result is a change in the volume of the enclosed gas (ΔV), as shown in the diagram below:

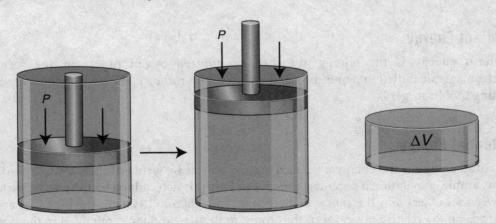

If the applied pressure is kept constant, it can be shown that the work done on or by the gas is given by the relationship:

$$w = -P\Delta V$$

where P is the applied pressure, and ΔV is the change in the volume of the gas. The negative sign is inserted in the equation in order to reconcile the sign conventions between work and volume change. For example, when work is done on the gas, the work is a positive quantity, but the volume change is negative since the gas is compressed. (Even if the pressure is not held constant, the performance of work requires that there be some change in the volume of the gas, although the amount of work done will not be equal to $P\Delta V$.)

ENERGY

Energy is an abstract concept that is recognized by the effects it has on objects. It is intimately related to the performance of work. For example, the energy released by falling water can be harnessed to drive electrical machinery. As with work, the SI unit of energy is the joule.

Kinetic Energy

Kinetic energy (E_k) is the energy possessed by an object by virtue of its *motion*, and is given by the equation:

$$E_k = \frac{1}{2}mv^2$$

where m is the mass of the object, and v is its speed. We are well aware of the destructive work a bullet can do by virtue of its large kinetic energy.

Radiant Energy

Radiant energy is the energy absorbed or emitted by an object in the form of electromagnetic radiation. For example, the visible light produced by a laser can be used as a cutting tool in surgery.

Potential Energy

Potential energy is the energy possessed among objects by virtue of their relative *positions*. For example, gravitational potential energy (on Earth) is determined by the relative distance between an object and the center of the Earth. In a chemical system, the potential energy depends on the relative positions of the atoms, molecules, and ions contained in the system.

Internal Energy

Internal energy (*E*) is the *total* kinetic and potential energies contained *within* an object. Internal energy is a state function and we will have more to say about this important quantity in the sections that follow.

Thermal Energy

Thermal energy is the portion of internal energy that is affected by changes in temperature. Do NOT equate thermal energy and temperature: they are different concepts. Temperature is related to the average kinetic energy of the particles within an object; it does not depend on the mass of the object. Since thermal energy is a portion of internal energy, it does depend on the mass of the object. For example, a lit match has a higher temperature than an ocean, but the ocean has far more thermal energy than the lit match.

Heat

Heat (*q*) is the part of thermal energy that is transferred between objects when a *difference in temperature* exists. The direction of heat transfer is from the hotter to the colder object. As the transfer occurs, the temperature of the hotter object will be lowered and the temperature of the colder object will be raised. (This assumes that nothing unusual, such as a phase change, is occurring.) When the objects reach the same temperature, known as the **equilibrium temperature**, the net transfer ceases. When an object *absorbs* heat, the process is known as **endothermic** and *q* is assigned a *positive* value; when an object *releases* heat, the process is known as **exothermic** and *q* is assigned a *negative* value.

THE FIRST LAW OF THERMODYNAMICS

The **first law of thermodynamics** describes the relationships among *internal energy*, *heat*, and *work*. The law is based on the law of conservation of energy, which states that the internal energy of the universe is a constant. It follows that any change in the internal energy of a system is accompanied by an *opposite* change in the internal energy of its surroundings. For example, a process that causes the internal energy of a system to *decrease* by 200 kilojoules, will cause the internal energy of the surroundings to *increase* by 200 kilojoules. These statements can be summarized by the following equations:

$$E_{universe} = \text{constant}$$

$$\Delta E_{universe} = \Delta E_{system} + \Delta E_{surroundings} = 0$$

$$\Delta E_{system} = -\Delta E_{surroundings}$$

While it is not possible to measure internal energy directly, it is possible to measure *changes* in internal energy. Rather than concentrate on a system and its surroundings, it is more useful to focus exclusively internal energy changes that occur within the system. In this context, the change in internal energy depends only on two properties: the heat transferred in or out of the system and the work done on or by the system. Mathematically, the first law takes the form:

$$\Delta E = q + w$$

Sample Problem

A system absorbs 35 joules of heat from its surroundings and does 22 joules of work on its surroundings. Calculate the internal energy change of the system.

Solution

Recalling the sign conventions for work and heat given earlier:

$$\Delta E = q + w = (+35 \text{ J}) + (-22 \text{ J}) = +13 \text{ J}$$

That is, the internal energy of the system *increases* by 13 joules.

The First Law and Constant Volume Processes

If the volume of a gaseous system is held constant, no work can be done on or by the system. In this case the first law takes the form:

$$\Delta E = q_v$$

where q_v is the heat transferred at constant volume. This provides a simple way of measuring the internal energy change: measure the heat transferred at constant volume.

The First Law and Constant Pressure Processes

If the pressure of a gaseous system is held constant, the first law takes the form:

$$\Delta E = q_p - P\Delta V$$

where q_p is the heat transferred at constant pressure. Under these conditions, it is very useful to create a new state function called **enthalpy** (H), which is defined as:

$$H = E + PV$$

A change in enthalpy (ΔH) is given by the relationship:

$$\Delta H = \Delta E + \Delta(PV)$$

At constant pressure, this simplifies to:

$$\Delta H = \Delta E + P\Delta V$$

If we compare this equation with the equation for the first law at constant pressure, it follows that:

$$\Delta H = q_p$$

When we measure the heat transferred at constant pressure, we measure the change in the enthalpy of the system. In practice, ΔH is slightly larger than ΔE since the heat transferred at constant pressure is used for pressure-volume work as well as for changing the internal energy.

Relationship of ΔH and ΔE at Constant Temperature

As we saw above, the enthalpy change in a gaseous system is given by the equation: $\Delta H = \Delta E + \Delta(PV)$. If we assume that all gases present obey the ideal gas law ($PV = nRT$), then we can rewrite the equation as: $\Delta H = \Delta E + \Delta(nRT)$. At constant temperature, this becomes:

$$\Delta H = \Delta E + RT\Delta n$$

where Δn is the change in the number of moles of gas within the system. The following problem applies this relationship.

Sample Problem

Consider the reaction: $N_2(g) + 3H_2(g) \rightarrow 2NH_3(g)$ at 298.15 K. If ΔH for this reaction is -92.38 kJ, what is the value of ΔE for this reaction. ($R = 8.315$ J/mol·K)

Solution

We obtain Δn from the coefficients of the equation: $\Delta n = n_{products} - n_{reactants} = 2 - (3 + 1) = -2$.

$$\Delta H = \Delta E + RT\Delta n$$

$$\Delta E = \Delta H - RT\Delta n$$

$$\Delta E = -92.38 \text{ kJ} - \left(8.315 \, \frac{\text{J}}{\text{mol} \cdot \text{K}}\right) \cdot (298.15 \text{ K}) \cdot (-2 \text{ mol}) \cdot \left(\frac{1 \text{ kJ}}{1,000 \text{ J}}\right)$$

$$\Delta E = \mathbf{-87.42 \text{ kJ}}$$

As we can see, ΔE is indeed smaller than ΔH.

CALORIMETRY: THE EXPERIMENTAL MEASUREMENT OF HEAT

Heat Capacity and Specific Heat

Objects differ in their abilities to transform heat transfer into temperature change. The **heat capacity** (**C**) of an object is one way of measuring this ability and is defined as:

$$C = \frac{q}{\Delta T}$$

where q is the quantity of heat transferred and ΔT is the temperature change. Since the algebraic sign of the temperature change and the heat transferred are the always the same, heat capacity is a positive quantity. A commonly used unit of heat capacity is the joule per kelvin (J/K) or, equivalently, the joule per Celsius degree (J/C°). Heat capacity is an extensive property, i.e., it depends on the mass of the object. It is most useful when the mass of the object is not the prime consideration.

Sample Problem

An object has a heat capacity of 57.5 J/K. If its temperature changes from 150.4°C to 121.8°C, how much heat is transferred?

Solution

$$\Delta T = 121.8°C - 150.4°C = -28.6C° = -28.6 \text{ K}$$

$$C = \frac{q}{\Delta T}$$

$$q = C\Delta T = \left(57.5 \frac{J}{K}\right) \cdot (-28.6 \text{ K}) = -\textbf{1640 J}$$

That is, 1640 joules of heat are released by the object.

When mass becomes an important factor, is more useful to use the **specific heat** (c_p), which is defined as the heat capacity per unit mass:

$$c_p = \frac{C}{m} = \frac{q}{m\Delta T}$$

A commonly used unit of specific heat is the joule per gram·kelvin (J/g·K). The subscript "p" in the symbol for specific heat, indicates that it is usually used in conditions where the pressure is constant. Specific heat is an intensive property that depends on the nature of the substance, rather than on its mass. A short table of specific heats follows:

THERMODYNAMICS I: THERMOCHEMISTRY

Specific Heats of Selected Substances and Mixtures

Substance	c_p (J/g·K)	Substance	c_p (J/g·K)	Substance	c_p (J/g·K)
Ag(s)	0.235	Cu(s)	0.385	Li(s)	3.58
Al(s)	0.897	Fe(s)	0.449	Mg(s)	1.02
AlF_3(s)	0.895	Glass(ℓ)	0.753	Mn(s)	0.479
As(s)	0.329	He(g)	5.19	Na(s)	1.23
Au(s)	0.129	Hg(ℓ)	0.140	NaCl(s)	0.858
Ca(s)	0.647	H_2O(s)	2.06	Si(s)	0.714
$CaCO_3$(s)	0.920	H_2O(ℓ)	4.19	SiO_2(s)	0.740
$CaSO_4$(s)	0.732	H_2O(g)	2.02	Sn(s)	0.228
CH_3CH_2OH(ℓ)	2.42	K(s)	0.757	Zn(s)	0.388

Note that the specific heat of a metallic substance, such as Ag, is smaller than the specific heat of a nonmetallic substance such as H_2O. A relatively small specific heat indicates that the substance translates a heat transfer into a relatively large temperature change, and vice versa.

Sample Problem

Calculate the heat absorbed by 50.0 grams of Cu(s) as it changes its temperature from 300. K to 500. K.

Solution

$$c_p = \frac{C}{m} = \frac{q}{m\Delta T}$$

$$q = c_p m\Delta T = \left(0.385\ \frac{J}{g \cdot K}\right) \cdot (50.0\ g) \cdot (200.\ K) = \textbf{3,850 J}$$

Sample Problem

A 100.-gram block of Au(s) initially at 300. K is brought into contact with a 50.0-gram block of Ag(s) initially at 400. K in an insulated environment. Calculate the equilibrium temperature of the two blocks.

Solution

This is an example of an *isolated system*, in which no heat is transferred to or from the surroundings. This can be expressed mathematically as:

$$q_{system} = q_{Au} + q_{Ag} = 0$$

$$q_{Au} = -q_{Ag}$$

$$(mc_p\Delta T)_{Au} = -(mc_p\Delta T)_{Ag}$$

$$\left[(100.\ g) \cdot \left(0.129\ \frac{J}{g \cdot K}\right) \cdot (T_{eq} - 300.\ K)\right]_{Au} = -\left[(50.0\ g) \cdot \left(0.235\ \frac{J}{g \cdot K}\right) \cdot (T_{eq} - 400.\ K)\right]_{Ag}$$

$$T_{eq} = \textbf{348 K}$$

Heat measurements for chemical processes are made in devices called **calorimeters**. There are two basic types of calorimeter: the *constant-volume calorimeter* (also known as a *bomb calorimeter*), and the *constant-pressure calorimeter*.

The Constant-Volume (Bomb) Calorimeter

The **bomb calorimeter** is a device for measuring the heat produced by a combustion reaction. A substance is placed in a steel container (the *bomb*), which is then filled with high-pressure oxygen. The container is surrounded by water and the entire apparatus is insulated from the external environment. The substance in the bomb is ignited electrically, and the temperature change of the water (and the bomb) is measured at the conclusion of the reaction.

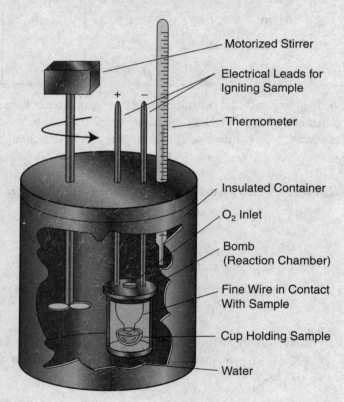

Since the bomb calorimeter is an isolated system, we can write:

$$q_{system} = q_{rxn} + q_{water} + q_{bomb} = 0$$

$$q_{rxn} = -(q_{water} + q_{bomb})$$

where q_{rxn} represents the heat produced by the reaction. The heat absorbed by the water is calculated from its mass, its specific heat, and the temperature change. The heat absorbed by the bomb is calculated from the temperature change and the heat capacity of the bomb, a quantity that is determined by combusting a known amount of a standard substance (such as benzoic acid: 1.000 gram liberates 26.38 kilojoules of heat) within the bomb.

The heats of combustion measured by a bomb calorimeter reflect the internal energy change (ΔE) of the reaction, since this is a constant-volume process. In general, heats of reaction are reported as changes in enthalpy (ΔH), a quantity that can be calculated from ΔE.

Sample Problem

A 1.92-gram sample of methanol (CH_3OH; $\mathcal{M}$ = 32.04 g/mol) was combusted in a bomb calorimeter whose heat capacity was 2.02 kJ/K. If the calorimeter contained 2.00 kilograms of water and the temperature change was 4.00 K, calculate the change in internal energy when 1 mole of methanol undergoes combustion.

Solution

$$q_{rxn} = -(q_{water} + q_{bomb})$$

$$q_{rxn} = -(m_{water} \cdot c_{Pwater} \cdot \Delta T) - (C_{bomb}\Delta T)$$

$$q_{rxn} = -\left((2.00 \times 10^3 \text{ g}) \cdot \left(4.19 \frac{J}{g \cdot K}\right) \cdot (4.00 \text{ K})\right) - \left(\left(2,020 \frac{J}{K}\right) \cdot (4.00 \text{ K})\right)$$

$$q_{rxn} = -25,400 \text{ J} = -25.4 \text{ kJ}$$

$$\Delta E = \frac{q_{rxn}}{n} = -\frac{-25.4 \text{ kJ}}{\left((1.92 \text{ g}) \cdot \left(\frac{1 \text{ mol}}{32.04 \text{ g}}\right)\right)} = -424 \frac{kJ}{mol}$$

Why was a c_p value used in a constant-*volume* calorimeter calculation? Only the *combustion reaction inside the bomb* took place at constant volume. The heat produced by the combustion was absorbed by the water and the bomb at approximately constant *pressure*.

An alternative, and simpler, method is to treat the bomb and the water as a single unit. The heat capacity of the calorimeter is determined and then an unknown substance is combusted. The temperature change produced by the unknown is used to determine the heat liberated.

Sample Problem

The heat capacity of a bomb calorimeter is 5.341 kilojoules per kelvin. When a 1.000-gram sample of a well-known cookie is combusted in the calorimeter, the temperature rises 3.513 kelvin.

(A) Calculate the heat liberated by the sample.

(B) If the mass of the cookie is 14.50 grams, how much heat would be liberated by the whole cookie?

(C) The nutritional Calorie (Cal) is equal to 4.186 kilojoules. What is the caloric value of the cookie?

Solution

(A) $q_{calorimeter} = C\Delta T = \left(5.341 \, \frac{kJ}{K}\right) \cdot (3.513 \text{ K}) = \mathbf{18.76 \ kJ}$

(B) $q_{cookie} = \left(18.76 \, \frac{kJ}{g}\right) \cdot (14.50 \text{ g}) = \mathbf{272.1 \ kJ}$

(C) $(272.1 \text{ kg}) \cdot \left(\dfrac{1.000 \text{ Cal}}{4.186 \text{ kJ}}\right) = \mathbf{64.99 \ Cal}$

The Constant-Pressure Calorimeter

As its name implies, the **constant-pressure calorimeter** is used to measure the heat liberated—or absorbed—by reactions taking place at constant pressure. It is not used for combustion reactions, but rather for reactions such as heat of (acid-base) neutralization and heat of solution. The calorimeter is easily constructed by using two nested Styrofoam® coffee cups, hence the name *coffee-cup calorimeter*.

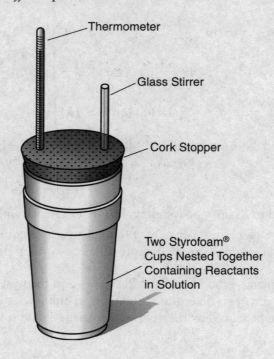

Thermometer

Glass Stirrer

Cork Stopper

Two Styrofoam®
Cups Nested Together
Containing Reactants
in Solution

Since the heat measurement is made at constant pressure, the heat of the reaction is equal to the enthalpy change of the reaction ($q_{rxn} = \Delta H_{rxn}$). As with the bomb calorimeter, the coffee-cup calorimeter is considered to be an isolated system:

$$q_{system} = 0 = q_{rxn} + q_{calorimeter}$$

$$q_{rxn} = \Delta H_{rxn} = -q_{calorimeter}$$

The heat transferred to or from the calorimeter is determined from its heat capacity and the temperature change. The following example shows how a coffee-cup calorimeter can be used to measure the heat of neutralization.

Sample Problem

When 100. mL of 0.500 M HCl is neutralized by 100. mL of 0.500 M NaOH in a coffee-cup calorimeter, the temperature of the resulting solution increases by 2.40 K. The reaction that takes place is:

$$HCl\ (aq) + NaOH(aq) \rightarrow NaCl(aq) + H_2O(\ell)$$

The heat capacity of the calorimeter is 335 joules per kelvin.

Calculate the heat of neutralization in kilojoules per mole.

Solution

Since the final solution is sufficiently dilute (0.250 M NaCl), we make the following assumptions:

- The volumes of the acid and base are additive, i.e., the final volume is 200. mL.

- The density and specific heat of the solution have the same values as those for pure water: 1.00 gram per milliliter and 4.186 joules per gram·kelvin, respectively.

$$q_{system} = 0 = -q_{rxn} + q_{solution} + q_{calorimeter}$$

$$q_{rxn} = -(q_{solution} + q_{calorimeter})$$

$$q_{rxn} = -(c_{solution}\, m_{solution}\, \Delta T + C_{calorimeter}\, \Delta T)$$

$$q_{rxn} = -\left(\left(4.186\ \frac{J}{g \cdot K} \cdot 200.\ g \cdot 2.40K\right) + \left(335\ \frac{J}{K} \cdot 2.40K\right)\right) \cdot \left(\frac{1\ kJ}{1,000\ J}\right)$$

$$q_{rxn} = -2.81\ kJ$$

100. mL of 0.500M HCl (and NaOH) corresponds to 0.0500 mole of acid (and base):

$$\Delta H_{rxn} = \left(\frac{-2.81\ kJ}{0.0500\ mol}\right) = -56.3\ \frac{kJ}{mol}$$

THERMOCHEMICAL EQUATIONS

Chemical equations obey the law of conservation of energy as well as the conservation of mass. The inclusion of an energy term in the equation produces a **thermochemical equation**. Since most common reactions are carried out at constant pressure, it is customary to show the *enthalpy change* following the equation, as shown below:

$$C_3H_8(g) + 5O_2(g) \rightarrow 3CO_2(g) + 4H_2O(\ell) \qquad \Delta H = -2,219\ kJ$$

This means that when the reaction takes place as written, and at constant pressure, 2,219 kilojoules of heat are released to the surroundings. We can also interpret the thermochemical equation to mean that the enthalpy of the products is 2,219 kilojoules *less* than the enthalpy of the reactants.

Since heat is an extensive variable, it follows that the enthalpy change for this reaction depends directly on the coefficients of the balanced equation. For example, if *2* moles of C_3H_8 had reacted with *10* moles of O_2 under the conditions described above, *twice* the amount of heat (4,438 kJ) would have been released at constant pressure.

Sample Problem

Consider the thermochemical equation: $2C(s) + 2H_2(g) \rightarrow C_2H_2(g) \quad \Delta H = +227.4 \text{ kJ}$

Calculate the enthalpy change for this system if 4.50 grams of C(s) reacts with an excess of $H_2(g)$.

Solution

$$(4.50 \text{ g C(s)}) \cdot \left(\frac{1 \text{ mol C(s)}}{12.01 \text{ g C(s)}} \right) \cdot \left(\frac{+227.4 \text{ kJ}}{2 \text{ mol C(s)}} \right) = \textbf{+42.6 kJ}$$

Standard State Conditions

In order to work with quantities such as enthalpy, it is convenient to define a set of conditions known as the **standard state**, in which the pressure is taken to be 100 kilopascals (exactly) and all substances are assumed to be in their pure state. Any temperature can be chosen, but the usual reference temperature is 298.15 K. An enthalpy change that is measured under standard state conditions is denoted by the symbol $\Delta H°$; the superscript "°" always indicates that standard state conditions are operative.

Hess's Law

Chemical reactions obey the laws of conservation of mass. As a result, it is possible to "add" a series of reactions to obtain a composite reaction:

$$CH_4(g) + 2O_2(g) \rightarrow CO_2(g) + 2H_2O(g)$$

$$2H_2O(g) \rightarrow 2H_2O(\ell)$$

$$\overline{CH_4(g) + 2O_2(g) + 2H_2O(g) \rightarrow CO_2(g) + 2H_2O(g) + 2H_2O(\ell)}$$

$$CH_4(g) + 2O_2(g) + \rightarrow CO_2(g) + 2H_2O(\ell) \text{ [net composite reaction]}$$

Note that substances that are duplicated on opposite side of the equation arrow do not appear in the final reaction.

Since enthalpy is a state function, the enthalpy change for a composite reaction is simply the sum of the enthalpy changes for the individual reactions; the nature of the individual steps is irrelevant. This is known as **Hess's law**. In developing a composite reaction, it may be necessary to reverse one of the individual reactions or multiply it by a common multiplier. In these cases, the following rules are employed:

- If a reaction is reversed, the sign of ΔH is reversed.
- If a common multiplier is used, ΔH is also multiplied by that multiplier.

The following example demonstrates how Hess's law is used:

Sample Problem

Consider the following thermochemical equations:

$$NO(g) + O_3(g) \rightarrow NO_2(g) + O_2(g) \qquad \Delta H° = -198.9 \text{ kJ}$$

$$O_3(g) \rightarrow \frac{3}{2}O_2(g) \qquad \Delta H° = -142.3 \text{ kJ}$$

$$O_2(g) \rightarrow 2O(g) \qquad \Delta H° = -495.0 \text{ kJ}$$

Calculate $\Delta H°$ for the reaction: $NO(g) + O(g) \rightarrow NO_2(g)$

Solution

- Since the first reaction and the composite reaction contain one mole of $NO(g)$ and $NO_2(g)$ on the same side of the equations, we leave the first reaction unchanged.

- The second reaction needs to be reversed in order to remove the presence of $O_3(g)$ from the composite:

$$\frac{3}{2}O_2(g) \rightarrow O_3(g) \quad \Delta H° = +142.3 \text{ kJ}$$

- The third reaction needs to be reversed *and* multiplied by $\frac{1}{2}$ in order to insure that one mole of $O(g)$ appears as a reactant in the composite:

$$O(g) \rightarrow \frac{1}{2}O_2(g) \quad \Delta H° = +247.5 \text{ kJ}$$

Adding the three "adjusted" reactions, we obtain:

$$NO(g) + O_3(g) \rightarrow NO_2(g) + O_2(g) \qquad \Delta H° = -198.9 \text{ kJ}$$

$$\frac{3}{2}O_2(g) \rightarrow O_3(g) \qquad \Delta H° = +142.3 \text{ kJ}$$

$$O(g) \rightarrow \frac{1}{2}O_2(g) \qquad \Delta H° = +247.5 \text{ kJ}$$

$$\overline{\phantom{NO(g) + O(g) \rightarrow NO_2(g) \qquad \Delta H° = +190.1 \text{ kJ}}}$$

$$NO(g) + O(g) \rightarrow NO_2(g) \qquad \Delta H° = +190.1 \text{ kJ}$$

Standard Enthalpies of Formation

A **formation reaction** is one in which one mole of a compound is formed from its elements under standard conditions and at a usual reference temperature of 298.15 K. A further requirement is that the elements be in their stable states. For example, the formation of $Al_2O_3(s)$ entails the reaction of $Al(s)$ and $O_2(g)$, the stable states of aluminum and oxygen at 298.15 K. The enthalpy change associated with a formation reaction is known as the **standard enthalpy of formation** (ΔH_f°). The unit of this quantity is kilojoule per mole of compound (kJ/mol). The thermochemical equation for the formation of $Al_2O_3(s)$ is shown below:

$$Al(s) + \frac{3}{2}O_2(s) \rightarrow Al_2O_3(s) \qquad\qquad \Delta H_f^\circ = -1,675.7 \text{ kJ/mol}$$

This means that 1675.7 kilojoules of heat is released when one mole of $Al_2O_3(s)$ is formed under standard conditions and at 298.15 K. In order to decompose one mole of $Al_2O_3(s)$ into its elements, 1675.7 kilojoules would have to be absorbed by the compound.

All elements in their stable states at 298.15 K are assigned a standard enthalpy of formation of *zero* since they are already considered to be formed. A listing of standard enthalpies of formation is given in Appendix B.

Standard Enthalpies of Formation and Reaction Enthalpies

It is possible to use standard enthalpies of formation to calculate the standard enthalpy change of a reaction. The steps for making such a calculation for the equation: $4NH_3(g) + 5O_2(g) \rightarrow 4NO(g) + 6H_2O(\ell)$ are given below.

- Multiply the standard enthalpies of formation of the products by their respective coefficients and add them together:

$4 \cdot \Delta H_f^\circ [\, NO(g)] + 6 \cdot \Delta H_f^\circ [H_2O(\ell)] = 4 \text{ mol} \cdot (+91.3 \text{ kJ/mol}) + 6 \text{ mol} \cdot (-285.8 \text{ kJ/mol}) = -1,350 \text{ kJ}$

- Repeat this procedure for the *reactants*:

$4 \cdot \Delta H_f^\circ [NH_3(g)] + 5 \cdot \Delta H_f^\circ [O_2(g)] = 4 \text{ mol} \cdot (-45.9 \text{ kJ/mol}) + 5 \text{ mol} \cdot (0 \text{ kJ/mol}) = -183.6 \text{ kJ}$

- Subtract the number obtained for the reactants from the number obtained for the products to obtain the standard enthalpy change for the overall reaction:

$$\Delta H^\circ = (-1,350 \text{ kJ}) - (-183.6 \text{ kJ}) = -1,166 \text{ kJ}$$

This procedure can be written symbolically as:

$$\Delta H^\circ = \left(\sum n \cdot \Delta H_f^\circ\right)_{products} - \left(\sum m \cdot \Delta H_f^\circ\right)_{reactants}$$

The symbols n and m represent the respective coefficients of the products and reactants.

Sample Problem

Using the appropriate standard enthalpies of formation, calculate the standard enthalpy change for the reaction: $Fe_2O_3(s) + 6HCl(g) \rightarrow 2FeCl_3(s) + 3H_2O(g)$

Solution

$$\Delta H° = (2 \cdot \Delta H_f° \, [FeCl_3(s)] + 3 \cdot \Delta H_f° \, [\, H_2O(g)]) - (1 \cdot \Delta H_f° \, [Fe_2O_3(s)] + 6 \cdot \Delta H_f° \, [HCl(g)])$$

$$\Delta H° = (2 \text{ mol} \cdot (-399.5 \text{ kJ/mol}) + 3 \text{ mol} \cdot (-241.8 \text{ kJ/mol})) -$$
$$(1 \text{ mol} \cdot (-824.2 \text{ kJ/mol}) + 6 \text{ mol} \cdot (-92.3 \text{ kJ/mol}))$$

$$\Delta H° = -146.4 \text{ kJ}$$

Chapter 7: Practice Questions

Multiple-Choice Questions

1. Lead(II) oxide is reduced with carbon to produce metallic lead, with the thermochemical equation:

   $PbO(s) + C(s) \rightarrow Pb(s) + CO(g)$
   $\Delta H= +106.8$ kJ/mol

   This is _____ reaction, which _____ heat.

   A. a formation; absorbs
   B. an exothermic; absorbs
   C. an exothermic; releases
   D. an endothermic; absorbs
   E. an endothermic; releases

2. Lead has been known and used for centuries. The first step to obtain the metal is to roast the mineral containing lead (II) sulfide in air to form lead(II) oxide:

   $PbS(s) + \frac{3}{2} O_2(g) \rightarrow PbO_s(s) + SO_2(g)$
   $\Delta H= -414$ kJ/mol

   What is the enthalpy change for the burning of 1.00 kilogram of solid lead (II) sulfide?

   A. −1,730 kJ
   B. −414 kJ
   C. −1.73 kJ
   D. +414 kJ
   E. +1,730 kJ

3. Solid sulfur can be burned to produce sulfur dioxide gas, with the following thermochemical equation:

   $S(s) + O_2(g) \rightarrow SO_2(g)$ $\Delta H= -296$ kJ/mol

   The burning of 1.00 kilogram of solid sulfur is used to heat 250.0 L of water. Assuming all of the energy is used to heat the water, and none is lost to the surroundings, what will the final temperature of the water be in kelvins, if the initial temperature is 25.0°C?

   A. 33.8 K
   B. 239.4 K
   C. −239.4 K
   D. 25.0 K
   E. 307.0 K

4. As a solid carbon sample burns completely it produces carbon dioxide gas, with the thermochemical equation:

$$C(s) + O_2(g) \rightarrow CO_2(g) \qquad \Delta H = -293.5 \text{ kJ/mol}$$

What is relationship between the enthalpy change and the heat of reaction for the burning of solid carbon at atmospheric pressure?

A. The enthalpy change is less than the heat of reaction at atmospheric pressure.

B. The enthalpy change is equal to the heat of reaction at atmospheric pressure.

C. The enthalpy change is greater than the heat of reaction at atmospheric pressure.

D. The enthalpy change is equal to the internal energy change, but not to the heat of reaction.

E. Any thermodynamic relationship must be determined at constant volume, not at constant pressure.

5. Solid sulfur is burned to produce sulfur dioxide gas:

$$S(s) + O_2(g) \rightarrow SO_2(g) \qquad \Delta H = -296 \text{ kJ/mol}$$

The heat released by burning 1.00 kilogram of solid sulfur is captured by 250.0 L of water. Assuming all of the energy is used to heat the water, and none is lost to the surroundings, what will the final temperature of the water be in degrees Celsius if the initial temperature is 25.0°C? The specific heat of water is 4.18 J/g·K.

A. 16.2°C
B. 25.0°C
C. 33.8°C
D. 154.0°C
E. 179.0°C

6. Enthalpy changes are determined experimentally for these reactions:

$$Pb(s) + 2\ Cl_2(g) \rightarrow PbCl_4(\ell) \qquad \Delta H = -329.3 \text{ kJ}$$

$$PbCl_2(s) + Cl_2(g) \rightarrow PbCl_4(\ell)\ \Delta H = +30.1 \text{ kJ}$$

What is the enthalpy change for the reaction of lead with chlorine to give lead(II) chloride, $PbCl_2$?

A. −359.4
B. −329.3
C. −299.2
D. +30.1
E. +359.4

7. The specific heat capacity of ethanol is 2.4 J/g·K and its density is 0.789 g/mL. A 250.0 mL sample of ethanol is heated from 25.0°C to 47.0°C at constant pressure. How much energy was added to the sample of ethanol?

A. 1.3×10^4 J
B. 1.0×10^4 J
C. 2.2×10^4 J
D. 2.8×10^4 J
E. 4.3×10^3 J

8. Hydrogen gas is burned to form steam:

$$H_2(g) + \frac{1}{2} O_2(g) \rightarrow H_2O(g)\ \Delta H° = -241.8 \text{ kJ}$$

The heat capacity, c_p, of aluminum is 0.90 J/g·K. How many grams of hydrogen gas is burned in the production of enough energy to raise the temperature of a 15.0 kg sample of aluminum by 6.5°C?

A. 0.36 g
B. 0.73 g
C. 0.88 g
D. 0.90 g
E. 2.8 g

9. The combustion of gaseous diborane, B_2H_6, proceeds according to the equation:

$B_2H_6(g) + 3 O_2(g) \rightarrow B_2O_3(s) + 3 H_2O(g)$
$\Delta H° = -1,941$ kJ

What is the enthalpy change for the reaction of boron oxide with water vapor to produce diborane gas and oxygen gas?

A. −3,882 kJ
B. −1,941 kJ
C. +970.5 kJ
D. +1,941 kJ
E. +3,882 kJ

10. The thermochemical equation for the combustion of hydrogen gas to form water vapor is:

$H_2(g) + \frac{1}{2} O_2(g) \rightarrow H_2O(g) \, \Delta H° = -241.8$ kJ

How many grams of hydrogen gas are consumed in the release 3580 kJ of energy?

A. 7.33 g
B. 7.40 g
C. 14.8 g
D. 15.0 g
E. 29.8 g

Free-Response Questions

1. Putting a small sealed bag containing water inside a larger sealed bag containing some ammonium nitrate crystals makes an "instant ice pack." The dissolving of ammonium nitrate (NH_4NO_3) is an endothermic process and the mass of ammonium nitrate required to chill the "ice pack" from room temperature (25.0°C) to 3.0°C can be determined. The heat of solution $(\Delta H_{solution})$ is known to be +25.69 kJ/mol. It is reasonable to assume that the specific heat capacity of the solution is the same as that of water, 4.184 J/g•°C.

(A) Distinguish between a constant volume calorimeter and a constant pressure calorimeter. Which directly measures the enthalpy change (ΔH) and which measures the internal energy change (ΔE)?

(B) Describe the calculations needed to compute the mass of NH_4NO_3 required to chill 135 grams of water.

(C) List the data which to be collected in an experiment required to verify the heat of solution.

(D) What are at least two likely sources of experimental error which can account for a high average experimental result for the mass of ammonium nitrate needed to chill the "ice pack"?

2. Ammonia, NH_3, is burned in the manufacture of nitric acid to produce NO_2. Benzene, C_6H_6, is burned in high-octane engines.

(A) Write a balanced equation for the complete combustion of ammonia gas.

(B) Determine ΔH_{rxn} for the combustion of ammonia using Hess's law and the heat of reaction data:

$$\frac{1}{2} N_2(g) + \frac{3}{2} H_2(g) \rightarrow NH_3(g) \qquad \Delta H^\circ = -46 \text{ kJ/mol}$$

$$\frac{1}{2} N_2(g) + O_2(g) = NO_2(g) \qquad \Delta H^\circ = +34 \text{ kJ/mol}$$

$$\frac{1}{2} O_2(g) + H_2(g) = H_2O(\ell) \qquad \Delta H^\circ = -286 \text{ kJ/mol}$$

(C) What is the difference between a molar heat of formation, a molar heat of reaction, and a molar heat of combustion?

(D) Determine the molar heat of combustion of benzene (C_6H_6).

Heats of formation (ΔH_f°):

$C_6H_6(\ell) = +46.0$ kJ/mol; $CO_2(g) = -393.5$ kJ/mol; $H_2O(\ell) = -286$ kJ/mol

The balanced chemical equation is: $C_6H_6(\ell) + \frac{15}{2} O_2(g) = 6CO_2(g) + 3H_2O(g)$

Multiple-Choice Answers

1. D   2. A   3. E   4. B   5. C   6. A   7. B   8. B   9. D   10. E

1. D

The sign for ΔH is positive, which signifies that energy is being gained by the system from the surroundings. The energy is absorbed in the form of heat. This type of reaction is called an *endothermic reaction*.

2. A

The molar mass of lead(II) sulfide is 239.3 g/mol. There are 4.18 moles in 1.00 kg of solid PbS. Since 414 kJ of energy is released for each mole of sulfur reacted, a total of 1730 kJ of energy is released. Since energy is released, the enthalpy change is negative.

3. E

To convert from degrees Celsius to kelvins, just add 273.15. The final temperature in Celsius is 33.8°C, so the temperature in kelvins is 307.0 K. See the explanation to question 5 for more information.

4. B

Since the reaction occurs at atmospheric pressure, the pressure is constant. At constant pressure, the enthalpy change of a reaction is equal to the heat of the reaction.

5. C

The specific heat capacity of water is 4.18 J/g·K. The density of water is 1.0 g/mL, so a 250.0 L sample has a mass of $2,500 \times 10^5$ g. The heat of the reaction, q, is −9,230 kJ. Using the equation $q = cm\Delta T$, the temperature change is 8.8 kelvins, which is equal to 8.8°C. Since heat is released by the reaction, the temperature of the water will increase by 8.8 degrees Celsius. The final temperature is 25.0°C + 8.8°C, or 33.8°C.

6. A

Use Hess's Law, keeping the first reaction as given, and reverse the second reaction. When the reactions are added the desired reaction, $Pb(s) + Cl_2(g) \rightarrow PbCl_2(s)$ results. This means when −329.3 kJ and −30.1 kJ are added the enthalpy change of −359.4 kJ results.

7. B

The 250.0 mL sample has a mass of 197.3 g. The temperature change was 22.0°C. Using the equation $q = cm\Delta T$, the quantity of heat absorbed can be determined. Since this reaction is occurring at constant pressure, the heat absorbed is equal to the change in internal energy of the ethanol.

8. B

Using the equation $q = cm\Delta T$, it can be calculated that 87.75 kJ are required to raise the temperature of the sample of aluminum by 6.5°C. 0.363 mole of hydrogen gas must be burned to produce this much energy. The mass of 0.363 moles of hydrogen gas is 0.73 g.

9. D

The enthalpy change for the reverse reaction is the same magnitude, but opposite in sign.

10. E

241.8 kJ is produced for each mole of hydrogen gas burned. For 3,580 kJ to be produced, 14.8 moles of hydrogen gas will have to be burned. The 14.8 moles of H_2 has a mass of 29.8 g

Free-Response Answers

1. (A) A constant volume calorimeter, sometimes called a bomb calorimeter, is a sealed container and is used to measure ΔE. A constant pressure calorimeter is generally open to atmospheric pressure, and is used to measure ΔH.

   (B) The solution process causes the temperature change. The heat is coming from the water. While it is probably not true, it is generally assumed that the heat the water loses is *all* gained during the solution process $[NH_4NO_3(s) \rightarrow NH_4NO_3(aq)]$. Thus, in calorimetry experiments it is common to generalize that heat loss = heat gain. Make sure the energy units are the same on both sides of the equation.

   Heat gained in solution process = Heat lost by the water

   The heat gained in solution process is the molar heat of solution multiplied by the molar mass and the unknown mass of NH_4NO_3. The heat lost by the water, q, is equal to $mc_p\Delta T$, with all values being known.

(C) Data to collect: initial temperature of the water just prior to reaction and maximum final temperature after mixing the ammonium nitrate with the water; quantity (mass or volume) of both the water and ammonium nitrate used. To get the *molar* heat of solution, divide the heat produced by the moles of solution used.

(D) Experimental errors: heat loss to calorimeter or air; incomplete transfer of solute; incomplete mixing; temperature not allowed to reach maximum before recording final temperature (about anything that will give a lower than needed final temperature).

2. (A) $4 NH_3(g) + 7 O_2(g) \rightarrow 4 NO_2(g) + 6 H_2O(\ell)$

(B) Equation of interest: $4 NH_3(g) + 7 O_2(g) \rightarrow 4 NO_2(g) + 6 H_2O(\ell)$

Rearrange the given equations for reactants and products and change the sign on ΔH if equation is reversed:

$$NH_3(g) = \frac{1}{2} N_2(g) + \frac{3}{2} H_2(g) \qquad\qquad \Delta H° = +46 \text{ kJ/mol}$$

$$\frac{1}{2} N_2(g) + O_2(g) = NO_2(g) \qquad\qquad \Delta H° = +34 \text{ kJ/mol}$$

$$\frac{1}{2} O_2(g) + H_2(g) = H_2O(\ell) \qquad\qquad \Delta H° = -286 \text{ kJ/mol}$$

Multiply each equation by the appropriate factor, including ΔH:

$$[4]\ NH_3(g) = \frac{1}{2} N_2(g) + \frac{3}{2} H_2(g) \qquad\qquad \Delta H° = +46 \text{ kJ/mol}$$

$$[4]\ \frac{1}{2} N_2(g) + O_2(g) = NO_2(g) \qquad\qquad \Delta H° = +34 \text{ kJ/mol}$$

$$[6]\ \frac{1}{2} O_2(g) + H_2(g) = H_2O(\ell) \qquad\qquad \Delta H° = -286 \text{ kJ/mol}$$

Add the resulting equations and heats of reaction to get the equation of interest and its heat of reaction:

$$4 NH_3(g) = 2 N_2(g) + 6 H_2(g) \qquad\qquad \Delta H° = +184 \text{ kJ/mol}$$

$$2 N_2(g) + 4 O_2(g) = 4 NO_2(g) \qquad\qquad \Delta H° = +136 \text{ kJ/mol}$$

$$3 O_2(g) + 6 H_2(g) = 6 H_2O(\ell) \qquad\qquad \Delta H° = -1,716 \text{ kJ/mol}$$

You get: $4 NH_3(g) + 7 O_2(g) \rightarrow 4 NO_2(g) + 6 H_2O(\ell)$ $\quad \Delta H_{rxn} = -1,396 \text{ kJ}$

(C) The molar heat of formation is the heat involved in producing a mole of a compound from its elements in their standard states. The heat of reaction for one mole of reactant in a stated reaction, and the heat of combustion is for the complete reaction of one mole of fuel with oxygen to produce $CO_2(g)$ and $H_2O(\ell)$.

(D) $\Delta H_c = \Sigma \Delta H(\text{products}) - \Sigma \Delta H(\text{reactants})$

$\Delta H_c = \{(6 \times -393.5 \text{ kJ/mol}) + (3 \times -286 \text{ kJ/mol})\} - \{(1 \times +46.0 \text{ kJ/mol}) + (15/2 \times 0 \text{ kJ/mol}\}$

$\Delta H_c = -3,270 \text{ kJ/mol } C_6H_6$

Atomic Structure and Periodic Properties

In order to understand the structure of atoms, one must first understand the properties of light.

THE WAVE NATURE OF LIGHT

In the classical sense, all light is composed of **waves**, which are periodic disturbances that spread energy throughout a medium. Since the disturbances are perpendicular to the direction of the energy flow, the waves are known as **transverse waves**. The diagram below represents a typical transverse wave:

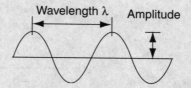

The **amplitude** of the wave is a measure of its intensity; for example, a visible light wave with a large amplitude would be a bright light, while one with a small amplitude would be a dim light. Since a wave is a periodic disturbance, the length of one complete cycle is known as the **wavelength** (λ). The wave also has a characteristic frequency (ν) that measures the number of complete cycles that pass a point in a unit of time. Both the wavelength and the frequency are related to the speed (v) of the wave by the equation:

$$v = \nu\lambda$$

In the SI, v is measured in meters per second, ν is measured in second^{-1}, which is also known as **hertz** (Hz), and λ is measured in meters. It is generally customary to measure the wavelengths of light in **nanometers** (nm, 1×10^{-9} m).

In particular, all light waves comprise the **electromagnetic spectrum** and the emission of light in any form is known as **electromagnetic radiation**. The electromagnetic spectrum, of which visible light is a part, is shown in the diagram below:

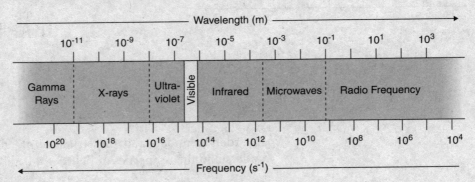

All electromagnetic radiation travels in space with a constant speed (c) whose approximate value is 3.00×10^8 meters per second. The frequency (or the wavelength) of an electromagnetic wave determines in which part of the spectrum it belongs.

Sample Problem

What is the frequency of red light whose wavelength is 720 nanometers?

Solution

$$c = \nu\lambda$$

$$\nu = \frac{c}{\lambda} = \frac{3.00 \times 10^8 \frac{m}{s}}{(720 \text{ nm}) \cdot \left(\frac{1 \times 10^{-9} \text{ m}}{1 \text{ nm}}\right)} = \mathbf{4.17 \times 10^{14} \text{ Hz}}$$

THE PARTICLE NATURE OF LIGHT

By the end of the nineteenth century, classical physicists were encountering serious problems with electromagnetic radiation, and the solutions to these problems necessitated a reformulation of the nature of light.

Black Body Radiation

When solid and liquid objects are heated they emit light. The familiar red-hot element of a toaster oven is but one example. As the temperature of the object is increased, the color of the emitted light changes from red-hot to white-hot as more blue light is added to the spectrum. It was known that the frequencies of the vibrating atoms in a heated object were responsible for the emitted radiation, but classical physics was unable to explain the *distribution* of this radiation. (Classical physics assumed that the radiation of energy was a continuous process.) In order to explain the distribution of the emitted radiation, the German physicist Max Planck made a revolutionary assumption: Energy absorbed or emitted by vibrating atoms occurred only in discrete multiples of the frequency of vibration and was given by the formula:

$$E = nh\nu \ (n = 1, 2, 3, \dots)$$

The minimum value of energy ($h\nu$) was known as a **quantum of energy**. The constant h is known as **Planck's constant** and has the approximate value of 6.63×10^{-34} joule·second. The extremely small value of Planck's constant explains why we do not sense the quantum nature of energy in our macroscopic world.

Sample Problem

Calculate the quantum of energy associated with red light whose frequency is 4.17×10^{14} Hz.

Solution

$$E = h\nu = (6.63 \times 10^{-34} \, \text{J} \cdot \text{s}) \cdot (4.17 \times 10^{14} \text{s}^{-1}) = \mathbf{2.76 \times 10^{-19} J}$$

The Photoelectric Effect

Another aspect of light that troubled nineteenth-century physicists was its behavior on photoemissive surfaces. A **photoemissive surface** is one that can release electrons when exposed to light, a phenomenon known as the **photoelectric effect**. The atypical behavior of light included the following observations:

- Each photoemissive surface would release electrons only if the incident light was at or above a specific *minimum frequency*. For example, the minimum frequency for the emission of photoelectrons from a cesium surface is 4.60×10^{14} hertz (orange light). Using light below this frequency will fail to release electrons from a cesium surface.

- The intensity of the incident light did not affect the maximum kinetic energy of the emitted photoelectrons. However, the rate at which photoelectrons were released depended on the intensity of the light.

- The maximum kinetic energy of the emitted photoelectrons depended on the frequency of the incident light (provided it was above the minimum frequency for the surface).

None of these behaviors could be explained by the classical notion of light as a wave. In 1905, Albert Einstein used and extended Planck's quantum theory in order to explain the photoelectric effect. Einstein suggested that in addition to possessing wave properties, light also possessed particle properties. Each quantum of light, known as a **photon**, had an energy that was related to its radiation frequency by the Planck formula:

$$E_{photon} = h\nu = \frac{hc}{\lambda}$$

When a photon strikes an electron on a photoemissive surface, the electron will be released only if the photon possesses the minimum energy necessary to do so. Any additional energy will be translated into the kinetic energy of the emitted electron.

Line Spectra

When the visible "white" light emitted from an incandescent solid or liquid is passed through a prism, the light is separated into its component colors and forms a **continuous spectrum** in which red blends into orange, orange blends into yellow, and so on.

When a gas at low pressure is made incandescent, the emitted light is not white and, when passed through a prism, a **line spectrum** is formed in which only specific colors of light are displayed. The nature of these colors depends on the nature of the gas. For example, the line spectrum of atomic hydrogen gas (H) in entirely different from the line spectrum of neon gas (Ne).

In the mid-nineteenth century a Swiss schoolteacher named Balmer was able to deduce a formula for the four lines of the hydrogen spectrum that he observed:

$$\nu = C \cdot \left(\frac{1}{2^2} - \frac{1}{n^2} \right) \ [n = 3, 4, 5, 6] \ \ C = (3.29 \times 10^{15} \text{ Hz})$$

The four lines of the visible line spectrum of atomic hydrogen, along with their colors, frequencies, and wavelengths are illustrated below:

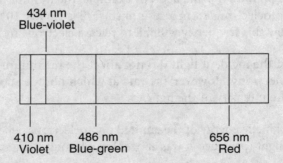

For example, if one inserts the value $n = 3$ into the Balmer equation, one obtains the frequency of the first (red) line of the hydrogen spectrum. The line spectra of hydrogen and other selected elements can be viewed at the following website: *www.colorado.edu/physics/PhysicsInitiative/Physics2000/quantumzone/*.

The Bohr Model of the Atom

Observing a line spectrum is one thing, explaining it is quite another. The classical nuclear model proposed by Rutherford, was utterly inadequate in providing an explanation for the existence of line spectra. Moreover, the rules of classical physics predicted that the orbiting electrons in his model should emit energy continuously, leading to the collapse of all atoms!

ATOMIC STRUCTURE AND PERIODIC PROPERTIES

In 1913, Neils Bohr, a Danish physicist, combined classical physics and Planck's quantum theory to arrive at a model of the hydrogen atom that was not only stable, but verified the Balmer formula for the line spectrum of hydrogen. Bohr made the following assumptions:

- The single electron of a hydrogen atom could orbit the nucleus in a circle that was associated with a specific radius and a specific energy. That is to say, he assumed that the orbits of the hydrogen atom were quantized.

- While in a specific orbit, or *energy level*, no radiation was emitted or absorbed by the atom. This condition is known as a **stationary state**.

- If an electron moved to a higher energy level, energy was absorbed; if it moved to a lower level, energy was emitted in the form of a photon of light, whose frequency (or wavelength) was given by the formula:

$$\Delta E = E_{final} - E_{initial} = h\nu = \frac{hc}{\lambda}$$

Note that ΔE represents the difference in the energies of the initial and final levels.

- The level with the lowest energy was known as the **ground state**; any higher level was called an **excited state**. The ground state was the closest that an electron could approach the hydrogen nucleus.

- Each energy level was assigned a *quantum number* (1, 2, 3, 4, . . . , n), with $n = 1$ designated as the ground state.

As a result of his calculations, Bohr arrived at a relationship that assigned a specific energy to each level:

$$\Delta E_n = -R_H \cdot \left(\frac{1}{n^2}\right) \ [n = 1, 2, 3 \dots] \ (R_H = 2.18 \times 10^{-18} \text{ J})$$

Sample Problem
Calculate the energy associated with the second energy level ($n = 2$).

Solution
$$\Delta E_n = -R_H \cdot \left(\frac{1}{n^2}\right) = -2.18 \times 10^{-18} \text{ J} \cdot \left(\frac{1}{2^2}\right) = -5.45 \times 10^{-19} \text{ J}$$

Combining the two equations given immediately above yields the Bohr equation for the hydrogen atom:

$$\Delta E = E_{final} - E_{initial} = h\nu = \frac{hc}{\lambda} = -R_H \cdot \left(\frac{1}{n_f^2} - \frac{1}{n_i^2}\right)$$

If we solve this equation for the frequency of the photon, ν, we obtain a *generalization* of the original Balmer formula:

$$\nu = \left| \frac{-R_H}{h} \cdot \left(\frac{1}{n_f^2} - \frac{1}{n_i^2} \right) \right|$$

The absolute value notation insures that the frequency will always be a *positive* number.

Sample Problem

Calculate the frequency and wavelength of the photon that is emitted when an electron moves from energy level 5 to energy level 2.

Solution

$$\nu = \left| \frac{-R_H}{h} \cdot \left(\frac{1}{n_f^2} - \frac{1}{n_i^2} \right) \right| = \left| \frac{-2.18 \times 10^{-18}\,\text{J}}{6.63 \times 10^{-34}\,\text{J} \cdot \text{s}} \cdot \left(\frac{1}{2^2} - \frac{1}{5^2} \right) \right| = \mathbf{6.90 \times 10^{-14}\,Hz}$$

$$\lambda = \frac{c}{\nu} = \left(\frac{3.00 \times 10^8\,\frac{\text{m}}{\text{s}}}{6.90 \times 10^{14}\,\text{Hz}} \right) \cdot \left(\frac{1\,\text{nm}}{10^{-9}\,\text{m}} \right) = \mathbf{434\,nm}$$

The color of this photon is blue.

Using the Bohr formulas we can draw the energy level diagram for a hydrogen atom, as shown below:

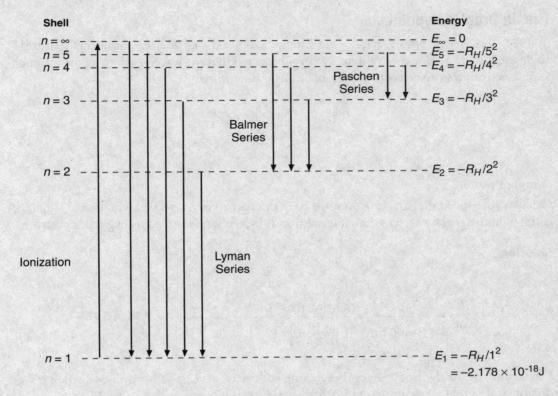

The energies are given in terms of the ground-state energy, $-R_H$. Note that as the energy-level number increases, the energy levels become closer to one another. The visible line spectrum, known as the *Balmer series*, is the result of electron transitions between energy level 2 and levels 3, 4, 5, ... Transitions between level 1 and levels 2, 3, 4, . . . are associated with photons found exclusively in the ultraviolet region of light; these transitions comprise the *Lyman series*. Transitions between level 3 and levels 4, 5, 6, . . . are associated with photons found exclusively in the infrared region of light. All of these spectra are *line spectra* since only certain specific transitions are allowed.

The topmost level, signified by ∞, has an energy of 0 joule; at this point, the electron is no longer part of the atom, and the atom is said to be *ionized*. The energy needed to ionize a single hydrogen atom in the ground state (2.18×10^{-18} joule) is known as the **ionization energy** of the hydrogen atom.

While the Bohr model explained the hydrogen atom, its use was limited to systems containing a single electron (such as He^+ and Li^{2+}). An atom with two electrons, such as helium, could not be explained adequately using Bohr's equations.

THE WAVE NATURE OF MATTER

The de Broglie Hypothesis

In 1924, the French physicist Louis de Broglie proposed that matter had a dual nature—analogous to the dual nature of light. De Broglie stated that the wavelength of a particle, such as an electron, was inversely proportional to its momentum:

$$\lambda = \frac{h}{p} = \frac{h}{mv}$$

Sample Problem

Calculate the speed that an electron must have in order to be associated with a wavelength of 0.100 nanometers (a wavelength that corresponds to X-rays in the electromagnetic spectrum).

Solution

$$\lambda = \frac{h}{mv}$$

$$v = \frac{h}{\lambda m} = \frac{(6.63 \times 10^{-34}\,\text{J} \cdot \text{s})}{(0.100\,\text{nm}) \cdot \left(\dfrac{1\,\text{m}}{10^9\,\text{nm}}\right) \cdot (9.11 \times 10^{-31}\,\text{kg})} = \mathbf{7.28 \times 10^6\,\dfrac{m}{s}}$$

Verification of the de Broglie hypothesis came within a few years of his proposal when it was demonstrated that electrons could be diffracted by crystals just as X-rays could.

The Uncertainty Principle

An object—in the classical sense—obeys Newton's laws of motion and, as a consequence, its position and speed can be measured simultaneously with complete accuracy. This is not the case with a wave, which spreads out through space. The German physicist Werner Heisenberg recognized that the wave nature of matter placed a fundamental limitation on an experimenter's ability to measure the position and momentum of a particle simultaneously. Heisenberg stated this limitation in a mathematical relationship known as the **uncertainty principle**:

$$\Delta p \cdot \Delta x \geq \frac{h}{4\pi}$$

If, for example, the uncertainty in the measurement of the momentum of a particle (Δp) is made very small, the measurement of the uncertainty in the particle's position (Δx) must be correspondingly larger. Simply stated: *the more we know about the momentum of a particle, the less we know about its position* (and *vice versa*). The uncertainty principle is not evident or important in our macroscopic world because of the smallness of Planck's constant. However, it plays a vital role in the domain of atoms and molecules.

THE QUANTUM-MECHANICAL MODEL OF THE HYDROGEN ATOM

In the 1920s a new method of dealing with atomic structure arose; it was based on the wave nature of matter and the uncertainty principle. The Austrian physicist Erwin Schrödinger, developed an equation that treated the electron on an atom of hydrogen as a wave. The solution to this equation yielded a series of **wave functions**, collectively denoted by the symbol ψ. Each wave function is known as an **orbital** and is associated with a characteristic energy.

While a wave function has no physical significance, the square of the wave function, ψ^2, is known as the **probability density** and is related to the probability of finding the electron a given distance from the nucleus. The following diagram illustrates one way of representing this probability in a ground-state hydrogen atom:

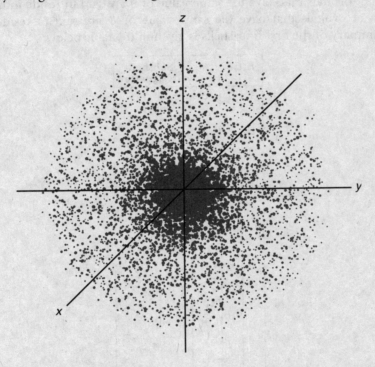

In regions where the probability of finding an electron is high, the region is said to have a high **electron density**. These regions correspond to the more intensely shaded portions of the diagram shown above.

The Schrödinger equation uses three **quantum numbers** to describe an orbital in a hydrogen atom:

- The **principal quantum number**, n, is an integer that describes the average distance of the electron from the nucleus; as n increases, the probability of finding an electron further away from the nucleus increases. As with the Bohr model of the hydrogen atom, the energy of the atom depends solely on the value of n and is given by the formula: $\Delta E_n = -R_H \cdot \left(\dfrac{1}{n^2}\right) [n = 1, 2, 3, \ldots]$.

- The **azimuthal quantum number**, ℓ, can have integral values between 0 and $n - 1$, and defines the *shape* of the orbital's electron density. The values of ℓ are usually associated with letter designations that are drawn from the descriptions of spectra that were developed in the past:

ℓ	0	1	2	3
letter	s	p	d	f

- The **magnetic quantum number**, m_ℓ, can have integral values between $+\ell$ and $-\ell$, and describes the *orientation* of the orbital in space.

The collection of orbitals that have the same value of n are said to reside in the same **shell**. The collection of orbitals that have the same value of ℓ are said to reside in the same **subshell**. A summary of the first four shells is given in the table below:

n	ℓ	m_ℓ	Subshell Designation
1	0	0	$1s$
2	0	0	$2s$
	1	1	$2p$
		0	
		−1	
3	0	0	$3s$
	1	1	$3p$
		0	
		−1	
	2	2	$3d$
		1	
		0	
		−1	
		−2	
4	0	0	$4s$
	1	1	$4p$
		0	
		−1	
	2	2	$4d$
		1	
		0	
		−1	
		−2	
	3	3	$4f$
		2	
		1	
		0	
		−1	
		−2	
		−3	

As we noted previously, the energy of a hydrogen atom (and all one-electron systems) depends solely on the value of n. Therefore, all of the subshells (and the orbitals) in a given shell have the same energy. We will see that this is *not* the case with many-electron atoms.

Orbital Shapes

s Orbitals

An *s* orbital can have *one* value of its magnetic quantum number (0), which corresponds to a *single* orientation in space. The shape of an *s* orbital is a spherical surface that is centered at the nucleus of the atom. The 1*s*, 2*s*, and 3*s* orbitals are progressively larger spheres. In cross-section, however, other differences become apparent, as shown in the following diagram:

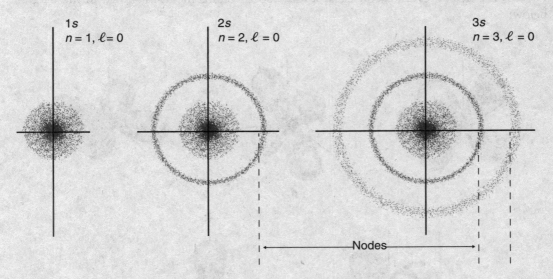

The 2*s* orbital consists of two distinct regions of electron density separated by a region in which the probability of finding an electron drops to zero. Any point in this region is known as a **node**. The 3*s* orbital consists of three distinct regions of electron density, each separated by a nodal region.

p Orbitals

A *p* orbital can have one of *three* values of its magnetic quantum number (+1, 0, −1). which corresponds to one of *three* possible orientations in space. The possible orientations of the 2*p* orbitals are shown below:

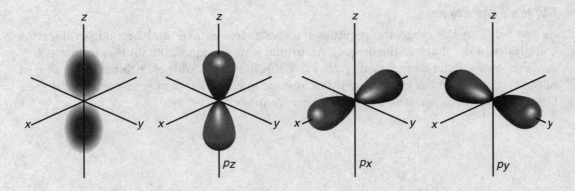

In each case, the electron density consists of two lobes aligned along one of the coordinate axes and each orbital is named according to its axis of alignment: $2p_x$, $2p_y$, and $2p_z$. The nodal regions of these orbitals are the two remaining planes; for example, the yz-plane is the nodal region for the $2p_x$ orbital.

d Orbitals

Four of the five $3d$ orbitals consist of a four-lobed arrangement, as shown in the diagram below:

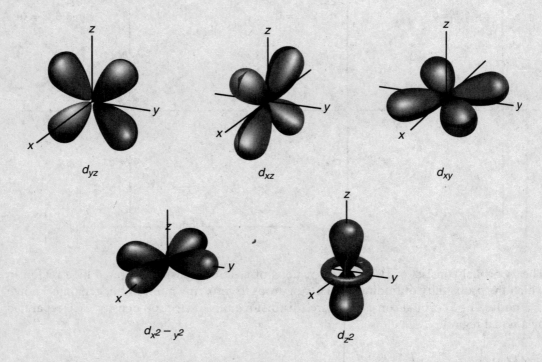

Three of these four are named for the plane in which they lie: $3d_{xy}$, $3d_{xz}$, and $3d_{yz}$; the fourth is named for the axes on which it is aligned: $3d_{x^2-y^2}$. The fifth orbital (d_{z^2}) is aligned along the z-axis and looks like a modified $2p_z$ orbital.

f Orbitals and Beyond

As the value of the magnetic quantum number increases to 3 and beyond, so does the complexity of the shape of the orbitals. An orbital whose magnetic quantum number is 4, 5, ... takes on the letter designations g, h, ... While the shapes of these orbitals are not given here, there are a number of websites that allow one to view their computer-generated shapes. One such website is: *www.uky.edu/~holler/html/orbitals_2.html*.

Many-Electron Atoms

The real power of the quantum-mechanical model lies in its ability to explain the structure and properties of all of the atoms in the Periodic Table. Many-electron atoms (that is, atoms whose atomic numbers are greater than or equal to 2) are composed of *hydrogen-like* orbitals. However, the energy of the atom does *not* depend solely on the value of the principal quantum number (n), but on the combined values of the principal and azimuthal quantum numbers (n and ℓ). The relative energies of these hydrogen-like orbitals can be determined from a device known as the *diagonal rule*, which is illustrated below:

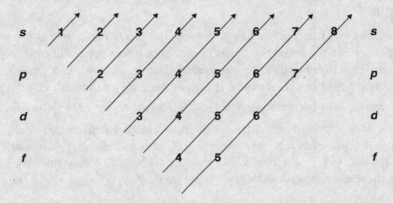

Using the diagonal rule is quite simple: follow the direction of the arrows from tail-to-head, and from left-to-right. The first ten orbitals in order of increasing energy are given below:

$$1s < 2s < 2p < 3s < 3p < 4s < 3d < 4p < 5s < 4d$$

Sample Problem

Use the diagonal rule to list the next six orbitals in order of increasing energy.

Solution

$$5p < 6s < 4f < 5d < 6p < 7s$$

This rather peculiar pattern of orbital energies is due to the fact that the electrons in a many-electron atom are attracted to the nucleus but repelled by each other. This makes it impossible to calculate the energy of any single electron. However, it is possible to estimate the energy of a single electron in this complex environment.

One important effect, known as the **screening effect**, is the shielding of outer-shell electrons by inner-shell electrons. As a result of the screening effect, the nuclear charge, Z, is reduced to a value known as the **effective nuclear charge**, Z_{eff}, which attracts the outer-shell electrons less strongly. The effective nuclear charge on an outer-shell electron can be estimated by subtracting the **screening constant**, σ, (which can be thought of as the average number of electrons that lie between the nucleus and the electron in question) from the nuclear charge:

$$Z_{eff} = Z - \sigma$$

One result of the screening effect is the ordering of energies of electrons within a single shell of a many-electron atom. As the value of the electron's magnetic quantum number, ℓ, increases, the average electron density is located further from the nucleus, and the electron experiences a larger screening effect. As a result, the energy of the electron is *increased*. Therefore, within a single shell, the following order is observed: $s < p < d < f$. This is quite unlike the situation in a hydrogen atom in which all of the orbitals within a single shell possess the same energy.

Electron Spin

In the mid-1920s, two Dutch physicists proposed that electrons in atoms possessed another intrinsic property known as **electron spin**. This proposal was the result of the observation of line spectra and experiments that demonstrated that an electron can generate a magnetic field in one of two possible orientations. It is now necessary to introduce a fourth quantum number, the **spin quantum number** (m_s), which has a value of $+\frac{1}{2}$ or $-\frac{1}{2}$. In 1925, the German physicist Wolfgang Pauli introduced his **exclusion principle**, which states that *no two electrons in an atom can have the same four quantum numbers*. As a result of the work done during this period, it was concluded that any orbital can contain a maximum of *two electrons*, and these electrons must have *opposite spins*.

The Electron Configurations of Atoms

Let us summarize what we have learned about shells, subshells, and orbitals in the table below:

Shell	Number of Subshells	Number of Orbitals	Maximum Number of Electrons
1	1 [s]	1	2
2	2 [s, p]	4 [1, 3]	8 [2, 6]
3	3 [s, p, d]	9 [1, 3, 5]	18 [2, 6, 10]
4	4 [s, p, d, f]	16 [1, 3, 5, 7]	32 [2, 6, 10, 14]
n	n [s, p, d, f, . . .]	n^2 [1, 3, 5, 7, . . .]	$2n^2$ [2, 6, 10, 14, . . .]

We note that every *s* subshell contains 1 orbital, every *p* subshell contains 3 orbitals, and so on. This can be verified by examining the table of quantum numbers found on page 134.

Let us adopt a shorthand notation for designating the number of electrons in a particular subshell:

The notation $3d^7$ means that the d subshell on the third shell contains 7 electrons. In order to designate the arrangement of electrons within orbitals, we need to adopt a final collection of rules:

- All of the orbitals in a subshell must contain one electron (that is, be *half-filled*) before any orbital in that subshell can be filled with two electrons. (This is known as the Hund rule.)

- All of the electrons in half-filled orbitals must have the same spin.

- Both electrons in a filled orbital must have opposite (paired) spins. (This is a direct result of the Pauli exclusion principle.)

In this book, we represent an orbital by a dash (underscore), and an electron by an arrow, whose direction indicates the value of the spin quantum number $\left(\uparrow \text{ for } +\frac{1}{2} \text{ and } \downarrow \text{ for } -\frac{1}{2}\right)$. The following diagrams represent empty, half-filled, and filled orbitals, respectively:

We can now represent the arrangement of electrons with a subshell notation of $3d^7$:

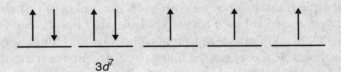

Note once again that the filling of an orbital occurs only after the entire subshell has been half-filled. Another way of designating the arrangement of these electrons is by assigning values to their four quantum numbers:

n	ℓ	m_ℓ	m_s
3	2	+2	$+\dfrac{1}{2}$
3	2	+1	$+\dfrac{1}{2}$
3	2	0	$+\dfrac{1}{2}$
3	2	−1	$+\dfrac{1}{2}$
3	2	−2	$+\dfrac{1}{2}$
3	2	+2	$-\dfrac{1}{2}$
3	2	+1	$-\dfrac{1}{2}$

Ground State Electron Configurations of the Elements

One of the powerful features of quantum mechanics is its ability to describe the electron configurations of the elements in the ground state. A technique known as the **aufbau** (building-up) **principle** is used: We begin with hydrogen and then add electrons—one at a time—into the lowest energy subshell that is available, that is, we follow the diagonal rule discussed on page 137. In addition, we must follow the rules for filling the orbitals within the subshell. The following table represents the filling patterns of the first twenty elements of the Periodic Table:

ATOMIC STRUCTURE AND PERIODIC PROPERTIES

Element	Atomic Number	Electron Configuration
H	1	$1s^1$
He	2	$1s^2$
	(Level 1 is complete.)	
Li	3	$1s^2\,2s^1$
Be	4	$1s^2\,2s^2$
B	5	$1s^2\,2s^2\,2p^1$
C	6	$1s^2\,2s^2\,2p^2$
N	7	$1s^2\,2s^2\,2p^3$
O	8	$1s^2\,2s^2\,2p^4$
F	9	$1s^2\,2s^2\,2p^5$
Ne	10	$1s^2\,2s^2\,2p^6$
	(Level 2 is complete.)	
Na	11	[Ne] $3s^1$
Mg	12	[Ne] $3s^2$
Al	13	[Ne] $3s^2\,3p^1$
Si	14	[Ne] $3s^2\,3p^2$
P	15	[Ne] $3s^2\,3p^3$
S	16	[Ne] $3s^2\,3p^4$
Cl	17	[Ne] $3s^2\,3p^5$
Ar	18	[Ne] $3s^2\,3p^6$
	(Level 3 is *not* complete, but it has 8 electrons.)	
K	19	[Ar] $4s^1$
Ca	20	[Ar] $4s^2$

The noble gases are unusually stable atoms and we use the [He] and [Ne] notations as shorthand ways of writing these configurations (also known as **noble gas cores**). Generally, it is accepted practice to write the subshell electron configuration of an atom or ion by using the noble gas core whose atomic number is closest to (but does not exceed) the atomic number of the atom or ion.

Sample Problem

(A) Draw the ground-state subshell configuration of an atom of titanium (Ti, atomic number = 22).

(B) Draw the orbital configuration of the $3d$ subshell of titanium.

Solution

There are two ways of writing the subshell configuration: in order of increasing energy, or in order of increasing shell number.

(A) [Ar]$4s^2 3d^2$ or   [Ar]$3d^2 4s^2$

In this book, we will write electron configurations in order of increasing shell number.

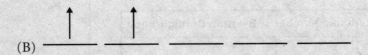

(B)

The rules that we use to predict filling patterns do have exceptions. When a subshell is half-filled or fully filled, the atom has additional stability (that is, lower energy). As a result, copper (Cu, atomic number 29) assumes an electron configuration of $[Ar]3d^{10}4s^1$, rather than the predicted electron configuration of $[Ar]3d^94s^2$. In the first electron configuration, all of the subshells are either fully filled or half-filled; in the second configuration, the $3d$ subshell is neither fully filled nor half-filled. Similarly, chromium (Cr, atomic number 24) assumes the electron configuration $[Ar]3d^54s^1$, rather than the configuration $[Ar]3d^44s^2$.

Valence Electrons and Lewis Symbols

The electrons in the outermost shell are known as **valence electrons**. We will see that valence electrons play an important role in the bonding of atoms. One way of designating the valence electrons in an atom is to employ a device known as a **Lewis symbol**, in which the symbol of the element is used to represent the nucleus and all of the *non*valence electrons of the atom. These are known collectively as the **kernel** of the atom. The *valence* electrons are represented by a series of "dots" and are placed around the kernel in a way that mimics their orbital arrangement: A filled orbital is represented by a pair of dots, and a half-filled orbital is represented by a single dot. The table below lists the Lewis symbols for the first ten elements.

H·
He:
Li·
Be:
:B·
Ċ·
:Ṅ·
:Ö·
:F̈·
:Ṅe:

The *order* of the placement of the dots is really a matter of convenience. For example, it does not matter where the dots for oxygen (O) are placed, as long as the dots represent the orbital configuration accurately: two pairs of dots and two single dots.

THE PERIODIC TABLE

In chapter 4 we learned that the physical and chemical properties of elements possess a considerable degree of regularity or *periodicity*. The British scientist Mosely discovered that this regularity was a function of atomic number. For this reason, the Periodic Table is an approximately rectangular array of elements arranged in order of increasing atomic number.

Recall that the (vertical) columns of the Table are known as **groups** (or **chemical families**), and are numbered from 1–18. Elements within a single group have similar physical and chemical properties. For example, the elements comprising Group 17 are all characteristically colored and they tend to gain a single electron in simple chemical reactions. Some of the groups have special names: Group 1 elements are named **alkali metals**; Group 2 metals are named **alkaline earth elements**; Group 17 elements are named **halogens**; and Group 18 elements are named **noble gases**.

The (horizontal) rows of the Table are known as **periods**, which are numbered from 1 to 7. Across a period, the properties of the elements change from **metallic** to **nonmetallic**. For example, in Period 3, the elements Na, Mg, and Al are metallic; the elements P, Cl, and Ar are nonmetallic; and the element Si is known as a **metalloid** (or **semimetal**) because it possess properties common to both metallic and nonmetallic elements.

The elements of Groups 1, 2, and 13–18 are known as the **representative elements**: their physical and chemical properties are the most easy to predict. The elements of Groups 3–11 are known as the **transition elements**, and display certain special properties such as color and multiple oxidation states. Another collection of elements is found as two rows placed at the bottom of the main body of the Table. These rows are named the **lanthanide** and **actinide** series, respectively.

The arrangement of the Periodic Table is more easily understood in terms of the electron configurations of the elements. The representative elements fill the *s* and *p* subshells; the transition elements and the elements of Group 12 fill the *d* subshells; and the lanthanides and actinides fill the *f* subshells. The valence shell electron configurations of every element within a single representative group are similar. For example, oxygen and sulfur are both members of Group 16: their valence shell electron configurations are, respectively, $2s^2 2p^4$ and $3s^2 3p^4$.

Sample Problem

What is the general valence shell electron configuration of the elements in Group 13?

Solution

$ns^2 np^1$, where n is the principal quantum number of the valence shell.

PERIODIC PROPERTIES

The table below lists a number of periodic properties for the representative elements. We will refer to this table in the discussion that follows.

Symbol	→	N
Atomic Radius (Calculated)/pm	→	56
Ionic Radius/pm (Ionic Charge)	→	171 (3–)
First Ionization Energy/(kJ/mol)	→	1402
Electron Affinity/(kJ/mol)	→	–7

	1	2	13	14	15	16	17	18
1	**H** 53 — 1312 –73							**He** 31 — 2372 0
2	**Li** 167 58 (1+) 520 –60	**Be** 112 27 (2+) 900 0	**B** 87 12 (3+) 801 –27	**C** 67 — 1086 –154	**N** 56 171 (3–) 1402 –7	**O** 48 140 (2–) 1314 –141	**F** 42 133 (1–) 1681 –328	**Ne** 38 — 2081 0
3	**Na** 190 102 (1+) 496 –53	**Mg** 145 72 (2+) 738 0	**Al** 118 53 (3+) 578 –44	**Si** 111 — 1086 –134	**P** 98 212 (3–) 1012 –72	**S** 88 184 (2–) 1000 –200	**Cl** 79 181 (1–) 1251 –349	**Ar** 71 — 1521 0
4	**K** 243 138 (1+) 419 –48	**Ca** 194 100 (2+) 590 0	**Ga** 136 62 (3+) 579 –29	**Ge** 125 — 762 –118	**As** 114 222 (3–) 947 –77	**Se** 103 198 (2–) 941 –195	**Br** 94 196 (1–) 1140 –325	**Kr** 88 — 1351 0
5	**Rb** 265 149 (1+) 403 –47	**Sr** 219 116 (2+) 550 0	**In** 156 72 (3+) 558 –29	**Sn** 145 — 709 –121	**Sb** 133 — 834 –101	**Te** 123 221 (2–) 869 –190	**I** 115 220 (1–) 1008 –295	**Xe** 108 — 1170 0
6	**Cs** 298 170 (1+) 376 –45	**Ba** 253 136 (2+) 503 0	**Tl** 156 88 (3+) 589 –30	**Pb** 154 — 716 –110	**Bi** 143 — 703 –110	**Po** 135 — 812 –183	**At** 127 — 920 –270	**Rn** 120 — 1037 0

Atomic Radius

There are a number of ways of estimating the size of a collection of atoms. If the atoms are bonded together, as in the H_2 molecule, for example, the **atomic radius** is defined to be one-half of the *internuclear distance* between the two bonded atoms. In a nonbonded atom, such as a noble gas atom, for example, the estimated atomic radius is defined to be the closest distance that a second atom can approach the nucleus of the first atom during a collision. The diagrams below illustrate these definitions:

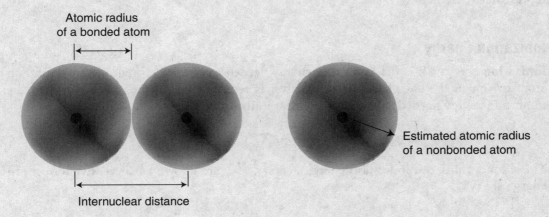

Atomic radius
of a bonded atom

Estimated atomic radius
of a nonbonded atom

Internuclear distance

Atomic radii are generally measured in picometers. In the table given above, the atomic radii are not measured values, but rather values that have been calculated using quantum-mechanical considerations. As we inspect the table, we find the atomic radii decrease left to right across a period, and increase from top to bottom down a group. We can explain the decrease in atomic radius across a period by recognizing that the nuclear charge increases steadily, but the amount of shielding does not because all of the additional electrons are added to the same shell. The increase in atomic radius down a group is the result of the reduction of Z_{eff} (see page 137) by the addition of additional shells of electrons.

Ionic Radius

The **ionic radius** is a measure of the size of a positive or negative ion. The ionic radius of a positive ion is *smaller* than its parent atom because the removal of one or more electrons decreases the electron-electron repulsion. The ionic radius of a negative ion is *larger* than its parent atom because the addition of one or more electrons increases the electron-electron repulsion.

As with neutral atoms, the ionic radius increases down a group. In comparing ionic radii across a period, however, we must be careful to compare ions that are **isoelectronic**, that is, have the same electron configurations. For example, in Period 3 of the Periodic Table, Na^+, Mg^{2+}, and Al^{3+} all have the [Ne] electron configuration, while S^{2-} and Cl^- have the [Ar] electron configuration. Across a period, the ionic radii of isoelectronic ions decrease.

Sample Problem

Which of these ions is smallest and which is largest in size: Na^+, Mg^{2+}, K^+, Ca^{2+}?

Solution

Mg^{2+} is the smallest ion; K^+ is the largest ion.

The table of periodic properties, given above, contains the ionic radii of selected representative elements.

Ionization Energy

Ionization energy (I) is the minimum energy needed to remove a single electron from an atom or ion in the gas phase. If the electron is removed from a neutral atom, then the minimum energy needed is known as the **first ionization energy** (I_1):

$$X(g) + I_1 \rightarrow X^+(g) + e^-$$

The second, third, . . . ionization energies (I_2, I_3, . . .) are known as **successive ionization energies**:

$$X^+(g) + I_2 \rightarrow X^{2+}(g) + e^-$$

$$X^{2+}(g) + I_3 \rightarrow X^{3+}(g) + e^-$$

$$\ldots$$

An atom can have as many successive ionization energies as it has electrons. The table of periodic properties given above lists the first ionization energies of the representative elements. In general the increase in atomic size and screening effects cause the ionization energy to decrease down a group.

Across a period, there is a general increase in the ionization energy due to the increase in nuclear charge and decrease in the size of successive atoms. There are notable exceptions, however. In Period 2, the first ionization energies of boron (B) and oxygen (O) are smaller than the ionization energies of the atoms that precede them. In boron, the last electron is added to a new, higher energy sublevel ($2p$) and, consequently, less energy is needed to remove it from the atom. In oxygen, the last electron fills the first $2p$ orbital and, as a result, the sublevel has less stability than the half-filled sublevel of its predecessor, nitrogen (N). Consequently, the first ionization energy of oxygen is less than that of nitrogen.

Successive Ionization Energies

As successive electrons are removed from an atom, its size decreases progressively and the energy needed to remove the next electron increases: that is, $I_1 < I_2 < I_3$ . . . . The table below lists the successive ionization energies for the first ten elements and illustrates this progressive increase.

Successive Ionization Energies/(kJ/mol)

Element	Configuration	I_1	I_2	I_3	I_4	I_5	I_6	I_7	I_8	I_9	I_{10}
H	$1s^1$	1,312									
He	$1s^2$	2,372	5,250								
Li	$[He]2s^1$	520	7,298	11,815							
Be	$[He]2s^2$	900	1,757	14,849	21,007						
B	$[He]2s^22p^1$	801	2,427	3,360	25,026	32,827					
C	$[He]2s^22p^2$	1,086	2,353	4,620	6,223	37,831	47,277				
N	$[He]2s^22p^3$	1,402	2,856	4,578	7,475	9,445	53,267	64,360			
O	$[He]2s^22p^4$	1,314	3,388	4,300	7,469	10,990	13,326	71,330	84,078		
F	$[He]2s^22p^5$	1,681	3,374	6,050	8,407	11,023	15,164	17,868	92,038	106,434	
Ne	$[He]2s^22p^6$	2,081	3,952	6,122	9,371	12,177	15,238	19,999	23,069	115,380	131,432

Examining the table more closely, we observe unusually large increases in ionization energies (in the shaded areas), beginning with the element lithium (Li). When all of the valence electrons of an atom have been removed, only the *noble gas core* remains. Given the great stability of a noble gas electron configuration, any attempt to remove the first electron from this configuration requires a significantly larger amount of energy. For example, nitrogen (N) has five valence electrons. Their removal requires 1,402, 2,856, 4,578, 7,475, and 9,445 kilojoules of energy per mole of nitrogen atoms. The removal of the sixth electron, which is part of the [He] core, requires 53,267 kilojoules of energy per mole—an almost six-fold increase in energy! The following graph illustrates the energies needed to remove the five electrons in boron.

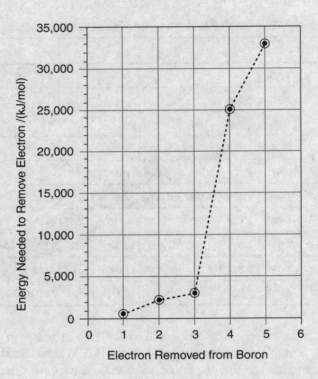

Electron Affinity

Electron affinity (*EA*) is the amount of energy that is involved with the *addition* of a single electron to an atom or ion. In general, the addition of an electron is accompanied by the release of energy:

$$Y + e^- \rightarrow Y^- + EA$$

In this book, we will represent such values as *negative* numbers. The larger the absolute value of the negative number, the more stable is the negative ion. If an electron affinity value is listed as zero, it means that the negative ion does not form. The table of periodic properties given above lists the electron affinities for a number of representative elements. The electron affinity depends on a number of factors: nuclear charge, atomic size, electron-electron repulsion, and the screening effect exerted on the incoming electron. In general, the tendency to form a negative ion increases as we move across a period, but there is not much change down a group. In Group 17, chlorine provides an interesting anomaly. It has the largest electron affinity of all the elements. Evidently, all of the factors responsible for determining electron affinity produce an ideal compromise in the chlorine atom and render it most hospitable to an incoming electron.

In Group 2 there is little, if any, tendency to form a negative ion, presumably because the incoming electron would reside in a new *p* subshell, and is especially well screened from the nuclear charge by the inner electrons.

In summary, the physical and chemical properties of substances depend on the structure and periodic properties of their constituent atoms. We will have occasion to examine these properties in the chapters that follow.

Chapter 8: Practice Questions

Multiple-Choice Questions

1. The photoelectric effect shows that a minimum energy is needed to eject an electron from a piece of metal. This supports the idea of quantized energy because:

   A. increasing the brightness of the light makes the electrons move faster.

   B. changing the color, or wavelength, of the light keeps the electrons quantized at the same level.

   C. quantized energy is the only way that energy can be explained at the time of its discovery.

   D. it is shown that a minimum frequency of light is needed. Making the light brighter doesn't help.

   E. all changes in brightness (intensity) and color (frequency) have no effect on the ejection of electrons from a metal.

2. What is the energy (in Joules) of a photon that has a frequency of 4.00×10^{10} Hz? (The Planck Constant has a value of 6.626×10^{-34} J•s.)

   A. 1.99×10^{-25}
   B. 2.65×10^{-23}
   C. 7.50×10^{-3}
   D. 1.20×10^{19}
   E. 6.02×10^{23}

3. The location of an electron is identified by the atomic orbital it is in. Which properties of an atomic orbital are specified by the first three quantum numbers?

   A. Length, width, and height
   B. Mass, speed, and direction
   C. Size, shape, and orientation
   D. Size, exact location, and velocity
   E. Color, frequency, and wavelength

4. What are the four quantum numbers identifying the outermost electron in a gallium ($_{31}$Ga) atom in the ground state?

   A. (3, 0, 0, −1/2)
   B. (3, 1, 1, +1/2)
   C. (4, 0, 0, +1/2)
   D. (4, 1, 1, +1/2)
   E. (4, 2, −2, −1/2)

5. Which of the following is the ground-state electron configuration of an oxide ion?

   A. $1s^2 2s^2 2p^4$

   B. $1s^2 2s^2 2p^5$

   C. $1s^2 2s^2 2p^6$

   D. $1s^2 2s^2 2p^6 3s^1$

   E. $1s^2 2s^2 2p^6 3s^2$

6. Position on the Periodic Table gives information regarding an element's electron configuration. Which is the same as the period number and the group number?

	Period number is the:	Group number is the:
A.	number of *p* orbitals;	number of elements in the group
B.	number of shells of electrons in the atom;	number of valence electrons
C.	total number of electrons in the outer shell;	total number of electrons in the atom
D.	number of subshells in the last energy level;	number of metal atoms in the group
E.	number of electrons in the outer subshell (*s*, *p*, *d*, or *f*);	charge on the most stable ion

7. Why does phosphorus have a lower electron affinity than silicon? This is explained by an electron being added to phosphorus to:

   A. a filled orbital.
   B. a new subshell.
   C. an empty orbital.
   D. a half-filled orbital.
   E. a new valence shell.

8. Which atom has the largest covalent radius?

   A. Argon
   B. Arsenic
   C. Phosphorus
   D. Selenium
   E. Sulfur

9. Fluorine is the most active nonmetal. What explains this high attraction for electrons?

   A. Fluorine is the most negative ion of any element.

   B. All of the fluorine electrons are in the outer shell.

   C. The fluorine nucleus is larger than any other nonmetal.

   D. Fluorine has atomic and mass numbers which are odd numbers.

   E. The fluorine atom has little shielding of its nucleus from its outer electrons.

10. What is the correct orbital diagram for the outer electrons of chromium in the ground state?

   A. 3*s*: ↑↓ 2*d*: ↑ ↑ ↑ ↑ __

   B. 4*s*: ↑↓ 3*d*: ↑↓ ↑↓ __ __ __

   C. 4*s*: ↑↓ 3*d*: ↑ ↑ ↑ ↑ __

   D. 4*s*: ↑ 3*d*: ↑ ↑ ↑ ↑ ↑

   E. 5*s*: ↑ 4*d*: ↑ ↑ ↑ ↑ ↑

Free-Response Questions

1. Physicists of the early 20th century contributed to the present quantum mechanical model of electronic configurations.

   (A) Werner Heisenberg is noted for his Uncertainty Principle. What is the Uncertainty Principle, and how does it apply to our current view of the location of electrons in atoms?

   (B) Wolfgang Pauli is noted for his Exclusion Principle. Draw the orbital diagram for sulfur, define the Exclusion Principle, and describe where the principle is applied to the electron structure of the valence electrons of sulfur.

   Sulfur: 3s: __ 3p: __ __ __ (Use arrows ↑↓)

   (C) Hund's Rule is applied when electrons are assigned to p, d, or f sublevels. Draw the orbital diagram for phosphorus, state Hund's Rule, and describe where this rule is applied in diagramming the electron structure of the valence electrons of this element.

   Phosphorus: 3s: __ 3p: __ __ __ (Use arrows ↑↓)

2. Trends in atomic radius, ionization energy, and electron affinity give experimental validity to the arrangement of elements in the Periodic Table.

   (A) What is the underlying cause of the decrease in atomic radius from left to right in a period?

   (B) Moving from rubidium to xenon, when there is an increase in +17 in the charge of the nucleus, causes a sharp increase in the ionization energy. Why then is there a rather large decrease in ionization energy when there is a +18 (related to rubidium) increase in the nuclear charge when cesium is considered.

   (C) Why is there a decrease in electron affinity correlated with atomic number moving within a period, yet a decrease when we move down a group?

Multiple-Choice Answers

1. D 2. B 3. C 4. D 5. C 6. B 7. D 8. B 9. E 10. D

1. D

The idea that photons are quantized helps explain why a threshold frequency is needed. Increasing the intensity of this light creates more photons, more electrons of equal energy, but does not increase the energy of the photons. Increasing the intensity of a lower frequency light which is below the threshold does not produce any electrons.

2. B

The correct answer is 2.65×10^{-23}. Using Planck's equation for the energy of a photon allows the computation of energy of a burst of light of a specific frequency.

$$E = h\nu = 6.626 \times 10^{-34} \text{ J} \cdot \text{s} \times 4.00 \times 10^{10} \text{ s}^{-1} = 2.65 \times 10^{-23} \text{ J}$$

3. C

The principal quantum number indicates the shell or relative distance from the nucleus, the angular momentum quantum number matches the subshell shape ($s, p, d, f, g, h, \ldots$), and the magnetic quantum number tells us about the orientation in three-dimensional space.

4. D

The four quantum numbers are: n, l, m_l, m_s. The atomic number of gallium is 31 and is in the 4th period and the Group IIIB of the Periodic Table. Because it is in Group 13 (IIIB), the outermost electron is in the p subshell. This gives it an l value of l. Possible values of m_l are $-1, 0$, and $+1$, and possible values of the spin quantum number, m_s, are $+1/2$ and $-1/2$.

5. C

The atomic number of oxygen is 8, and the oxide ion is O^{2-} has 2 more electrons than the oxygen atom giving a total of 10 electrons in this ion. The correct answer to this question is found by adding up the numbers present in the superscripts.

6. B

The period number [row] predicts the number of shells in a given electron configuration. The group number [column] tells how many valence electrons to expect in the elements found in the group.

7. D

Electron affinity represents the amount of energy liberated as an electron is added to the valence orbital with the lowest energy. A number of factors effect electron affinity—nuclear charge, atomic size, electron-electron repulsion, and the screening effect exerted on the incoming electron. In the case of phosphorus, an electron is being added to an orbital that is already occupied by an electron, which is less favorable than adding an electron to an empty orbital (this is the case with silicon).

8. B

Atomic radius increases for those elements in the Periodic Table that are down and to the left in the table, so As (arsenic) will have the largest size of the atoms listed.

9. E

Fluorine has 7 electrons in its outermost shell, one short of the ultimate objective—an octet of electrons. It is in only the second period. The outer shell of electrons (it only has 2) of fluorine is not far away from the nucleus, which decreases the shielding of the outer electrons from the nucleus. The other answer choices are all incorrect.

10. D

All of the answer choices have 6 outer electrons, so this is not an issue. Chromium is in Period 4, which predicts that the outer electrons will be in the $4s$ and $3d$ subshells. Answer (B) is a violation of Hund's Rule, which predicts that electron will all occupy empty before pairing. (D) is the best choice for this question because filled and half-filled subshells are more stable than other arrangements. In order to achieve this ground or lower energy state an electron is "promoted" from the $4s$ to the $3d$ subshell.

Free-Response Answers

1. (A) The Uncertainty Principle states that it is impossible to determine both the position and momentum of an electron in an atom simultaneously. The position is more important in the electron model, and the probability of an electron being in a certain area of the atom is emphasized currently.

   (B) Sulfur: $3s$: ↑↓ $3p$: ↑↓ ↑ ↑ The Exclusion Principle states that no two electrons in an atom can have the same four quantum numbers. This leads to the important conclusion that no atomic orbital can have more than 2 electrons, and the 2 electrons must have opposite spin.

   (C) Phosphorus: $3s$: ↑↓ $3p$: ↑ ↑ ↑ Hund's Rule states that the most stable arrangement of electrons is that with the maximum number of unpaired electrons. The three $3p$ electrons are in three orbitals, and all have parallel spin.

2. (A) The nuclear charge increases steadily with the increase in atomic number. The additional electrons are being added to the same energy level, so the shielding effect is minimal and there is no increase is the number of shells. Thus, the nucleus is more effective at pulling in the outer edge of the electron cloud.

   (B) Rubidium and xenon are in the same period, and thus have the same number of shells with the increase in nuclear charge. The outer electrons are more tightly held in xenon. Cesium is in the same family as rubidium, down a period, meaning that it has one more shell. Of the 18 additional electrons, 17 of them are placed into inner levels, thus contributing to the shielding of the outer electron. This combined effect creates a much more loosely held electron.

   (C) Similar explanation as that given in part (B) of this question. The increase in electron affinity is due to greater effectiveness of the nucleus, but a decrease is noted when the nuclear charge has less impact on outer electrons.

Chemical Bonding and Molecular Shape

A chemical bond involves a "link" between atoms. As a result of bond formation, the energy of the bonded atoms is lower than the energies of the separated atoms. For example, the energy of bonded magnesium and oxide ions in MgO is less than the total energy of separated magnesium and oxygen atoms. Similarly, nitrogen and hydrogen atoms bond to form NH_3 because the energy of the molecule is lower than the total energy of the separated atoms.

An understanding of chemical bonding allows us not only to predict the properties bonded substances, but also to create new substances with specific properties.

IONIC BONDS

An **ionic bond** is the result of the electrostatic attraction between positive and negative ions. When a representative metal atom loses electrons, the resulting positive ion generally has a noble gas configuration, that is, a completed *octet* (Na^+, Mg^{2+}, Al^{3+}) or *duplet* (Li^+, Be^{2+}) of electrons. When a nonmetal atom gains electrons, the resulting negative ion also has a completed octet (F^-, O^{2-}) or duplet (H^-) of electrons.

Some heavier metallic elements can form positive ions that have more than one charge. For example, while aluminum only forms Al^{3+}, indium, another member of Group 13, can form both In^{3+} and In^+. The explanation for this phenomenon is based on the differences in the energies of the *s* and *p* electrons in the valence shell. In larger atoms, the two *s* electrons in the valence shell are poorly shielded and have considerably lower energies than the *p* electrons. Consequently, they may remain with the atom when the *p* electrons are removed, yielding a positive ion that differs by two charge units. This is known as the **inert-pair effect**.

Sample Problem

Lead is a member of Group 14. Predict the two positive ions that it may form.

Solution

As a member of Group 14, the **Pb^{4+}** ion, with a completed octet, is expected. As a result of the inert-pair effect, the more common **Pb^{2+}** ion is usually formed. Once again, the two ions differ by two charge units.

Lattice Enthalpies

A common misconception is the notion that completed octets are solely responsible for the stability of ionic compounds. Consider the formation of Mg^{2+} and O^{2-} from their respective atoms: In forming the Mg^{2+} ion, 2188 kJ/mol is required (first and second ionization energies); in forming the O^{2-} ion, 703 kJ/mol is required (first and second electron affinities). Since a total of 2891 kilojoules is required, we should expect the formation of MgO to be highly unlikely.

The stabilizing factor in the formation of ionic compounds is the considerable release of energy that occurs when the oppositely charged ions arrange themselves in a solid crystal. The positions occupied by the ions are known as **lattice points** and the energy that is released is known as the **lattice enthalpy**. The larger the lattice enthalpy, the more stable is the ionic compound.

Lattice enthalpy depends:

- *directly* on the charge of the ions, and
- *inversely* on the distance between them.

For example, the lattice enthalpy of MgO is 3850 kJ/mol, while the lattice enthalpy for KCl is only 717 kJ/mol. This can be explained as follows:

- Mg^{2+} and O^{2-} are more highly charged than K^+ and Cl^-, leading to a stronger attraction between the Mg^{2+} and O^{2-} ions.

- Mg^{2+} and O^{2-} are smaller than K^+ and Cl^-. The smaller size of the Mg^{2+} and O^{2-} ions allows them to approach each other more closely and this also leads to a stronger attraction between the Mg^{2+} and O^{2-} ions.

The table below lists the lattice enthalpies (in kJ/mol) for a number of ionic compounds in order to show the variation of ionic size and charge. Note that lattice enthalpies are usually reported as positive numbers.

Halides							
LiF	1,046	LiCl	861	LiBr	818	LiI	759
NaF	929	NaCl	787	NaBr	751	NaI	700
KF	826	KCl	717	KBr	689	KI	700
Oxides							
MgO	3,850	CaO	3,461	SrO	3,283	BaO	3,114
Sulfides							
MgS	3,406	CaS	3,119	SrS	2,974	BaS	2,832

The Born-Haber Cycle

Lattice enthalpies cannot be measured directly; they are obtained by an indirect series of measurements, collectively known as a **Born-Haber cycle**. Since enthalpy is a state function (see chapter 7), the total enthalpy change for the entire cycle is 0. Let us construct a six-step hypothetical scheme that begins and ends with 1 mole of NaCl(s) at 298.15 K. The Born-Haber for this cycle is shown in the diagram that follows.

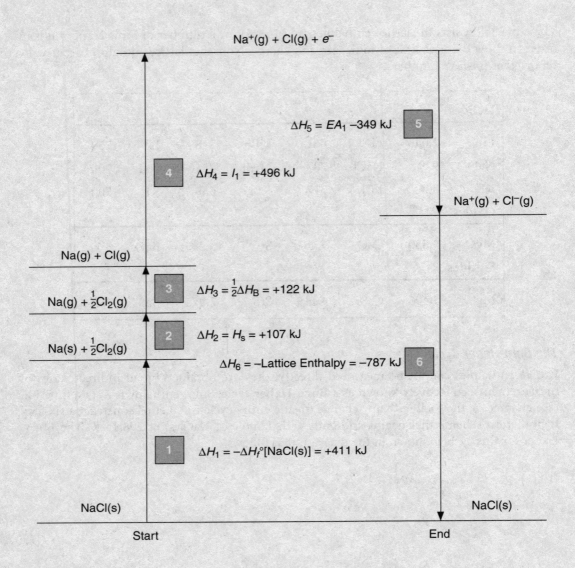

- If we examine Step 1 in reverse, we recognize that it is the standard heat of formation of NaCl(s), ΔH_f°, [−411 kJ]. In this step, 1 mole of NaCl(s) "forms" 1 mole of Na(s) atoms and one-half mole of Cl_2(g) molecules. **ΔH for Step 1 is +411 kJ.**

- In Step 2, Na(s) atoms sublime to form Na(g), and the enthalpy change is the heat of sublimation, H_s, of Na(s). **ΔH for Step 2 is +107 kJ.**

- In Step 3, one-half mole of Cl_2(g) molecules are dissociated into 1 mole of Cl(g) atoms. Therefore, the enthalpy change for this step equals one-half the bond enthalpy ΔH_B for Cl_2(g). [See pages 169–173.] **ΔH for Step 3 is +122 kJ.**

- In Step 4, the Na(g) atoms are ionized to Na^+(g) ions, and the enthalpy change for this step is the first ionization energy, I_1, for Na(g). **ΔH for Step 4 is +496 kJ.**

- In Step 5, the Cl(g) atoms are ionized to Cl⁻(g) ions, and the enthalpy change for this step is the electron affinity, EA_1, for Cl(g). **ΔH for Step 5 is −349 kJ.**

- Since ΔH_{cycle} = 0 kJ, we can calculate the value of the enthalpy change for Step 6, which is the *negative* of the *lattice enthalpy*. **ΔH for Step 6 is −787 kJ.**

As we can see, the large negative value of Step 6 is the stabilizing influence for NaCl(s).

Properties of Ionic Compounds

Ionic compounds are crystalline solids with high melting and boiling points, due to the strong interionic attractions. The rigidity of the crystal lattice causes ionic crystals to be quite brittle. When melted or dissolved in water, ionic compounds conduct electricity due to the mobility of the ions that are present. However, some ionic solids, such as AgCl and MgO, are insoluble in water because the very strong interionic attractions in these solids make the ions resistant to separation by water molecules.

COVALENT BONDS

When two atoms share one or more pairs of electrons, the result is a **covalent bond**. For example, the substances N_2, F_2, NH_3, and CH_4 all employ covalent bonding. In simple molecules, the goal of covalent bonding is to complete an octet of valence electrons. The notable exception is hydrogen, in which the valence shell is complete with two electrons.

Lewis Structures of Diatomic Molecules

The simplest covalent molecule is H_2. In this molecule, a single pair of electrons is shared by two hydrogen atoms and can be represented as follows: H : H or H—H. Since the pair of electrons belongs to *both* hydrogen atoms, each atom has a completed valence shell of 2 electrons. The second representation, in which a line is used to represent the shared pair of electrons, is known as a **Lewis structure**. Since only one pair is shared, the covalent bond is known as a **single bond**. Another molecule in which a single pair of electrons is shared is F_2. The Lewis structure for F_2 is:

$$\ddot{\underset{\cdot\cdot}{F}} - \ddot{\underset{\cdot\cdot}{F}}$$

Note that the sharing of a pair of electrons by the two fluorine atoms results in each atom having a completed octet. The unshared pairs of valence electrons surrounding each fluorine atom are known as **lone pairs** (or **nonbonded pairs**) of electrons.

Adjacent atoms can also share two or three pairs of electrons known, respectively, as **double** and **triple bonds** and represented by two or three parallel lines. For example, the Lewis structures for O_2 and N_2 are:

$$:\ddot{O}\!=\!\ddot{O}: \text{ and } :N\!\equiv\!N:$$

Note that each of the atoms in these molecules also has a completed octet of valence electrons. Strictly speaking, the Lewis structure for O_2 is not really correct. We will examine why this is so later on in the chapter.

How to Write Lewis Structures of Polyatomic Molecules and Ions

If we consider a polyatomic molecule such as NH_3, we know that the molecule contains three hydrogen atoms and one nitrogen atom. It is quite another matter to determine how the atoms are bonded within the molecule. In order to be able to write Lewis structures for polyatomic molecules and polyatomic ions, we will adopt a set of rules that will make the task easier. As we state each rule, we will apply it to the stepwise "construction" of the NH_3 molecule.

1. Count the total number of valence electrons in the molecule or ion.

   The nitrogen atom has 5 valence electrons and the three hydrogen atoms have 1 valence electron each. This yields a total of 8 valence electrons in the molecule.

2. If the species is a polyatomic ion, add 1 electron for each negative charge and subtract 1 electron for each positive charge.

   NH_3 is a neutral molecule; there are no electron additions or subtractions.

3. Draw a preliminary Lewis structure in which all adjacent pairs of atoms are connected by single bonds. A word of advice: It is a good idea to begin with a symmetrical structure, in which a single atom is centrally located in the molecule.

   In the NH_3 molecule, we will make nitrogen the central atom and surround it by the three hydrogen atoms.

$$\begin{array}{c} H-N-H \\ | \\ H \end{array}$$

4. If any atom does not have a complete valence shell, add one or more lone pairs to complete it.

   Since nitrogen has only six valence electrons, we add a lone pair to complete its octet.

$$\begin{array}{c} H-\ddot{N}-H \\ | \\ H \end{array}$$

5. Compare the total number of valence electrons in the Lewis structure with the number you counted in Steps 1 and 2 of these rules. If they agree, your Lewis structure is complete.

The Lewis structure has a total of 8 valence electrons and the structure for NH_3 is complete.

Sample Problem

Draw the Lewis structure for the $[OBr]^-$ ion.

Solution

There are a total of 14 valence electrons: 6 from oxygen, 7 from bromine, and 1 for the extra negative charge. The completed Lewis structure is:

$$\left[:\ddot{O} - \ddot{Br}: \right]^-$$

Note that Lewis structures for polyatomic ions are placed in brackets.

If a completed Lewis structure has too many valence electrons, it is an indication that multiple bonds are needed in the structure. If two adjacent atoms each contain a lone pair and they share a single bond between them, the valence-electron count can be lowered by two if the lone pairs and the single bond are replaced by a double bond. Similarly, replacing four lone pairs and a single bond by a triple bond will lower the valence-electron count by four. The next sample problem shows how this is done.

Sample Problem

Draw the Lewis structure for the NO_2^- ion

Solution

There are a total of 18 valence electrons: 5 from nitrogen, 12 from both oxygen atoms, and 1 for the single negative charge. The preliminary structure is:

$$:\ddot{N} - \ddot{O}:$$
$$|$$
$$:\ddot{O}:$$

The problem with this structure is that it has a total of 20 valence electrons. We need to reduce this number by two. Therefore, we remove a single bond and two lone pairs, producing the structure:

$$\left[\begin{array}{c} :N = \ddot{O}: \\ | \\ :\ddot{O}: \end{array} \right]^-$$

If we count the valence electrons, we see that they now total 18, and the structure is completed.

Comparing Ionic and Covalent Bonds

The ionic and covalent models of bonding represent two extremes, neither of which is realized completely: *No bond is purely ionic or purely covalent.* One way of determining the relative ionic or covalent character of a bond is to compare the electronegativities of the two bonded atoms. Roughly speaking, **electronegativity** is a quantitative measure of the ability of an atom to attract electrons in a chemical bond. The American chemist Linus Pauling first developed a scale of electronegativity. Here is a table of the electronegativities of the representative elements:

Electronegatives of Representative Elements (Pauling Scale)

H 2.20							He (−)
1	**2**	**13**	**14**	**15**	**16**	**17**	**18**
Li 0.98	Be 1.57	B 2.04	C 2.55	N 3.04	O 3.44	F 3.98	Ne (−)
Na 0.93	Mg 1.31	Al 1.61	Si 1.90	P 2.19	S 2.58	Cl 3.16	Ar (−)
K 0.82	Ca 1.00	Ga 1.81	Ge 2.01	As 2.18	Se 2.55	Br 2.96	Kr 3.00
Rb 0.82	Sr 0.95	In 1.78	Sn 1.96	Sb 2.05	Te 2.10	I 2.66	Xe 2.60
Cs 0.79	Ba 0.89	Tl 1.62	Pb 2.33	Bi 2.02	Po 2.00	At 2.20	Rn (−)
Fr 0.70	Ra 0.90	Uut	Uuq (−)	Uup	Uuh	Uus	Uuo

Note that the metallic elements generally have low electronegativities and nonmetallic elements generally have high electronegativities. Within a periodic group, electronegativity decreases with atomic number, since larger atoms attract electrons less strongly. Cesium and francium are the least electronegative elements. Across a period, the electronegativity increases due to the decrease in atomic size and increase in nuclear charge. Fluorine and oxygen are the most electronegative elements.

The relative ionic character of a bond can be approximated by calculating the electronegativity difference of two bonded atoms. A large difference in electronegativity indicates that the bond is predominantly ionic. For example, the electronegativity of Na is 0.93 and F is 3.98. The difference is 3.05, which suggests that the compound NaF is predominantly linked by ionic bonds. In contrast, we expect HI (electronegativity difference of 0.46) to be a predominantly covalently bonded compound.

Nevertheless, it is useful to have some rough guidelines to assess the nature of the bond between to atoms. In this book, an electronegativity difference of 1.70 or more will indicate that the bond is essentially ionic; an electronegativity difference between 0.00 and 0.40 will indicate that the bond is essentially covalent; an electronegativity difference between 0.41 and 1.69 will indicate that the bond has both ionic and covalent character.

Another way of viewing the nature of a chemical bond is to relate the electronegativity difference to a concept called *percent ionic character*, which indicates what percentage of the bond is ionic, the remainder being covalent. A graph of percent ionic character versus electronegativity difference is shown below:

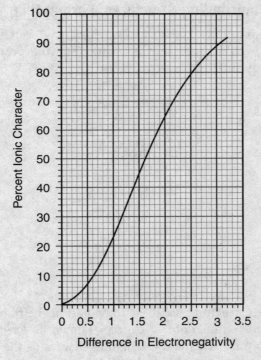

Based on this graph, we can conclude that NaBr is approximately 90% ionic and 10% covalent, while HI is only 5% ionic and 95% covalent.

Resonance and Resonance Hybrids

Actually, we can draw *two* equivalent Lewis structures for the NO_2^- ion, differing only in the positions of the single and double bonds. Both structures are shown below:

These two structures illustrate a concept known as **resonance**, a phenomenon that exists when more than one structure can be drawn for a species. These structures are known as a **contributing resonance structures**, which are usually illustrated together and separated by a double arrow. The actual structure of a species is a blend of its contributing resonance structures and is known as a **resonance hybrid**. The resonance hybrid of the NO_2^- ion might be represented as shown below:

$$\left[\begin{array}{c} N = O \\ \| \\ O \end{array} \right]^-$$

Selecting Suitable Contributing Structures: Formal Charge

The compound N_2O, commonly known as "laughing gas," can be associated with five possible Lewis structures, which are shown below. (For the present, ignore the signed numbers on top of each atom.)

$$\overset{2-}{:\ddot{N}} - \overset{2+}{O} \equiv \overset{0}{N:} \quad \text{(I)}$$

$$\overset{1-}{:\ddot{N}} = \overset{2+}{O} = \overset{1-}{\ddot{N}:} \quad \text{(II)}$$

$$\overset{1-}{:\ddot{N}} = \overset{1+}{N} = \overset{0}{\ddot{O}:} \quad \text{(III)}$$

$$\overset{0}{:N} \equiv \overset{1+}{N} - \overset{1-}{\ddot{O}:} \quad \text{(IV)}$$

$$\overset{2-}{:\ddot{N}} - \overset{1+}{N} \equiv \overset{1+}{O:} \quad \text{(V)}$$

How can we decide which structures are likely to be part of the resonance hybrid for N_2O? In order to make such determinations, a concept known as *formal charge* is employed. **Formal charge** is an arbitrary scheme for assigning charges to the atoms in a polyatomic molecule or ion. *These charges are not real*, but they do serve a purpose in "weeding out" unfavorable structures.

In determining formal charge, a number of electrons is assigned to each atom in the species according to specific rules. This number is then compared with the number of valence electrons present in the *unbonded* atom: If the assigned number exceeds the number of valence electrons, then the atom has a negative formal charge; if the assigned number is less, then the atom has a positive formal charge; if the two numbers are equal, then the atom has zero formal charge.

The rules for assigning a formal charge to an atom are as follows:

1. All lone pairs of electrons, L, are assigned solely to that atom.

2. The atom is also assigned one-half the number of shared pairs of electrons, S.

3. If V is the number of valence electrons in the unbonded atom, then the atom's formal charge, FC, is:

$$FC = V - \left(L + \frac{1}{2}S \right)$$

The assignment of one-half of the shared electrons to each atom in a bonded pair indicates that formal charge stresses the *covalent* nature of each of the bonds present in the molecule or ion.

Returning to the five N_2O structures given above, we note that the signed numbers appearing above each atom are the calculated formal charges of that atom. In order to select the most favorable structures we adopt the following guidelines:

1. The formal charges should be as close to 0 as possible.

2. The negative formal charges should reside on the most electronegative atoms whenever possible.

3. Adjacent atoms should not have formal charges with *like* signs.

Note that structures (III) and (IV) meet the first guideline, while structure (IV) meets the second guideline. These are the most favored contributing resonance structures for N_2O. Structure (V) is not at all favorable because it does not follow any of the three guidelines.

Sample Problem

Draw five possible resonance structures for CO_2 and select the most favorable resonance structures for this compound.

Solution

$$\overset{0}{:\ddot{O}} = \overset{0}{C} = \overset{0}{\ddot{O}:} \quad \text{(I)}$$

$$\overset{1+}{:O} \equiv \overset{0}{C} - \overset{1-}{\ddot{\underset{..}{O}}:} \quad \text{(II)}$$

$$\overset{2-}{:\ddot{C}} = \overset{2+}{O} = \overset{0}{\ddot{O}:} \quad \text{(III)}$$

$$\overset{1-}{:C} \equiv \overset{2+}{O} - \overset{1-}{\ddot{\underset{..}{O}}:} \quad \text{(IV)}$$

$$\overset{3-}{:\underset{..}{\ddot{C}}} - \overset{2+}{O} \equiv \overset{1+}{O:} \quad \text{(V)}$$

Clearly, structure (I) is best since all of the formal charges are 0. Structure (II) is second best and structure (V) is worst.

Oxidation Number: The Ionic Aspect of a Covalent Bond

We have met the term *oxidation number* previously, but we never established how these apparent charges were calculated. Oxidation numbers are determined in much the same way as formal charges except that all of the shared electrons are assigned to the more electronegative atom. In making such an assignment, we stress the *ionic* nature of the bond.

Sample Problem

Determine the oxidation numbers of carbon and oxygen in structure (I) of CO_2 given in the last sample problem.

Solution

All of the shared electrons are assigned to the oxygen atoms because oxygen is more electronegative than carbon. Therefore, carbon, with no electrons assigned to it, has an oxidation number of 4+, while each oxygen, with 8 electrons assigned to it, has an oxidation number of 2−.

Exceptions to the Octet Rule in Molecules

Central Atoms with Unpaired Electrons

Consider the molecules NO, NO_2, and ClO_2. The total number of valence electrons in each of these molecules is, respectively, 11, 17, and 19, and there is no way that the electrons in these molecules can be arranged to produce complete pairing. The preferred Lewis structure for the NO molecule is:

$$\cdot \ddot{N} = \ddot{O} :$$

Note that the nitrogen atom contains only seven electrons. The unpaired electron is associated with the nitrogen atom because it is less electronegative than oxygen. Molecules that contain an unpaired electron are very reactive and are known as **free radicals**.

Sample Problem

Draw the Lewis structure of ClO_2.

Solution

$$\overset{\displaystyle \cdot \ddot{C}l - \ddot{O}:}{\underset{:\ddot{O}:}{|}}$$

Central Atoms with Less Than 8 Electrons

Boron and beryllium are examples of atoms that contain less than eight electrons in some of their compounds. For example, in the compound BF_3, boron has only six electrons in its valence shell. In the compound $BeCl_2$, beryllium has only four electrons in its valence shell. The Lewis structures of these molecules are shown below. Note the symmetry of each molecule.

$$:\ddot{F} - B - \ddot{F}: \qquad :\ddot{C}l - Be - \ddot{C}l:$$
$$\underset{:\ddot{F}:}{\overset{\displaystyle |}{}}$$

Expanded Valence Shells: Central Atoms with More Than 8 Electrons

In compounds such as PCl_5 and SF_6, as well as in ions such as ICl_4^-, the central atoms have more than eight electrons in their valence shells. The Lewis structures of these species are shown below:

Note that in PCl_5 phosphorus has 10 electrons in its valence shell, while sulfur and iodine in SF_6 and ICl_4^- each have 12 electrons in their respective valence shells. Also note the periodic placement of each of these central atoms: Since they are located in Periods 3 (P and S) and 5 (I), they can use their available d subshells to accommodate the additional electrons. This idea is supported by the fact that a compound such as NCl_5 does not exist because nitrogen is located in Period 2 and does not have a d subshell.

Drawing Lewis structures for species that have central atoms with expanded valence shells is done using the same rules that were developed earlier in this chapter. For PCl_5 or SF_6, this is a no-brainer because the number of peripheral atoms are, respectively, five and six, leading to 10 and 12 electrons in the expanded valence shells of the central atom. For an ion such as ICl_4^- or a molecule such as XeF_2, the preliminary Lewis structures contain *too few* electron pairs:

- In ICl_4^-, we must account for a total of 36 valence electrons; the preliminary Lewis structure contains only 32 electrons. We complete the structure by *adding* two additional lone pairs of electrons. Therefore, the central atom (I) contains a valence shell with 12 electrons.

- In XeF_2, we must account for a total of 22 valence electrons; the preliminary Lewis structure contains only 20 electrons. We complete the structure by *adding* one additional lone pair of electrons. Therefore, the central atom (Xe) contains a valence shell with 10 electrons.

Sample Problem

Determine the Lewis structure of BrF_5.

Solution

We must account for a total of 42 valence electrons (seven electrons contributed by each of the six halogen atoms). Even when we place 5 fluorine atoms around the central bromine atom, the preliminary Lewis structure contains only 40 electrons. Therefore, we must add one additional lone pair of electrons to the valence shell of bromine. The completed Lewis structure is:

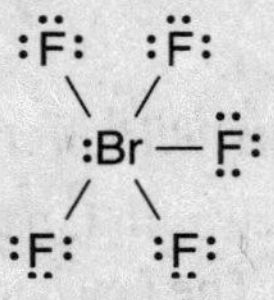

At times, it is possible to draw Lewis structures for a species that may or may not contain an expanded valence shell. The PO_4^{3-} ion is one such case.

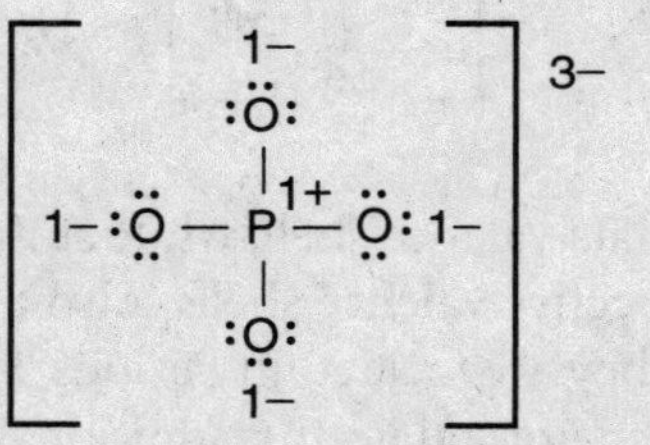

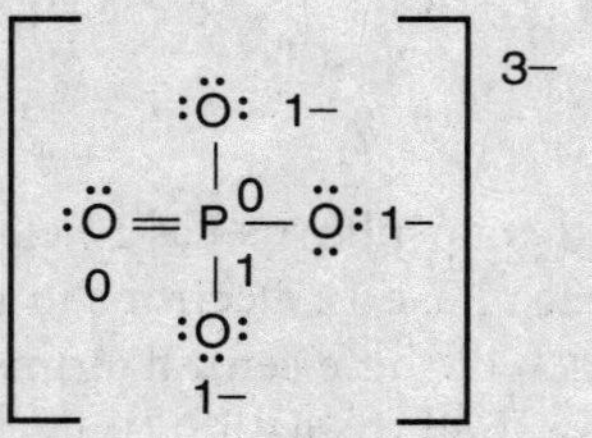

Based on formal charges (which are shown in the diagrams), it is reasonable to select the structure with the expanded valence shell. However, recent work has lent support to the concept that the structure without the expanded valence shell is the most favored one. In this book, if there is a choice between drawing a Lewis structure with or without an expanded valence shell, we will opt for the second choice: that is, *no* expanded valence shell.

STRENGTHS OF COVALENT BONDS

When two hydrogen atoms approach each other, a number of electrostatic forces are present between the atoms: each nucleus attracts the electron in the other atom, both nuclei repel each other, and both electrons repel each other. These forces govern the potential energy of the hydrogen-hydrogen system. Since the magnitudes of all of these forces depend on the distance between the atoms, the formation of a hydrogen molecule must represent a compromise that *minimizes* the potential energy of the system.

The diagram below represents the changes in potential energy of two hydrogen atoms as a function of the distance between their nuclei.

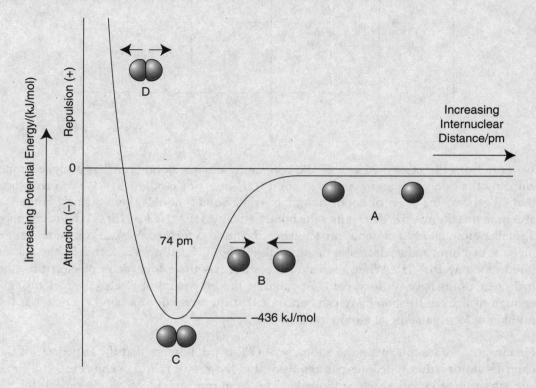

At large distances (Point *A*), the potential energy is nearly 0 kilojoules per mole, indicating that both atoms are separate and isolated. As the atoms approach (Point *B*), the energy of the system begins to drop, indicating that the attractive forces are the dominant interactions. The potential energy becomes a minimum (−436 kilojoules per mole) at an internuclear distance of 74 picometers (Point *C*). It is at this point that the hydrogen molecule is formed and stable. Attempts to decrease the internuclear distance further produces a steep rise in potential energy, indicating that the repulsive forces are now dominant (Point *D*). Needless to say, this is a highly unstable configuration.

Bond Enthalpies of Diatomic Molecules

One way of assessing the strength of a covalent bond in a diatomic molecule is to determine the energy needed to separate molecule into its constituent atoms, a process known as **bond dissociation**. This energy is known as the **bond enthalpy**, ΔH_B, and it is measured in kilojoules per mole of molecules. As we saw in the last section, the potential energy of the hydrogen molecule is –436 kJ/mol. Therefore, we can conclude that the bond enthalpy for H_2 is +436 kJ/mol. The following table lists the bond enthalpies of a number of diatomic molecules:

Molecule	ΔH_B /(kJ/mol)	Molecule	ΔH_B /(kJ/mol)
H_2	436	HF	565
N_2	944	HCl	431
O_2	496	HBr	366
F_2	158	HI	299
Cl_2	242	CO	1,074
Br_2	193		
I_2	151		

We note from the table that a molecule containing a triple bond (N_2) has a higher bond enthalpy than one containing a double bond (O_2) or a single bond (F_2). We can conclude that, in general, the order of bond strength is: triple bond > double bond > single bond. We also note that atomic size determines the bond enthalpy: HF > HCl > HBr > HI. Presumably, a smaller atom allows a closer approach when the bond is formed. We can conclude that an increase in atomic radius decreases the strength of a bond. Finally, we note that F_2 has a lower bond enthalpy that H_2. While each fluorine atom has three lone pairs of electrons, the hydrogen atoms have no lone pairs. Apparently, the lone pairs of electrons repel strongly enough to weaken the bond. We can conclude that the strength of a bond decreases as the number of lone pairs on an atom increases.

Finally, how are we to explain the anomaly of Cl_2? It would seem that the larger size of the chlorine atoms reduces the lone-pair repulsion, but the atomic radius of chlorine is still small enough to allow a relatively close approach. This compromise evidently is responsible for the higher bond enthalpy of Cl_2. Recall that we saw a similar situation when we studied the electron affinity of chlorine in chapter 8.

Average Bond Enthalpies of Polyatomic Molecules

Bond enthalpies in polyatomic molecules vary from one compound to another. For example, the C=O bond enthalpy in CO_2 is different than the C=O bond enthalpy in $(CH_3)_2CO$. As a result, a large number of bond enthalpies for a particular type of bond have been tabulated and averaged; they are known as **average bond enthalpies**, $\underline{\Delta H_B}$. The following table lists a number of average bond enthalpies.

Bond	$\overline{\Delta H_B}$/(kJ/mol)	Bond	$\overline{\Delta H_B}$/(kJ/mol)	Bond	$\overline{\Delta H_B}$/(kJ/mol)
C–H	412	C–F	484	N–O	210
C–C	348	C–Cl	338	N=O	630
C=C	612	C–Br	276	N–F	195
C≡C	837	C–I	238	N–Cl	381
C–O	360.	N–H	388	O–H	463
C=O	743	N–N	163	O–O	157
C–N	305	N=N	409		

Reaction Enthalpy from Bond Enthalpies

If we assume that a chemical reaction is the result of bond breaking (in the reactant molecules) and bond formation (in the product molecules), then we can use bond enthalpies to estimate the enthalpy change of a reaction under standard conditions, $\Delta H°$. Let us use bond enthalpies to estimate $\Delta H°$ for the reaction:

$$N_2(g) + 3H_2(g) \rightarrow 2NH_3(g)$$

In this reaction 1 mole of N_2 and 3 moles of H_2 are dissociated, while 2 moles of NH_3 are formed. The Lewis structure of NH_3 is:

$$H - \ddot{N} - H$$
$$|$$
$$H$$

and we can conclude that 6 moles of N–H bonds are formed. Since enthalpy is a state function, we can determine the enthalpy of the reaction by adding the individual bond enthalpies, remembering that the enthalpy change of bond *formation* is *negative*, while the enthalpy change of bond *dissociation* is *positive*:

$$\Delta H° = (1 \text{ mol}) \cdot \Delta H_B(N_2) + (3 \text{ mol}) \cdot \Delta H_B(H_2) - (6 \text{ mol}) \cdot \overline{\Delta H_B}(N-H)$$

$$\Delta H° = (1 \text{ mol}) \cdot (944 \text{ kJ/mol}) + (3 \text{ mol}) \cdot (436 \text{ kJ/mol}) - (6 \text{ mol}) \cdot (388 \text{ kJ/mol}) = -76 \text{ kJ}$$

The $\Delta H°$ for this reaction, calculated from standard enthalpies of formation is –92 kJ, a difference of 17%. Why the difference? Remember that we used an *average* bond enthalpy in the calculation.

Sample Problem

Estimate $\Delta H°$ for the reaction: $2CO(g) + O_2(g) \rightarrow 2CO_2(g)$

Solution

In this reaction, 2 moles of CO and 1 mole of O_2 are dissociated, while 2 moles of CO_2 are formed. Since the Lewis structure of CO_2 is O=C=O, we can conclude that 4 moles of C=O bonds are formed.

$$\Delta H° = (2 \text{ mol}) \cdot \Delta H_B(CO) + (1 \text{ mol}) \cdot \Delta H_B(O_2) - (4 \text{ mol}) \cdot \overline{\Delta H_B}(C=O)$$

$$\Delta H° = (2 \text{ mol}) \cdot (1074 \text{ kJ/mol}) + (1 \text{ mol}) \cdot (496 \text{ kJ/mol}) - (4 \text{ mol}) \cdot (743 \text{ kJ/mol}) = \mathbf{-328 \text{ kJ}}$$

The value of $\Delta H°$ calculated from standard heats of formation is −566 kJ, a difference of 42%. We can do better! If we use the *actual* ΔH_B for the C=O bond in CO_2 (−799 kJ/mol), we obtain −552 kJ, which is in much better agreement (2.4% difference) with the thermodynamic value.

Bond Lengths

If we examine the bond-formation curve (page 169) once again we note that the H_2 bond is considered formed at an internuclear distance of 74 picometers. The internuclear distance that is associated with a given bond is known as the **bond length**. As with bond enthalpies, bond lengths are specific for bond in diatomic molecules, and are averages for bonds in polyatomic molecules. The table below lists a number of bond lengths:

Molecule	Bond Length /pm	Bond	Average Bond Length /pm
H_2	74	C–H	109
N_2	110.	C–C	154
O_2	121	C=C	134
F_2	142	C≡C	120.
Cl_2	199	C–O	143
Br_2	228	C=O	112
I_2	268	O–H	96
		N–H	101

In general, the larger the atom, the longer the bond. Also note that a multiple bond is shorter than its corresponding single bond: single bond > double bond > triple bond, presumably because the additional shared electrons pull the atoms closer together in the molecule.

MOLECULAR AND (POLYATOMIC) IONIC SHAPE

The size and shape of a molecule or polyatomic ion, together with the strength of its bonds and the distribution of electrons within those bonds largely determine the properties of the molecule or ion. One way of determining shape is to employ a model of molecular structure known as the *VSEPR model*.

The VSEPR Model

The **Valence-Shell-Electron-Pair-Repulsion** (**VSEPR**) model is based on four simple premises:

- Both shared and lone pairs of electrons repel each other.

- Lone pairs repel more strongly than shared pairs.

- Multiple bonds are not distinguished from single bonds; they are considered equivalent for the purposes of determining shape.

- The overall shape of the molecule or ion is determined by placing bonded electrons and lone pairs *as far apart as possible*.

In employing this model we will focus on a molecule or ion that has *one central atom* bonded to a number of peripheral atoms. For example, in NH_3, nitrogen is the central atom and the three hydrogen atoms are the peripheral atoms. One important aspect of the model is the prediction of *bond angles*. A **bond angle** is the angle between the central atom and *two* of its peripheral atoms. For example, the angle between the central oxygen atom and the two peripheral hydrogen atoms in H_2O is approximately 105°, as shown below:

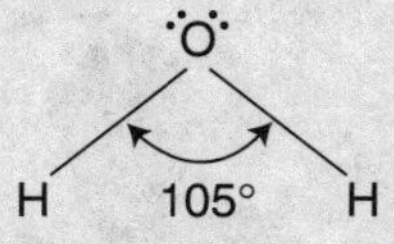

Note that the lone pairs are also placed at angles. Each pair of bonded and nonbonded electrons around the central atom defines a region known as an **electron domain**.

Molecules and Ions Without Lone Pairs on the Central Atom

When the central atom has no lone pairs, the bonded pairs solely determine the overall geometry of the electron domains. This, in turn, determines the shape of the molecule or ion. If we designate the central atom by C and each peripheral atom by P, we can prepare a table that relates the number of peripheral atoms to the shape of the molecule. We do not consider diatomic molecules and ions because they *must* be linearly shaped.

Number of Electron Domains	Formula	Geometry of Electron Domains	3-D Spatial Arrangement	Bond Angle(s)	Example
2	CP_2	Linear	180°	180°	$BeCl_2$
3	CP_3	Trigonal planar	120°	120°	BF_3
4	CP_4	Tetrahedral	109.5°	109.5°	CH_4
5	CP_5	Trigonal bipyramidal	90°, 120°	120° (equatorial), 90° (axial)	PCl_5
6	CP_6	Octahedral	90°, 90°	90°	SF_6

Since four out of the five shapes are associated with only one bond angle, all of the bonds in such molecules have equivalent energies. PCl_5, however, is associated with *two* bond angles: 120° and 90°. The 120° angles are found between the bonds that *surround* the molecule. These are called **equatorial bonds**. The 90° angles are associated with the bonds that point above and below the equatorial plane; they are called **axial bonds**. These atoms do *not* have equivalent energies, and we will see that they have a profound effect on central atoms that do possess lone pairs.

Sample Problem

Use the VSEPR model to determine the shape of the CO_2 molecule.

Solution

The Lewis structure of CO_2 is O=C=O. Since the carbon has no lone pairs, the molecule assumes a **linear shape** (bond angle = 180°).

Molecules and Ions with Lone Pairs on the Central Atom

When lone pairs are present on the central atom, the overall geometry of the electron domain is the same as the geometry when no lone pairs are present. However, the shape of the molecule or ion is another matter entirely. Molecular shapes are determined by the positions of the nuclei of the atoms within the molecule or ion; neither bonded nor lone pairs of electrons are visible to the instruments that perform such determinations. As a result, when a bonded pair is replaced by a lone pair, the shape of the molecule or ion will be quite different than the geometry of its electron domains.

The table below illustrates the effect of lone pairs on the shapes of molecules or ions:

Number of Electron Domains	Number of Lone Pairs Present	Geometry of Electron Domains	Spatial Arrangement of Domains	Shape of Molecule	Example
3	1	Trigonal planar		bent	$[NO_2]^-$
4	1	Tetrahedral		trigonal pyramidal	NH_3
	2			bent	H_2O
5	1	Trigonal bipyramidal		seesaw	SF_4
	2			t-shaped	ClF_3
	3			linear	XeF_2
6	1	Octahedral		square pyramidal	BrF_5
	2			square planar	XeF_4

When a molecule or ion with five electron domains has one or more lone pairs, the lone pairs will always occupy the *equatorial* positions. Apparently, the larger bond angles in the equatorial positions (120° versus 90°) minimize the extra repulsion produced by lone pairs of electrons.

The presence of lone pairs and/or multiple bonds tends to compress bond angles because they repel bonded pairs more strongly. Consider the following two examples:

- There are four electron domains in H_2O, which is a tetrahedral configuration (109.5°). However, two of these domains are lone pairs and, as a result, the H–O–H bond angle shrinks to approximately 105°.

- The molecule Cl_2CO (phosgene) has no lone pairs but the C and the O are joined by a double bond. There are three electron domains, which is a trigonal planar configuration (120°). The presence of a multiple bond shrinks the Cl–C–Cl bond angle to approximately 111°.

Sample Problem
Use the VSEPR model to predict the shape of the ICl_4^- ion.

Solution
The Lewis structure of ICl_4^- is shown on page 167. We see that there are six electron domains and two lone pairs on the iodine atom. If we refer to the table given on page 176, we predict that ICl_4^- will have a **square planar** structure.

Charge Distributions Within Molecules

The electronegativities of the atoms within a molecule determine how the electronic charge will be distributed among the atoms of that molecule. Regions that have a high electron density will be relatively negative and regions that have a low electron density will be relatively positive. This creates a condition known as **polarity**, which is the imbalance of electric charge. Polarity can be a property either of a bond or of the entire molecule. One way of assessing polarity is to measure a quantity known as a *dipole moment*.

Dipole Moments

A **dipole** is an imbalance of electric charge and a **dipole moment**, μ, measures the size of this imbalance. If two *equal and opposite* electric charges, Q, are separated by a distance, r, the dipole moment of this charge distribution is given by the formula:

$$\mu = Qr$$

The SI unit of dipole moment is the coulomb·meter, but dipole moments are usually reported in *debyes*, D ($1\ D = 3.34 \times 10^{-30}$ C·m).

If the charge distribution between two bonded atoms is entirely even, $\mu = 0$, and the bond is said to be *nonpolar*. If the dipole moment of a covalent bond is not 0, the bond is said to be *polar*. The unequal charge distribution in a polar bond is usually represented as shown below:

$$\overset{\delta+}{H}\!\!-\!\!\overset{\delta-}{Cl}$$

The Cl atom, with its higher electronegativity, has the larger share of the negative charge. The symbol "δ" indicates that the charges on the atoms are not whole charges, as in oppositely charged ions, but are "partial" charges around the covalent bond.

When the dipole moments of the individual bonds in a molecule are added together (in a vector-like fashion), the dipole moment of the entire molecule is obtained. If the dipole moment of the molecule is 0, the molecule is said to be *nonpolar*; if the dipole moment is not 0, the molecule is said to be *polar*.

If a molecule has great symmetry, it is possible for it to be nonpolar even though it contains polar bonds. For example, CO_2 and CCl_4 are nonpolar molecules with polar bonds: CO_2 is a linear molecule with no lone pairs on the carbon atom (O=C=O). Even though the C=O bonds are polar, the symmetry of the molecule as a whole renders the molecule nonpolar. Similarly, CCl_4 is tetrahedral and carbon has no lone pairs; the symmetry of the molecule causes $\mu = 0$, even though the four C–Cl bonds are polar.

The following sample problems illustrate how the dipole moment of a diatomic molecule is calculated, and how a measured dipole moment is used to calculate the partial positive and negative charges on each atom in the molecule.

Sample Problem

The diatomic molecule HBr has a bond length of 141 picometers. Assuming that the molecule consists of two oppositely charged ions, calculate the dipole moment of this molecule.

Solution

Since H and Br are assumed to have relative charges of 1+ and 1−, respectively, the magnitude of each of these elementary charges is 1.60×10^{-19} C.

$$\mu = Qr = (1.60 \times 10^{-19} \text{ C}) \cdot (141 \text{ pm}) \cdot \left(\frac{1 \text{ m}}{10^{12} \text{ pm}}\right) \cdot \left(\frac{1 \text{ D}}{3.34 \times 10^{-30} \text{ C} \cdot \text{m}}\right) = \mathbf{6.75 \text{ D}}$$

Sample Problem

The measured dipole moment of HBr is 0.82 D. Calculate the partial relative charges on the H and Br atoms in this molecule.

Solution

$$\mu = Qr$$

$$Q = \frac{\mu}{r} = \frac{(0.82 \text{ D}) \cdot \left(\dfrac{3.34 \times 10^{-30} \text{ C} \cdot \text{m}}{1 \text{ D}}\right) \cdot \left(\dfrac{1e}{1.60 \times 10^{-19} \text{ C}}\right)}{(141 \text{ pm}) \cdot \left(\dfrac{10^{-12} \text{ m}}{1 \text{ pm}}\right)} = \mathbf{0.12 \textit{ e}}$$

This should come as no surprise: the electronegativity difference between Br and H is 0.76. Using the graph on page 163, we see that this value corresponds to a bond with approximately 13% ionic character. Moreover, the measured dipole moment of 0.82 D is approximately 12% of the "ionic" dipole moment (6.75 D) calculated in the last problem. As we can see the symbols $\delta+$ and $\delta-$ really do represent partial charges.

BONDING THEORIES

Although we have written Lewis structures for covalent molecules and predicted their shapes, we have not considered *how* covalent bonding takes place. There are two principal approaches to bonding:

- *Valence-bond theory* takes a *local* view of bonding: it concentrates on how adjacent atoms share electrons within a molecule. The drawing of Lewis structures and the use of the VSEPR model are closely related to the valence-bond approach.
- *Molecular orbital theory* takes a *global* view of bonding: *all* of the electrons in a molecule are needed to describe how bonding occurs. There are situations in which the molecular orbital approach is the only way to describe a molecule.

Valence-Bond Theory (VB Theory)

In VB theory, bonds are formed when electrons in atomic orbitals of neighboring atoms are *paired*. In order for this event to occur the atomic orbitals of the neighboring atoms must *overlap*.

Orbital Overlap

The simplest molecule is H_2. When separated, each hydrogen atom has a single electron in its 1s atomic orbital. As the atoms approach to form a single H–H bond, **orbital overlap** occurs, leading to a merging of the two orbitals and the pairing of the electron spins. This merging of orbitals is known as **s–s overlap**.

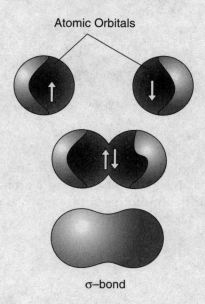

Atomic Orbitals

σ–bond

In HF, the unpaired valence electron in fluorine occupies the $2p_z$ orbital. As the H–F single bond is formed, the $1s$ orbital of the hydrogen atom and the $2p_z$ orbital of the fluorine merge, a condition known as **s–p_z overlap.**

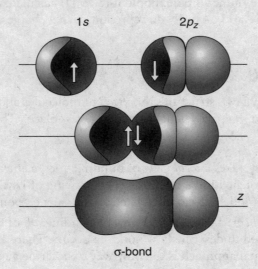

In F_2, **p_z–p_z overlap** is responsible for the formation of the F–F single bond.

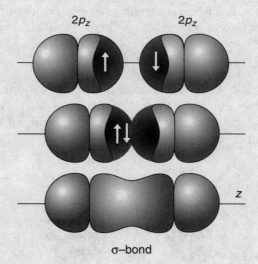

Sigma and Pi Bonding

In all three cases we just described, the single bond was formed by the *head-to-head overlap* of the two merging atomic orbitals. As a result, the merged orbitals have symmetry about the internuclear axis. When such a bond is formed, it is known as a **sigma bond (σ bond)**. All single bonds are composed of σ bonds.

Multiple bonds pose a slightly different problem. Consider the approach of two nitrogen atoms to form the diatomic molecule N_2. In each nitrogen atom, the three unpaired valence electrons lie, respectively, in the $2p_x$, $2p_y$, and $2p_z$ orbitals. As the atoms approach, the two $2p_z$ orbitals overlap head-to-head and form a σ bond. Since all three atomic orbitals are mutually at right angles, both $2p_x$ orbitals and both $2p_y$ orbitals can *not* overlap in a head-to-head fashion. Instead, they undergo **side-by-side overlap**, forming two merged orbitals known as **pi bonds (π bonds)**. Each π bond is double-lobed, lying above and below the internuclear axis. The π_x and π_y bonds lie at right angles to one another.

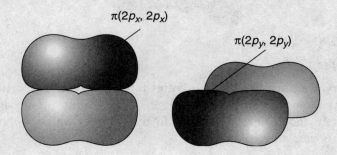

$\pi(2p_x, 2p_x)$

$\pi(2p_y, 2p_y)$

Since the Lewis structure for N_2 is drawn with a triple bond, we can conclude that triple bonds are composed of one σ bond and two π bonds. If the bond between atoms is a double bond, it consists of one σ bond and one π bond. In summary:

Single bond: one σ bond

Double bond: one σ bond and one π bond

Triple bond: one σ bond and two π bonds

Delocalized Pi Bonding

When the positions of multiple bonds in contributing resonance structures differ, the resonance hybrid contains a π bond that is "smeared out" over the entire structure. Such a phenomenon is known as **delocalized π bonding**. Consider the contributing resonance structures for the NO_3^- ion:

The delocalized π bond, involving the $2p_x$ orbitals of the nitrogen atom and the three oxygen atoms, is shown below:

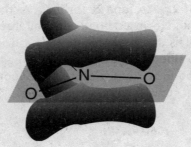

Hybridized Orbitals

In using VB theory, we must be able to account for a variety of properties of polyatomic molecules, such as molecular shape and the relative enthalpies of the bonds within the molecules. In certain molecules, we cannot do these things unless we employ a scheme known as *orbital hybridization.*

sp Hybridization

In the ground state, the element Be contains two paired valence electrons in a completed 2s orbital; its 2p orbitals are empty:

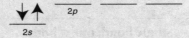

Based on this information, it is unclear why Be forms $BeCl_2$, a linear molecule with two single bonds that have equal bond enthalpies. The only way to make sense of these observations is to assume that, somehow, the valence electrons in Be *separated* and formed two equivalent σ bonds with the unpaired valence electrons of the Cl atoms. The separation of the valence electrons in Be can be considered to occur in stages:

1. One of the valence electrons is *promoted* to an empty 2p orbital:

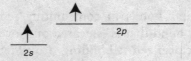

2. The two occupied orbitals blend to form two **sp hybrid orbitals**, so named because the hybrid comes from the blending of one *s* and one *p* orbital. The remaining two *p* orbitals are unchanged:

The angle between the two *sp* orbitals is 180°, that is, it is a *linear* arrangement. The electron density diagrams for the formation of *sp* orbitals is shown here:

s Orbital   *p* Orbital

Two *sp* Hybrid Orbitals   *sp* Hybrid Orbitals Shown Together
(Large Lobes Only)

Each of these hybrid *sp* orbitals forms a σ bond by overlapping (head-to-head) with a $2p_z$ orbital of a chlorine atom, and the linear molecule is formed with two energy-equivalent bonds.

sp² Hybridization

BF_3 is a trigonal planar molecule (bond angles = 120°) with three energy-equivalent bonds. In this molecule, B also hybridizes its orbitals, but in a slightly different fashion since B already has a single valence electron in one of the 2*p* orbitals. The promotion of a 2*s* electron to a 2*p* orbital results in a valence shell with three *unpaired* electrons. These valence electrons form three orbitals, known as ***sp²* hybrid orbitals**. The lobes of the orbitals are 120° apart. The electron density diagrams for this formation is shown below:

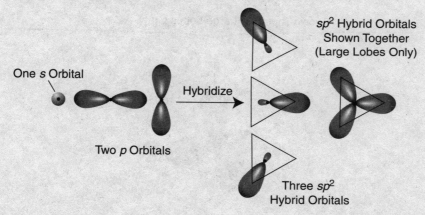

One *s* Orbital

Two *p* Orbitals

Hybridize

Three *sp²* Hybrid Orbitals

sp² Hybrid Orbitals Shown Together (Large Lobes Only)

Each of the *sp²* orbitals forms a σ bond with the $2p_z$ orbital of a chlorine atom, and the trigonal planar molecule is formed.

sp³ Hybridization

CH_4 is a tetrahedral molecule (bond angles =109.5°) with four energy-equivalent bonds. The promotion of a 2*s* electron to a 2*p* orbital results in a valence shell with four unpaired electrons. These valence electrons form four orbitals, known as ***sp³* hybrid orbitals**. The lobes of the orbitals are 109.5° apart. Here are the electron density diagrams for this formation:

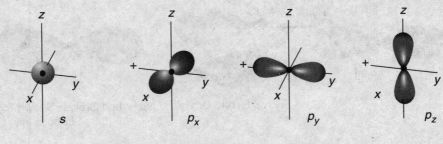

Hybridize to Form Four sp^3 Hybrid Orbitals

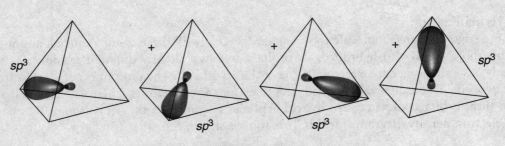

Shown Together (Large Lobes Only)

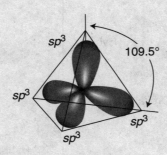

Each of the sp^3 orbitals forms a σ bond with the $1s$ orbital of a hydrogen atom, and the tetrahedral molecule is formed. Other species in Period 2 also employ sp^3 hybrid orbitals, including NH_3 (bond angles = 107°), H_2O (bond angle = 105°), and NH_4^+ (bond angle = 109.5°).

Note that in each of the cases considered the central atom was located in Period 2. In such cases the bond angles provide a clue as to the type of hybridization that is employed by the central angle.

Carbon and Hybrid Orbitals

Carbon is unique among all of the elements in that it bonds with itself (as well as with other elements) and forms an almost endless variety of compounds. We will explore the chemistry of carbon in chapter 16. Not only can carbon form sp^3 hybrid orbitals, but it can also form sp^2 hybrid orbitals (which are employed when carbon forms double bonds), and sp hybrid orbitals (which are employed when carbon forms triple bonds).

d-Orbital Hybridization

Central atoms located in Period 3 and above can use empty *d* orbitals to receive a promoted *s* electron. For example, when phosphorus promotes its $3s$ electron, its valence shell has the configuration $3s^1 3p^3 3d^1$, that is, five unpaired electrons in its valence shell. All five occupied orbitals form five ***sp³d* hybrid orbitals** whose lobes have bond angles of 90° and 120°. The PCl_5, SF_4, and BrF_3 molecules employ this type of hybridization.

Sulfur promotes its $3s$ electron and its paired $3p$ electron into two *d* orbitals. Its valence shell has the configuration $3s^1 3p^3 3d^2$, that is, six *unpaired* electrons in its valence shell. All six occupied orbitals form six ***sp³d²* hybrid orbitals** whose lobes have bond angles of 90°. The SF_6, XeF_4, and ClF_5 molecules employ this type of hybridization.

Molecular Orbital Theory (MO Theory)

When a species contains one or more unpaired electrons it is weakly *attracted* by a magnetic field, a property known as **paramagnetism**; when a species does not contain unpaired electrons it is *repelled* by a magnetic field, a property known as **diamagnetism**.

O_2 is a paramagnetic substance, yet its Lewis structure (page 160) contains no unpaired electrons. We can only conclude that the VB approach fails to account this observed property of O_2. There are other failures as well: The hydrogen molecule ion, H_2^+, contains a one-electron bond, a condition that is prohibited by VB theory. Similarly, the substance diborane (B_2H_6) has 14 valence electrons, not enough to bond its eight atoms in a way that pairs all of the bonding electrons.

Molecular orbital theory was introduced in the 1920s to correct the failures of VB theory. It begins with the premise that two atomic orbitals with similar energies overlap in a way that produces two molecular orbitals. One of the molecular orbitals has a *lower energy* than its corresponding atomic orbitals, a condition that promotes the formation of a stable bond. For this reason, it is called a **bonding molecular orbital**. The other molecular orbital has a *higher energy* than its corresponding atomic orbitals, a condition that destabilizes the formation of a bond; it is known as an **antibonding molecular orbital**.

When two atomic orbitals overlap in a head-to-head fashion, the molecular orbitals are called σ **molecular orbitals**. For example, the overlap of two $1s$ atomic orbitals produces the bonding molecular orbital designated σ_{1s}, and the antibonding molecular orbital designated σ^*_{1s}. Other σ molecular orbitals are formed from the overlap of two $2s$ or two $2p_z$ atomic orbitals. The formation of molecular orbitals from 1s atomic orbitals are represented by the following diagram:

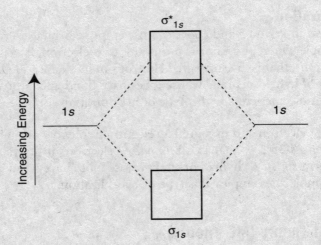

The formation of σ molecular orbitals from either $2s$ or $2p_z$ atomic orbitals could be represented by a diagram very similar to the one given above.

When two atomic orbitals overlap in a side-by-side fashion, the molecular orbitals are called **π molecular orbitals**. The overlap of two $2p_x$ and two $2p_y$ atomic orbitals produces the two bonding molecular orbitals designated π_{2p}, and two antibonding molecular orbitals designated π^*_{2p}.

The following diagram represents the formation of the six molecular orbitals that result from the overlap of the six corresponding $2p$ atomic orbitals in the Period 2 elements O, F, and Ne:

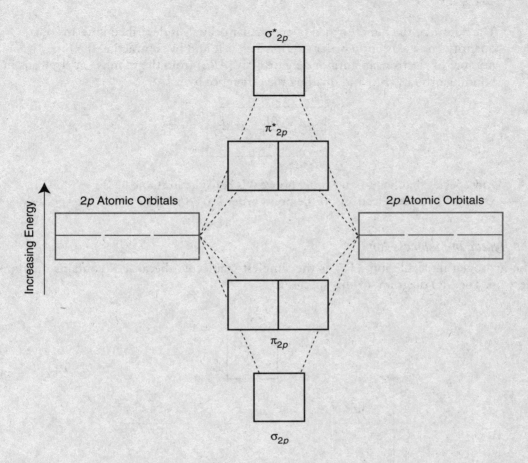

In the Period 2 elements Li through N, the π_{2p} bonding orbitals lie at a lower energy than the σ_{2p} bonding orbital. This is the result of the larger interactions between the 2s and 2p atomic orbitals in the first five elements of Period 2.

The filling of molecular orbitals with electrons follows rules that are similar to the filling of atomic orbitals:

- Molecular orbitals can contain 0, 1, or 2 electrons (empty, half-filled, or filled).

- Lower energy molecular orbitals fill first.

- The two electrons in a filled molecular orbital must have paired spins. (Pauli exclusion principle)

- The two π molecular orbitals (bonding or antibonding) have equal energies: each orbital must be half-filled before any orbital can be filled with two electrons. (Hund's rule)

- An electron in an antibonding orbital, "cancels" its corresponding electron in a bonding orbital.

- The nature of the bond formed by molecular orbitals is described in terms of a quantity known as **bond order (*BO*)**, which is found by subtracting the total number of electrons in antibonding orbitals (*ABE*) from the number of electrons in bonding orbitals (*BE*) and dividing the difference by 2:

$$BO = \frac{BE - ABE}{2}$$

Any species that has a bond order above 0 is considered stable because it has an excess of bonding electrons. If the bond order is 0 (*BE* = *ABE*), then the species does not exist.

The Hydrogen-Molecule Ion

The hydrogen molecule ion, H_2^+, is the simplest "molecule" because it contains only one electron. The MO diagram for this species is:

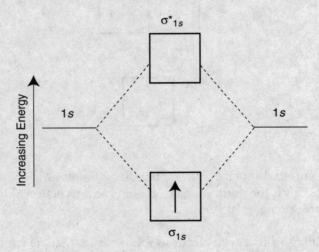

This species has 1 bonding and 0 antibonding electrons. It is stable and has a bond order of $\frac{1}{2}$. Since H_2^+ has one unpaired electron, it is paramagnetic.

Molecular Orbitals in Period 1 Molecules

The H_2 molecule has two electrons and both of these are located in the σ_{1s} molecular orbital. Since there are 2 bonding electrons and 0 antibonding electrons, the molecule is stable and has a bond order of 1. Note that this corresponds to the single bond predicted by VB theory. Also note that H_2 has no unpaired electrons; it is diamagnetic.

He_2, on the other hand, has four electrons, two in the σ_{1s} orbital and two in the σ^*_{1s} orbital. The bond order is 0 and we conclude that He_2 does not exist.

Molecular Orbitals in Period 2 Diatomic Molecules

The Period 2 diatomic molecules begin with Li_2 and end with Ne_2. Li_2 has a total of six electrons: four of these electrons mimic the He_2 configuration and, as a result, they make no contribution to bond formation. These four electrons are called **nonbonding electrons**. The remaining two electrons are located in the σ_{2p} orbital. Consequently, there is an excess of 2 bonding electrons, and Li_2 is stable, diamagnetic, and has a bond order of 1.

Sample Problem

Does Be_2 exist? Use MO theory to justify your answer.

Solution

Be_2 does *not* exist. Be_2 has a total of eight electrons. The first six mimic the configuration of Li_2. The remaining two electrons are located in the σ^*_{2p} orbital and the bond order of the molecule is 0.

The following is a summary of the placement of electrons in molecular orbitals in B_2 through Ne_2, together with the properties of each molecule, when it exists:

Orbital	B_2	C_2	N_2
σ^*_{2p}	☐	☐	☐
π^*_{2p}	☐ ☐	☐ ☐	☐ ☐
σ_{2p}	☐	☐	↑↓
π_{2p}	↑ ↑	↑↓ ↑↓	↑↓ ↑↓
σ^*_{2s}	↑↓	↑↓	↑↓
σ_{2s}	↑↓	↑↓	↑↓

Orbital	O_2	F_2	Ne_2
σ^*_{2p}	☐	☐	↑↓
π^*_{2p}	↑ ↑	↑↓ ↑↓	↑↓ ↑↓
π_{2p}	↑↓ ↑↓	↑↓ ↑↓	↑↓ ↑↓
σ_{2p}	↑↓	↑↓	↑↓
σ^*_{2s}	↑↓	↑↓	↑↓
σ_{2s}	↑↓	↑↓	↑↓

	B_2	C_2	N_2	O_2	F_2	Ne_2
Bond Order	1	2	3	2	1	0
Bond Enthalpy (kJ/mol)	290	620	941	495	155	—
Bond Length (A)	1.59	1.31	1.10	1.21	1.43	—
Magnetic Behavior	Paramagnetic	Diamagnetic	Diamagnetic	Paramagnetic	Diamagnetic	—

The O_2 molecule is of particular interest; it has a bond order of 2 (a "double" bond) and it is also paramagnetic because of its two unpaired electrons. MO theory can describe the structure of O_2 in a way that VB theory cannot. The disadvantage: we cannot draw an accurate Lewis structure for O_2.

Metallic Bonding

Metallic substances possess a variety of unique properties:

- Excellent electrical and thermal conductivity
- Metallic luster
- **Ductility**: the ability to be drawn into wire
- **Malleability**: the ability to undergo a change in shape

Two principal models are used to account for metallic properties: the *electron-sea model* and *band theory*.

The Electron-Sea Model

In the **electron-sea model** of metallic structure, the crystal lattice of the metal is occupied by positive ions, namely, the nuclei plus the noble gas core of electrons. The valence electrons are delocalized and move freely about the lattice. In response to an applied electric field, the electrons move opposite to the direction of the field, resulting in an electric current. All of the properties given above given above can be explained using this model. For example, the freely moving valence electrons are able to transfer kinetic energy to the ions of the crystal lattice, resulting in thermal conductivity.

However, not all of the properties (such as the periodic trend in melting points) can be explained by this model. Moreover, the electron-sea model fails to explain the conductive behavior of *metalloids*, which show an *increase* in conductivity with an increase in temperature. This is opposite to the behavior of metals, which show a decrease in conductivity as the temperature rises.

Another model known as the *band theory* is able to account for all of the properties of metals, metalloids, and nonmetals.

The Band Theory of Solids

The **band theory** is an outgrowth of MO theory. Let us use Li as an example: A single atom of Li has 1 valence electron located in the 2s atomic orbital. We saw that the overlap of two 2s atomic orbitals resulted in the creation of two molecular orbitals, one at a lower energy (the bonding orbital), and one at a higher energy (the antibonding orbital). The large difference in energy makes it highly unlikely that an electron would be able to bridge this energy gap.

However suppose that one mole of 2s atomic orbitals in Li overlapped. This would lead to the creation of 1 mole of molecular orbitals. In this case, the energy difference between any two adjacent molecular orbitals would be negligible. This large collection of molecular orbitals is known as an *electron band*, in particular, a **valence band** (because it contains valence electrons). Since each orbital can accommodate a maximum of 2 electrons, it follows that the valence band of Li is only half-filled, allowing the valence electrons free movement within the band. This leads to appreciable electrical conductivity.

Nonmetals contain a large number of valence electrons and, as a result, the valence band is nearly filled, restricting the movement of electrons. To be sure, there are other empty bands from higher energy molecular orbitals, but there is a large energy gap between these bands. As a result, the movement of electrons by promotion to the higher bands is highly unlikely, accounting for the poor conductive properties of nonmetals.

In metalloids, the situation is somewhat different: the energy gap between two adjacent bands is not nearly as large as in the nonmetal. While a metalloid such as silicon is a poor conductor at room temperature, an increase in temperature enables the more energetic electrons to be promoted to the next band, and the conductivity of the metalloid increases. The bands of metals, nonmetals, and metalloids are illustrated below:

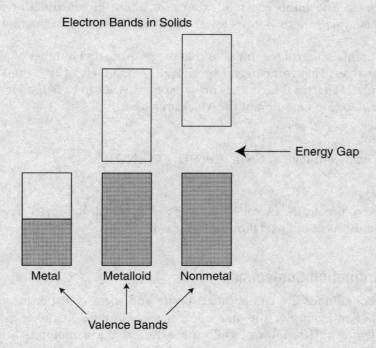

Electron Bands in Solids

A (VERY) BRIEF INTRODUCTION TO COORDINATION COMPOUNDS

Structures of Coordination Complexes

A species that contains a central metal ion bonded to a surrounding group of molecules or ions is known as a **complex**. If the complex has a net charge, the species is known as a **complex ion**. $[Ag(NH_3)_2]^+$ is one example of a complex ion. Compounds that contain complexes are known collectively as **coordination compounds**. An example of a coordination compound is $[Co(NH_3)_5Cl](NO_3)_2$, in which the complex ion is $[Co(NH_3)_5Cl]^{2+}$. Notice that the complex is set off in brackets to isolate it from the rest of the compound.

The molecules or ions that surround the central metal atom are known as **ligands** or *complexing agents*. For example, the ligands in $[Co(NH_3)_5Cl]^{2+}$ are the five NH_3 molecules and the single Cl^- ion. The metal ions and the ligands within these brackets are called the **coordination sphere of the metal**.

Every ligand is normally either a negative ion or a polar molecule. A further requirement of ligands is that they contain at least one lone pair of electrons. As such, they can serve as *electron-pair donors* or **Lewis bases** (see chapter 14). Metal ions (particularly transition metal ions) have vacant valence orbitals and these can serve as electron-pair acceptors or **Lewis acids**. The part of the ligand that bonds with the metal is known as the **donor atom**. In the complex ion $[Co(NH_3)_5Cl]^{2+}$, the five nitrogen atoms and the one chloride ion serve as donor atoms for Co. The number of donor atoms constitute the **coordination number** of the metal. In this case, there are six donor atoms, therefore, Co has a coordination number of 6.

A useful, but not infallible, rule of thumb is that the coordination number of a metal is equal to twice its charge. The complexes $[Ag(NH_3)_2]^+$, $[Co(NH_3)_5Cl]^{2+}$, and $[Ni(CN)_4]^{2-}$ illustrate this rule. (Whereas this rule has many exceptions, it can be helpful it completing an equation in the Reaction Question of the AP examination.)

Sample Problem

Determine the oxidation state of Co in $[Co(NH_3)_5Cl](NO_3)_2$.

Solution

NH_3 is a neutral molecule; the two NO_3^- ions, as well as the Cl^- ion each carry a (1–) charge. Therefore, Co must have and oxidation state of (**3+**).

Naming Coordination Compounds

Coordination compounds are complicated species and there is a specific set of rules for naming them. For example, the name of the compound $[Co(NH_3)_5Cl](NO_3)_2$ is *pentamminechlorocobalt(III) nitrate*, while the name of the compound $Na_2[MoOCl_4]$ is *sodium tetrachlorooxymolybdate(IV)*—quite a mouthful!

- In the first compound, *pentamminechloro–* represents the five NH_3 and the one Cl^- ligand; *–cobalt(III)* represents the metal ion, together with its oxidation state; finally, *nitrate* represents the negative ion that is *not* part of the complex.

- In the second compound, *sodium* represents the positive ion that is *not* part of the complex; *tetrachlorooxo–* represents the four Cl^- and the one O^{2-} ligands; *–molybdate(IV)* represents the metal ion, together with its oxidation state. The *–ate* ending indicates that the complex is a *negative ion*.

- In each compound, note that the name of the positive ion preceded the name of the negative ion.

Within the complex, the following rules control:

- The ligands are named—in alphabetical order—before the metal is named.

- Prefixes (di–, tri–, tetra–, . . .) are used to indicate the number of ligands of each type present, but they do *not* affect the alphabetical ordering of the ligands. For example *triammine* would be placed before *dichloro*.

- Negative ions that are ligands end in the letter *o*, such as *cyano*, *chloro*, *oxo*, ...

- Neutral molecules that are ligands take their own names, with the notable exceptions NH_3 (*ammine*) and H_2O (*aqua*).

- The metal ion appears last; its oxidation state is indicated by a Roman numeral in parentheses. If the complex is a negative ion, the metal takes the ending *–ate*.

Sample Problem

Name the following species:

(A) $[Ni(NH_3)_6]^{2+}$

(B) $K_2[CoCl_4]$

(C) $[Co(H_2O)_4Br_2]Br$

Solution

(A) hexaamminenickel(II)

(B) potassium tetrachlorocobaltate(VI)

(C) tetraaquadibromocobalt(III) bromide

Chapter 9: Practice Questions

Multiple-Choice Questions

1. Which does not make a good electron donor in the formation of a coordinate covalent bond?

   A. Cl^-
   B. H_2O
   C. PF_3
   D. CH_4
   E. NH_3

2. In the molecule shown, in which bond is the electron density shifted furthest *away* from the carbon atom? (Electronegativity values: C = 2.5; H = 2.1; Cl = 3.0; Br = 2.8; O = 3.5)

   $$HO - \overset{\overset{\displaystyle H}{|}}{\underset{\underset{\displaystyle Cl}{|}}{C}} - Br$$

   A. The C–H bond only
   B. The C–O bond only
   C. The C–Cl bond only
   D. The C–Br bond only
   E. The C–Cl and the C–Br bond only

3. The formate ion, HCO_2^-, has two equivalent bonds from the carbon atom to each oxygen atom. Each bond is intermediate in length between a single bond and a double bond. Resonance theory is used to describe bonding in this ion. Which is an acceptable resonance structure for this ion?

   I. $\left[H-C\!\!\begin{smallmatrix} \ddot{O} \\ \ddot{\ddot{O}}\!: \end{smallmatrix} \right]^-$ III. $\left[H-C\!\!\begin{smallmatrix} \ddot{O}\!: \\ \ddot{\ddot{O}} \end{smallmatrix} \right]^-$

   II. $\left[H-C\!\!\begin{smallmatrix} \ddot{O} \\ \ddot{\ddot{O}}\!: \end{smallmatrix} \right]^-$ IV. $\left[H\!=\!C\!\!\begin{smallmatrix} \ddot{O}\!: \\ \ddot{\ddot{O}}\!: \end{smallmatrix} \right]^-$

   A. II only
   B. I and III only
   C. II and IV only
   D. I, II, and III only
   E. I, II, III, and IV

4. Which statement is true of a chemical species that donates electrons to a coordinate covalent bond?

   A. It is a metal ion.
   B. It is a noble gas atom.
   C. It violates the octet rule.
   D. It has only nonpolar bonds.
   E. It has a lone pair of valence electrons.

5. Which of these atoms doesn't always obey the octet rule?

H	Cl	O

   A. H only
   B. Cl only
   C. O only
   D. Both H and Cl
   E. Both H and O

6. What are the formal charges respectively of N, C, and O in this resonance form of the cyanate ion (NCO^-)?

   $$\left[:\ddot{N}\!=\!C\!=\!\ddot{O}: \right]^-$$

   A. 0, 0, 0
   B. 0, 0, –1
   C. –1, 0, 0
   D. –1, +1, –1
   E. –2, 0, +1

7. The molecule thionyl bromide, $SOBr_2$, has three skeleton structures. Using the formal charges and electronegativity, which is the most preferred? (Electronegativity values: Br: 2.8; O: 3.5; S: 2.5)

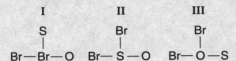

I	II	III
S	Br	Br
|	|	|
Br—Br—O	Br—S—O	Br—O—S

A. I
B. II
C. III
D. I and II are equally preferred over III
E. II and III are equally preferred over I

8. The structures of three nitrogen–containing compounds are shown. Match the name of the compound to the correct C-to-N bond length.

Compounds:

H_3C-NH_2 (methylamine)

$H_3CC \equiv N$ (acetonitrile)

$CH_3CH_2CH=NCH_2CH_3$
     (N-ethylpropanalimine)

Bond Lengths:				
85 pm	116 pm	124 pm	147 pm	200 pm
A.	116 pm		H_3C-NH_2	
	124 pm	$CH_3CH_2CH=NCH_2CH_3$		
	147 pm		$H_3CC \equiv N$	
B.	116 pm		$H_3CC \equiv N$	
	124 pm	$CH_3CH_2CH=NCH_2CH_3$		
	147 pm		H_3C-NH_2	
C.	85 pm		$H_3CC \equiv N$	
	116 pm	$CH_3CH_2CH=NCH_2CH_3$		
	124 pm		H_3C-NH_2	
D.	124 pm		$H_3CC \equiv N$	
	147 pm	$CH_3CH_2CH=NCH_2CH_3$		
	200 pm		H_3C-NH_2	
E.	85 pm		H_3C-NH_2	
	24 pm	$CH_3CH_2CH=NCH_2CH_3$		
	200 pm		$H_3CC \equiv N$	

9. How much energy is needed to completely dissociate one mole of H_3C-CN (acetonitrile) into its atoms?

Bond Energies in kJ:			
C≡N	C–C	N–H	C–H
89	347	391	413

A. 1,891 kJ
B. 2,130 kJ
C. 1,651 kJ
D. 2,201 kJ
E. 2,477 kJ

10. Which molecule has a Lewis structure that is an expanded octet?

A. IF_5

B. BF_4^-

C. SO_3^{2-}

D. BrF_2^+

E. Cl_2SO

11. Combining one *s* and two *p* orbitals creates:
    A. one sp^2 hybrid orbital.
    B. one *sp* hybrid orbital.
    C. two sp^3 hybrid orbitals.
    D. three *sp* hybrid orbitals.
    E. three sp^2 hybrid orbitals.

12. What are the electron pair geometry and the molecular geometry of SF_4?

A. Trigonal bipyramidal; seesaw
B. Trigonal bipyramidal; T-shaped
C. Trigonal bipyramidal; trigonal bipyramidal
D. Octahedral; square planar
E. Octahedral; square pyramidal

13. Which of the statements about the IF_5 molecule is (are) true?

    I. All its bond angles are identical.

    II. It has no net dipole moment.

    III. It contains polar bonds.

    A. I only
    B. II only
    C. III only
    D. I and II only
    E. I and III only

14. The boron atom in boron tribromide has sp^2 hybridization. What are the angles between the bromine atoms in BBr3?

    A. All angles are 90°.
    B. Two angles are 90° and the other is 120°.
    C. All angles are 120°.
    D. Two angles are 120° and the other is 180°.
    E. All angles are 180°.

15. What is the hybridization of the central selenium atom of selenium hexafluoride, SeF_6?

    A. sp^5

    B. s^2p^3d

    C. spd^2

    D. sp^3d

    E. sp^3d^2

16. The diagram below shows the arrangement of electron groups in a molecule with octahedral electron pair geometry and a square pyramidal molecular geometry. Where will the next lone pair most likely be located if one more external atom is absent?

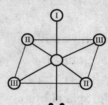

    A. Position I only
    B. Position II only
    C. Position III only
    D. Position I or Position II
    E. Position I, Position II, or Position III

Free-Response Questions

1. Consider the thiocyanate ion, SCN^-.

   (A) Draw an acceptable Lewis electron dot structure for the thiocyanate ion. Show the bonds using a line, and the nonbonding electrons with dots.

   (B) The thiocyanate ion has three resonance structures. Draw the two additional (to the answer in (A) resonance structures for this ion.

   (C) Would you classify the C–S bonds in the carbonate ion as pure covalent, polar covalent, coordinate covalent, or ionic? The electronegativity values of C and S are 2.5 and 2.5, respectively.

   (D) Estimate the CN bond lengths in the thiocyanate ion, given that the average CN single bond length is 147 pm, the average CN double bond length is 127 pm, and the average CN triple bond length is 115 pm.

2. Ionic compounds consist of ions arranged in a crystal structure or lattice.

   (A) What is lattice energy, and what is the generic equation for the reaction of the solid MX in a reaction involving lattice energy?

   (B) What do the sign and magnitude of lattice energy signify?

   (C) Describe the steps in the Born-Haber cycle.

   (D) Calculate the lattice energy of lithium fluoride given the following information:

Reaction	ΔH
1. $Li(s) \rightarrow Li(g)$	161 kJ/mol
2. $\frac{1}{2} F_2(g) \rightarrow F(g)$	77 kJ/mol
3. $Li(g) \rightarrow Li^+ (g) + e^-$	520 kJ/mol
4. $F(g) + e \rightarrow F^-(g)$	−328 kJ/mol
5. $Li^+(g) + F^-(g) \rightarrow LiF(s)$	?
6. $Li(s) + \frac{1}{2} F_2(g) \rightarrow LiF(s)$	−617 kJ/mol (LiF)

3. A molecule has the formula PF_5.

   (A) Draw an acceptable Lewis structure for this molecule.

   (B) What is the molecular geometry of PF_5?

   (C) Describe the bond angles between the central atom and peripheral atoms.

   (D) The dipole moment of this molecule is zero. What can you conclude about the arrangement of the peripheral atoms around the central atom? Explain your reasoning.

Multiple-Choice Answers

1. D   2. B   3. B   4. E   5. D   6. C   7. B   8. B   9. E   10. A   11. E   12. A   13. E
14. C   15. E   16. A

1. D

Methane (CH_4) is not a good electron donor in the formation of a coordinate covalent bond because it doesn't have a lone pair of electrons to donate to the bond.

2. B

The C–O bond will be the one in which the electron density is shifted the furthest away from the carbon atom. The bond density is always shifted towards the atom with the *highest* electronegativity difference between it and the central carbon atom. The difference between the C and Cl is 0.5; the difference between the C and H is –0.4; the difference between the C and Br is 0.3; and the difference between the C and O is 1.0.

3. B

These are sound Lewis structures that are equivalent except for the switching of the double and single bonds between C and O. By considering a hybrid of these two structures, we have an explanation for C–O bonds that are intermediate in length between single and double bonds.

4. E

Having a lone pair of electrons is the most important characteristic of electron donors to coordinate covalent bonds.

5. D

Hydrogen always violates the octet rule because it can have at most two electrons. Chlorine and all other elements in the third period or below on the Periodic Table can also violate the octet rule. This is because it's possible for them to use unfilled *d* orbitals in bonding, thereby exceeding the eight-electron limit for atoms that can only use *s* and *p* orbitals in bonding.

6. C

Formal charge equals the number of valence electrons minus the number of electron plus half the bonding electrons. Free nitrogen has 5 valence electrons, and this nitrogen has 4 lone electrons plus 2 bonding electrons giving a formal charge of –1. The carbon is 4 – (0 + 4), so its formal charge is 0. The oxygen is 6 valence electrons – (4 lone + 2 bonding), so its formal charge is also 0.

7. B

II is preferred. The formal charge of the central S atom is +1, the O is –1, and the Br is 0. Oxygen has the highest electronegativity and sulfur the lowest which correlates nicely with this structure.

Structure I is not preferred. The formal charge of the central Br atom is +2, the O and S are –1, and the external Br is 0. The Br–S bond is upside down, the bromine should have a negative formal charge and the sulfur a positive charge to correlate with their electronegativities. When the O and the S are switched in Structure III, formal charge predicts the O is positive (+1) and the S negative (–1), the opposite of what their electronegativies indicate should happen.

8. B

116 pm = acetonitrile ($H_3CC\equiv N$)

124 pm = N–ethylpropanalimine ($CH_3CH_2CH=NCH_2CH_3$)

147 pm = methylamine ($H_3C–NH_2$)

The higher the bond order, the shorter the bond. The two outside values on this list, 85 pm and 200 pm, are way too far from the average C–N bond lengths to be probable answers to this question.

9. E

Three moles of C–H bonds (413 kJ each), 1 mole of C–C bonds (347 kJ) and one mole of CN bonds (891 kJ) are broken for a total of 2,477 kJ.

10. A

IF_5 has 7 valence electrons contributed by the I, and 5 bonding electrons contributed by the five F atoms for a total of 12 electrons. The structure of IF must involve 6 pairs of bonding and unbonding electrons which means an expanded octet is involved.

11. E

When atomic orbitals are hybridized, the number of hybrid orbitals formed is the same as the number of contributing atomic orbitals. One s and two p orbitals hybridize to form three sp^2 hybrid orbitals.

12. A

SF_4 has 4 bonding pairs and 1 nonbonding pair for a total of 5 electron pairs. This gives a trigonal bipyramidal electron structure. According to the VSEPR theory, a molecule with 4 atoms bonded to the central atom in a trigonal bipyramidal geometry has a seesaw molecular geometry.

13. E

According to the VSEPR theory, the electron pair geometry of IF_5 is octahedral and the molecular geometry is square pyramidal. The bond angles in an octahedral molecule are all 90 degrees from each other. Because of its lone pair of electrons, it is a dipole and the I–F bonds have an electronegativty difference and are all polar.

14. C

The bromine atoms will be evenly spaced around the central boron atom. The three bromine atoms form a triangle with the boron atom at the center. All four atoms are in a single plane.

15. E

There are six peripheral atoms, so there must be six hybrid orbitals. The hybrid orbitals are created from the combination of one s, three p, and two d orbitals.

16. A

If two pairs of electrons in an octahedral arrangement are lone pairs, these pairs attempt to be as far apart as possible. So, the next lone pair will be in position I, and the result is a square planar molecule.

Free-Response Answers

1.

(A) and (B) (in any order)

$$\left[\ddot{\text{S}}=\text{C}=\ddot{\text{N}}\right]^- \rightleftarrows \left[:\text{S}\equiv\text{C}-\ddot{\text{N}}:\right]^- \rightleftarrows \left[:\ddot{\text{S}}-\text{C}\equiv\text{N}:\right]^-$$

(C) Polar covalent, despite the fact the electronegativity difference is 0 the only true pure covalent bond is when both atoms are the same element.

(D) Bond order = 2.00 as the average, so the bond length will be about that of a double bond, which is 127 pm.

2. (A) The lattice energy is the amount of energy required to change an ionic solid into isolated gaseous phase ions. Lattice energy can be described by the equation:

$$MX(s) \rightarrow M^+(g) + X^-(g)$$

(B) Since the definition of lattice energy involves the breaking of ionic bonds, energy is required or absorbed, and the lattice energy will have a positive heat value. The larger the lattice energy, the more energy required to break the ionic bonds.

(C) The Born-Haber cycle is based on Hess's law, which states that the enthalpy change for a reaction is the sum of the enthalpy changes for individual steps in the reaction. The Born-Haber cycle is composed of five steps:
Step 1 is the sublimation of the metal.
Step 2 is the dissociation of the diatomic nonmetal molecules.
Step 3 is the ionization of the gaseous metal atoms.
Step 4 is the formation of the gaseous nonmetal anions.
Step 5 is the formation of the ionic solid from gaseous metal and nonmetal ions.

(D) The sum of the energy changes for steps 1–5 will be equal to the overall energy change found in 6. Using this information, the value for step 5 is –1,047 kJ/mol. The reaction depicted in step 5 is the reverse of the reaction that defines lattice energy. So the lattice energy of LiF is equal in magnitude, but opposite in sign to the value of step 5, or +1,047 kJ/mol LiF.

3. (A)

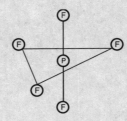

(B) Five groups of bonding electrons gives trigonal pyramidal electron pair and molecular geometries. There are no lone pairs and the molecular geometry will be the same.

(C) Bond angles: All will be 90° since it is octahedral.

(D) Since the dipole moment is zero, the molecule is nonpolar. This verifies the molecule is symmetrically arranged.

Intermolecular Forces and the Condensed Phases of Matter

In this chapter, we examine the attractive (and repulsive) forces between molecules as well as the behavior of real gases, liquids, and solids.

INTERMOLECULAR FORCES

If we cool any collection of gas molecules sufficiently, we observe that it becomes a liquid and then, with more cooling, a solid. In addition, molecular substances have the ability (within certain limitations) to form solutions. These behaviors arise from the fact that molecules exert attractive forces on one another. There are three recognized types of *intermolecular attractions*: London dispersion forces, dipole attractions, and hydrogen bonding. These attractive forces are known collectively as **van der Waals forces**.

London Dispersion Forces

London dispersion forces are present in *all* molecular substances and constitute the main form of attraction between molecules. In neighboring molecules, the electrons of one molecule repel the electrons of the neighboring molecule and an *instantaneous dipole* (i.e., a *temporary dipole*) is induced in *both* molecules. The London force is the mutual attraction of the opposite ends of these neighboring dipoles. London forces are most easily observable in nonpolar substances since no other type of van der Waals forces exists between the molecules. The strength of London forces depends on the mass and shape of the molecule, and is reflected in the melting and boiling points of the substances: the stronger the London force, the higher the melting and boiling points. The following table relates molar mass to the boiling point of the elements of Group 17:

Substance	Molar Mass/(g mol^{-1})	Boiling Point /K
F_2	38.0	85.1
Cl_2	71.0	238.6
Br_2	159.8	332.0
I_2	253.8	457.6

Molecular shape also determines the strength of London forces. In molecules that are more linear in shape, there is greater contact between the molecules than between more spherically shaped molecules. This is summarized in the table below, which relates the boiling point to the shapes of three hydrocarbons known as the *pentanes*:

Substance	Structure (2-dimensional)	Boiling Point /K
n-pentane	$CH_3CH_2CH_2CH_2CH_3$	309.2
isopentane	CH_3 | $CH_3CHCH_2CH_3$	301.0
neopentane	CH_3 | $CH_3 - C - CH_3$ | CH_3	282.6

London dispersion forces are also responsible for the mutual solubility of nonpolar molecules, such as Br_2 in CCl_4.

Dipole Attractions

A polar molecule has a *permanent* positive and negative end. For this type of molecule, an additional type of intermolecular attraction exists: **dipole attraction**, in which the negative end of one molecule is attracted to the positive end of another. This type of attraction between two HCl molecules is illustrated below.

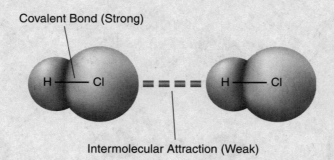

Within polar compounds having similar molar masses, the London forces are of comparable strength; the strength of the dipole attractions, however, is closely related to the polarity of its constituent molecules. This relationship is illustrated in the table below:

Substance	Molar Mass/ (g mol^{-1})	Dipole Moment /D	Boiling Point /K
n-propane ($CH_3CH_2CH_3$)	44	0.1	231
methoxymethane (CH_3OCH_3)	46	1.3	248
ethanal (CH_3CHO)	41	2.7	294

Dipole attractions are also responsible for the mutual solubility of polar molecules, such as NH_3 in H_2O.

In comparing the intermolecular forces between molecules, molar mass is the first consideration:

- When the molar masses are comparable, differences in intermolecular attractions are mainly due to differences in molecular polarity (that is, the strengths of the dipole attractions).

- When the molar masses differ greatly, differences in intermolecular attractions are mainly due to differences in the molar masses themselves (that is, the strength of the London forces).

Ion-Dipole Forces

An ionic substance such as KNO_3 is highly soluble in H_2O, a liquid composed of polar molecules. The ability of solid KNO_3 to dissolve in H_2O is due to the presence of **ion-dipole forces**, in which the positive ions are effectively surrounded by the negative ends of the H_2O molecules, and the negative ions by the positive ends. This is shown in the diagram below:

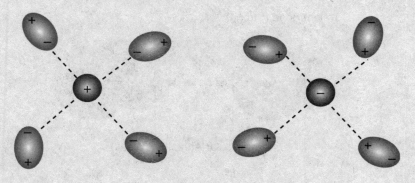

Hydrogen Bonding

The graph below relates boiling point to molar mass for the hydrogen compounds of the elements in Groups 14 and 16.

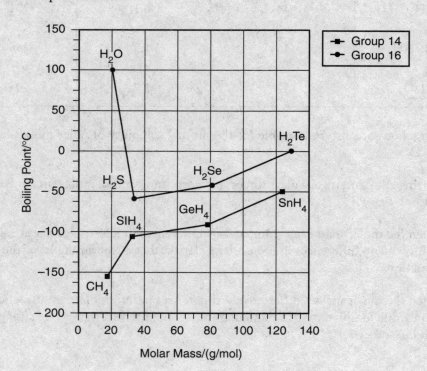

INTERMOLECULAR FORCES & THE CONDENSED PHASES OF MATTER

The graph of Group 14 is as predicted: as the molar mass increases, the boiling point increases. The graph of Group 16, however, displays an anomaly: H_2O, the compound with the smallest molar mass, has the highest boiling point. A graph of the hydrogen compounds in Groups 15 and 17 would yield the same type of result: NH_3 and HF, the compounds with the smallest molar masses, have the highest boiling points. Evidently, H_2O, NH_3, and HF have unusually strong intermolecular forces. These forces are known as **hydrogen bonds** and they are a unique type of dipole attraction. A molecule containing hydrogen bonded to a small, highly electronegative atom (such as N, O, or F) is highly polar. The small size of all of these atoms allows adjacent molecules to approach each other closely, resulting in a strong interaction among the molecules. The diagram below illustrates some examples of hydrogen bonding in H_2O and NH_3:

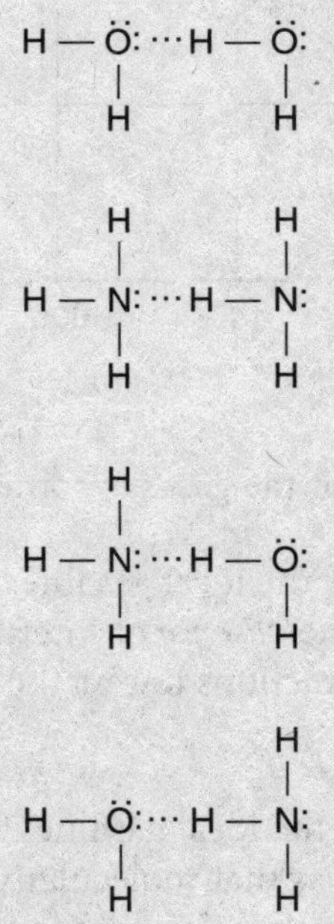

$$
\begin{array}{c}
\text{H} - \ddot{\text{O}}\text{:} \cdots \text{H} - \ddot{\text{O}}\text{:} \\
| \qquad\qquad | \\
\text{H} \qquad\qquad \text{H}
\end{array}
$$

$$
\begin{array}{c}
\text{H} \qquad\qquad \text{H} \\
| \qquad\qquad | \\
\text{H} - \text{N:} \cdots \text{H} - \text{N:} \\
| \qquad\qquad | \\
\text{H} \qquad\qquad \text{H}
\end{array}
$$

$$
\begin{array}{c}
\text{H} \\
| \\
\text{H} - \text{N:} \cdots \text{H} - \ddot{\text{O}}\text{:} \\
| \qquad\qquad | \\
\text{H} \qquad\qquad \text{H}
\end{array}
$$

$$
\begin{array}{c}
\text{H} \\
| \\
\text{H} - \ddot{\text{O}}\text{:} \cdots \text{H} - \text{N:} \\
| \qquad\qquad | \\
\text{H} \qquad\qquad \text{H}
\end{array}
$$

THE BEHAVIOR OF REAL GASES

In chapter 6, we explored the behavior of ideal gases and we developed an equation ($PV = nRT$) to quantify this behavior. *Real gases*, however, do not obey this equation under certain conditions of pressure and temperature. A useful quantity for examining gas behavior is known as **compressibility** (Z) and is defined as follows:

$$
Z = \frac{PV}{nRT}
$$

For an ideal gas, Z must equal 1. The graph below relates the compressibility of various gases at 300 K (313 K for CO_2) as a function of pressure:

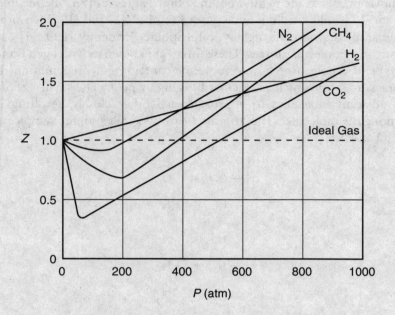

At low pressures, the behavior of all of the gases becomes more ideal (that is, $Z \rightarrow 1$).

• For those points that fall *below* the ideal gas line, $PV < nRT$. One of the assumptions made about an ideal gas was the *absence* of intermolecular forces. In real gases, the presence of intermolecular attractions lowers the observed pressure from its ideal value and, as a result, $Z < 1$.

• For those points that lie *above* the ideal gas line, $PV > nRT$. Another assumption that was made about an ideal gas was that molecular volume is *negligible*. Consequently, ideal gas molecules move throughout the entire volume of the container. In real gases, however, the molecules do have volume and, as a result, the "free" volume in which they move is considerably *less* than the volume of the container. When we measure the volume that a gas occupies, we are really measuring the container volume and not the volume in which it is free to move. As a result, the V term in the equation is too large and $Z > 1$.

The deviation of a gas from ideal behavior also depends on the temperature. The graph below traces the compressibility of N_2 (as a function of pressure) at various temperatures:

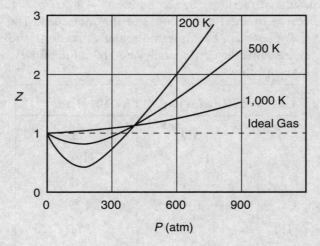

At higher temperatures, the increased average kinetic energy of the molecules reduces the effectiveness of the intermolecular forces and the gas exhibits behavior that is more closely ideal: at 1,000 K, the graph no longer lies *below* the ideal line. Nevertheless, deviations still remain and these are attributable to the finite nature of the molecular volume.

The van der Waals Equation

In order to predict the behavior of real gases more accurately, a number of equations have been developed. One of the best known equations was developed by the Dutch physicist Johannes van der Waals:

$$\left(P + \frac{n^2 a}{V^2}\right)(V - nb) = nRT$$

The first term $\left(+\dfrac{n^2 a}{V^2}\right)$ corrects the observed pressure, which is too *small* because of the presence of intermolecular attractions; the second term $(-nb)$ corrects the observed volume, which is too *large* because of the finite volume of the molecules. The constants a and b, therefore, represent the roles played by intermolecular attractions and molecular volume, respectively. They are determined experimentally and are unique for each gas. A table of van der Waals constants is given below:

Gas	$a\,/(L^2\ atm\ mol^{-2})$	$b\,/(10^{-2}\ L\ mol^{-1})$
Ar	1.317	3.20
C_6H_6	18.57	11.93
Cl_2	6.260	5.42
CO_2	3.610	4.29
H_2	0.2420	2.65
H_2O	5.465	3.05
NH_3	4.170	3.71

Sample Problem

Use the van der Waals equation to estimate the pressure in a 10.0-liter tank containing 20.0 moles of CO_2, if the temperature is 320. K.

Solution

Rearrange the van der Waals equation to solve for pressure:

$$\left(P + \frac{n^2 a}{V^2}\right)(V - nb) = nRT$$

$$P = \frac{nRT}{(V - nb)} - \frac{n^2 a}{V^2}$$

$$= \frac{(20.0\ mol)\cdot(0.0821\ L\ atm\ mol^{-1}\ K^{-1})\cdot(320.\ K)}{\big((10.0\ L) - \big((20.0\ mol)\cdot(4.29\times10^{-2}\ L\ mol^{-1})\big)\big)} - \frac{(20.0\ mol)^2\cdot(3.610\ L^2\ atm\ mol^{-2})}{(10.0\ L)^2}$$

$$= 43.0\ atm$$

If the ideal gas equation $\left(P = \dfrac{nRT}{V}\right)$ had been used in the calculation, the calculated pressure would have been 52.5 atmospheres. The smaller value produced by the van der Waals equation reflects the dominant role played by intermolecular attractions inside the tank of gas.

LIQUIDS: SURFACE TENSION AND VISCOSITY

Intermolecular forces also determine the magnitudes of two familiar properties of liquids: surface tension and viscosity.

Surface Tension

If we compare the intermolecular forces on the molecules entirely *within* a liquid to the molecules on the *surface* of a liquid, we find that the molecules within the liquid experience a greater net attractive force, as shown in the diagram below:

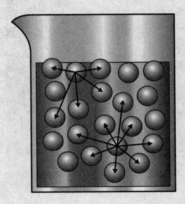

As a result, the surface molecules are drawn inward and the area of the surface of the liquid is reduced. (Falling droplets of a liquid assume a spherical shape because a sphere has the *lowest* surface area to volume ratio.) The smaller surface area causes the liquid to behave as though it has a "skin" on surface, and this *surface tension* enables insects to walk on the surface of water. **Surface tension** is formally defined as the energy needed to increase the surface area of a liquid by 1 unit. The most commonly used units of surface tension are joules per square meter (J/m^2). The surface tensions of water and mercury at 20°C are, respectively, 7.3×10^{-2} J/m^2 and 4.6×10^{-1} J/m^2.

When a liquid is confined within a container, such as a glass tube, two types of forces are present on the molecules of the liquid:

- **Cohesive forces**, which bind the molecules of the liquid

- **Adhesive forces**, which bind the liquid molecules to the molecules to the surface of the container

In the case of water, the adhesive forces are greater than the cohesive forces, and the curved upper surface, known as a **meniscus**, is concave; in the case of mercury, the opposite is true and the meniscus is convex. This is shown in the diagram below:

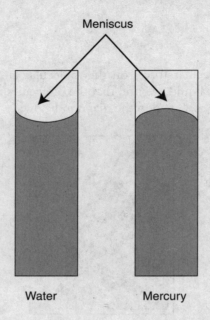

Viscosity

Viscosity is the resistance of a liquid to flow. Honey and motor oil are examples of particularly viscous fluids. Viscosity is measured by determining the time it takes a given volume of a liquid to pass through a thin tube under the influence of gravity. Since this property is related to the relative ease with which liquid molecules can move past each other, it is dependent in intermolecular attractions. Viscosity is temperature dependent: placing a jar of honey in boiling water turns the honey into a relatively free-flowing liquid. As the honey cools, its viscosity increases once again. The decrease in viscosity with increasing temperature is attributed to the increase in the average molecular kinetic energy of the liquid, which overcomes the attractive forces between the molecules.

CRYSTALLINE SOLIDS

A solid in which its atoms, ions, or molecules are present in a state of relative disorganization is said to be **amorphous**. In many solids, however, the structural particles are arranged in a highly ordered fashion known as a **crystal**.

INTERMOLECULAR FORCES & THE CONDENSED PHASES OF MATTER

The properties of crystalline solids depend on the type of structural particles that make up the *crystal lattice*. The general properties of crystalline solids, studied in chapter 9, are reviewed in the table below:

Crystal Type	Structural Particles	Principal Atrractive Forces Between Particles	Characteristics of the Crystal	Examples
Ionic Crystals	Positive and negative ions	Electrostatic attractions between ions	Hard, brittle; moderate to high melting points; nonconductors as solids, but conductors as liquids; many dissolve in water	KCl, CaF_2, CsBr, MgO, $BaCl_2$, NaCl
Covalent Network Crystals	Atoms	Covalent bonds	Very hard; insoluble in most ordinary liquids; sublime or melt at high temperatures; most are nonconductors	C (diamond), C (graphite), SiC, BN, SiO_2
Metallic Crystals	Positive ions and delocalized, mobile valence electrons	Metallic bonds	Hardness varies from very soft to very hard; melting point varies from very low to very high; lustrous; malleable; ductile; high thermal and electrical conductivity	Na, Mg, Cu, Al, Zn, Fe, Li, W, Pt, Cr, Ag, Pb, Au, Hg
Polar Molecular Crystals	Polar molecules	Dispersion forces and dipole–dipole attractions	Low to moderate melting points; soluble in some polar and nonpolar liquids	HCl, $CHCl_3$, H_2S
Hydrogen-bonded Molecular Crystals	Molecules with H bonded to O, N, F	Hydrogen bonds	Low to moderate melting points; soluble in some hydrogen-bonded and some polar liquids	H_2O, NH_3, HF, CH_3OH
Nonpolar Molecular Crystals	Atoms or nonpolar molecules	Dispersion forces	Extremely low to moderate melting points; soluble in nonpolar solvents	Ar, Cl_2, H_2, CH_4, I_2, CO_2, CCl_4

A detailed study of crystal structure is beyond the scope of this book; the reader should refer to any standard general chemistry text for more information.

CHANGE OF PHASE

The transitions among the solid, liquid, and gaseous phases, and the energy changes accompanying these phase changes, are illustrated below:

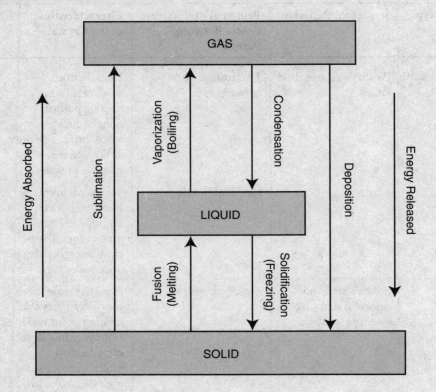

The solid phase is the most orderly and the gas phase, the least. Any transition from a more orderly to a less orderly phase (such as liquid → gas) requires the absorption of energy in order to overcome the attractive forces between the particles. Associated with each phase transition is a corresponding enthalpy change: The **enthalpy of vaporization** (ΔH_{vap}) is the heat absorbed by one mole of liquid when it changes to gas at constant pressure. For example, ΔH_{vap} for H_2O is 40.67 kJ/mol. Similarly, the **enthalpy of fusion** (ΔH_{fus}) is the heat absorbed by one mole of solid when it changes to liquid at constant pressure. For example, ΔH_{fus} for ice is 6.01 kJ/mol. The large difference between the enthalpy of fusion and the enthalpy of vaporization reflects the additional energy that is needed to overcome the intermolecular attractions in vaporizing a liquid.

In addition, there is an **enthalpy of sublimation** (ΔH_{sub}) corresponding to the energy changes in a solid–gas transition. Since enthalpy is a state function, it follows that: $\Delta H_{sub} = \Delta H_{fus} + \Delta H_{vap}$. In the case of ice, the enthalpy of sublimation is 40.67 kJ/mol + 6.01 kJ/mol = 46.68 kJ/mol.

Heating and Cooling Curves

A heating curve traces the temperature and phase changes in a substance as heat is absorbed by it at constant pressure. The graph below represents an idealized heating curve for 1 mole of H_2O as it changes from ice at −25°C to steam at 120°C:

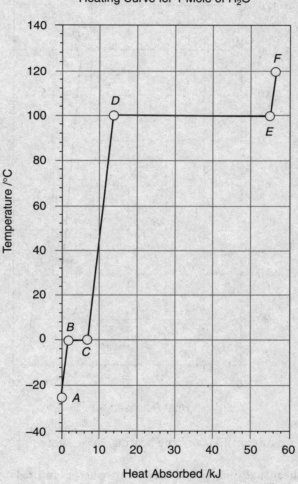

Heating Curve for 1 Mole of H_2O

- During segment *AB*, ice is warmed from −25°C to 0°C; during segment *CD*, liquid water is warmed from 0°C to 100°C; during segment *EF*, steam is warmed from 100°C to 120°C. The *slopes* of these segments are dependent on the specific heats of ice, liquid water, and steam, respectively.

- During segment *BC*, ice melts to liquid water at its melting point, 0°C; during segment *DE*, liquid water vaporizes to steam at its boiling point, 100°C. The *lengths* of these segments are dependent on the heats of fusion of ice and vaporization of liquid water, respectively. That a phase change occurs at constant temperature indicates that the heat absorbed by the system is used to overcome intermolecular attractions rather than to increase the average kinetic energy of the molecules.

An idealized cooling curve for 1 mole of H_2O between 120°C and –25°C is given below:

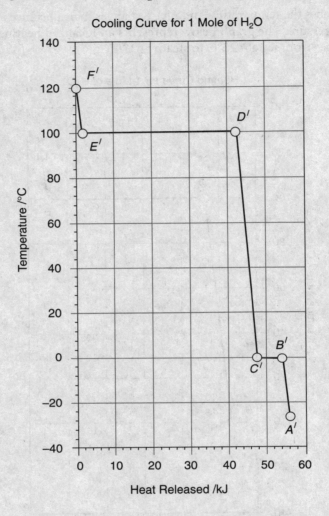

Cooling Curve for 1 Mole of H_2O

Note that this graph is a mirror image of the heating curve shown page 215.

- A quantity of heat that is absorbed between two temperatures on the heating curve is equal to the quantity of heat released between the same two temperatures on the cooling curve.

- Freezing and condensation take place at the same respective temperatures as melting and vaporization. If a mixture of ice and water at 0°C and 1 atmosphere were isolated from its surroundings, the masses of each phase would remain unchanged. This is not a static situation, but rather a **dynamic phase equilibrium**, in which the rates of the opposing processes of melting and freezing are equal. Similarly, in an isolated mixture of water and steam at 100°C and 1 atmosphere, the rates of vaporization and condensation will also be equal.

Vapor Pressure

The escape of molecules from the surface of a liquid into the gas phase is known as **evaporation**. The rate of evaporation depends on the nature of the liquid, the temperature, and the area of the exposed surface:

• Ethanol evaporates more readily than water; the application of ethanol to the skin has a dramatic cooling effect (as it absorbs the heat from the skin).

• A clothes dryer brings wet clothes in contact with heated air (to raise the temperature of the water), and tumbles the clothes (to expose more of the wet surfaces), in order to promote rapid and efficient drying.

The effect of temperature on evaporation is illustrated in the diagram below, which plots the kinetic energies of the surface molecules of a hypothetical liquid at two temperatures:

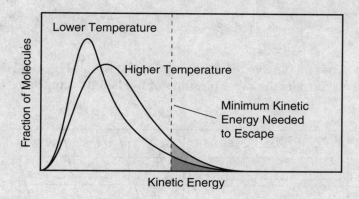

For every liquid, there is a minimum kinetic energy that the surface molecules need to escape the liquid. Increasing the temperature shifts the distribution curve so that more of the surface molecules have this minimum energy and the rate of evaporation is increased.

If a liquid is placed in a container that is connected to a manometer, the pressure produced by the evaporated liquid will cause a change in the levels of the manometer. As evaporation proceeds, however, condensation is also taking place and, eventually, the rates of these opposing processes will become equal and a dynamic equilibrium will exist between the liquid and gas phases. At this point, the pressure of the evaporated liquid (or *vapor*) is known as the (**equilibrium**) **vapor pressure**. This process is illustrated in the diagram below:

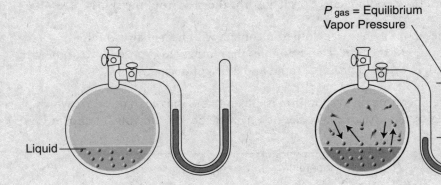

Vapor pressure depends on the *volatility* of the liquid (that is, on ΔH_{vap}) and the temperature. The diagram below illustrates the vapor pressures of five liquids as functions of temperature:

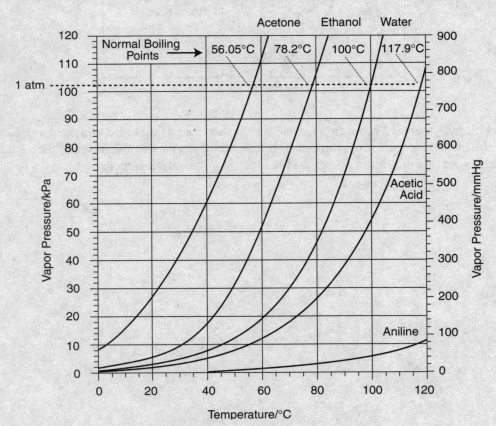

INTERMOLECULAR FORCES & THE CONDENSED PHASES OF MATTER

At a given temperature, acetone has the highest the vapor pressure: it is the most volatile liquid. In contrast, aniline ($C_6H_5NH_2$) has the lowest vapor pressure. As the temperature of a liquid is raised, its vapor pressure increases until the vapor pressure equals the *external pressure above the liquid*: at that point the liquid *boils*. Therefore, a liquid has many boiling points: for example, if the external pressure above a container of ethanol were adjusted to 18 kilopascals (135 mmHg), the ethanol would boil at an approximate temperature of 40°C. Usually, a liquid is characterized by its **normal boiling point**, the temperature at which the vapor pressure of the liquid equals one atmosphere. The normal boiling points of the first four liquids are displayed on the vapor pressure curves shown above. Note that the liquids with the *highest* vapor pressures have the *lowest* normal boiling points.

A table of the equilibrium vapor pressure of water from 0°C to 220°C (in mmHg and kPa) is given in Appendix B.

Mathematically, the relationship between vapor pressure and temperature is given by the **Clausius-Clapeyron equation:**

$$\ln P = \frac{-\Delta H_{vap}}{RT} + C$$

This equation is based on the following assumptions:

- The volume of the vaporized liquid is negligible compared to the volume of the vapor.
- The vapor behaves as an ideal gas.
- ΔH_{vap} is constant over the temperature interval of the data.

Under these conditions, a plot of $\ln P$ versus $1/T$ will yield a straight line whose slope is equal to $-\Delta H_{vap}/R$. The table and the graph shown below depict how the Clausius-Clapeyron equation is applied to vapor pressure data for ethanol (C_2H_5OH) over the temperature range 10°C – 60°C (283.15 K – 333.15 K):

T/K	P/Pa	$1/T/(K^{-1})$	$\ln P$
283.15	3,146.4	0.0035317	8.0540
288.15	4,293.0	0.0034704	8.3647
293.15	5,852.9	0.0034112	8.6747
298.15	7,866.0	0.0033540	8.9703
303.15	10,506	0.0032987	9.2597
308.15	13,826	0.0032452	9.5343
313.15	18,039	0.0031934	9.8003
318.15	23,198	0.0031432	10.052
323.15	29,624	0.0030945	10.296
328.15	37,410	0.0030474	10.530
333.15	47,023	0.0030017	10.758

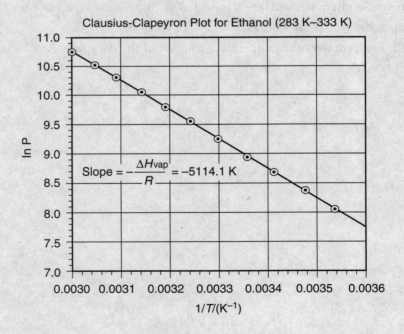

Clausius-Clapeyron Plot for Ethanol (283 K–333 K)

Based on the slope of the best-fit straight line, the calculated value of ΔH_{vap} for ethanol over this temperature range is 42.521 kJ/mol.

Phase Diagrams

All phase changes are both pressure and temperature dependent. A **phase diagram** is a graph that relates pressure and temperature of the system to the phases of a substance that are present within the system. The general phase diagram shown below is typical of many substances:

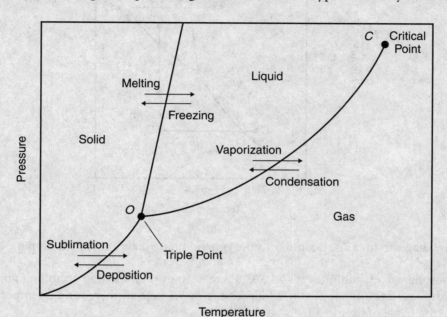

- Each line signifies a pressure-temperature combination in which the two adjacent phases are in dynamic equilibrium.

- A pressure-temperature combination that does *not* lie on a line indicates that a single phase is present.

- Point O is known as the **triple point**, in which all three phases coexist in dynamic equilibrium. Note that there is only one triple point in this diagram, and it is associated with a unique pressure and temperature.

- Point C is known as the **critical point** and it is associated with a unique temperature and pressure known, respectively, as the **critical temperature** (T_C) and **critical pressure** (P_C). The critical region of a substance will be discussed in the next section.

The Phase Diagram for Carbon Dioxide (CO_2)

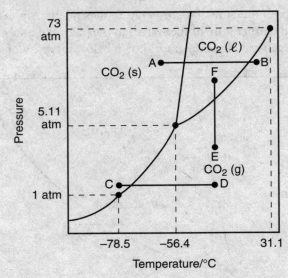

The phase diagram for CO_2 is similar to the general phase diagram shown on page 221.

At 1 atmosphere, CO_2 sublimes at $-78.5°C$; there is no normal melting or boiling point since the liquid phase does not appear until a pressure of 5.11 atmospheres and a temperature of $-56.4°C$ (the triple point of CO_2) is attained.

- The horizontal line segments *AB* and *CD* represent temperature changes at constant pressure: As the temperature increases over *AB*, CO_2 is converted directly from solid to gas; as the temperature increases over *CD*, CO_2 is converted from solid to liquid and from liquid to gas.

- The vertical line segment *EF* represents a change in pressure at constant temperature: As the pressure increases over *EF*, CO_2 is converted from gas to liquid and from liquid to solid. The positive slope of the solid–liquid line indicates that solid CO_2 is more dense than liquid CO_2 at the same temperature.

INTERMOLECULAR FORCES & THE CONDENSED PHASES OF MATTER

The Phase Diagram for Water (H₂O)

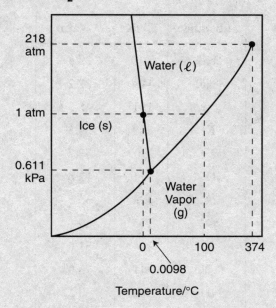

The phase diagram for H_2O differs from the general phase diagram in that the slope of the solid–liquid line is *negative*.

At 1 atmosphere, H_2O melts at 0°C and boils at 100°C. Ice sublimes only at very low pressures (below 0.611 kilopascal). The triple point of water occurs at 0.611 kilopascal and a temperature that is now defined exactly as 273.16 K. This temperature, T_{tr}, is one of the fixed points on the Kelvin scale of temperature (the other being 0 K, absolute zero).

- The negative slope of the solid–liquid line implies that increasing the pressure of solid ice at 0°C will transform the ice into liquid water, a phenomenon that occurs because ice at 0°C is less dense than liquid water at 0°C.

Critical Temperature and Pressure

At the critical temperature and pressure, the liquid–gas line in a phase diagram terminates. That is to say, at any temperature above the critical temperature, it is impossible to liquefy a gas, no matter how much pressure is applied to it. The critical temperature depends on the intermolecular attractions within a substance: the greater the intermolecular forces, the higher the critical temperature.

As the applied pressure is increased to a gas above its critical temperature, the density of the gas can approach the density of the liquid phase. Although the substance is still technically a gas, it is more appropriately called a **supercritical fluid**. Supercritical fluids can function as solvents, as liquids do, and they are capable of dissolving a wide range of substances. For example, supercritical CO_2 is used to dissolve the caffeine in coffee beans without removing the other components necessary for flavor and aroma.

The table below lists the critical temperatures and pressures for a number of substances:

Substance	T_C /K	P_C /atm
He	5.19	2.24
Ne	44.4	27.2
Ar	151	48.3
Kr	209	54.3
Xe	290.	57.7
H_2	33.0	12.8
N_2	126	33.5
O_2	155	49.8
F_2	144	51.0
Cl_2	417	78.9
NH_3	406	112
H_2O	647	218
CO_2	304	72.3

At the critical point, a gas is also associated with a molar **critical volume** (V_C). It is of interest that all three critical constants can be calculated by using the constants (a, b, and R) associated with the van der Waals equation:

$$T_C = \frac{8a}{27Rb} \qquad P_C = \frac{a}{27b^2} \qquad V_C = 3b$$

Sample Problem

Estimate the critical constants for NH_3 using the van der Waals relationships given above and the values given in the table found on page 210.

Solution

$$T_C = \frac{8a}{27Rb} = \frac{8 \cdot (4.170 \text{ L}^2 \text{ atm mol}^{-2})}{27 \cdot (0.0821 \text{ L atm mol}^{-1} \text{ K}^{-1}) \cdot (0.0371 \text{ L mol}^{-1})} = \textbf{407 K}$$

$$P_C = \frac{a}{27b^2} = \frac{(4.170 \text{ L}^2 \text{ atm mol}^{-2})}{27 \cdot (0.0371 \text{ L mol}^{-1})^2} = \textbf{113 atm}$$

$$V_C = 3b = 3 \cdot (0.0371 \text{ L mol}^{-1}) = \textbf{0.111 L mol}^{-1}$$

Note that T_C and P_C vary by less than 1 percent from the measured values given in the table.

Chapter 10: Practice Questions

Multiple-Choice Questions

1. Which is true?

   A. London forces are also called dispersion forces.

   B. London forces occur between permanent dipoles.

   C. Hydrogen bonds are weaker than dipole-dipole forces.

   D. Hydrogen bonds occur only between nonpolar molecules.

   E. Dispersion forces account for the high boiling point of water.

2. Which hydrocarbon has the strongest dispersion forces between its molecules?

   A. CH_4
   B. C_2H_6
   C. C_5H_{12}
   D. C_6H_6
   E. C_8H_{18}

3. Which best explains why water is a liquid at ordinary earth conditions while all other molecules with comparable molecular weights on earth are gases? Water:

   A. is polar.
   B. exhibits hydrogen bonding.
   C. is the most common liquid on earth.
   D. has extremely strong dispersion forces.
   E. takes the shape of its container but has a fixed volume.

4. Why is it possible to float a pin on the surface of water?

   A. The surface tension of the water is very low.

   B. The density of the pin is less than that of the water surface tension.

   C. Only the sharp tip of the pin will break the surface tension of the water.

   D. The surface molecules of the water change bonding as they dry out.

   E. The mass of the pin does not exert enough force to increase the surface area needed to sink it.

5. What type of solid has a high melting point, does not conduct heat and electricity in either the solid or liquid state, and does not dissolve in water or organic solvents?

   A. Ionic
   B. Network
   C. Molecular
   D. Amorphous
   E. High viscosity solution

6. What statement correctly justifies two correct reasons why MgS has a higher melting point than CsI?

   A. Mg has higher metallic character and S is a more reactive nonmetal.

   B. MgS has a greater ionic charge and its ions are smaller.

   C. MgS has larger dispersion forces and a less structured crystal lattice.

   D. MgS has greater electron mobility and forms an amorphous solid.

   E. MgS has a high solubility in water and forms solutions with a high electrical conductivity.

7. Select the choice that correctly matches the Structural Unit, the Attractive Force between them, and an Example of the type of solid.

	Structural Unit	Attractive Force	Example
A.	molecules	covalent bonds	iodine
B.	diatomic molecules	covalent bonds	helium
C.	metallic anions	electrostatic forces	gold
D.	metallic cations	covalently bonded ions	diamond
E.	oppositely charged anions and cations	electrostatic forces	sodium chloride

8. Which causes in increase in vapor pressure?

   A. Increased pressure
   B. Decreased pressure
   C. Increased intermolecular attractions
   D. Increased temperature
   E. Decreased temperature

9. When a solid melts, the temperature of the solid:

   A. first decreases, then increases after all the solid melts.

   B. remains the same and the potential energy increases.

   C. decreases due to a decrease in average molecular kinetic energy.

   D. increases at a characteristic slope related to the strength of attractive forces.

   E. increases at a positive rate when there is a density increase, and at a rate slope when there is a density decrease.

10. Which can be used to predict the strength of the intermolecular attractions between molecules?

   I.   The viscosity of a liquid
   II. The pressure on a liquid
   III. The molar mass of a liquid
   IV. The boiling point of a liquid

   A. I only
   B. I and IV only
   C. II and III only
   D. I, II, and IV only
   E. II, III, and IV only

Free-Response Questions

1. The phase change diagram for carbon dioxide, CO_2, is shown in the diagram.

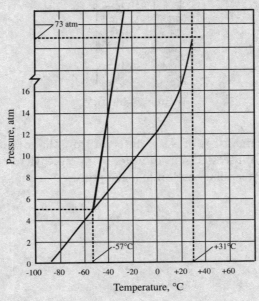

(A) Give the temperature and pressure of the:
  (i)   triple point.
  (ii)   critical point.
  (iii)   normal sublimation point.

(B) The temperature of a quantity of solid carbon dioxide is increased at a pressure of 12 atm. State the:
  (i)   melting temperature.
  (ii)   boiling temperature.

(C) Will this solid float or sink in liquid CO_2? What characteristic of this phase diagram enables this prediction?

2. What type of solid will these substances form? Choose from network, molecular, metallic, or ionic.

(A) C
(B) $CaCO_3$
(C) CO_2
(D) I_2
(E) NaOH
(F) K
(G) SiO_2
(H) U
(I) Xe

3. Test tubes containing water and mercury are viewed side-by-side in glass tubes.

Water   Mercury

(A) Why is the shape of the meniscus of water different from that of mercury?
(B) What will be the shape of the meniscus of water in a polyethylene tube?
(C) Will water rise higher by capillary action in a glass tube or in a polyethylene tube of the same diameter?

Multiple-Choice Answers

1. A   2. E   3. B   4. C   5. B   6. B   7. E   8. D   9. B   10. B

1. A

London forces are the attractive forces between all molecules due to instantaneous dipoles, and are also called dispersion forces.

2. E

As the number of electrons in a molecule increases, the strength of the dispersion forces also increases because there is a higher likelihood for the formation of temporary, instantaneous dipoles due to the lack of symmetry in the distribution of electrons around the molecule. This is a quick way to pick the molecule with the highest number of electrons and molar mass.

3. B

When asked a question about why water has certain unusual properties, you can make a pretty safe bet that hydrogen bonding will be the reason behind it. Hydrogen bonds help hold the water molecules more tightly together than otherwise would be expected, allowing it to remain in the liquid phase.

4. C

The mass of the pin is not great enough to distort the surface of the water to the degree needed to surround the pin. This only works for things of light enough total mass, such as staples, pins, and water bugs.

5. B

Different crystalline solids have different characteristics. An amorphous solid has molecules present in a state of relative disorganization and do not have high melting points. Ionic solids conduct in the liquid phase and are often soluble in water. Molecular crystals have low melting points. Viscosity is the resistance of liquid to flow.

6. B

The strength of attractive forces between molecules depends on the mass and shape of the molecule. As the attractive forces grow stronger, the melting and boiling points of a substance increase. Due to the ±2 charge of the ions in MgS, the bond between Mg^{2+} and S^{2-} is stronger than the bond between the ±1 charge of the Cs^+ and I^- ions. Ionic radius increases from upper right to lower left of the Periodic Table. The smaller ions of MgS move the ions closer, also leading to the prediction of a higher melting point than that of CsI.

7. E

The attractive force between two oppositely charged particles is known as electrostatic force. A good example of this would be sodium chloride, NaCl.

8. D

Vapor pressure is affected by temperature alone. An increase in temperature increases the kinetic energy of the liquid molecules, allowing more of them to have sufficient energy to escape to the vapor state.

9. (B)

Phase change diagrams show that as a substance is changing states, the temperature does not change—all the energy added to the system goes towards increasing the potential energy of the substance to move the molecules further apart. All of the other choices are incorrect because they imply there a change in temperature with the melting of a solid.

10. B

The correct answer is: I and IV only. The viscosity and boiling points of a liquid increase as the intermolecular forces increase. The external pressure on a liquid affects other properties, such as boiling point, but it does not allow the prediction of attraction strength. The molar mass of a substance is also not dependent upon the strength of the intermolecular forces.

Free-Response Answers

1. (A) (i)   The triple point temperature and pressure is 5 atm and –57°C.

    (ii)   The critical point is 73 atm and 31°C.

    (iii)   The normal sublimation point is: 1 atm and –80°C

   (B) (i)   The melting temperature is –40°C.
    (ii)   The boiling temperature is 0°C.

   (C) This solid will sink in liquid CO_2 since the solid-liquid equilibrium line has a positive slope, indicating the solid phase is denser than the liquid.

2. (A) C   network
   (B) $CaCO_3$ ionic
   (C) CO_2 molecular
   (D) I_2 molecular
   (E) NaOH   ionic
   (F) K   metallic
   (G) SiO_2 network
   (H) U   metallic
   (I) Xe   molecular

3. (A) The water meniscus seems to creep up the walls of the tube, primarily because of hydrogen bonding between the water and the wall. The mercury has more attraction for other mercury atoms than for the glass walls.

   (B) The shape of the water meniscus in a polyethylene tube is similar to the mercury in the glass tube. Water has no attraction for polyethylene.

   (C) Water will rise higher by capillary action in a glass tube being assisted by the hydrogen bonding with the walls of the tube.

Solutions and Their Properties

CHAPTER ELEVEN

Homogeneous mixtures are uniform and have the same composition throughout. A homogeneous mixture is classified according to the size of the particles contained within the mixture:

- If any of the particles are greater than 500 nanometers in diameter, the mixture is classified as a **suspension**. Suspensions are only momentarily homogeneous because they separate with time. Paint is an example of a suspension.

- Particles sizes of 2–500 nanometers form mixtures known as **colloids** or **colloidal dispersions**. While colloids may appear murky or even opaque, they do not separate. A light beam that is passed through a non-opaque colloid is clearly visible, a phenomenon known as the **Tyndall effect**. The spreading of automobile headlight beams in a fog is an example of the Tyndall effect. The following table lists a number of colloid types:

Type of Colloid	Formation Criteria	Example
Aerosol	A liquid is dispersed in a gas	Fog
Aerosol	A solid is dispersed in a gas	Smoke
Foam	A gas is dispersed in a liquid	Whipped cream
Solid foam	A gas is dispersed in a solid	Marshmallow
Emulsion	A liquid is dispersed in a liquid	Milk
Solid emulsion	A liquid is dispersed in a solid	Butter
Sol	A solid is dispersed in a liquid	Gelatin

- If the particle diameter is in the range 0.1–2 nanometers (the size of ions and small molecules), the mixture is classified as a **solution**. Solutions that are not solid have a degree of transparency and do not exhibit the Tyndall effect. The component that is *dispersed* in a solution is known as the **solute**; the other component, the *dispersing agent*, is known as the **solvent**. Generally, the solvent is the component that is present in greater quantity. Vinegar is 5% acetic acid and 95% water: water is the solvent of this solution. Rum that is 151 proof contains approximately 75% grain alcohol and 25% water: alcohol is the solvent in this solution. The table below lists a number of types of solutions:

Solute	Solvent	Example
Gas	Gas	Air; all gas mixtures
Gas	Liquid	Carbonated water; ammonia solution
Gas	Solid	Hydrogen in palladium metal
Liquid	Liquid	Vinegar; rubbing alcohol
Liquid	Solid	Mercury-silver dental amalgam
Solid	Liquid	Sugar in water
Solid	Solid	Metallic alloys: brass; 14 karat gold

THE SOLUTION PROCESS

In forming a solution, solute-solvent attractions must replace solute-solute and solvent-solvent attractions. Therefore, these attractions must be similar in nature and strength. Chemists have a familiar saying that summarizes this principle: *"like dissolves like."* For example, ethanol (C_2H_5OH) and water dissolve freely in one another because the hydrogen bonding present among the ethanol molecules and among the water molecules is replaced by the hydrogen bonding between the ethanol and water molecules. Similarly, NaCl dissolves in water because the interionic attractions among Na^+ and Cl^- ions are replaced by ion-dipole attractions between water molecules and Na^+ and Cl^- ions. In contrast, oil does not dissolve in water because the London forces present among the nonpolar oil molecules and the hydrogen bonding present among the polar water molecules are too dissimilar. We would, however, expect oil to dissolve in benzene (C_6H_6, another nonpolar liquid) due to the ability of the benzene and oil molecules to form London forces with one another.

From an energy point of view, the solution process can be divided into three steps:

- *Separation of solute*: In this step, the intermolecular attractions among the solute particles (ionic attractions, dipole forces, London forces) must be overcome. This is an endothermic process.

- *Separation of solvent*: In this step, the (dipole forces, London forces) among the solvent particles must be overcome. This is also an endothermic process.

- *Solvation*: In this step, attractions are formed between solute and solvent particles (ion-dipole attractions, dipole forces, London forces) that stabilize the solution. This is an exothermic process: $\Delta H_{solvation}$ <0. When solute particles form attractions with water molecules, we say that the solute is **hydrated**, and the energy associated with this process is known as the **hydration energy** of the solute. Hydration energy depends on the charge and size of the solute particles: Larger charges on solute particles attract the water molecules more strongly (Ba^{2+} has a larger hydration energy than K^+); smaller particles allow the water molecules to approach the solute more closely (Na^+ has a larger hydration energy than Rb^+).

The **heat of solution** ($\Delta H_{solution}$) is the algebraic sum of these processes: If more energy is absorbed (during separation of solute and solvent) than is released (during solvation), the heat of solution is *endothermic* and the temperature of the solution will be lowered. NH_4NO_3 has an endothermic heat of solution ($\Delta H_{solution}$ = +25.7 kJ / mol) and it is used for "cold packs." If less energy is absorbed (during separation of solute and solvent) than is released (during solvation), the heat of solution is *exothermic* and the temperature of the solution will be raised. $MgSO_4$ has an exothermic heat of solution ($\Delta H_{solution}$ = −91.2 kJ / mol) and it is used for "heat packs."

THE CONCENTRATION OF SOLUTIONS

The labeling of a solution as *dilute* or *concentrated* provides no information on the quantities of components present within a solution. Chemists have devised a number of ways of expressing the concentration of solutions in a quantitative way.

In chapter 5, we introduced **molarity (M)** as a measure of solution concentration: you should review this material now. There are, in addition, other ways of expressing the concentration of a solution.

Mass Percent (Mass %)

The **mass percent** of a component of a solution is defined as follows:

$$\text{mass \%} = \frac{\text{mass of component}}{\text{mass of solution}} \times 100\%$$

It is essential that the masses of the component and the solution be expressed in the same units.

Sample Problem

Calculate the mass percent of ethanol in an ethanol-water solution if 25.0 grams of ethanol are dissolved in 125 grams of water.

Solution

$$\text{mass \%} = \frac{\text{mass of component}}{\text{mass of solution}} \times 100\% = \frac{25.0 \text{ g}}{25.0 \text{ g} + 125 \text{ g}} \times 100\% = \textbf{16.7\%}$$

Parts per Million (ppm) and Parts per Billion (ppb)

These methods of expressing concentration are quite similar to mass percent and are very useful for describing concentrations in very dilute solutions. Their definitions are:

$$\text{ppm} = \frac{\text{mass of component}}{\text{mass of solution}} \times 10^6$$

$$\text{ppb} = \frac{\text{mass of component}}{\text{mass of solution}} \times 10^9$$

Sample Problem

A 3.50-gram sample of groundwater is found to contain 4.71 micrograms (μg, 10^{-6} g) of Pb^{2+} ion. What is the concentration of the Pb^{2+} ion in ppm?

Solution

This solution is so dilute that we can take the mass of the solution to be the mass of the water, 3.50 grams.

$$\text{ppm} = \frac{\text{mass of component}}{\text{mass of solution}} \times 10^6 = \frac{4.71 \times 10^{-6} \text{ g}}{3.50 \text{ g}} \times 10^6 \text{ ppm} = \textbf{1.35 ppm}$$

Mole Fraction (*X*)

The **mole fraction** of a component of a solution is based entirely of the number of moles of particles present in the solution. It is defined as:

$$X_i = \frac{\text{moles of particles of component } i}{\text{total moles of particles present in the solution}}$$

In a solution consisting of two components, A and B, the mole fraction of component A (X_A) is defined as:

$$X_A = \frac{n_A}{n_A + n_B}$$

Mole fraction has no units assigned to it.

Sample Problem

Calculate the mole fraction of glucose ($C_6H_{12}O_6$, $M = 180.2$ g/mol) in a solution that contains 50.0 grams of glucose dissolved in 200.0 grams of water ($M = 18.02$ g/mol).

Solution

$$X_{glucose} = \frac{n_{glucose}}{n_{glucose} + n_{water}} = \frac{\left(50.0 \text{ g} \cdot \frac{1 \text{ mol}}{180.2 \text{ g}}\right)}{\left(\left(50.0 \text{ g} \cdot \frac{1 \text{ mol}}{180.2 \text{ g}}\right) + \left(200. \text{ g} \cdot \frac{1 \text{ mol}}{18.02 \text{ g}}\right)\right)} = 0.0244$$

On occasion, the mole fraction is multiplied by 100% and the answer is expressed as *mole percent* (mol %). In this problem, the concentration of the glucose is 2.44 mol %.

Molality (*m*)

The **molality** of a solution is defined as the number of moles of solute per kilogram of *solvent*:

$$m = \frac{\text{moles of solute}}{\text{kilograms of solvent}} = \frac{n_{solute}}{kg_{solvent}}$$

Sample Problem

Calculate the molality of the glucose solution given in the last sample problem.

Solution

$$m = \frac{n_{glucose}}{kg_{water}} = \frac{\left(50.0 \text{ g} \cdot \frac{1 \text{ mol}}{180.2 \text{ g}}\right)}{200. \text{ g} \cdot \frac{1 \text{ kg}}{1,000 \text{ g}}} = 1.39 \text{ m}$$

Converting between Molarity and Molality

Since molarity is based on the volume of the solution, while molality is based on the mass of the solvent, one needs to know the density of the solution in order to convert between these two expressions of concentration. The next problem illustrates how this is accomplished.

Sample Problem

A 0.944 M solution of glucose has a density of 1.0624 g/mL. Calculate the molality of this solution.

Solution

The mass of 1.0000 L (1000.0 mL) of this solution is 1062.4 g, and contains 0.944 mole of glucose.

$$180.2 \frac{g}{mol} \cdot 0.944 \frac{mol}{L} = 170.1 = \frac{\text{g glucose}}{\text{L of solution}}$$

$$\text{mass of water} = 1,062.4 \text{ g} - 170.1 = 892.3 \text{ g of water} = 0.8923 \text{ kg}$$

$$m = \frac{0.944 \text{ mol}}{0.8923 \text{ kg}} = 1.06 \text{ } m$$

SOLUBILITY

All gas mixtures and certain combinations of liquids are able to form solutions with an infinite variety of compositions. For example, a solution of ethanol (C_2H_5OH) and water can vary from 0% ethanol–100% water to 100% ethanol–0% water. Such combinations of liquids are said to be **miscible**.

Most solutions, however, have a limit on how much solute will dissolve in a given amount of solvent. This limit is known as the **solubility** of the solute. When a solid such as NaCl is first dissolved in water, the solution is said to be **unsaturated** because its limit has not been reached. When the limit is finally reached, additional solid NaCl will not dissolve, but will appear at the bottom of the solution. This solution is now said to be **saturated**, and a dynamic equilibrium exists between the solid phase and the aqueous phase:

$$NaCl(s) \rightleftharpoons NaCl(aq)$$

The "split" arrow means that the *rate of NaCl(s) that is dissolving into the aqueous phase is equal to the rate of NaCl (aq) that is crystallizing into the solid phase.*

Solubility can be reported in a variety of ways: The solubility of common solids in water is usually reported as *grams of solute per 100 grams of water*. The solubility of common gases in water is usually reported in terms of *molarity*, *ppm*, or as a *mole fraction*.

Solubility and Temperature

The solubility of a substance is generally temperature dependent. The solubilities of most—but not all—solids increase with increasing temperature. The graph below illustrates the solubilities of a number of ionic solids in water as a function of temperature:

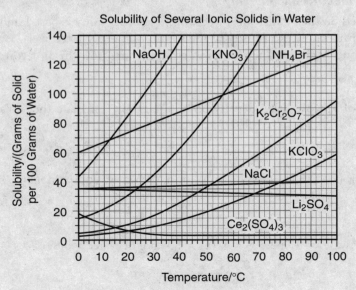

Solubility of Several Ionic Solids in Water

Note that the rates of increase (or decrease) in solubility vary widely: while the solubility curve of KNO_3 rises sharply, the solubility curve of NaCl shows hardly any change over the 100° temperature range.

If we prepare a saturated solution of NH_4Br at 10°C and then raise the temperature to 50°C, the solution will become unsaturated. Conversely, if we prepared the saturated solution at 50°C, and then cooled it to 10°C, we would observe excess solid crystallizing from the solution. However, some solutions, such as sodium acetate ($NaC_2H_3O_2$) in water, will not crystallize on cooling: they form an unstable type of solution that is termed **supersaturated**. Generally, substances that form viscous solutions, such as sugar syrups, form supersaturated solutions fairly readily. If the solution is disturbed, or allowed to stand over time, the excess solute will crystallize.

Gases, however, are a different story: As the temperature of a gas-liquid solution rises, the increase in the average kinetic energy of the gas molecules allows them to escape more readily from the solution: that is, the solubility of a gas *decreases* with increasing temperature. The solubility curve shown below illustrates the solubility of CO_2 in water as a function of temperature when the pressure of the gas above the water is 100 kilopascals:

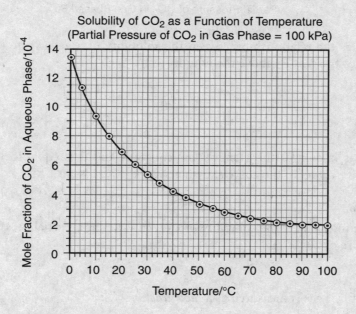

Solubility of CO_2 as a Function of Temperature
(Partial Pressure of CO_2 in Gas Phase = 100 kPa)

Solubility and Pressure

The solubility of gases in a liquid are also pressure-dependent: As the partial pressure of the gas above the liquid increases at constant temperature, the solubility of the gas in the liquid increases *linearly*, as illustrated in the following graph of the solubility of CO_2 in water as a function of the partial pressure of the CO_2 above the water (at a temperature of 20°C).

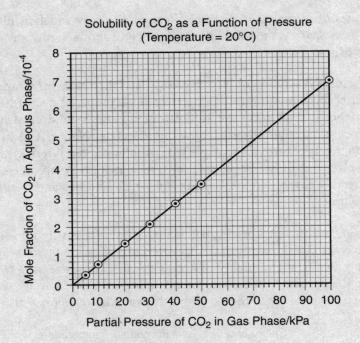

Solubility of CO_2 as a Function of Pressure
(Temperature = 20°C)

Henry's law states the relationship between the solubility of a gas (S) in a liquid and the partial pressure of the gas (P) above the liquid:

$$S = kP$$

The constant k is known as the **Henry's law constant** and it is equal to the *slope* of the pressure-solubility curve. In the curve shown above: $k = 7.04 \times 10^{-6}$ kPa^{-1}.

Sample Problem

Use Henry's law to calculate the solubility of CO_2 in water at a partial pressure of 60.0 kPa and a temperature of 20°C.

Solution

The solubility of this gas is measured as a *mole fraction* (X):

$$S = kP$$

$$X_{CO_2} = (7.04 \times 10^{-6} \text{ kPa}^{-1}) \cdot 60.0 \text{ kPa} = \mathbf{4.24 \times 10^{-4}}$$

Note that this is equal to the value given on the graph above.

THE VAPOR PRESSURE OF LIQUID-LIQUID SOLUTIONS

Ideal Solutions

We began the study of gas behavior in chapter 6 by defining the concept of an *ideal gas*. In chapter 10, we continued the exploration of gases by describing how *real gases* deviated from ideal behavior. In order to begin the study of solutions we need to define the concept of the *ideal solution*. We will assume that our ideal solution consists of two miscible liquids that are *volatile* (that is, each liquid has a measurable vapor pressure). In an ideal solution the solute-solute, solvent-solvent, and solute-solvent interactions are identical with one another. A solution is considered ideal if it meets four criteria:

1. The volume of the solution is equal to the sum of the volumes of the individual liquids. (In the real world, this is not always the case: 1 liter of ethanol and 1 liter of water do not produce 2 liters of solution when mixed.)

2. The enthalpy of solution is zero ($\Delta H_{solution} = 0$).

3. The vapor pressure of each component in the solution is found from a relationship known as **Raoult's law**:

$$P_i = X_i P_i^\circ$$

   where P_i is the vapor pressure of the component in the solution, X_i is the mole fraction of the component, and P_i° is the vapor pressure of the *pure* liquid component at the same temperature. The total vapor pressure of the solution is found by applying Raoult's law to each component, and then adding the individual vapor pressures together. Raoult's law describes the statistical behavior of the components of a solution: the vapor pressure of each component depends on the relative number of component molecules present in the solution. For example, if the mole fraction of component A in an ideal solution is 0.5, then the vapor pressure that A produces will be one-half the vapor pressure of pure A.

4. The *vapor above the solution is ideal*: In particular, Dalton's law of partial pressures is obeyed (see chapter 6).

A graph illustrating the application of Raoult's law to a solution of methanol and ethanol at 20°C is below. The graph is based on the assumption that the solution is ideal over the entire range of composition.

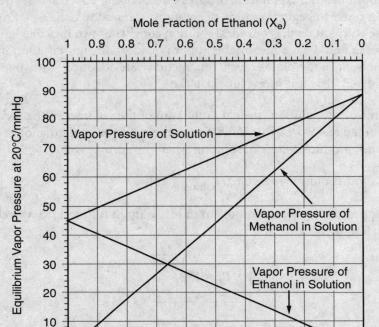

Vapor Pressure of an Ideal Methanol-Ethanol Solution at 20°C
(Raoult's Law)

Inspection of the graph shows that when the mole fraction of methanol is 0.80 the vapor pressure produced by the methanol is approximately 70.2 mmHg; similarly, the mole fraction of ethanol is 0.20 the vapor pressure produced by the methanol is approximately 9.8 mmHg. The total vapor of the solution is approximately 80.0 mm Hg (70.2 mmHg + 9.8 mmHg).

Sample Problem

At 20°C, a nearly ideal solution is prepared by mixing 0.400 mole of methanol and 0.600 mole of ethanol. The vapor pressures of the pure liquids at 20°C are: $P°_{methanol} = 88.7$ mmHg; $P°_{ethanol} = 44.5$ mmHg.

(A) Use Raoult's law to calculate the equilibrium vapor pressure of the solution.

(B) Use Dalton's law of partial pressures to calculate the composition of the *vapor above the solution*.

Solution

In the solution:

(A) $P_{solution} = P_{methanol} + P_{ethanol} = X_{methanol} P^{\circ}_{methanol} + X_{ethanol} P^{\circ}_{ethanol}$

$$= (0.400) \cdot (88.7 \text{ mmHg}) + (0.600) \cdot (44.5 \text{ mmHg})$$

$$= 35.5 \text{ mmHg} + 26.7 \text{ mm Hg} = \textbf{62.2 mmHg}$$

In the vapor:

(B) $P_{methanol} = X_{methanol} P_{total}; P_{ethanol} = X_{ethanol} P_{total}$

$$X_{methanol} = \frac{P_{methanol}}{P_{total}} = \frac{35.5 \text{ mmHg}}{62.2 \text{ mmHg}} = \textbf{0.571}$$

$$X_{ethanol} = \frac{P_{ethanol}}{P_{total}} = \frac{26.7 \text{ mmHg}}{62.2 \text{ mmHg}} = \textbf{0.429}$$

Even though there are more ethanol molecules than methanol molecules present in the solution ($X_{ethanol} = 0.600$), the fraction of the total vapor pressure produced by the ethanol (26.7 mmHg) is *less* than the fraction of the total vapor pressure produced by the methanol (35.5 mmHg). This is easily understood when one considers that pure methanol (VP = 88.7 mmHg) is far more **volatile** than pure ethanol (VP = 44.5 mmHg). As a result of this difference in volatility, the *vapor* has more molecules of methanol than ethanol ($X_{methanol} = 0.571$).

Now suppose that this vapor were *condensed* into a liquid and then allowed to evaporate and come to equilibrium. What would happen is explored in the next problem.

Sample Problem

The newly condensed liquid from the last problem now has the following composition: $X_{methanol} = 0.571$; $X_{ethanol} = 0.429$. The vapor pressures of the pure liquids at 20°C are: $P^{\circ}_{methanol} = 88.7$ mmHg; $P^{\circ}_{ethanol} = 44.5$ mmHg.

(A) Calculate the equilibrium vapor pressure of the solution.

(B) Calculate the composition of the vapor above the solution.

Solution

In the solution:

(A) $P_{solution} = P_{methanol} + P_{ethanol} = X_{methanol} P^{\circ}_{methanol} + X_{ethanol} P^{\circ}_{ethanol}$

$$= (0.571) \cdot (88.7 \text{ mmHg}) + (0.429) \cdot (44.5 \text{ mmHg})$$

$$= 50.6 \text{ mmHg} + 19.1 \text{ mm Hg} = \textbf{69.7 mmHg}$$

In the vapor:

(B) $P_{methanol} = X_{methanol}P_{total}$; $P_{ethanol} = X_{ethanol}P_{total}$

$$X_{methanol} = \frac{P_{methanol}}{P_{total}} = \frac{50.6 \text{ mmHg}}{69.7 \text{ mmHg}} = \mathbf{0.726}$$

$$X_{ethanol} = \frac{P_{ethanol}}{P_{total}} = \frac{19.1 \text{ mmHg}}{69.7 \text{ mmHg}} = \mathbf{0.274}$$

We note that the vapor pressure of the liquid methanol-ethanol mixture has risen from 62.2 mmHg to 69.7 mmHg. We also note that the new vapor is even richer in methanol (the mole fraction has risen from 0.571 to 0.726).

This suggests a way to *separate* liquids based on differences in their volatilities. The graph below compares the composition of a methanol-ethanol mixture in the liquid and vapor phases as a function of vapor pressure:

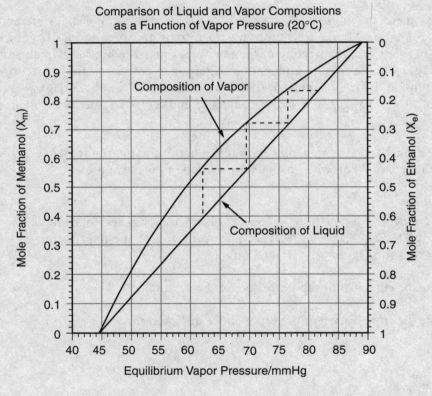

Comparison of Liquid and Vapor Compositions as a Function of Vapor Pressure (20°C)

We note that at a given equilibrium vapor pressure, the vapor phase is always richer in the more volatile component (in this case, methanol). The dashed vertical lines represent these comparisons. The dashed horizontal lines represent the condensation of the vapor into a liquid. The total vapor pressure rises because the liquid is also richer in the more volatile component. The technique of **fractional distillation** involves *repeated* vaporization and condensation, carried out at the normal boiling points of the liquid fractions. As the liquid becomes richer in the more volatile component, its boiling point drops. The number of vaporization-condensation cycles determines the purity of the liquids that are separated.

All of the material in this section has been based on the assumption that the solution in question is ideal. Obviously, that is not the case in the real world. The differences among solute, solvent, and solute-solvent interactions cause many solutions to deviate markedly from Raoult's law: In some cases the total vapor pressure of a solution is less than the sum of the vapor pressures of its components, and in other cases it is greater. Even the technique of fractional distillation has its limits: It is impossible to purify an ethanol-water mixture beyond 95%–5% by mass, and other means must be employed in order to effect the separation.

SOLUTIONS WITH NONVOLATILE SOLUTES

Let us consider the case of a volatile solvent that has a nonvolatile solute ($VP_{solute} = 0$) dissolved in it. The vapor pressure of the solution is determined solely by the vapor pressure of the solvent. Since the mole fraction of the solvent must be less than 1, *the vapor pressure of the solution must be less than the vapor pressure of the pure solvent*. The graph below compares the vapor pressure of pure water (the solvent) with the vapor pressure of an ideal solution containing a nonvolatile solute ($X_{solute} = 0.2$) as a function of temperature.

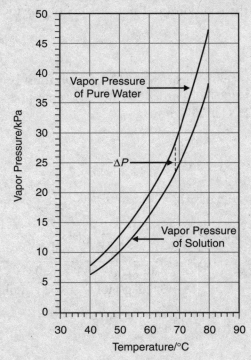

Lowering of Vapor Pressure with a Nonvolatile Solution

We can use Raoult's law to calculate the ΔP, the amount that the vapor pressure is lowered:

$$P_{solution} = P_{solute} + P_{solvent} = 0 + P_{solvent}$$

$$P_{solution} = P_{solvent} = X_{solvent}P^{\circ}_{solvent} \text{ (Raoult's law)}$$

$$\Delta P = P^{\circ}_{solvent} - P_{solution} = P^{\circ}_{solvent} - X_{solvent}P^{\circ}_{solvent} = P^{\circ}_{solvent}(1 - X_{solvent})$$

However, $(1 - X_{solvent}) = X_{solute}$

$$\therefore \Delta P = P^{\circ}_{solvent}X_{solute}$$

The result of these calculations, which appears inside the box, is most important: It says that the vapor pressure lowering depends on *the relative number of solute particles and the vapor pressure of the pure solvent.* We must know the identity of the solvent in order to know its vapor pressure, but we *do not need to know the identity of the solute*, only the number of particles present in the solution. Any property of a solution that depends on the number of solute particles present in the solution (and not the identity of the solute) is known as a **colligative property.**

Sample Problem

How many grams of sucrose (table sugar, $M = 342.3$ g/mol) must be dissolved in 320 grams of water ($M = 18.02$ g/mol) at 25°C in order to lower the vapor pressure by 1.5 mmHg? The vapor pressure of water at 25°C is 23.8 mmHg.

Solution

$$\Delta P = P^{\circ}_{water}X_{sucrose}$$

$$1.5 \text{ mmHg} = (23.8 \text{ mmHg})X_{sucrose}$$

$$X_{sucrose} = 0.063 = \frac{\dfrac{m_{sucrose}}{342.3 \text{ g/mol}}}{\dfrac{m_{sucrose}}{342.3 \text{ g/mol}} + \dfrac{320 \text{ g}}{18.02 \text{ g/mol}}}$$

$$m_{sucrose} = \textbf{410 g}$$

COLLIGATIVE PROPERTIES OF SOLUTIONS

Boiling Point Elevation and Freezing Point Depression

The diagram below superimposes a phase diagram for a solution on a phase diagram for a pure liquid (the solvent) in order to show the effects of vapor pressure lowering.

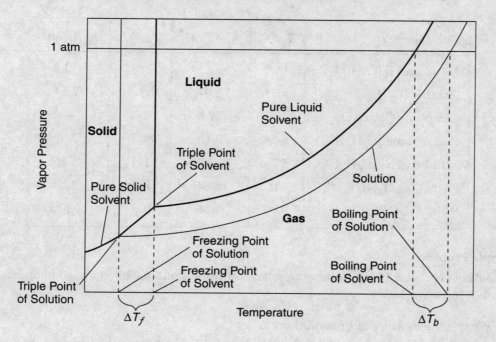

If we examine this diagram, we observe the following:

- The liquid-vapor curve of the solution lies below the liquid-vapor curve for the pure solvent since the vapor pressure of the solution is lower than the vapor pressure of the pure solvent.

- The triple point of the solution has a lower temperature than the triple point of the solvent and, as a result, the solid-liquid line of the solution lies to the left of the solid-liquid line of the solvent.

- The normal boiling point of the solution is *higher* than the normal boiling point of the pure liquid.

- The normal freezing point of the solution is *lower* than the normal freezing point of the pure liquid.

How can we calculate the amount of these colligative properties? Both boiling point elevation and freezing point depression depend on similar relationships:

$$\Delta T_b = k_b m$$
$$\Delta T_f = k_f m$$

The term ΔT is the amount that the boiling point is elevated or the freezing point is lowered; k is a constant that depends on the solvent that is used; m is the molality of the solute. These relationships are not too different from the relationship we derived above for vapor pressure lowering (ΔP): Molality is closely related to mole fraction, and—for small sections of the phase diagram—ΔP and ΔT are nearly proportional. A table of various solvents is given below. The table contains the normal freezing and boiling points (T_f, T_b), as well as k_f and k_b.

Substance	$T_f/°C$	$k_f/C° \, m^{-1}$	$T_b/°C$	$k_b/C° \, m^{-1}$
Acetic acid [$HC_2H_3O_2$]	16.7	3.63	118	3.22
Benzene [C_6H_6]	5.53	5.07	80.1	2.64
Camphor [$C_{10}H_{16}O$]	178	37.8	207	---
Cyclohexane [C_6H_{12}]	6.54	20.8	80.7	2.92
Nitrobenzene [$C_6H_5NO_2$]	5.76	6.87	211	5.20
Phenol [C_6H_5OH]	40.9	6.84	182	3.54
Water [H_2O]	0.00	1.86	100.	0.513

Sample Problem

A solution is prepared by dissolving an unknown compound in 22.0 grams of benzene. The solution freezes at 4.25°C.

(A) What is the molality of this solution?

(B) At what temperature would this solution boil?

(C) What is the molar mass of the compound?

Solution

(A) $\Delta T_f = 5.53°C - 4.25°C = 1.28 \, C°$

$\Delta T_f = k_f m \Rightarrow 1.28 C° = (5.07 \, C°/m) \cdot m$

$m = \mathbf{0.252m}$

(B) $\Delta T_b = k_b \, m = (2.64 \, C°/m) \cdot 0.252 \, m = 0.667 \, C°$

$\Delta T_{b_{solution}} = T_{b_{solvent}} + \Delta T_b = 80.1°C + 0.667 \, C° = \mathbf{80.8°C}$

(C) The *molal* composition of the solution is: $\dfrac{0.252 \, mol_{solute}}{1 \, kg \, benzene}$

The *mass* composition of the solution is: $\dfrac{1.45 \, g_{solute}}{22.0 \, g \, benzene \cdot \left(\dfrac{1 \, kg}{1000 \, g}\right)} = \dfrac{65.9 \, g_{solute}}{1 \, kg \, benzene}$

$\therefore \mathcal{M} = \dfrac{65.9 \, g_{solute}}{0.252 \, mol_{solute}} = \mathbf{262 \, g/mol}$

Osmotic Pressure

The diagram below represents a u-tube in which an aqueous solution of glucose is separated from water by means of a *semipermeable membrane* that allows the passage of water (solvent) molecules, but not glucose (solute) molecules. Initially, the liquid levels on both sides of the tube are equal.

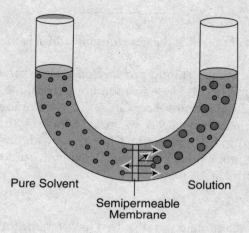

The rate at which water molecules pass through the membrane is dependent on the rate of molecular collisions with the membrane. Since the concentration of water molecules is greater in the *pure solvent*, the water molecules collide more frequently with the water side of the membrane and there is a net flow of water into the solution, a process known as **osmosis**. As a result, the level of the liquid on the solution side begins to rise and the level on the water side falls. The difference in the height of the liquid levels creates a difference in pressure that eventually is sufficient to stop the osmosis. This pressure difference is known as **osmotic pressure** (π) and is shown in the diagram below:

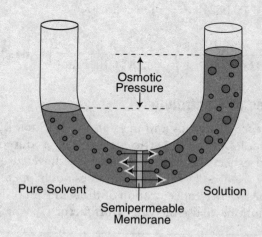

Alternatively, we could think of osmotic pressure as the *additional* pressure needed on the solution side to prevent osmosis from occurring. The osmotic pressure that can be produced by a solution is given by the relationship:

$$\pi = MRT$$

M is the molarity of the solution, R is the gas constant, and T is the absolute temperature.

If two solutions have the same osmotic pressure and are separated by a semipermeable membrane, osmosis will not occur. These two solutions are said to be **isotonic**. The average osmotic pressure of blood is 7.7 atmospheres at 25°C. An isotonic solution of NaCl (known as *normal saline*) is approximately 0.9% by mass. Solutions that have relatively higher osmotic pressures are said to be **hypertonic**, while those with lower osmotic pressures are designated **hypotonic**. Osmotic pressures are particularly useful for calculating large molar masses such as those of proteins.

Sample Problem

When 3.50×10^{-3} gram of a protein was dissolved in water, the total volume of the solution was 5.00×10^{-3} liter. The osmotic pressure of the solution was 1.54 mmHg at 298.15 kelvin. Calculate the molar mass of the protein.

Solution

$$\pi = MRT$$

$$M = \frac{\pi}{RT} = \frac{1.54 \text{ mmHg} \cdot \left(\dfrac{1 \text{ atm}}{760 \text{ mmHg}}\right)}{\left(0.0821 \dfrac{\text{L atm}}{\text{mol K}}\right) \cdot (298.15 \text{ K})} = 8.28 \times 10^{-5} \text{ mol/L}$$

$$\mathcal{M} = \left(\frac{3.50 \times 10^{-3} \text{ g}}{5.00 \times 10^{-3} \text{ L}}\right) \cdot \left(\frac{1 \text{ L}}{8.28 \times 10^{-5} \text{ mol}}\right) = \textbf{8460 g/mol}$$

Colligative Properties of Electrolytes

An **electrolyte** is a substance that yields ions in solutions. If 1 mole of NaCl were dissolved in 1 kilogram of water, 2 moles of solute particles would be produced (1 mole of Na^+ and 1 mole of Cl^-). We would expect the freezing point to be lowered by 2×1.86 C°, and the boiling point elevated by 2×0.513 C°. In fact, this does not happen. A phenomenon known as *ion-pairing* reduces the *effective* concentration of the solute. In order to assess the behavior of *real* electrolytes in solution, we use the **van't Hoff factor** (i), which is defined as follows:

$$i = \frac{\text{observed colligative property of electrolyte}}{\text{colligative property of nonelectrolyte of the same concentration}}$$

For example, if 1.0 m NaCl (aqueous) were a nonelectrolyte, its freezing point would be lowered by 1.86C°. The observed lowering is 3.37 C°. Therefore, the van't Hoff factor is: $i = \dfrac{3.37\,C^\circ}{1.87\,C^\circ} = 1.81$. If NaCl behaved ideally, we would have expected $i = 2.00$. The chart below lists a number of van't Hoff factors for various aqueous solutions as functions of concentration:

van't Hoff Factors (i) and Concentration

Solute	1.00 m	0.100 m	0.0100 m	0.00100 m	Ideal Value
Glucose	1.00	1.00	1.00	1.00	1.00
NaCl	1.81	1.87	1.94	1.97	2.00
$MgSO_4$	1.09	1.21	1.53	1.82	2.00
$Pb(NO_3)_2$	1.31	2.13	2.63	2.89	3.00

We observe that glucose, a nonelectrolyte, maintains its ideal value of 1.00 at all concentrations. The remaining solutes deviate increasingly from the ideal value as the concentration of the solution increases, because the amount of ion pairing increases. Note that $MgSO_4$ shows greater deviation than NaCl, presumably because the doubly charged Mg^{2+} and SO_4^{2-} ions attract more strongly than the singly charged Na^+ and Cl^- ions.

In order to adjust for deviations from ideality, we can incorporate the van't Hoff factor into each of the equations describing colligative properties:

$$\Delta T_f = i k_f m$$
$$\Delta T_b = i k_b m$$
$$\pi = iMRT$$

Sample Problem

Calculate the freezing point depression in a 0.0100 m aqueous solution of $MgSO_4$.

Solution

We use the table to obtain the van't Hoff factor for 0.0100 $MgSO_4$.

$$\Delta T_f = i k_f m = (1.53) \cdot (1.86\ C^\circ/m)(0.0100\ m) = \textbf{0.0285 } C^\circ$$

Chapter 11: Practice Questions

Multiple-Choice Questions

1. Which pair of liquids will be miscible?

   I. C_6H_6 (ℓ) and CCl_4 (ℓ)
   II. C_4H_{10} (g) and H_2O (ℓ)
   III. CH_3COOH (l) and H_2O (ℓ)

   A. I only
   B. II only
   C. III only
   D. I and III only
   E. I, II, and III

2. Arrange these solutes in order of increasing solubility in benzene (C_6H_6): NaI (s); C_2H_5OH (ℓ); C_6H_{14}.

   A. C_2H_5OH (ℓ) < C_6H_{14} < NaI (s)

   B. C_6H_{14} (ℓ) < C_2H_5OH (l) < NaI (s)

   C. NaI (s) < C_6H_{14} < C_2H_5OH (ℓ)

   D. NaI (s) < C_2H_5OH (ℓ) < C_6H_{14}

   E. C_2H_5OH (ℓ) < NaI (s) < C_6H_{14}

3. Rank the solutions in order of decreasing freezing point:

   I.   0.03 molal sodium iodide
   II. 0.04 molal sucrose
   III. 0.04 molal potassium sulfate

   A. I > II > III
   B. II > I > III
   C. II > III > I
   D. III > II > I
   E. III > I > II

4. In calculating the freezing point depression of a potassium carbonate solution, the van't Hoff factor is used. What is the *ideal* value of the van't Hoff factor for potassium carbonate as the solute?

   A. 0
   B. 1
   C. 2
   D. 3
   E. 4

5. Henry's Law constant for O_2 in water at 25°C is 1.26×10^{-3} M/atm. Which is a reasonable constant when the temperature is 50°C?

   I. 6.69×10^{-4} M/atm
   II. 1.26×10^{-3} M/atm
   III. 6.38×10^{-2} M/atm

   A. I
   B. II
   C. III
   D. I and II only
   E. II and III only

6. What factor(s) causes a definite, large number of water molecules to hydrate a cation such as Fe^{3+}?

   I.   Small cations are more effective than large ones.
   II. Large cations are more effective than small ones.
   III. Cations with high charges strongly attract the O atoms of the H_2O molecules.
   IV. Cations with low charges strongly attract the O atoms of the H_2O molecules.

   A. I only
   B. I and III only
   C. I and IV only
   D. II and III only
   E. II and IV only

7. A solution containing 3.86 g of an unknown solute which will not dissociate dissolved in 150. g of ethyl acetate boils at 78.21°C. The normal boiling point of ethyl acetate is 77.06°C, and K_b for ethyl acetate is +2.77°C/m. What is the molar mass of the solute?

   A. 9.30
   B. 16.0
   C. 22.4
   D. 62.0
   E. 71.3

8. No solid is present in a beaker holding a saturated solution of NaCl at 25°C. What can be done to increase the amount of dissolved NaCl in this solution?

   A. Add more solid NaCl.

   B. Raise the temperature of the solution.

   C. Lower the temperature of the solution.

   D. Raise the temperature of the solution and add more solid NaCl.

   E. Lower the temperature of the solution and add more solid NaCl.

9. Which measurement is most suitable for the determination of the molecular weight of oxyhemoglobin, an organic molecule with a molar mass of many thousands?

   A. The osmotic pressure
   B. The vapor pressure lowering
   C. The elevation of the boiling point
   D. The depression of the freezing point
   E. Any of these four are equally as good

10. A solution of CS_2 and acetone, CH_3COCH_3, in which the mole fraction of the CS_2 is 0.25, has a total vapor pressure of 600 mm Hg at 35°C. The vapor pressures of CS_2 and acetone at this temperature are, respectively, 512 mmHg and 344 mmHg. Which is true about CS_2 and acetone solutions?

    A. A mixture of 100.0 mL CS_2 and 100.0 mL CH_3COCH_3 has a volume of 200.0 mL.

    B. A mixture of 100.0 mL CS_2 and 100.0 mL CH_3COCH_3 has a volume significantly less than 200.0 mL.

    C. Raoult's Law is obeyed by both the CS_2 and CH_3COCH_3 in this solution.

    D. When CS_2 and CH_3COCH_3 are mixed, heat is released.

    E. When CS_2 and CH_3COCH_3 are mixed, heat must be supplied in order to produce a solution

Free-Response Questions

1. Knowledge of the solution process allows the explanation of faster chemical reaction rates, the computation of molar masses of experimental substances, and the explanation of common experiences.

   (A) Name three conditions that affect the solubility of a gas such as CO_2 in the cola solution of "soda."

   (B) Why does a cucumber shrivel up (and eventually form a pickle) when it is immersed in a concentrated solution of brine?

   (C) This statement is *false*. Change it to make it true. "Colligative properties such as freezing point depression depend on the nature of the solvent, the nature of the solute, and the concentrations of the solute."

2. Dimethylglyoxime {DMG, $(CH_3CNOH)_2$ (s)} is used as a reagent in qualitative analysis to precipitate the nickel(II) ion. A solution is made by dissolving 57.0 grams of DMG in 430.0 g of ethanol, C_2H_5OH (ℓ).

   (A) What is the mole fraction of DMG in this solution?

   (B) What is the molality of this solution?

   (C) What is the vapor pressure over the solution at ethanol's normal boiling point, 78.4°C?

   (D) What is the boiling point of this solution (DMG does not dissociate)? K_b (ethanol) = +1.22°C/m.

Multiple-Choice Answers

1. D   2. D   3. B   4. D   5. A   6. B   7. D   8. D   9. A   10. E

1. D

A key phrase to remember for this problem is "like dissolves like." Polar compounds are soluble with other polar compounds, and nonpolar compounds are soluble with other nonpolar compounds. Roman numeral II mixes nonpolar C_4H_{10} (g) with polar H_2O (l), and they are immiscible.

2. D

Benzene is a nonpolar hydrocarbon. Hexane, the only nonpolar compound of the three choices, will be the most soluble. The ionic salt NaI is not at all soluble in benzene or other hydrocarbons. The C_2H_5OH is intermediate since it has a polar end, and more importantly, a nonpolar end.

3. B

Increasing the number of particles in a solution causes the freezing point to decrease. The 0.03 m solution of sodium iodide, NaI, has two ions per mole, giving an ion concentration of 0.06 m. This causes a larger decrease in the freezing point than would a 0.04 molal solution of sucrose, since the sucrose does not dissociate. A 0.04 m solution of potassium sulfate, K_2SO_4, contains three ions per mole 2 K^+ and 1 SO_4^-, and an effective ion concentration of 0.12 m.

4. D

The van't Hoff factor is used to account for the number of ions the solute breaks into when dissolving. For example: K_3PO_4 has a van't Hoff factor of 4 (3 K plus 1 PO_4^{3-} ions). Glucose does not dissociate and has a van't Hoff factor of 1. K_2CO_3 has a theoretical van't Hoff factor of 3 since it breaks into 3 ions when dissolved.

5. A

The solubility of a gas *decreases* as the temperature is increased. Henry's law relates the relationship between the solubility of a gas (S) in a liquid and the partial pressure of the gas (P) above the liquid as $S = kP$. The constant, k, is dependent on temperature and decreases with increasing pressure.

6. B

Cations with strong positive charges strongly attract the negatively charged oxygen atom of the polar water molecule. Small ions are more effective than large ones because the charge is more highly concentrated in small ions.

7. D

The boiling point elevation equation is $\Delta T = K_b m$. Substituting the given values into this equation gives $(78.21 - 77.06)\ °C = 2.77°C/m \times \dfrac{(3.86\ g \div (molar\ mass))}{0.150\ kg}$. Solving for the molar mass leads to the answer of 62.0 g/mol. Another approach is to compute the required molality from ΔT and K_b, and multiply this molality by 0.150 kg to give the moles of solute. Then the 3.86 g can be divided by the this number of moles to get the molar mass.

8. D

The solubility of a solid in a liquid *increases* as the temperature increases. Answers (B) and (D) are therefore possibilities, but in order to increase the amount of NaCl you must also remember to add some NaCl.

9. A

Large molecules tend to decompose when subjected to extreme changes in temperature. They also have extremely low vapor pressures, making such measurements inaccurate at best. Osmosis is the movement of solvent molecules from low solute concentration to high solute concentration, and the osmotic pressure, π, is proportional to the molar concentration, M, of the oxyhemoglobin. The molar mass of the solute in a solution containing a known number of grams of solute and giving a measured osmotic pressure is accomplished using the equation $\pi = MRT$.

10. E

The preparation of a solution of a nonpolar solute and a polar solvent is not spontaneous. Heat must be supplied to increase the solubility of the CS_2. When the components are mixed, the volume is usually less, but not significantly less than their total. The solute molecules occupy part of the empty space between the differently shaped and sized solvent molecules. If Raoult's Law is obeyed in this example, the total pressure would have been 386 mmHg, not 600 mmHg. $P_T = (0.25 \times 512) + (0.75 \times 344) = 386$.

Free-Response Answers

1. (A) Pressure, temperature, and the shape and polarity of the solute.

   (B) The water in the cucumber moves to the brine in an effort to dilute it in an osmosis effect.

   (C) "Colligative properties such as freezing point depression depend on the nature of the solute (*does or does not dissociate, van't Hoff constant*), and on the concentrations of the solute (*molality*)."

2. (A) mole fraction $= x = \dfrac{\text{(moles of solute)}}{\text{(moles of solution)}}$

   $$\dfrac{(57.0/116.13)}{(57.0/116.13)} + (430.0/46.07) = \dfrac{0.490}{(0.490 + 9.33)} = 0.0499$$

   (B) molality $= m = \dfrac{\text{moles of solute}}{\text{kg of solvent}} = \dfrac{0.490 \text{ g}}{0.430 \text{ kg}} = 1.14 \text{ m}$

   (C) The vapor pressure is lowered because of the presence of the solute. It is assumed the DMG will contribute an insignificant number of molecules to the vapor phase. Using Raoult's Law:

   $$P_{ethanol} = X_{ethanol} \, P^\circ_{ethanol} = 0.950 \times 760 \text{ mmHg} = 722 \text{ mmHg}$$

   (D) $\Delta T = K_b m = +1.22°C/m \times 1.14 \text{ m} = 1.30°C$

   $T_{final} = 78.4 \,°C + 1.30°C = 79.7°C$

Chemical Kinetics

The area known as **chemical kinetics** focuses on the study of the rates of reactions and the mechanisms by which these reactions occur. In general, a number of factors affect the rate of a reaction:

- The nature of the reactants
- The concentrations of the reactants
- The temperature at which the reaction occurs
- The presence of a catalyst

MEASUREMENT OF REACTION RATES

The rate of a chemical reaction is usually measured by calculating changes in concentration over time. Consider a hypothetical reaction in which a reactant A is converted to a product B: $A \rightarrow B$. The table below provides sample data for this reaction:

Time /s	[A] /M	[B] /M
0	1.70	0.00
50	1.20	0.498
100	0.850	0.850
150	0.601	1.10
200	0.425	1.275
250	0.301	1.99
300	0.212	1.49
350	0.150	1.55
400	0.106	1.59
450	0.0751	1.62
500	0.0531	1.65

The symbols [A] and [B] are used to represent the respective molar concentrations of A and B. Although we used molar concentrations in this example, we could have used any other quantity that is related to the concentration. For example, in a reaction involving gases, we could have tracked the *pressure* in the reaction container over time.

The graph below follows the changes in these concentrations:

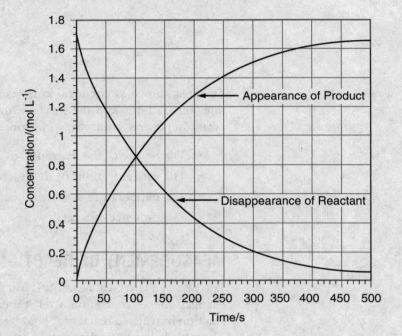

In this reaction, we can measure the rate by considering the rate at which A disappears or the rate at which B appears. Since the coefficients of the reaction are both 1, the numerical values of these rates are equal. For this reaction, we define the **average rate** as:

$$\text{Rate}_{average} = -\frac{\Delta[A]}{\Delta t} = +\frac{\Delta[B]}{\Delta t}$$

The negative sign is used to signify that the reactant (A) disappears, and the positive sign is used to signify that the product (B) appears. The units of reaction rate are generally: concentration/time (such as M/s).

Sample Problem

Calculate the average rate of the disappearance of A between 100 and 400 seconds.

Solution

$$\text{Rate}_{average} = -\frac{\Delta[A]}{\Delta t} = -\frac{0.106 \text{ M} - 0.850 \text{ M}}{400. \text{ s} - 100. \text{ s}} = \textbf{0.00248 M/s}$$

A more useful way to measure the rate of a reaction is to calculate the **instantaneous rate**, which is defined (for this reaction) as:

$$\text{Instantaneous rate} = -\frac{d[A]}{dt} = +\frac{d[B]}{dt}$$

The symbol $\frac{d[\ ]}{dt}$ is known as a *derivative*, and it indicates that the rates have been measured over time intervals that approach 0 second.

Both the average and instantaneous rates of a reaction can be calculated from a concentration versus time graph. The graph below tracks only the disappearance of A as a function of time and displays average and instantaneous rates.

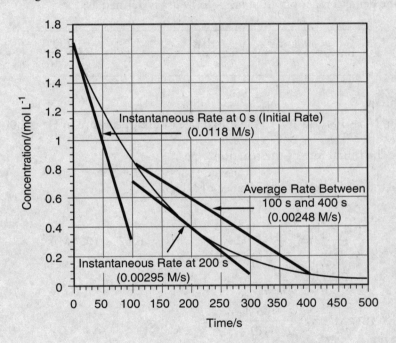

In order to calculate the average rate over a period of time, a straight line is drawn between two points on the graph, in this case between 100 and 400 seconds. The *absolute value* of the slope of this line equals the average rate. In order to calculate the instantaneous rate at a specific moment in time, a tangent line is drawn to the curve at this time. The *absolute value* of the slope of the tangent line is equal to the instantaneous rate. Instantaneous rates are shown for 200 seconds and for 0 seconds. The instantaneous rate at 0 seconds is known as the **initial rate**: it is the rate that is most often used, since the appearance of products can interfere with the study of a reaction.

Reaction Rates and Stoichiometry

Consider the reaction $N_2 + 3H_2 \rightarrow 2NH_3$. In this case the rate of disappearance of N_2 is different from the rate of disappearance of H_2, and both rates are different from the rate of formation of NH_3. This is the result of the stoichiometry of the reaction: For every **1** mole of N_2 that reacts, **3** moles of H_2 react, and **2** moles of NH_3 are formed. In order to provide consistency, the rate of a reaction is defined so that the rate will be the same regardless of the substance whose concentration is measured. In this case:

$$\text{Rate} = -\frac{1}{1}\frac{d[N_2]}{dt} = -\frac{1}{3}\frac{d[H_2]}{dt} = +\frac{1}{2}\frac{d[NH_3]}{dt}$$

That is to say, the rate of this reaction is equal to:

- The rate of disappearance of N_2
- One-third the rate of disappearance of H_2
- One-half the rate of formation of NH_3

The rate of the general reaction $aA + bB \rightarrow cC + dD$ is defined as:

$$\text{Rate} = -\frac{1}{a}\frac{d[A]}{dt} = -\frac{1}{b}\frac{d[B]}{dt} = +\frac{1}{c}\frac{d[C]}{dt} = +\frac{1}{d}\frac{d[D]}{dt}$$

Sample Problem

In the reaction $N_2 + 3H_2 \rightarrow 2NH_3$, the initial rate of disappearance of H_2 is 2.00×10^{-3} M/s.

(A) Calculate the initial rate of the reaction.

(B) Calculate the initial rate of appearance of NH_3.

Solution

(A) $\quad \text{Rate} = -\frac{1}{3}\frac{d[H_2]}{dt} = -\frac{1}{3}(-2.00 \times 10^{-3} \text{ M/s}) = \mathbf{6.67 \times 10^{-4} \text{ M/s}}$

(B) $\quad \text{Rate} = +\frac{1}{2}\frac{d[NH_3]}{dt}$

$\quad\quad 6.67 \times 10^{-4} \text{ M/s} = +\frac{1}{2}\frac{d[NH_3]}{dt}$

$\quad\quad \frac{d[NH_3]}{dt} = \mathbf{1.33 \times 10^{-3} \text{ M/s}}$

Concentration and Reaction Rate: Rate Laws

For every chemical reaction it is possible to write a **rate law** that relates the reaction rate to the concentrations of the reactants. The rate law for the hypothetical reaction $2A + 3B + 4C \rightarrow$ products, takes the form:

$$\text{Rate} = k[A]^x[B]^y[C]^z$$

The terms x, y, and z are known as **reaction orders**. The reaction order refers to the reactant with which it is associated; it may be zero, positive, negative, an integer, or a fraction. We will limit our discussion to reaction orders that are 0, 1, or 2.

- If a reaction is *zero* order with respect to a reactant, it means that the rate does *not* depend on the concentration of that reactant: for example, doubling the concentration will not increase or decrease the rate.

- If a reaction is *first* order with respect to a reactant, it means that the rate is directly proportional to the concentration of that reactant: for example, doubling the concentration will double the rate.

- If a reaction is *second* order with respect to a reactant, it means that the rate is directly proportional to the *square* of the concentration of that reactant: for example, doubling the concentration will quadruple the rate.

- The **overall order** of a reaction is the *sum* of the individual reaction orders: for the reaction shown above, the overall order is $x + y + z$.

The term k is known as the **rate constant**; it relates the concentrations, and their orders, to the rate of the reaction.

It must be emphasized that *all rate laws are experimentally determined*: they cannot as a rule be deduced from the coefficients of the reaction. Let us examine how a rate law is actually determined and a rate constant is calculated. The table below lists data for the initial rate of reaction in the reaction $2NO(g) + O_2(g) \rightarrow 2NO_2(g)$:

Trial	[NO] /M	[O_2] /M	Initial Rate /(M/s)
1	0.0126	0.0125	7.05×10^{-3}
2	0.0252	0.0250	5.64×10^{-2}
3	0.0252	0.0125	2.82×10^{-2}

We begin by writing a tentative rate law: Rate = $k[NO]^x[O_2]^y$. In order to calculate the order of the NO (x), we choose two trials in which the concentration of the other reactant (O_2) is *constant* (trials 1 and 3) and proceed with the following steps:

- Write the rate laws for trials 1 and 3 by substituting the values given in the table (omitting the units for the sake of simplicity):

$$\text{Rate} = k[NO]^x[O_2]^y$$
Trial 1: $7.05 \times 10^{-3} = k(0.0126)^x(0.0125)^y$
Trial 3: $2.82 \times 10^{-2} = k(0.0252)^x(0.0125)^y$

- Divide the rate law for trial 3 (the larger values) by the rate law for trial 1 and solve for x:

$$\frac{2.82 \times 10^{-2}}{7.05 \times 10^{-3}} = \frac{k(0.0252)^x(0.0125)^y}{k(0.0126)^x(0.0125)^y}$$

$$4.00 = \frac{(0.0252)^x}{(0.0126)^x} = \left(\frac{0.0252}{0.0126}\right)^x = 2.00^x$$

$$\therefore x = 2$$

- Choose two trials in which the concentration of NO is constant (trials 2 and 3) and repeat the calculations:

$$\text{Rate} = k[NO]^2[O_2]^y$$
$$\text{Trial 2: } 5.64 \times 10^{-2} = k(0.0252)^2(0.0250)^y$$
$$\text{Trial 3: } 2.82 \times 10^{-2} = k(0.0252)^x(0.0125)^y$$

$$\frac{5.64 \times 10^{-2}}{2.82 \times 10^{-3}} = \frac{k(0.0252)^2(0.0250)^y}{k(0.0252)^2(0.0125)^y}$$

$$2.00 = \frac{(0.0250)^y}{(0.0125)^y} = \left(\frac{0.0250}{0.0125}\right)^y = 2.00^y$$

$$\therefore y = 1$$

The completed rate law for the reaction is: **Rate = $k[NO]^2[O_2]^1$**; the overall order is 3.

In order to determine the value of the rate constant k, substitute the values and units for any single trial into the rate law:

$$\text{Rate} = k[NO]^2[O_2]^1$$
$$5.64 \times 10^{-2} \text{ M/s} = k(0.0252 \text{ M})^2(0.0250 \text{ M})^1$$

$$k = \frac{5.64 \times 10^{-2} \text{ M/s}}{(0.0252 \text{ M})^2(0.0250 \text{ M})^1} = \textbf{3.58} \times \textbf{10}^\textbf{3} \textbf{ M}^{\textbf{-2}}\textbf{s}^{\textbf{-2}}$$

Sample Problem

Determine the rate law and rate constant for the reaction $S_2O_8{}^{2-}$(aq) + $3I^-$(aq) $\rightarrow$ $2SO_4{}^{2-}$ + $I_3{}^-$(aq), given the following data:

Trial	$[S_2O_8{}^{2-}]$ /M	$[I^-]$ /M	Initial Rate /(M/s)
1	0.018	0.036	2.6×10^{-6}
2	0.027	0.036	3.9×10^{-6}
3	0.036	0.054	7.8×10^{-6}
4	0.050	0.072	1.4×10^{-5}

Solution

$$Rate = k[S_2O_8{}^{2-}]^x[I^-]^y$$

Trial 1: $2.6 \times 10^{-6} = k(0.018)^x(0.036)^y$

Trial 2: $3.9 \times 10^{-6} = k(0.027)^x(0.036)^y$

$$\frac{3.9 \times 10^{-6}}{2.6 \times 10^{-6}} = \frac{k(0.027)^x(0.036)^y}{k(0.018)^x(0.036)^y}$$

$$1.5 = \frac{(0.027)^x}{(0.018)^x} = \left(\frac{0.027}{0.018}\right)^x = 1.5^x$$

$$\therefore x = 1$$

$$Rate = k[S_2O_8{}^{2-}]^1[I^-]^y$$

Trial 2: $3.9 \times 10^{-6} = k(0.027)^1(0.036)^y$

Trial 3: $7.8 \times 10^{-6} = k(0.036)^1(0.054)^y$

$$\frac{7.8 \times 10^{-6}}{3.9 \times 10^{-6}} = \frac{k(0.036)^1(0.054)^y}{k(0.027)^1(0.036)^y}$$

$$2.0 = \left(\frac{0.036}{0.027}\right)^1\left(\frac{0.054}{0.036}\right)^y = (1.333)(1.5^y)$$

$$1.5 = 1.5^y$$

$$\therefore y = 1$$

$$\textbf{Rate} = k[S_2O_8{}^{2-}]^1[I^-]^1$$

Trial 4: 1.4×10^{-5} M/s $= k(0.050 \text{ M})^1(0.072 \text{ M})^1$

$$k = \frac{1.4 \times 10^{-5} \text{ M/s}}{(0.050 \text{ M})^1(0.072 \text{ M})^1} = \textbf{3.9} \times \textbf{10}^{-3} \textbf{ M}^{-1}\textbf{s}^{-1}$$

REACTION ORDER: CONCENTRATION-TIME RELATIONSHIPS

For each reaction order there is a unique relationship between the concentration of the reactant and the elapsed time. These relationships allow us to examine a set of concentration-time data and determine the order of the reaction.

A quantity that we shall also examine is known as the **half-life** ($t_{0.5}$) of a reaction: it is the time needed for the concentration of a reactant to decrease to one-half of its initial value. We will see that the mathematical expressions we derive for this quantity depend on the order of a reaction.

First Order Reactions

A **first order reaction** is one whose rate equation has the form: Rate $= k_1[\text{Reactant}]^1$. The rate constant k_1 has the units: time^{-1} (for example, s^{-1}).

If reactant A undergoes a first order reaction, the mathematical form of the rate equation is:

$$-\frac{d[A]}{dt} = k_1[A]$$

This differential equation can be solved by the calculus technique known as *integration*. (If you have not had calculus, skip down to the bottom line in which the solved equation is given in boldface type.)

$$-\frac{d[A]}{dt} = k_1[A]$$

$$\int_{[A]_0}^{[A]_t} \frac{d[A]}{[A]} = \int_0^t k_1 dt$$

$$\ln[A]_t - \ln[A]_0 = -k_1 t$$

$$\boxed{\ln[A]_t = -k_1 t + \ln[A]_0}$$

The quantity $[A]_0$ is the initial concentration of A, and the quantity $[A]_t$ is the concentration of A at time t.

Sample Problem

A first order reaction has a rate constant of 1.00×10^{-3} s^{-1} and the initial concentration of the reactant is 1.50 M. What is the concentration of the reactant at 1000. seconds?

Solution

We omit the concentration units in the calculations since the logarithm of a quantity is never associated with a unit.

$$\ln[A]_1 = -k_1 t + \ln[A]_0 = -(1.00 \times 10^{-3} \text{ s}^{-1}) \cdot (1000. \text{ s}) + \ln(1.50)$$

$$\ln[A]_t = -1.00 + 0.41 = -0.59$$

$$[A]_t = e^{-0.59} = \textbf{0.554 M}$$

We can express the answer with concentration units since the final exponentiation step ($e^{-0.59}$) changes $\ln[A]_t$ back into $[A]_t$.

A graph of [A] versus t yields a curve known as an *exponential decay curve*, but a graph of ln[A] versus t yields a straight line in which the slope $= -k_1$. The data table and graphs presented below illustrate this point:

Time /s	[A] /M	ln[A]
0	1.70	0.531
50	1.20	0.182
100	0.850	−0.163
150	0.601	−0.509
200	0.425	−0.856
250	0.301	−1.20
300	0.212	−1.55
350	0.150	−1.90
400	0.106	−2.24
450	0.0751	−2.59
500	0.0531	−2.94

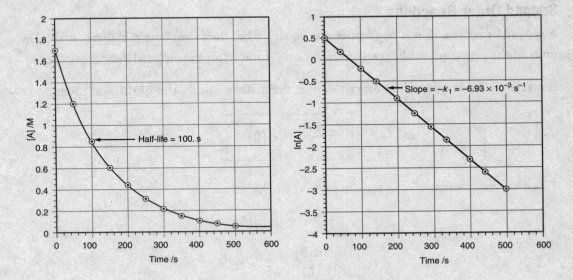

The calculated value of the slope is -6.93×10^{-3}; therefore, the value of $k_1 = 6.93 \times 10^{-3}\ \text{s}^{-1}$.

An important feature of the half-life of a first order reaction is its *constancy*. We can derive an expression for the half-life of a first order reaction by using the integrated rate equation:

$$\ln[A]_t - \ln[A]_0 = -k_1 t$$

$$\ln\left(\frac{[A]_0}{2}\right) - \ln[A]_0 = -k_1 t_{0.5}$$

$$-\ln\frac{1}{2} = -k_1 t_{0.5} \Rightarrow \ln 2 = k_1 t_{0.5}$$

$$\boxed{t_{0.5} = \frac{\ln 2}{k_1}}$$

If we use this relationship with the value of k_1 calculated from the slope of the graph of $\ln[A]$ versus t, we obtain: $t_{0.5} = 100.$ seconds. If we examine the exponential decay curve shown above, we see that the concentration of A is reduced to one-half of the previous value every 100. seconds.

A particularly important application of first order reactions is the study of the decay of radioactive isotopes. This will be examined in chapter 17.

Second Order Reactions

A **second order reaction** is one whose rate equation has the form: Rate = $k_2[\text{Reactant}]^2$. The units of the rate constant k_2 are: concentration^{-1} time^{-1} (for example, $M^{-1}\,s^{-1}$).

If reactant B undergoes a first order reaction, the mathematical form of the rate equation is:

$$\boxed{-\frac{d[B]}{dt} = k_2[B]^2}$$

This differential equation can also be solved by integration:

$$-\frac{d[B]}{dt} = k_2[B]^2$$

$$\int_{[B]_0}^{[B]_t} \frac{d[B]}{[B]^2} = -\int_0^t k_2 dt$$

$$\frac{1}{[B]_0} - \frac{1}{[B]_t} = -k_2 t$$

$$\boxed{\frac{1}{[B]_t} = -k_2 t + \frac{1}{[B]_0}}$$

Sample Problem

A second order reaction has a rate constant of 2.00×10^{-2} $M^{-1}s^{-1}$ and the initial concentration of the reactant is 0.400 M. What is the concentration of the reactant at 500. seconds?

Solution

$$\frac{1}{[B]_t} = k_2t + \frac{1}{[B]_0} = (2.00 \times 10^{-2} \, M^{-1} \, s^{-1}) \cdot (500. \, s) + \frac{1}{0.400 \, M}$$

$$[B]_t = 8.00 \times 10^{-2} \, M$$

A graph of [B] versus t yields a curve, but a graph of $\frac{1}{[B]}$ versus t yields a straight line in which the *slope* $= +k_2$. The data table and graphs presented below illustrate these points:

Time /10^3 s	[B]/M	$\frac{1}{[B]}$ /M^{-1}
0.00	0.0100	100.
1.00	0.00625	160.
1.80	0.00476	210.
2.80	0.00370	270.
3.60	0.00313	320.
4.40	0.00270	370.

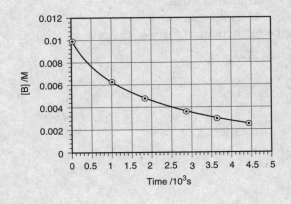
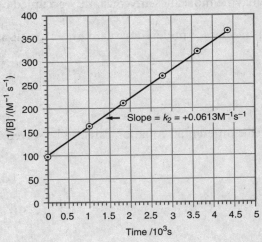

The calculated value of the slope equals $k_2 = 6.13 \times 10^{-2}$ M^{-1} s^{-1}.

We can also derive an expression for the half-life of a second order reaction by using the integrated rate equation:

$$\frac{1}{[B]_t} = +kt + \frac{1}{[B]_0}$$

$$\frac{1}{\dfrac{[B]_0}{2}} = +kt_{0.5} + \frac{1}{[B]_0}$$

$$\frac{2}{[B]_0} - \frac{1}{[B]_0} = kt_{0.5}$$

$$\boxed{t_{0.5} = \frac{1}{k[B]_0}}$$

We observe that the half-life of a second order reaction is dependent on the concentration of the reactant B and, therefore, changes as time progresses. Although we will not do so here, it can be shown that each succeeding half-life of a second order reaction doubles in value.

Zero Order Reactions

A **zero order reaction** is one whose rate is constant and whose rate equation has the form: Rate $= k_0$. The units of the rate constant k_0 are: concentration/time (for example, M/s). Zero order reactions occur in enzyme-catalyzed biological processes in which the active sites of the enzyme molecules are constantly occupied by substrate molecules.

If reactant Z undergoes a zero order reaction, the mathematical form of the rate equation is:

$$\boxed{-\frac{d[Z]}{dt} = k_0}$$

This differential equation can also be solved by integration:

$$-\frac{d[Z]}{dt} = k_0$$

$$\int_{[Z]_0}^{[Z]_t} d[Z] = -\int_0^t k_0 dt$$

$$[Z]_t - [Z]_0 = -k_0 t$$

$$\boxed{[Z]_t = -k_0 t + [Z]_0}$$

Sample Problem

A zero order reaction has a rate constant of 5.00×10^{-4} M/s. If the initial concentration of the reactant is 0.0800 M, at what time does the concentration reach 0.0300 M?

Solution

$$[Z]_t = -k_0 t + [Z]_0$$

$$0.0300 \text{ M} = -(5.00 \times 10^{-4} \text{ M/s})t + (0.0800 \text{ M})$$

$$t = \frac{-0.0500 \text{ M}}{-(5.00 \times 10^{-4} \text{ M/s})} = \textbf{100. s}$$

A graph of [Z] versus t yields a straight line in which the slope = $-k_0$. The data table and graph presented below illustrate these points:

Time / s	[Z] /M
0	0.0800
20.0	0.0700
40.0	0.0600
60.0	0.0500
80.0	0.0400
100.	0.0300
120	0.0200

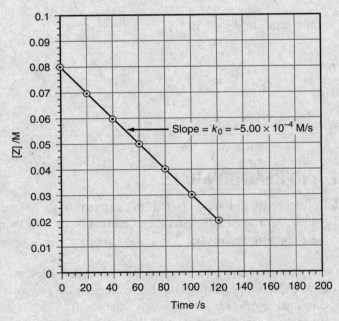

The calculated value of the slope equals $k_0 = 5.00 \times 10^{-4}$ M/s.

We can also derive an expression for the half-life of a second order reaction by using the integrated rate equation:

$$[Z]_t = -kt + [Z]_0$$

$$\frac{[Z]_0}{2} = -kt_{0.5} + [Z]_0$$

$$\frac{[Z]_0}{2} - [Z]_0 = -kt_{0.5}$$

$$\boxed{t_{0.5} = \frac{[Z]_0}{2k}}$$

The half-life of a zero order reaction is also dependent on the concentration of the reactant Z and, therefore, changes as time progresses. Although we will not do so here, it can be shown that each succeeding half-life of a zero order reaction is reduced by a factor of 2.

The table below summarizes first, second, and zero order reactions:

Property	First order	Second order	Zero order
Rate law	Rate = $k_1[A]$	Rate = $k_2[B]^2$	Rate = k_0
Integrated rate law	$\ln[A]_t = -k_1 t + \ln[A]_0$	$\frac{1}{[B]_t} = k_2 t + \frac{1}{[B]_0}$	$[Z]_t = -k_0 t + [Z]_0$
Plot that yields a straight line	$\ln[A]$ versus t	$\frac{1}{[B]}$ versus t	$[Z]$ versus t
Slope of straight line	$-k_1$	$+k_2$	$-k_0$
Half-life ($t_{0.5}$)	$\frac{\ln 2}{k_1}$	$\frac{1}{k_2[B]_0}$	$\frac{[Z]_0}{2k_0}$

TEMPERATURE AND REACTION RATE

The **collision model** of chemical kinetics is based on the premise that *molecules must collide in order to react*. Therefore, an increase in reaction rate must be related to an increase in the *collision frequency* of the reacting molecules.

It is also a well-known fact that the rate of a chemical reaction increases with increasing temperature. In order to explain the relationship between reaction rate and temperature, we need to examine the kinetic-molecular theory of gases (chapter 6): an increase in temperature causes an increase in the speeds and the collision frequency of the molecules.

However, it has been found that the increase in reaction rate is much smaller than the predicted increase in collision frequency. In the latter part of the 19th century, Svante Arrhenius of Sweden proposed that reacting molecules needed to possess a minimum energy, known as the **activation energy** (E_a). The **reaction enthalpy diagram** shown below follows the progress of a hypothetical *endothermic* reaction:

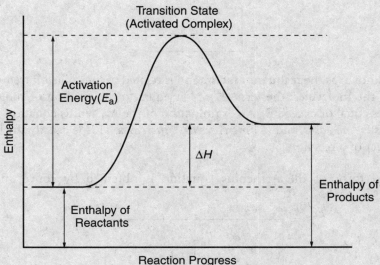

At the left of the diagram, we find the reactants. Those reactant molecules whose kinetic energies equal or exceed E_a form an intermediate **transition state** or **activated complex**. As the energy decreases, the transition state is converted to products. The difference between the final level of the products and the initial level of the reactants is equal to ΔH of the reaction. In this case, the enthalpy of the products is greater than the enthalpy of the reactants and ΔH is positive.

If we examine the molecular distribution diagram shown below, which was first introduced in chapter 6, we observe that an increase in temperature increases the number of molecules whose kinetic energies equal or exceed E_a:

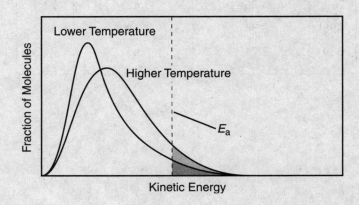

Since any rate law takes the form of: Rate = k[Reactant]^x, we can conclude that *a change in temperature must also affect the rate constant* k. Arrhenius suggested that the rate constant must depend on four factors: (1) the temperature, (2) the activation energy, (3) the collision frequency, and (4) the fraction of molecules whose collisions were *oriented* properly. The **Arrhenius equation** has the form:

$$k = Ae^{-E_a/RT}$$

The term A is known as the **frequency factor** and is related to the collision frequency and the orientation of the molecules; the term $e^{-E_a/RT}$ is equal to the fraction of molecules whose kinetic energies equal or exceed E_a. An examination of the Arrhenius equation reveals that k (1) *decreases as* E_a *increases*, and (2) *increases as* T *increases*. This is exactly the behavior that we would expect of a reaction.

A more useful form of the Arrhenius equation is obtained by converting it into its logarithmic form:

$$\ln k = -\frac{E_a}{R}\left(\frac{1}{T}\right) + \ln A$$

A graph of ln k versus $(1/T)$ yields a straight line whose slope equals $-E_a/R$ and whose y-intercept equals ln A. This is shown in the table and graph for the reaction NO + $O_3 \rightarrow$ NO_2 + 0_2 that follow:

T/K	$1/T/(10^{-3}\,K^{-1})$	$k/(10^9\,M^{-1}\,s^{-1})$	ln k
195	5.13	1.08	20.8
230.	4.35	2.95	21.8
260.	3.85	5.42	22.4
298	3.36	12.0	23.2
369	2.71	35.5	24.3

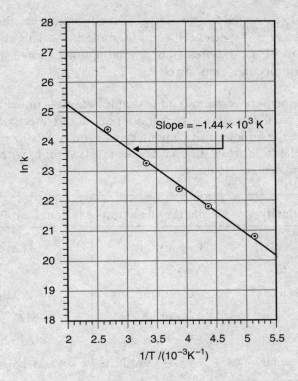

Sample Problem

The slope of the graph shown is -1.44×10^3 K and its y-intercept is 28.1. (The y-intercept is *not* shown on the graph.) Calculate E_a and A.

Solution

The slope of the graph $= -E_a/R$:

$$-1.44 \times 10^3 \text{ K} = -\frac{E_a}{8.315 \text{ J mol}^{-1}\text{K}^{-1}}$$

$$E_a = 1.20 \times 10^4 \text{ J/mol} = \textbf{12.0 kJ/mol}$$

The y-intercept $= \ln A$:

$$A = e^{\ln A} = e^{28.1} = \textbf{1.60} \times \textbf{10}^{\textbf{12}} \textbf{ M}^{-1}\textbf{s}^{-1}$$

Note that A has the same units as the rate constant k.

REACTION MECHANISMS

The **mechanism of a reaction** is a detailed description of the interactions among the reacting molecules. Most reactions occur as a result of molecular collisions and often involve a series of simpler steps known as **elementary reactions**. The number of species that participate in each elementary reaction is known as the **molecularity** of that step. For example, the reaction: $NO(g) + O_3(g) \rightarrow NO_2(g) + O_2(g)$ actually occurs as the result of a single elementary step involving two species (NO and O_3 molecules). This elementary reaction is said to be **bimolecular**. Elementary reactions involving one or three species are said to be **unimolecular** and **termolecular**, respectively.

The rate law for an elementary reaction is quite simple because—unlike a "regular" chemical reaction—no experimental data is necessary: it can be deduced directly from the elementary reaction. For example, the bimolecular elementary reaction $A + B \rightarrow$ products has the rate law: Rate $= k[A][B]$. The table below summarizes the rate laws for hypothetical unimolecular, bimolecular, and termolecular reactions:

Molecularity	Elementary Reaction	Rate Law: Rate =
Unimolecular	$A \rightarrow$ products	$k[A]$
Bimolecular	$2A \rightarrow$ products	$k[A]^2$
Bimolecular	$A + B \rightarrow$ products	$k[A][B]$
Termolecular	$3A \rightarrow$ products	$k[A]^3$
Termolecular	$2A + B \rightarrow$ products	$k[A]^2[B]$
Termolecular	$A + B + C \rightarrow$ products	$k[A][B][C]$

Multistep Mechanisms

It was noted at the beginning of this section that a reaction usually occurs as the result of a series of elementary reactions. This series of simpler steps is known as the **multistep mechanism** of the overall reaction. Let us examine a multistep mechanism that has been proposed for the gas-phase reaction:

$$4HBr + O_2 \rightarrow 2H_2O + 2Br_2$$

Step 1: $HBr + O_2 \rightarrow HOOBr$ \qquad (slow step)

Step 2: $HOOBr + HBr \rightarrow 2HOBr$ \qquad (fast step)

Step 3: $HOBr + HBr \rightarrow H_2O + Br_2$ \qquad (fast step)

The molecules HOOBr and HOBr are known as **intermediates**. Since the overall reaction does not contain any intermediates, these molecules must be consumed as the multistep mechanism progresses. For example, HOOBr is consumed in Step 2, and HOBr is consumed in Step 3.

The three steps must also "add" to produce the overall reaction, including the correct coefficients. This is accomplished if Step 3 is multiplied by 2 before the reactions are "added" together.

The final requirement for a plausible mechanism is that it must produce the experimental rate law of the overall reaction. In this case, the experimental rate law is: Rate $= k[HBr]^1[O_2]^1$.

The Rate-Determining Step

In a multistep mechanism, each step proceeds at a different rate. This is indicated in the mechanism shown above. The slowest step determines the overall rate because, in effect, it creates a "traffic jam." The slowest step of a mechanism is known as the **rate-determining step**. In the mechanism shown above, Step 1 is the rate-determining step. If we write the rate law for this bimolecular elementary reaction, we obtain: Rate = $k[HBr]^1[O_2]^1$, which is in complete agreement with the experimental rate law for the overall reaction..

The rate law of the overall reaction given above was easily reconciled with the proposed mechanism because the first elementary reaction was the rate-determining step.

Now, let us examine a gas-phase reaction that is more complicated:

$$2NO + Br_2 \rightarrow 2NOBr$$

The experimental rate law for this reaction is: Rate = $k[NO]^2[Br_2]^1$. It is tempting to propose that his reaction occurs in a single elementary step: the rate laws certainly agree with one another. However, this would involve a termolecular collision between two NO molecules and a single Br_2 molecule. The probability that this would occur at a reasonable rate is very small. Instead, consider the following two-step mechanism in which the second step is the rate-determining step:

Step 1: $NO + Br_2 \underset{k_{-1}}{\overset{k_1}{\rightleftharpoons}} NOBr_2$ (fast step)

Step 2: $NOBr_2 + NO \xrightarrow{k_2} 2NOBr_2$ (slow step)

If we write the rate law for the slow step, we obtain: Rate = $k[NOBr][NO]$. There are three things wrong with this rate law: (1) It is first order with respect to NO; the experimental rate law is second order with respect to NO. (2) The rate law contains an intermediate, which is not allowed since intermediates never appear in the overall reaction. (3) The rate law does *not* contain a $[Br_2]$ term, which appears in the experimental rate law for the overall reaction.

Let us consider the first step of the mechanism. Note that a split double arrow is associated with it. This means that the rate at which $NOBr_2$ is formed (from NO and Br_2) is equal to the rate at which decomposes (back to NO and Br_2). That is, a state of **rapid equilibrium** exists for this step. Therefore, we can write two rate laws for this step:

$$Rate_{forward} = k_1[NO][Br_2]$$

$$Rate_{reverse} = k_{-1}[NOBr_2]$$

Since the rates are equal, we can equate both terms and rearrange them to obtain:

$$k_1[NO][Br_2] = k_{-1}[NOBr_2]$$

$$[NOBr_2] = \frac{k_1}{k_{-1}}[NO][Br_2]$$

When we insert this result into the rate law for the second elementary step, we obtain:

$$Rate = k_2[NO][NOBr_2]$$

$$[NOBr_2] = \frac{k_1}{k_{-1}}[NO][Br_2]$$

$$Rate = k_2[NO]\frac{k_1}{k_{-1}}[NO][Br_2] = \frac{k_1k_2}{k_{-1}}[NO]^2[Br_2]$$

This result agrees with the rate law for the overall reaction and the rate constant k is given by the combined constants $\frac{k_1k_2}{k_{-1}}$.

Sample Problem

The following two-step mechanism has been proposed for a gas-phase reaction:

Step 1: $H_2 + ICl \rightarrow HI + HCl$ (slow step)

Step 2: $HI + ICl \rightarrow I_2 + HCl$ (fast step)

(A) What is the overall reaction?

(B) What intermediates are present?

(C) What is the molecularity of each elementary step?

(D) Based on the mechanism, what is the expected rate law?

Solution

(A) $H_2 + 2ICl \rightarrow I_2 + 2HCl$

(B) HI is the only intermediate.

(C) Both steps are bimolecular.

(D) Rate = $k[H_2][ICl]$

CATALYSIS

A **catalyst** is a substance that increases the rate of a reaction without being consumed in the reaction. Catalysts are generally classified as either homogeneous or heterogeneous. A **homogeneous catalyst** appears in the same phase as the overall reaction. For example, if a reaction takes place in aqueous solution, a **homogeneous catalyst** will also be present in the aqueous phase. A heterogeneous catalyst exists in a different phase than the reactants. For example, the gas-phase reaction in which ethene (C_2H_4) reacts with H_2 to produce ethane (C_2H_6) is catalyzed with a finely divided metal such as platinum. The metal provides an expanded surface that increases the likelihood of favorable H_2–C_2H_4 collisions.

Even though catalysts are not consumed in a reaction, they most definitely play a part in the mechanism of a reaction. Let us consider the decomposition of H_2O_2, an aqueous-phase reaction that is catalyzed by I^- ions. The two-step mechanism of the catalyzed reaction is:

Step 1: $H_2O_2 + I^- \rightarrow H_2O + IO^-$

Step 2: $H_2O_2 + IO^- \rightarrow H_2O + O_2 + I^-$

Overall: $2H_2O_2 \rightarrow 2H_2O + O_2$

The intermediate is the IO^- ion, which is formed in the first step and consumed in the second. The catalyst, I^-, reacts in the first step, but is regenerated in the second. The first step is the rate-determining step and the rate of the reaction is given by the rate law: Rate = $k[H_2O_2][I^-]$. This is consistent with experimental observations.

Catalysts work by providing an alternative pathway that has the effect of *lowering the activation energy of a reaction*. The diagram below tracks the progress of a hypothetical *exothermic* reaction.

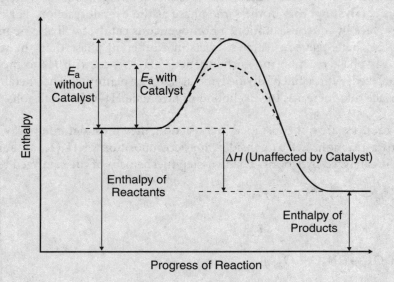

The catalyzed pathway is denoted by the dashed curve and the uncatalyzed pathway, by the solid curve. Note that the enthalpy change of the reaction (ΔH) is unaffected by the presence of the catalyst.

Chapter 12: Practice Questions

Multiple-Choice Questions

1. Nitric oxide reacts with chlorine gas to form nitrosyl chloride. The rate of the reaction doubles when the concentration of chlorine doubles, and quadruples when the concentration of nitric oxide doubles. What is the rate law for the reaction? The balanced equation is:

   $2 NO (g) + Cl_2 (g) \rightarrow 2 NOCl (g)$

   A.   rate = k $[NO][Cl_2]$
   B.   rate = k $[NO]^2[Cl_2]$
   C.   rate = k $[NO]^2[Cl_2]^2$
   D.   rate = k $[NO]^4[Cl_2]$
   E.   rate = k $[NO]^4[Cl_2]^2$

2. A mechanism for the reaction of nitrogen dioxide gas with carbon monoxide to produce nitric oxide gas and carbon dioxide gas is:

   Step 1: $2 NO_2 (g) \rightarrow NO_3 (g) + NO (g)$
   slow

   Step 2: $NO_3 (g) + CO (g) \rightarrow NO (g) + CO_2 (g)$ *fast*

   The first step in this reaction is:

   A.   a catalyst.
   B.   rate-limiting.
   C.   reversible.
   D.   unimolecular.
   E.   zero order.

3. N_2O_5 gas decomposes to nitrogen dioxide and oxygen gases at 55°C. The rate of decomposition of N_2O_5 gas in the first 100.0 s is 3.1×10^{-5} M/s. What is the rate of formation of nitrogen dioxide in the first 100.0 s?

   A. 7.75×10^{-6} M/s
   B. 1.55×10^{-5} M/s
   C. 3.10×10^{-5} M/s
   D. 6.20×10^{-5} M/s
   E. 1.24×10^{-4} M/s

4. In the overall reaction from question 2, at what rate will carbon dioxide gas be produced relative to the rate of consumption of nitrogen dioxide?

   A.   At the same rate
   B.   Twice as fast
   C.   Half as fast
   D.   Four times as fast
   E.   One fourth as fast

5. N_2O_5 gas decomposes to nitrogen dioxide and oxygen gases at 55°C. The rate of the reaction is directly proportional to the concentration of N_2O_5 (g). What is the rate law for the reaction?

   A.   rate = k $[N_2O_5]$
   B.   rate = k $[N_2O_5]^2$
   C.   rate = k $[N_2O_5][NO_2][O_2]$
   D.   rate = k $[N_2O_5]^2[NO_2]^4[O_2]^2$
   E.   The rate law can only be determined experimentally.

6. Which statement about the order of reaction is true?

   A.   A second order reaction is also bimolecular.

   B.   The order of a reaction must be a positive integer.

   C.   The order of a reaction is only determined by experiment.

   D.   The order of a reaction increases with increasing temperature.

   E.   The order of a reaction can be determined from the correctly balanced net ionic equation for the reaction.

7. Which of the statements is (are) true?

    I. Molecules can collide in any orientation, and a reaction occurs if there is enough energy.

    II. Activation energy is the difference in energy between the reactants and the products.

    III. Increasing the temperature decreases the activation energy.

    IV. The Arrhenius equation can be used to determine the activation energy of a reaction.

    A.   IV only
    B.   II and IV only
    C.   I and III only
    D.   III and IV only
    E.   I, II, and III only

8. Which statement is true?

    A.   There is a single rate-determining step in any reaction mechanism.

    B.   Endothermic reactions have higher activation energies than exothermic reactions.

    C.   The rate of a catalyzed reaction is independent of the concentration of the catalyst.

    D.   The specific rate constant of a reaction is independent of the concentrations of the reacting species.

    E.   The rate law for a reaction depends on the concentrations of all reactants that appear in the stoichiometric equation.

9. Nitrogen dioxide gas reacts with carbon monoxide gas to form nitric oxide gas and carbon dioxide gas. The frequency factor in the Arrhenius equation, A, for the reaction is 1.3×10^{10}, and at 700.0 K, the rate constant, k, is $1.3/(M \cdot s)$. What is the activation energy, E_a, for the reaction?

    A.   58 kJ/mol
    B.   15 kJ/mol
    C.   7.6 kJ/mol
    D.   1,600 kJ/mol
    E.   130 kJ/mol

10. In an "iodine-clock" experiment, the time, t, it takes for the blue color of the starch-iodine complex to appear is a measure of the initial rate of formation of 3. The rate of the reaction is studied using the iodine-clock technique, and the time needed to produce the blue color is recorded.

$$3\ I^- + S_2O_8^{2-} \rightleftharpoons I^- + 2SO_4^{2-}$$

Run Number	$[I^-]$ (M)	$[S_2O_8^{2-}]$ (M)	t (S)
1	0.0800	0.0400	44.0
2	0.0800	0.0800	22.1
3	0.1600	0.0200	43.9
4	0.0400	0.0400	88.0

The order of this reaction with respect to $S_2O_8^{2-}$ is:

    A.   0.
    B.   1/2.
    C.   1.
    D.   2.
    E.   3.

11. A reaction is first order in the reactant 1 and second order in reactant 2. When the concentration of reactant 2 is doubled, the rate:

    A.   halves.
    B.   stays the same.
    C.   is increased by a factor of 2.
    D.   is increased by a factor of 4.
    E.   is increased by a factor of 8.

Free–Response Questions

1. Hydrogen peroxide, H_2O_2, decomposes to water and oxygen.

   (A) Given the following data, write the rate law for the decomposition of hydrogen peroxide.

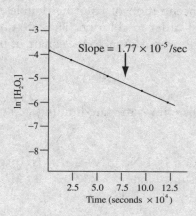

   (B) What is the rate constant for the decomposition of hydrogen peroxide in this experiment?

   (C) What is the half-life of hydrogen peroxide in our experiment?

2. At high temperature, nitric oxide reacts with hydrogen gas to form nitrogen gas and steam.

Experiment	[NO] initial, M	[H_2] initial, M	Rate (M/s)
1	0.25	0.10	0.039
2	0.10	0.10	0.0063
3	0.25	0.30	0.12

   (A) From the data, determine the rate law for the reaction.

   (B) What is the rate constant for the reaction?

   (C) What is the rate of the reaction when the initial concentration of nitrogen monoxide is 0.15 mol/L and the initial concentration of hydrogen is 0.070 mol/L?

   (D) Explain why this reaction is not an elementary reaction.

   (E) A proposed reaction mechanism for this reaction is given. Is this mechanism consistent with the rate law?

   $$2\,NO\,(g) + H_2\,(g) \rightarrow N_2O\,(g) + H_2O\,(g) \qquad slow$$
   $$N_2O\,(g) + H_2\,(g) \rightarrow N_2\,(g) + H_2O\,(g) \qquad fast$$

Multiple-Choice Answers

1. B   2. B   3. D   4. C   5. A   6. C   7. A   8. D   9. E   10. C   11. D

1. B

Since the rate doubles when the concentration of chlorine doubles, rate and concentration are directly proportional, and the reaction is first order in chlorine. The rate quadruples when the concentration of nitric oxide doubles, making the rate is proportional to the square of the NO concentration. The reaction is second order in nitric oxide.

2. B

The slow step in a two-step reaction like this is the rate-limiting step of the reaction.

3. D

The balanced equation for the reaction is $2\ N_2O_5\ (g) \rightleftharpoons 4\ NO_2\ (g) + O_2\ (g)$. The nitrogen dioxide is produced twice as fast as N_2O_5 is consumed. rate $= 2 \times (3.0 \times 10^{-5}\ M/s) = 6.20 \times 10^{-5}\ M/s$

4. C

The nitrogen dioxide is consumed in step 1 and the carbon dioxide is produced in the second step. The molar ratio of $NO_2:CO_2$ is 2:1. This means that for every 2 moles of NO_2 consumed, 1 mole of CO_2 will be produced. CO_2 is therefore being produced half as fast as NO_2 is being consumed.

5. A

Since the rate of the reaction is directly proportional to the concentration of N_2O_5, the reaction is first order in N_2O_5.

6. C

The order of a chemical reaction is determined by experiment, and the order does not necessarily follow the stoichiometry of the reaction. Orders of reaction can have negative values, particularly in the instance of $[OH^-]$. Temperature affects only the rate constant, k, in the rate equation.

7. A

The Arrhenius equation relates the activation energy to the rate constant, the temperature, and the frequency factor, and it can be used to determine the activation energy of a reaction. Molecules must collide in an orientation that allows the atoms to rearrange. The activation energy is the energy difference between the reactants and the energy of the activate complex, not the products. Increasing the temperature increases the number of molecules that have enough energy to react, but the activation energy itself doesn't change

8. D

The rate constant of a reaction is dependent upon the activation energy and the temperature only.

9. E

The correct answer is 130 kJ/mol. Using the Arrhenius equation, it is possible to solve for the activation energy E_a. $\ln(1.3) = -\left(\frac{E_a}{R}\right)\left(\frac{1}{700.0\ K}\right) + \ln(1.3 \times 10^{10})$, so E_a is 130 kJ/mol. Make sure to use the energy form of R = 0.008314 kJ / mol · K.

10. C

In Runs 1 and 2, when the $[S_2O_8^{2-}]$ is doubled, the time of reaction is halved. This means the rate of reaction has doubled, indicating the reaction is first order in $[S_2O_8^{2-}]$. Further investigation of this data will indicate the reaction is also first order in $[I^-]$.

11. D

"Second order" means that whatever change occurs with the concentration has a "squared" effect on the rate. For example, if it were third order, and the concentration increases by a factor of 2, the reaction rate is increased by a factor of $8 = (2)^3$.

Free-Response Answers

1. (A) Since a plot of $\ln[H_2O_2]$ versus time gives a straight line, this is a first order reaction. Therefore rate = k $[H_2O_2]$.

   (B) The integrated first order reaction equation is $\ln [A]_t = -kt + \ln [A]_0$. Therefore the slope of the line equals $-k$. The slope of the line is -1.77×10^{-5} s^{-1}, so the rate constant is 1.77×10^{-5} s^{-1}.

   (C) The half-life of a first order reaction equals $\ln 2 / k$. The rate constant is 1.77×10^{-5} s^{-1}, so the half-life is calculated to be 3.9×10^4 s. This is the same as 653 min, or 10.9 hr.

2. (A) When the concentration of nitric oxide is held constant and the concentration of hydrogen is tripled, the rate triples. Therefore the reaction is first order in hydrogen. When the concentration of nitric oxide is increased by a factor of 2.5, the rate is increased by a factor of 6.3. When we square 2.5, we get 6.3, so the reaction is second order in nitric oxide. This gives a rate law of rate = k $[NO]^2 [H_2]$.

   (B) To find the rate constant, use the data from any of the three experiments, but only the first experiment is used here. The rate, 0.039 mol/L · s, equals the rate constant times (0.25 mol/L)^2 (0.10 mol/L). When solved for the rate constant, k = 6.2 mol^2 / L^2 · s.

   (C) The rate law and rate constant is used to determine the rate at a given concentration of reactants. Rate = (6.2 mol^2 / L^2 · s)(0.015 mol/L)^2(0.0070 mol/L), or rate = 9.8×10^{-3} mol/L · s.

   (D) For an elementary reaction, the order of the reactants is equal to their stoichiometric coefficients. This is because an elementary reaction is a single molecular event. If this is an elementary reaction, the rate law will be rate = $k [NO]^2[H_2]^2$. Since this is not the correct rate law, there must be a different reaction mechanism.

   (E) The rate law for this reaction mechanism is: rate = k $[NO]^2 [H_2]$, which is consistent with the rate law. Both steps of the reaction mechanism add up to the overall reaction, which is essential.

Chemical Equilibrium

CHAPTER THIRTEEN

We begin this chapter with the study of reactions in which the reactants and products are either in the gas phase or dissolved in solution (*homogeneous reactions*). Later, we will study reactions in which pure solids or liquids are also present (*heterogeneous reactions*).

In chapter 12, we examined the changes in concentration of reactants and products over time. Let us consider a simple hypothetical reaction $A(g) \rightarrow B(g)$, which we assume is first order in A. The following table and graph track the disappearance of A and the appearance of B over time.

Time /s	[A] /M	[B] /M
0	0.2000	0.0000
50	0.1213	0.07869
100	0.07358	0.1264
150	0.04463	0.1554
200	0.02707	0.1729
250	0.01642	0.1836
300	0.009957	0.1900
350	0.006039	0.1940
400	0.003663	0.1963
450	0.002222	0.1978
500	0.001348	0.1987
550	0.0008174	0.1992
600	0.0004958	0.1995
650	0.0003007	0.1997
700	0.0001824	0.1998

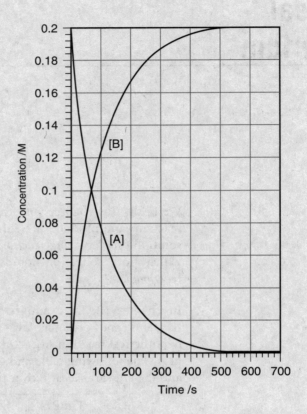

Examination of the table and graph shows that the concentration of *A* approaches 0.0000 M, while the concentration of *B* approaches 0.2000 M.

In most reactions, however, this does *not* occur. The next table and graph illustrate a more likely scenario:

Time /s	[A] /M	[B] /M
0	0.2000	0.00000
50	0.1296	0.07035
100	0.09642	0.1036
150	0.08072	0.1193
200	0.07330	0.1267
250	0.06980	0.1302
300	0.06815	0.1319
350	0.06737	0.1326
400	0.06700	0.1330
450	0.06682	0.1332
500	0.06674	0.1333
550	0.06670	0.1333
600	0.06668	0.1333
650	0.06667	0.1333
700	0.06667	0.1333

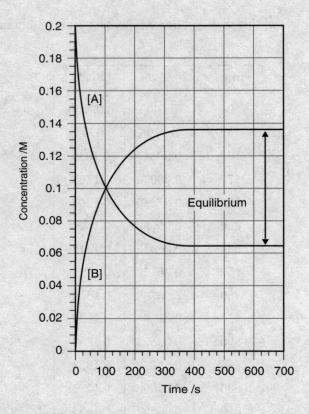

We note that the concentrations of *A* and *B* approach constant nonzero values, beginning at about 400 seconds. When the concentrations of reactants and products become constant over time, we say that the reaction has come to a state of **equilibrium**, and there are no further apparent changes in the system.

Why does this occur? Strictly speaking, equilibrium is the province of *thermodynamics* and complex energy concentrations are needed to describe equilibrium systems. We will begin, however, by taking a *kinetic* approach.

KINETICS AND EQUILIBRIUM

In the last section, we assumed that the reaction A $\rightarrow$ B proceeded in the *forward* direction only. This is not the case: Most chemical reactions are *reversible*, at least to some extent, and we now write the reversible reaction using the "split arrow" notation:

$$A(g) = \underset{k_{-1}}{\overset{k_1}{\rightleftharpoons}} B(g)$$

The rate constants k_1 and k_{-1} are written above and below the split arrow. As time progresses, the rate of *disappearance* of *A* is impeded because the reverse reaction is regenerating this substance. Similarly, the rate of *appearance* of *B* is impeded because it is being converted to *A*.

The following table and graph illustrate how the forward and reverse rates change over a period of time ($k_1 = 0.01000$ s^{-1}; $k_{-1} = 0.005000$ s^{-1}).

Time /s	$Rate_{forward}$ /(M/s)	$Rate_{reverse}$ /(M/s)
0	0.002000	0.0000000
50	0.001296	0.0003518
100	0.0009642	0.0005179
150	0.0008072	0.0005964
200	0.0007330	0.0006335
250	0.0006980	0.0006510
300	0.0006815	0.0006593
350	0.0006737	0.0006632
400	0.0006700	0.0006650
450	0.0006682	0.0006659
500	0.0006674	0.0006663
550	0.0006670	0.0006665
600	0.0006668	0.0006666
650	0.0006667	0.0006666
700	0.0006667	0.0006666

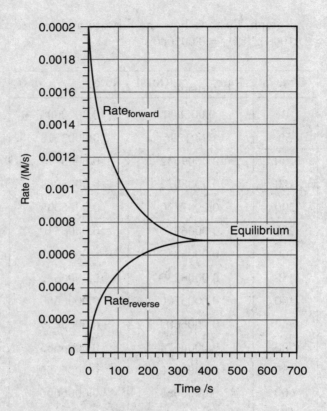

If we examine the table and graph, we see that *when a state of equilibrium is attained, the rates of the forward and reverse reactions are equal*. This occurs at approximately 400 seconds. Therefore, a reaction that is at equilibrium is in a state of **dynamic equilibrium**.

THE EQUILIBRIUM CONSTANT

The Equilibrium Constant and Concentration

Let us examine the table of page 285, once again. We observe that the *ratio of the equilibrium concentrations of B to A is constant*:

$$\frac{[\text{B}]_{eq}}{[\text{A}]_{eq}} = \text{constant}(= 2.00)$$

For the reaction: $w\text{W(g)} + x\text{X(g)} \rightleftharpoons y\text{Y(g)} + z\text{Z(g)}$, we define the **equilibrium constant with respect to concentration** (K_c) as:

$$K_c = \frac{[Y]^y[Z]^z}{[W]^w[X]^x}$$

The ratio on the right side of the equation is known as an **equilibrium constant expression**. Notice that it takes the form of "products" divided by "reactants" and the *coefficients* of the equation appear as *exponents* in this expression.

Sample Problem

What is the equilibrium constant expression for the reaction: $N_2(g) + 3H_2(g) \rightleftharpoons 2NH_3(g)$?

Solution

$$K_c = \frac{[NH_3]^2}{[N_2][H_2]^3}$$

Calculation of K_c

Experiments have shown that the equilibrium constant for a reaction is constant regardless of the initial composition of the chemical system. The only restriction is that the temperature must be held constant during all such determinations. In practice, the units of the reactants, the products, and the equilibrium constant itself are omitted from any calculations.

The table below presents data for the reaction given in the sample problem at a temperature of 773K (500°C):

	$[N_2]$ /M	$[H_2]$ / M	$[NH_3]$ / M	$K_c = \dfrac{[NH_3]^2}{[N_2][H_2]^3}$
Initial	1.00	1.00	0.000	———
Equilibrium	0.921	0.763	0.157	6.02×10^{-2}
Initial	0.000	0.000	1.00	———
Equilibrium	0.399	1.20	0.203	6.02×10^{-2}
Initial	2.00	1.00	3.00	———
Equilibrium	2.59	2.77	1.82	6.02×10^{-2}

In the first experiment, N_2 and H_2 are present initially, but not NH_3; in the second experiment, NH_3 is the only substance present initially; in the third experiment, all three substances are present initially. In every case, equilibrium concentrations are different but the calculated value of the equilibrium constant is the same. Experiments 1 and 2 also show us that equilibrium can be approached from either direction.

Sample Problem

The following equilibrium concentrations were compiled for the reaction: $2NO(g) + O_2(g) \rightleftharpoons 2NO_2(g)$ at 230°C:

$$[NO] = 0.0542 \text{ M}; [O_2] = 0.127 \text{ M}; [NO_2] = 15.5 \text{ M}$$

Calculate the equilibrium constant K_c for this reaction.

Solution

$$K_c = \frac{[NO_2]^2}{[NO]^2[O_2]} = \frac{(15.5)^2}{(0.0542)^2(0.127)} = \textbf{6.44} \times \textbf{10}^5$$

If the coefficients of a reaction are all *multiplied* by the same number n, the new equilibrium constant $K_c{}'$ is related to the original equilibrium constant K_c by the following expression:

$$\boxed{K_c{}' = K_c{}^n}$$

Sample Problem

Use the data of the last problem to calculate the equilibrium constant for the reaction: $4NO(g) + 2O_2(g) \rightleftharpoons 4NO_2(g)$.

Solution

$$K_c{}' = K_c{}^2 = (6.44 \times 10^5)^2 = 4.15 \times 10^{11}$$

If a reaction is reversed, the new equilibrium constant $K_c{}'$ is related to the original equilibrium constant K_c by the following expression:

$$\boxed{K_c{}' = \frac{1}{K_c}}$$

Sample Problem

Calculate the equilibrium constant for the reaction: $2NO_2(g) \rightleftharpoons 2NO(g) + O_2(g)$.

Solution

$$K_c{}' = \frac{1}{K_c} = \frac{1}{6.44 \times 10^5} = \textbf{1.55} \times \textbf{10}^{-6}$$

Suppose that two reactions have respective equilibrium constants of K_1 and K_2. If the two reactions can be added together, then the equilibrium constant for the composite reaction K_c is related to K_1 and K_2 by the expression:

$$\boxed{K_c = K_1 \cdot K_2}$$

Sample Problem

Given the following data:

$$C_2H_2O_4 \rightleftharpoons H^+ + C_2HO_4^- \qquad K_1 = 6.5 \times 10^{-2}$$

$$C_2HO_4^- \rightleftharpoons H^+ + C_2O_4^{2-} \qquad K_2 = 6.1 \times 10^{-5}$$

Calculate the equilibrium constant K_c for the composite reaction: $C_2H_2O_4 \rightleftharpoons 2H^+ + C_2O_4^{2-}$.

Solution

$$K_c = K_1 \cdot K_2 = (6.5 \times 10^{-2}) \cdot (6.1 \times 10^{-5}) = \mathbf{3.97 \times 10^{-6}}$$

Equilibrium Constants and Pressure

When a reaction involves gases, it is possible to express the equilibrium constant in terms of the partial pressures of the gases present. For example, the **equilibrium constant with respect to partial pressure** (K_p) for the reaction $N_2(g) + 3H_2(g) \rightleftharpoons 2NH_3(g)$ is:

$$K_P = \frac{P_{NH_3}^2}{P_{N_2} P_{H_2}^3}$$

We can write this expression involving gases because the molar concentration of an ideal gas is *directly proportional* to its partial pressure:

$$P_i V = n_i RT$$

$$P_i = \frac{n_i RT}{V} = \left(\frac{n_i}{V}\right) RT$$

$$\boxed{P_i = [i]RT}$$

Sample Problem

When the reaction $PCl_5(g) \rightleftharpoons PCl_3(g) + Cl_2(g)$ reaches equilibrium at 250°C, the equilibrium partial pressures of PCl_5, PCl_3, and Cl_2 are 0.875 atm, 0.463 atm, and 1.98 atm, respectively. Calculate K_p for this reaction.

Solution

$$K_P = \frac{P_{PCl_3} P_{Cl_2}}{P_{PCl_5}}$$

$$K_P = \frac{(0.463)(1.98)}{(0.875)}$$

$$K_P = \mathbf{1.05}$$

Relationship Between K_P and K_c

Consider the reaction $wW(g) + xX(g) \rightleftharpoons yY(g) + zZ(g)$ for which:

$$K_P = \frac{P_Y^y P_Z^z}{P_W^w P_X^x} \quad \text{and} \quad K_c = \frac{[Y]^y[Z]^z}{[W]^w[X]^x}$$

The relationship between K_P and K_c is derived as follows:

$$P_i = [i]RT$$

$$K_P = \frac{P_Y^y P_Z^z}{P_W^w P_X^x} = \frac{([Y]RT)^y([Z]RT)^z}{([W]RT)^w([X]RT)^x} = \frac{[Y]^y[Z]^z}{[W]^w[X]^x}(RT)^{y+z-w-x}$$

$$\boxed{K_P = K_c(RT)^{\Delta n}}$$

where Δn ($y+z-w-x$) is the difference in the number of moles between reactants and products.

Sample Problem

Given the following data at 500. K:

(A) $N_2(g) + 3H_2(g) \rightleftharpoons 2NH_3(g)$ $K_c = 62$

(B) $H_2(g) + I_2(g) \rightleftharpoons 2HI(g)$ $K_c = 54$

Calculate K_P for each reaction.

Solution

$K_P = K_c(RT)^{\Delta n}$

$R = 0.0821 \dfrac{\text{L atm}}{\text{mol K}}$

(A) $\Delta n = 2 - 1 - 3 = -2$

$K_P = 62 \cdot (0.0821 \cdot 500.)^{-2} = \mathbf{3.7 \times 10^{-2}}$

(B) $\Delta n = 2 - 1 - 1 = 0$

$K_P = 54 \cdot (0.0821 \cdot 500.)^0 = \mathbf{54}$

Note that in reaction (B) $K_c = K_P$. This is always true in homogeneous reactions when the number of moles of reactants and products are equal.

Heterogeneous Equilibria

When a reaction contains pure solids or pure liquids, these substances never appear as part of the equilibrium constant expression. The concentration of a pure liquid or a pure solid is its mass divided by its own volume (that is, its *density*), and density is *never* altered during a reaction if the temperature remains constant. In practice the concentration of the pure liquid or pure solid is considered to be equal to 1, and only gases and dissolved solutes appear in the equilibrium constant expression.

Sample Problem

Determine the equilibrium constant expressions (K_c) for the following reactions:

(A) $H_2(g) + Br_2(\ell) \rightleftharpoons 2HBr(g)$

(B) $C(s) + \frac{1}{2}O_2(g) \rightleftharpoons CO(g)$

(C) $HC_2H_3O_2(aq) + H_2O(\ell) \rightleftharpoons H_3O^+(aq) + C_2H_3O_2^-(aq)$

(D) $CaCO_3(s) \rightleftharpoons CaO(s) + CO_2(g)$

Solution

(A) $K_c = \dfrac{[HBr]^2}{[H_2]}$

(B) $K_c = \dfrac{[CO]}{[O_2]^{\frac{1}{2}}}$

(C) $K_c = \dfrac{[H_3O^+][C_2H_3O_2^-]}{[HC_2H_3O_2]}$

(D) $K_c = [CO_2]$

THE PROGRESS OF A REACTION TOWARD EQUILIBRIUM

Suppose that some N_2, H_2, and NH_3 are mixed in a container. How can we predict the *direction* in which the chemical reaction $N_2(g) + 3H_2(g) \rightleftharpoons 2NH_3(g)$ moves in order to reach equilibrium? In order to answer this question, we introduce the **reaction quotient** (Q), which is the equilibrium constant expression into which any set of data may be substituted. The difference between Q and the equilibrium constant K is that the expression for K can only contain concentrations or partial pressures that exist at equilibrium, while Q does not have this restriction.

In practice, data values are substituted into the expression for Q and the expression is evaluated. The value of Q is then compared with the value of K. In effect, Q gives us an instantaneous "snapshot" of the system, while K describes the system at equilibrium. The following example shows how the comparison of Q with K is made.

Sample Problem

The equilibrium constant K_c for the reaction $N_2(g) + 3H_2(g) \rightleftharpoons 2NH_3(g)$ is 62 at 500. K. A container contains the following concentrations at 500. K: $[N_2] = 0.070$ M; $[H_2] = 0.12$ M; $[NH_3] = 0.27$ M.

(A) Calculate the value of Q.

(B) Compare the value of Q with the value of K_c ($= 62$).

Solution

$$Q = \frac{[NH_3]^2}{[N_2][H_2]^3}$$

(A) $Q = \dfrac{(0.27)^2}{(0.070)(0.12)^3} = \mathbf{6.0 \times 10^2}$

(B) $\mathbf{Q > K_c}$

What information does this comparison provide? Since Q is greater than K_c, the numerator of the expression for Q is relatively larger than the numerator for the expression for K_c. We can conclude that the system contains too much product (NH_3) to be at equilibrium. If the system is to reach equilibrium, some of the NH_3 will have to be converted to N_2 and H_2, and the reverse reaction will have to be favored over the forward reaction. In the lingo of the chemist, this is known as a *shift to the left*. This shift will continue until equilibrium is attained.

Let us summarize the three possibilities that can occur when Q is compared with K:

- If $Q < K$, the reaction will shift to the *right* in order to produce more *products* until equilibrium is attained.

- If $Q > K$, the reaction will shift to the *left* in order to produce more *reactants* until equilibrium is attained.

- If $Q = K$, there will be *no* shift because the reaction is already at equilibrium.

LE CHÂTELIER'S PRINCIPLE

If a system at equilibrium is disturbed, the system will react in a way that will restore the system to a new equilibrium state. This principle was stated by Henri Le Châtelier during the nineteenth century and is known as **Le Châtelier's principle**. The disturbance to the system is known as a *stress*, and the stress results in a change known as a *shift*. In a chemical reaction, the only possible shifts are to the right (more products are produced), and to the left (more reactants are produced).

Among the possible stresses to a chemical reaction at equilibrium are: changes in concentrations, changes in container volume (of reactions in which reactants or products are gases), dilution (of reactions which take place in solution), changes in temperature, and the addition of a catalyst.

Concentration Changes

There are four possible scenarios:

- The concentration of at least one reactant is increased
- The concentration of at least one reactant is decreased
- The concentration of at least one product is increased
- The concentration of at least one product is decreased

Let us consider a hypothetical reaction at equilibrium: $A(g) + B(g) \rightleftharpoons C(g) + D(g)$. If the equilibrium concentration of reactant A is *increased* to twice its equilibrium value, the system will shift to the *right* in order to *remove* some of the A from the system. We can see why this must be so by comparing the equilibrium constant expression before the application of the stress to the reaction quotient expression after the application of the stress:

$$\text{Before stress: } K_c = \frac{[C]_{eq}[D]_{eq}}{[A]_{eq}[B]_{eq}}$$

$$\text{After stress: } Q_c = \frac{[C]_{eq}[D]_{eq}}{(2[A]_{eq})[B]_{eq}} = \frac{K_c}{2}$$

Since $Q_c < K_c$, the reaction will shift to the right.

When the reaction has returned to equilibrium, it will be a *new* equilibrium point ("You can't go home again!"): There will be increases in [C], [D], and [A] (since not all of the added A will have been removed); only [B] will have decreased. The following table is a summary of possible changes in the equilibrium concentration of reactant A or product D, and the effects these changes have on the reaction given above. (Note: ↑ means increased, and ↓ means decreased.)

Stress	Reaction shifts:	When equilibrium is re-established:
[A]↑	to the right	[A]↑, [B]↓, [C]↑, [D]↑
[A]↓	to the left	[A]↓, [B]↑, [C]↓, [D]↓
[D]↑	to the left	[A]↑, [B]↑, [C]↓, [D]↑
[D]↓	to the right	[A]↓, [B]↓, [C]↑, [D]↓

In every case, however, the value of the equilibrium constant K_c remains *unchanged* if the temperature is held constant.

Volume Changes

When a system containing gases is at equilibrium, a change in the volume of the container is a stress on the system and will result in a shift of the equilibrium point. Let us consider a system at equilibrium:

$$2O_3(g) \rightleftharpoons 3O_2(g)$$

Suppose that the container is compressed so that the volume is reduced to one-half of the original volume. It follows that the *concentrations of both gases will be doubled* $\left(\dfrac{n}{V/2} = 2\dfrac{n}{V}\right)$. Let us now compare the equilibrium constant expression before the volume reduction to the reaction quotient expression after the volume reduction:

$$\text{Before volume reduction: } K_c = \frac{[O_2]^3_{eq}}{[O_3]^2_{eq}}$$

$$\text{After volume reduction: } Q_c = \frac{(2[O_2]_{eq})^3}{(2[O_3]_{eq})^2} = 2K_c$$

Since $Q_c > K_c$, the reaction will shift to the *left*. If we inspect the reaction we see that the left side has fewer gas particles (3 moles) than the right side (2 moles). Therefore, we can draw the following general conclusion: *A reduction in the volume of the container will shift reaction so that the number of gas particles will be reduced. Similarly, an increase in volume will lead to the production of more gas molecules.* This rule cannot be applied to liquids or solids since their concentrations are unaffected by a change in the volume of the container.

Sample Problem

In which direction will a shift occur for the following:

(A) $H_2(g) + I_2(g) \rightleftharpoons 2HI(g)$ Volume is increased

(B) $H_2(g) + I_2(s) \rightleftharpoons 2HI(g)$ Volume is decreased

(C) $N_2(g) + 3H_2(g) \rightleftharpoons 2NH_3(g)$ Volume is decreased

(D) $CaCO_3(s) \rightleftharpoons CaO(s) + CO_2(g)$ Volume is increased

Solution

(A) There is **no shift**. The number of particles is the same on both sides of the equation.

(B) There is a **shift to the left**. There are fewer gas particles on the left side of the equation.

(C) There is a **shift to the right**. There are fewer gas particles on the right side of the equation.

(D) There is a shift **to the right**. There are more gas particles on the right side of the equation.

Addition of Solvent to an Equilibrium System in Solution

Consider the following equilibrium system that exists in aqueous solution at 298 K:

$$HC_2H_3O_2(aq) + H_2O(\ell) \rightleftharpoons H_3O^+(aq) + C_2H_3O_2^-(aq)$$

If the initial concentration of the $HC_2H_3O_2$ is 1.0 M, 0.42% of the molecules are ionized at equilibrium; if the initial concentration is 0.10 M, 1.3% of the molecules are ionized; if the initial concentration is 0.010 M, 4.2% of the molecules are ionized. Why should the addition of H_2O, a pure liquid, shift this reaction to the right?

This effect is analogous to the effect of increasing the container volume in a system containing gas particles. The general rule is that *the addition of pure solvent shifts the system in a direction that will increase the number of dissolved particles*. In this case, there are more dissolved particles on the right side of the equation.

Temperature and Equilibrium

The standard enthalpy change (ΔH°) for the reaction $N_2(g) + 3H_2(g) \rightleftharpoons 2NH_3(g)$ at 298.15 K is −91.8 kJ. Heat is released in the forward reaction and absorbed in the reverse reaction.

We can rewrite this reaction so that it includes the enthalpy change:

$$N_2(g) + 3H_2(g) \rightleftharpoons 2NH_3(g) + 91.8 \text{ kJ}$$

How would this system behave if the temperature were changed? An increase in temperature means that heat was added to the system. The system must shift in a direction that relieves the stress, that is, heat must be absorbed by the system. We conclude that *an increase in temperature favors the endothermic reaction*. A similar argument leads us to the conclusion that *a decrease in temperature favors the exothermic reaction*.

We know that an increase in temperature increases the rates of *all* reactions, including the forward and reverse reactions in the N_2–H_2–NH_3 system, but not to the same extent: Increasing the temperature increases the rate of the endothermic reaction more than it increases the rate of the exothermic reaction. As a result, the shift is toward the endothermic reaction.

If the temperature is decreased, the rates of the forward and reverse reactions will decrease, but the rate of the endothermic reaction will decrease *more* than the rate of the exothermic reaction. As a result, the shift is toward the exothermic reaction.

A change in temperature must, of necessity, also change the equilibrium constant of the reaction. In the N_2–H_2–NH_3 system, an increase in temperature favors the breakdown of NH_3 to N_2 and H_2. When a new equilibrium point is established, the system will contain more N_2 and H_2, and less NH_3 than it did before the temperature increase. By inserting these changes into the equilibrium constant expression:

$$\frac{[NH_3\downarrow]^2}{[N_2\uparrow][H_2\uparrow]^3}$$

we can conclude that the equilibrium constant for this system will *decrease* with an increase in temperature. This reasoning is supported by the data table shown below:

Temperature /K	K_c
298	4.2×10^8
400	4.5×10^4
500	6.2×10^1

Catalysts and Equilibrium

A catalyst increases the rate of a reaction by lowering its activation energy. However, the addition of a catalyst increases the forward and reverse rates *equally*. As a result, a system at equilibrium will remain at equilibrium. (If a system is *not* at equilibrium, the addition of a catalyst will shorten the time needed to attain equilibrium.)

CALCULATIONS INVOLVING EQUILIBRIUM SYSTEMS

In solving equilibrium problems, we must be aware of the balanced equation that represents the system, and we must be able to keep track of changes that occur within the system as it progresses toward equilibrium. A very useful tool is a template that lists all of the information we will need to solve the problem. A sample template for a hypothetical reaction is shown below:

$K_c = 2.0 \times 10^{-4}$	$A(g) + 2B(g) \rightleftharpoons C(g) + 3D(g)$			
Substance:	[A]	[B]	[C]	[D]
Initial concentration:				
Change in concentration:				
Equilibrium concentration:				

An equilibrium problem is solved by completing the template and inserting the equilibrium concentrations into the equilibrium constant expression for the reaction. If the expression contains unknown variables, it will be necessary to solve the (linear, quadratic, ...) equation that arises.

Sample Problem

The reaction $H_2(g) + I_2(g) \rightleftharpoons 2HI(g)$ is carried out at 690. K with the following initial concentrations: $[H_2] = 1.000 \times 10^{-3}$ M; $[I_2] = 2.000 \times 10^{-3}$ M; $[HI] = 0.000$ M. When the reaction reaches equilibrium, $[HI] = 1.880 \times 10^{-3}$ M. Calculate the equilibrium constant K_c for this reaction.

Solution

The initial template is shown below:

K_c = ???	$H_2(g) + I_2(g) \rightleftharpoons 2HI(g)$		
Substance:	$[H_2]$	$[I_2]$	$[HI]$
Initial concentration:	1.000×10^{-3}	2.000×10^{-3}	0.000
Change in concentration:			
Equilibrium concentration:			1.880×10^{-3}

Inspection of the reaction coefficients reveals that for every 1 mole of H_2 (or I_2) reacted, 2 moles of HI will be formed. Therefore, the change in the concentration of HI will always be equal to twice the change in the concentration of H_2 (or I_2). In this case, the change in the concentration of HI is $+1.880 \times 10^{-3}$ ($1.880 \times 10^{-3} - 0.000$). It follows that the change in the concentrations of H_2 and I_2 will each be one-half this quantity: 9.40×10^{-4}. We now proceed to complete the template:

K_c = ???	$H_2(g) + I_2(g) \rightleftharpoons 2HI(g)$		
Substance:	$[H_2]$	$[I_2]$	$[HI]$
Initial concentration:	1.000×10^{-3}	2.000×10^{-3}	0.000
Change in concentration:	-9.40×10^{-4}	-9.40×10^{-4}	$+1.880 \times 10^{-3}$
Equilibrium concentration:	6.0×10^{-5}	1.060×10^{-3}	1.880×10^{-3}

We now calculate the equilibrium constant by substituting the equilibrium constant expression:

$$K_c = \frac{[HI]^2}{[H_2][I_2]} = \frac{(1.880 \times 10^{-3})^2}{(6.0 \times 10^{-5})(1.060 \times 10^{-3})} = \textbf{56}$$

Sample Problem

$PCl_5(g)$ undergoes the following dissociation reaction at 500°C: $PCl_5(g) \rightleftharpoons PCl_3(g) + Cl_2(g)$.

1.000 mole of PCl_5 is placed in a 2.000-liter container. At equilibrium, 19.04% of the PCl_5 molecules have dissociated. Calculate the equilibrium constant for this reaction.

Solution

We will not use the template in solving this problem. At equilibrium, 0.1904 mole of PCl_3 and 0.1904 mole of Cl_2 will have been formed. Therefore, $(1.000 - 0.1904) = 0.810$ mole of PCl_5 will remain at equilibrium. However, *these are not concentrations*: We must divide the number of moles by the container volume (2.00 liters) in order to obtain the concentrations: $[PCl_5] = 0.405$ M; $[PCl_3] = [Cl_2] = 0.0952$ M.

$$K_c = \frac{[PCl_3][Cl_2]}{[PCl_5]} = \frac{(0.0952)^2}{(0.405)} = \textbf{0.0224}$$

Sample Problem

At 690. K, the reaction $H_2(g) + I_2(g) \rightleftharpoons 2HI(g)$ has an equilibrium constant $K_c = 55.6$. A 255.8-gram sample of HI is placed in a 2.000 liter container at 690 K. When equilibrium is attained, what are the equilibrium concentrations of H_2, I_2, and HI?

Solution

We need to find the initial concentration of the HI. First we convert the mass in grams to moles using the molar mass of HI (127.9 g/mol), and we obtain 2.000 moles. Then we divide the number of moles by the volume of the container (2.000 L) to obtain an initial concentration of 1.000 M.

Although the changes in the concentrations of the three substances are not known, we do know (from the equation coefficients) that the change in [HI] must be *twice* the change in $[H_2]$ and $[I_2]$. We assign $-2x$ to the change in [HI] (since HI is breaking down) and $+x$ to the changes in $[H_2]$ and $[I_2]$ (since they are being formed). The completed template is shown below:

$K_c = 55.6$	$H_2(g) + I_2(g) \rightleftharpoons 2HI(g)$		
Substance:	$[H_2]$	$[I_2]$	[HI]
Initial concentration:	0.000	0.000	1.000
Change in concentration:	$+x$	$+x$	$-2x$
Equilibrium concentration:	x	x	$1.000 - 2x$

$$K_c = \frac{[HI]^2}{[H_2][I_2]}$$

$$55.6 = \frac{(1.000 - 2x)^2}{x^2} = \left(\frac{1.000 - 2x}{x}\right)^2$$

$$\sqrt{55.6} = \sqrt{\left(\frac{1.000 - 2x}{x}\right)^2}$$

$$7.46 = \frac{1.000 - 2x}{x}$$

$$x = [H_2] = [I_2] = \textbf{0.106 M}$$

$$1.000 - 2x = [HI] = \textbf{0.789 M}$$

We were able to avoid solving a quadratic equation because the right side of line 2 of the solution was a perfect square. Usually, this will not be the case.

Sample Problem

At a certain temperature, the reaction $2N_2(g) + O_2(g) \rightleftharpoons 2N_2O(g)$ has an equilibrium constant $K_c = 2.0 \times 10^{-13}$. A 3.00-liter container at this temperature contains 0.150 mole of N_2 and 0.270 mole of O_2 initially. Calculate the equilibrium concentrations of N_2, O_2, and N_2O.

Solution

We divide the number of moles by 3.00 liters to obtain the initial concentrations: $[N_2] = 0.050$ M; $[O_2] = 0.090$ M; $[N_2O] = 0.000$ M.

The completed template is shown below:

$K_c = 2.0 \times 10^{-13}$	$2N_2(g) + O_2(g) \rightleftharpoons 2N_2O(g)$		
Substance:	$[N_2]$	$[O_2]$	$[N_2O]$
Initial concentration:	0.050	0.090	0.000
Change in concentration:	$-2x$	$-x$	$+2x$
Equilibrium concentration:	$0.050 - 2x$	$0.090 - x$	$2x$

$$K_c = \frac{[N_2O]^2}{[N_2]^2[O_2]}$$

$$2.0 \times 10^{-13} = \frac{(2x)^2}{(0.050 - 2x)^2(0.090 - x)}$$

When the terms are rearranged, the result is a cubic equation, which can be solved exactly by means of a graphing calculator. However, we can use the very small size of K_c to solve this problem more easily. Since K_c is so small, we can assume that the terms containing x can be neglected in the denominator: $(0.050 - 2x)^2 \approx (0.050)^2$ and $(0.090 - x) \approx (0.090)$. We continue:

$$K_c = \frac{[N_2O]^2}{[N_2]^2[O_2]}$$

$$2.0 \times 10^{-13} = \frac{(2x)^2}{(0.050)^2(0.090)}$$

$$4.50 \times 10^{-17} = 4x^2$$

$$x = 3.35 \times 10^{-9}$$

$$[N_2] = 0.050 - 2x = 0.050 - 6.70 \times 10^{-9} \approx \mathbf{0.050 \ M}$$

$$[O_2] = 0.090 - x = 0.090 - 3.35 \times 10^{-9} \approx \mathbf{0.090 \ M}$$

$$[N_2O] = 2x = \mathbf{6.71 \times 10^{-9} \ M}$$

How do we know when the equilibrium constant is small enough to use approximation methods? A handy rule of thumb is to employ the "5% rule":

- If the equilibrium constant *seems* small to you, assume it is small enough and solve the problem by approximation.

- Compare the value you calculated for x with the value from which x was *neglected*.

- If $x < 5\%$ of this value, your approximation was correct. If not, you must solve the resulting equation exactly.

In this problem, $x = 3.35 \times 10^{-9}$ M. In the term $(0.050 - 2x)$, $x << 5\%$ of 0.050 and it was perfectly acceptable to neglect x.

Sample Problem

At 523 K, the reaction $PCl_5(g) \rightleftharpoons PCl_3(g) + Cl_2(g)$ has an equilibrium constant $K_P = 78.3$. A 0.150-mole sample of PCl_5 is placed in a 5.00-liter container and allowed to come to equilibrium at 523 K. Calculate the total pressure in the container.

Solution

First, we need to calculate the initial partial pressure of the PCl_5, in atmospheres, using the ideal gas law:

$$PV = nRT$$

$$P = \frac{nRT}{V} = \frac{(0.150 \text{ mol})(0.0821 \text{ L atm/ mol K})(523 \text{ K})}{5.00 \text{ L}} = \textbf{1.29 atm}$$

Next, we complete the template and solve the resulting quadratic equation:

$K_P = 78.3$	$PCl_5(g) \rightleftharpoons PCl_3(g) + Cl_2(g)$		
Substance:	P_{PCl_5}	P_{PCl_3}	P_{Cl_2}
Initial partial pressure:	1.29	0.0	0.0
Change in partial pressure:	$-x$	$+x$	$+x$
Equilibrium partial pressure:	$1.29 - x$	x	x

$$K_p = \frac{[P_{PCl_3}][P_{Cl_2}]}{[P_{PCl_5}]}$$

$$78.3 = \frac{x^2}{1.29 - x}$$

$$x = \left\{ {}^{+1.27}_{-79.6} \right\} \Rightarrow \text{Reject the negative root}$$

$$x = P_{PCl_3} = P_{Cl_2} = 1.27 \text{ atm}$$

$$1.29 - x = P_{PCl_5} = 1.29 - 1.27 = 0.02 \text{ atm}$$

$$P_{total} = P_{PCl_5} + P_{PCl_3} + P_{Cl_2} = x + x + (1.29 - x) = 1.29 + x = \textbf{2.56 atm}$$

SOLUBILITY EQUILIBRIA

In chapter 11 we learned that a saturated solution is a system in which the undissolved and dissolved solute are in dynamic equilibrium. In this section, we will study the equilibria that exist in saturated solutions of *sparingly soluble ionic compounds*.

Let us consider the equilibrium that exists in a saturated solution of AgCl, whose equation is:

$$AgCl(s) \rightleftharpoons Ag^+(aq) + Cl^-(aq)$$

Since the AgCl is in the solid phase it does not appear in the equilibrium constant expression:

$$K_{sp} = [Ag^+][Cl^-]$$

The equilibrium constant K_{sp} is known as the **solubility product** because the concentrations of the dissolved ions are multiplied together. The equilibrium constant expression on the right side of the equation is known as the **ion product**.

Sample Problem

Write the equation and the K_{sp} expression for a saturated solution of $Mg(OH)_2$.

Solution

$$Mg(OH)_2(s) \rightleftharpoons Mg^{2+}(aq) + 2OH^-(aq)$$
$$K_{sp} = [Mg^{2+}][OH^-]^2$$

If K_{sp} values of compounds with similar formula types ($AgCl$, CuI, $CaSO_4$) are compared, we can conclude that the smaller the K_{sp} value, the lower the solubility. Care must be taken *not* to compare the K_{sp} values of compounds with *different* formula types ($AgCl$, $Cu(OH)_2$). A table of K_{sp} values is given in Appendix B.

An interesting comparison of solubility equilibria is observed in the following experiment:

A saturated solution of AgCl ($K_{sp} = 1.77 \times 10^{-10}$) contains a white precipitate of solid AgCl. When a solution of I^- ions is added, the white precipitate disappears and is replaced by a yellow precipitate of AgI ($K_{sp} = 8.52 \times 10^{-17}$).

How can this observation be explained? We need to recognize that (1) AgI is less soluble than AgCl, and (2) there are two *simultaneous equilibria* in the solution:

$$AgCl(s) \rightleftharpoons Ag^+(aq) + Cl^-(aq)$$

$$AgI(s) \rightleftharpoons Ag^+(aq) + I^-(aq)$$

As the I^- ions are added to the solution, AgI(s) forms using the Ag^+ ions that are already present in the aqueous phase. Le Châtelier's principle states that the removal of Ag^+ ions will shift the AgCl equilibrium to the right, causing the AgCl(s) to dissolve.

Calculating K_{sp} from Solubility Data

We can use the mass solubility (grams per liter), or the molar solubility (moles per liter) of a saturated solution to calculate the K_{sp} of a sparingly soluble ionic compound. We will assume that all solutions are at 298.15 K. The next two sample problems show how this is done:

Sample Problem

The mass solubility of $CaSO_4$ ($\mathcal{M} = 136.04$ g/mol) is 0.955 g/L. Calculate the K_{sp} of $CaSO_4$.

Solution

$$CaSO_4(s) \rightleftharpoons Ca^{2+}(aq) + SO_4^{2-}(aq)$$

The solution involves three steps:

- Convert mass solubility to molar solubility using the molar mass of the compound.
- Use the balanced equation to recognize that the molar solubility is also the concentration of both ions in solution.
- Substitute the ion concentrations into the K_{sp} expression.

$$0.955 \frac{g}{L} \cdot \left(\frac{1 \text{ mol}}{136.04 \text{ g}} \right) = 7.02 \times 10^{-3} \frac{\text{mol}}{L} = \left[Ca^{2+} \right] = \left[SO_4^{2-} \right]$$

$$K_{sp} = \left[Ca^{2+} \right]\left[SO_4^{2-} \right] = (7.02 \times 10^{-3})^2 = \mathbf{4.93 \times 10^{-5}}$$

Sample Problem

The molar solubility of Ag_2SO_4 is 1.44×10^{-2} mol/L. Calculate the K_{sp} of this compound.

Solution

$$Ag_2SO_4(s) \rightleftharpoons 2Ag^+(aq) + SO_4^{2-}(aq)$$

We can see from the balanced equation that 1 mole of $Ag_2SO_4(s)$ produces 2 moles of $Ag^+(aq)$ and 1 mole of $SO_4^{2-}(aq)$. Therefore, in a saturated solution:

$$[Ag^+] = 2(1.44 \times 10^{-2} \text{ mol/L}) = 2.88 \times 10^{-2} \text{ mol/L}$$

$$[SO_4^{2-}] = 1.44 \times 10^{-2} \text{ mol/L}$$

We now substitute these concentration values into the K_{sp} expression:

$$K_{sp} = [Ag^+]^2[SO_4^{2-}]$$

$$K_{sp} = (2.88 \times 10^{-2})^2(1.44 \times 10^{-2}) = \mathbf{1.19 \times 10^{-5}}$$

Comparison of the solutions of these two sample problems illustrates the dangers of comparing the K_{sp} values of two compounds that have dissimilar formula types: The K_{sp} of Ag_2SO_4 is *smaller* than the K_{sp} of $CaSO_4$, yet the molar solubility of Ag_2SO_4 is *larger* than the molar solubility of $CaSO_4$.

Calculating Molar Solubilities from K_{sp} Values

If we start with the K_{sp}, we can work backwards and calculate the molar (or mass) solubility of the compound.

Sample Problem

Calculate the molar solubility of CuI ($K_{sp} = 1.27 \times 10^{-12}$).

Solution

$$CuI(s) \rightleftharpoons Cu^+(aq) + I^-(aq)$$

We complete the template, assuming that the initial concentrations of the *ions* are 0.00 M. For each 1 mole of CuI that dissolves, 1 mole of Cu^{2+} and 1 mole of I^- are formed. The concentration of solid CuI remains constant throughout the dissolving process:

$K_{sp} = 1.27 \times 10^{-12}$	$CuI(s) \rightleftharpoons Cu^+(aq) + I^-(aq)$		
Substance:	[CuI]	$[Cu^+]$	$[I^-]$
Initial concentration:	constant	0.00	0.00
Change in concentration:	constant	$+x$	$+x$
Equilibrium concentration:	constant	x	x

$$K_{sp} = [Cu^+][I^-]$$
$$1.27 \times 10^{-12} = x^2$$
$$x = [Cu^+] = [I^-] = 1.16 \times 10^{-6} \text{ M}$$
$$\therefore \text{ The molar solubility of CuI is } \mathbf{1.13 \times 10^{-6} \text{ M}}$$

Sample Problem

Calculate the molar solubility of $Zn(OH)_2$ ($K_{sp} = 3.00 \times 10^{-17}$).

Solution

$$Zn(OH)_2(s) \rightleftharpoons Zn^{2+}(aq) + 2OH^-(aq)$$

$K_{sp} = 3.00 \times 10^{-17}$	$Zn(OH)_2(s) \rightleftharpoons Zn^{2+}(aq) + 2OH^-(aq)$		
Substance:	$[Zn(OH)_2]$	$[Zn^{2+}]$	$[OH^-]$
Initial concentration:	constant	0.00	0.00
Change in concentration:	constant	$+x$	$+2x$
Equilibrium concentration:	constant	x	$2x$

$$K_{sp} = [Zn^{2+}][OH^-]^2$$
$$3.00 \times 10^{-17} = (x) \cdot (2x)^2 = 4x^3$$

Solving for x:
$$x = [Zn^{2+}] = 1.96 \times 10^{-6} \text{ M}$$
$$2x = [OH^-] = 3.92 \times 10^{-6}$$

The molar solubility of $Zn(OH)_2 = [Zn^{2+}] = \mathbf{1.96 \times 10^{-6} \text{ M}}$

Why is the molar solubility of the compound equal to the concentration of the Zn^{2+} ion? Refer to the reaction and note that $Zn(OH_2)$ and Zn^{2+} are in a 1:1 mole ratio.

Predicting Whether Precipitation Will Occur

Just as we can write a reaction quotient for other equilibrium reactions, we can write one for solubility equilibria. For example, the reaction quotient for the AgCl equilibrium is:

$$Q_{sp} = [Ag^+][Cl^-]$$

- If $Q_{sp} < K_{sp}$, the solution contains fewer ions than would be present at equilibrium. That is, the solution is *unsaturated*.

- If $Q_{sp} > K_{sp}$, the solution contains more ions than would be present at equilibrium. The solute would crystallize until the ion product equals K_{sp}.

- If $Q_{sp} = K_{sp}$, the system is at equilibrium.

Sample Problem

A 200.-mL solution of 4.00×10^{-3} M $BaCl_2$ is added to a 600.-mL solution of 8.00×10^{-3} M K_2SO_4. Assuming that the volumes are additive, will $BaSO_4$ ($K_{sp} = 1.08 \times 10^{-10}$) precipitate from the solution?

Solution

Calculate the number of mmoles of Ba^{2+} and SO_4^{2-} in the uncombined solutions (volume $\times$ molarity): $n(Ba^{2+}) = 0.800$ mmol; $n(SO_4^{2-}) = 4.80$ mmol.

The volume of the combined solution is 800. mL. Therefore, the ion concentrations are now:

$$[Ba^{2+}] = \frac{0.800 \text{ mmol}}{800. \text{ mL}} = 1.00 \times 10^{-3} \text{ M}$$

$$[SO_4^{2-}] = \frac{4.80 \text{ mmol}}{800. \text{ mL}} = 6.00 \times 10^{-3} \text{ M}$$

We now calculate Q_{sp} and compare it with K_{sp}:

$$Q_{sp} = [Ba^{2+}][SO_4^{2-}] = (1.00 \times 10^{-3}) \cdot (6.00 \times 10^{-3}) = 6.00 \times 10^{-6}$$

Since $Q_{sp} > K_{sp}$, **$BaSO_4$ will precipitate from the solution.**

Sample Problem

Will a precipitate of $Ca(OH)_2$ ($K_{sp} = 5.02 \times 10^{-6}$) form if 2.00 mL of 0.200-M NaOH is added to 1.00×10^3 mL of 0.100-M $CaCl_2$?

Solution

The uncombined solutions contain 0.400 mmol of OH^- and 100. mmol of Ca^{2+}, respectively.

Since we are working with *three* significant figures, we can consider the volume of the combined solution to be 1.00×10^3 mL.

$$[OH^-] = \frac{0.400 \text{ mmol}}{1.00 \times 10^3 \text{ mL}} = 4.00 \times 10^{-4} \text{ M}$$

$$[Ca^{2+}] = \frac{100. \text{ mmol}}{1.00 \times 10^3 \text{ mL}} = 1.00 \times 10^{-1} \text{ M}$$

$$Q_{sp} = [Ca^{2+}][OH^-]^2 = (1.00 \times 10^{-1}) \cdot (4.00 \times 10^{-4})^2 = 1.60 \times 10^{-8}$$

Since $Q_{sp} < K_{sp}$, **$Ca(OH)_2$ will not precipitate from the solution.**

Fractional Precipitation

Fractional precipitation involves the separation of ions using differences in the solubility of their compounds. For example, a solution containing Br^- and Cl^- ions can be separated by adding Ag^+ ions to the solution. This procedure is possible because the K_{sp} values of AgCl and AgBr are 1.77×10^{-10} and 5.35×10^{-13}, respectively. As Ag^+ ion is added to the solution, AgBr, the compound with the smaller K_{sp} begins to precipitate first. After a time, AgCl will begin to precipitate. The sample problem below shows how this is accomplished:

Sample Problem

A solution contains: $[Cl^-] = 2.00 \times 10^{-2}$ M and $[Br^-] = 2.00 \times 10^{-2}$ M. Solid $AgNO_3$ is slowly added to the well-stirred solution.

(A) What is $[Ag^+]$ when AgBr begins to precipitate?

(B) What is $[Ag^+]$ when AgCl begins to precipitate?

(C) What is $[Br^-]$ when AgCl begins to precipitate?

Solution

(A) AgBr begins to precipitate when $[Ag^+][Br^-] = K_{sp}$:

$[Ag^+] \cdot (2.00 \times 10^{-2}) = 5.35 \times 10^{-13}$

$[Ag^+] = \mathbf{2.68 \times 10^{-11} \text{ M}}$

(B) $[Ag^+] \cdot (2.00 \times 10^{-2}) = 1.77 \times 10^{-10}$

$[Ag^+] = \mathbf{8.85 \times 10^{-9} \, M}$

(C) When AgCl begins to precipitate, the solution is still saturated with respect to AgBr ($[Ag^+][Br^-] = K_{sp}$):

$(8.85 \times 10^{-9}) \cdot [Br^-] = 5.35 \times 10^{-13}$

$[Br^-] = \mathbf{6.05 \times 10^{-5} \, M}$

The $[Br^-]$ in part (C) corresponds to 0.30% of the original bromide ion concentration. Therefore, fractional precipitation allowed 99.70% of the bromide ions to be separated from the chloride ions in solution.

Solubility and the Common-Ion Effect

Consider the solubility equilibrium: $AgI(s) \rightleftharpoons Ag^+(aq) + I^-(aq)$ ($K_{sp} = 8.52 \times 10^{-17}$).

In pure water, the solubility of AgI is determined solely by its K_{sp} value. If a *common ion* such as $I^-(aq)$ was added to this solution, the equilibrium would shift *to the left*, causing more AgI(s) to precipitate. This is equivalent to saying that the solubility of AgI would decrease. This phenomenon is known as the **common-ion effect**. We will use it again in chapter 14.

Sample Problem

(A) Calculate the solubility of AgI in pure water.

(B) Calculate the solubility of AgI in a solution containing 1.00×10^{-3} M NaI.

Solution

(A)

$K_{sp} = 8.52 \times 10^{-17}$	$AgI(s) \rightleftharpoons Ag^+(aq) + I^-(aq)$		
Substance:	[AgI]	$[Ag^+]$	$[I^-]$
Initial concentration:	constant	0.00	0.00
Change in concentration:	constant	$+x$	$+x$
Equilibrium concentration:	constant	x	x

$$K_{sp} = [Ag^+][I^-]$$
$$8.52 \times 10^{-17} = x^2$$
$$x = [Ag^+] = [I^-] = 9.23 \times 10^{-9} \, M$$
$$\therefore \text{The molar solubility of AgI is } \mathbf{9.23 \times 10^{-9} \, M}$$

(B) Since the addition of NaI adds *external* I^- ions to the solution, the solubility of AgI is determined solely by $[Ag^+]$.

$K_{sp} = 8.52 \times 10^{-17}$	$AgI(s) \rightleftharpoons Ag^+(aq) + I^-(aq)$		
Substance:	[AgI]	$[Ag^+]$	$[I^-]$
Initial concentration:	constant	0.00	0.00100
Change in concentration:	constant	$+y$	$+y$
Equilibrium concentration:	constant	y	$0.00100 + y$

We know from the value of K_{sp} that y will be very small. Therefore, $0.00100 + y \approx 0.00100$.

$$K_{sp} = [Ag^+][I^-]$$
$$8.52 \times 10^{-17} = y \cdot (0.00100)$$
$$y = [Ag^+] = 8.52 \times 10^{-14} \text{ M}$$

$\therefore$ The molar solubility of AgI is $\mathbf{8.52 \times 10^{-14}}$ **M**

In the presence of 0.00100-M NaI, the solubility of AgI is 108,000 times less than its solubility in pure water!

Complex-Ion Equilibria

In chapter 9 we introduced coordination chemistry and complex ions. The solubility of a substance can be enhanced greatly when a complex ion forms in solution. For example, AgCl is sparingly soluble in water, but it can be made to dissolve when moderately concentrated NH_3 is added to the solution. The overall reaction for this process is:

$$AgCl(s) + 2NH_3(aq) \rightleftharpoons [Ag(NH_3)_2]^+(aq) + Cl^-(aq)$$

We can view this overall equilibrium reaction as being composed of two simultaneous equilibria:

$$AgCl(s) \rightleftharpoons Ag^+(aq) + Cl^-(aq)$$

$$Ag^+(aq) + 2NH_3(aq) \rightleftharpoons [Ag(NH_3)_2]^+(aq)$$

The formation of the complex ion has a very large equilibrium constant and the NH_3 removes the Ag^+ ion from the aqueous phase. As a result, the AgCl equilibrium is shifted to the right and the solid AgCl dissolves.

The equilibrium constant for the formation of a complex ion is known as a **formation constant** (K_f). For example, the overall reaction and formation constant for the diamminesilver(I) ion ($[Ag(NH_3)_2]^+$) is:

$$Ag^+(aq) + 2NH_3(aq) \rightleftharpoons [Ag(NH_3)_2]^+(aq)$$

$$K_f = \frac{[[Ag(NH_3)_2]^+]}{[Ag^+][NH_3]^2}$$

In general, the larger the value of K_f, the more stable the complex ion. The table below lists the formation constants for selected complex ions:

Overall Formation Reaction	K_f
$Ag^+(aq) + 2CN^-(aq) \rightleftharpoons [Ag(CN)_2]^-(aq)$	5.6×10^8
$Ag^+(aq) + 2NH_3(aq) \rightleftharpoons [Ag(NH_3)_2]^+(aq)$	1.6×10^7
$Au^+(aq) + 2CN^-(aq) \rightleftharpoons [Au(CN)_2]^-(aq)$	2.0×10^{38}
$Cu^{2+}(aq) + 4NH_3(aq) \rightleftharpoons [Cu(NH_3)_4]^{2+}(aq)$	1.2×10^{13}
$Hg^{2+}(aq) + 4Cl^-(aq) \rightleftharpoons [Hg(Cl)_4]^{2-}(aq)$	1.2×10^5
$Fe^{2+}(aq) + 6CN^-(aq) \rightleftharpoons [Fe(CN)_6]^{4-}(aq)$	7.7×10^{36}
$Ni^{2+}(aq) + 6NH_3(aq) \rightleftharpoons [Ni(NH_3)_6]^{2+}(aq)$	5.6×10^8

The following sample problem explores the equilibrium state of a complex ion.

Sample Problem

A solution is prepared that has the following initial concentrations: $[Cu^{2+}] = 0.20$ M; $[NH_3] = 1.20$ M.

The relevant reaction is: $Cu^{2+}(aq) + 4NH_3(aq) \rightleftharpoons [Cu(NH_3)_4]^{2+}(aq)$. When equilibrium is reached, what is $[Cu^{2+}]$ in the solution?

Solution

$K_f = 1.2 \times 10^{13}$	$Cu^{2+}(aq) + 4NH_3(aq) \rightleftharpoons [Cu(NH_3)_4]^{2+}(aq)$		
Substance:	$[Cu^{2+}]$	$[NH_3]$	$[[Cu(NH_3)_4]^{2+}]$
Initial concentration:	0.20	1.20	0.00
Change in concentration:	$-x$	$-4x$	$+x$
Equilibrium concentration:	$0.20 - x$	$1.20 - 4x$	x

$$Cu^{2+}(aq) + 4NH_3(aq) \rightleftharpoons [Cu(NH_3)_4]^{2+}(aq)$$

$$K_f = \frac{\left[[Cu(NH_3)_4]^{2+}\right]}{[Cu^{2+}][NH_3]^4}$$

$$1.2 \times 10^{13} = \frac{x}{(0.20 - x)(1.20 - 4x)^4}$$

We can estimate the answer of this fifth degree equation without actually solving it: The very large value of K_f infers that the forward reaction goes essentially to completion. If $[Cu^{2+}] \approx 0$ at equilibrium, then $x \approx 0.20$ M (and $[NH_3] \approx 1.20$ M $- 4(0.20$ M$) = 0.40$ M). This is not a very satisfying answer: we should be able to provide a better estimate of the $[Cu^{2+}]$ at equilibrium.

In order to accomplish this, we use an old trick. We assume that the reaction goes to completion, and then we work backwards, that is, we assume that the reaction proceeds to the *left*: In this case, the "initial" concentrations are as follows: $[Cu^{2+}] = 0$; $[NH_3] = 0.40$ M; $[[Cu(NH_3)_4]^+] = 0.20$ M.

$K_f = 1.2 \times 10^{13}$	$Cu^{2+}(aq) + 4NH_3(aq) \rightleftharpoons [Cu(NH_3)_4]^{2+}(aq)$		
Substance:	$[Cu^{2+}]$	$[NH_3]$	$[[Cu(NH_3)_4]^{2+}]$
Initial concentration:	0.00	0.40	0.20
Change in concentration:	$+y$	$+4y$	$-y$
Equilibrium concentration:	y	$0.40 + 4y$	$0.20 - y$

$$Cu^{2+}(aq) + 4NH_3(aq) \rightleftharpoons [Cu(NH_3)_4]^{2+}(aq)$$

$$K_f = \frac{\left[[Cu(NH_3)_4]^{2+}\right]}{[Cu^{2+}][NH_3]^4}$$

$$1.2 \times 10^{13} = \frac{0.20 - y}{(y)(0.40 + 4y)^4}$$

This fifth degree equation certainly doesn't look any better than the first one, but it is!!! The variable y now represents the $[Cu^{2+}]$ at equilibrium, and we know from above that it must be very small. Therefore, we can neglect y in the other two terms:

$$1.2 \times 10^{13} = \frac{0.20 - y}{(y)(0.40 + 4y)^4} = \frac{0.20}{(y)(0.40)^4}$$

$$y = [Cu^{2+}] = 6.5 \times 10^{-13} \text{ M}$$

Why does this method work? Recall that approaching equilibrium from either direction is an entirely equivalent operation.

Complex Ions and Solubility

Complexing agents such as NH_3 (known also as **ligands**) can be used to control the precipitation of sparingly soluble ionic compounds.

Sample Problem

What is the minimum $[NH_3]$ necessary to prevent precipitation of AgCl in a solution that contains $[Ag^+] = 0.10$ M and $[Cl^-] = 0.010$ M?

$$AgCl(s) \rightleftharpoons Ag^+(aq) + Cl^-(aq); K_{sp} = 1.77 \times 10^{-10}$$

$$Ag^+(aq) + 2NH_3(aq) \rightleftharpoons [Ag(NH_3)_2]^+(aq); K_f = 1.6 \times 10^7$$

Solution

If no precipitation is to occur, $[Cl^-]$ must be maintained at 0.010 M.

$$K_{sp} = [Ag^+][Cl^-]$$

$$1.77 \times 10^{-10} = [Ag^+](0.010)$$

$$[Ag^+] = 1.77 \times 10^{-8} \text{ M}$$

We use the same approach as we did in the last problem, only the initial $[NH_3]$ is not known and the equilibrium concentration of $[Ag^+]$ must be 1.77×10^{-8} M:

$K_f = 1.6 \times 10^7$	$Ag^+(aq) + 2NH_3(aq) \rightleftharpoons [Ag(NH_3)_2]^+(aq)$		
Substance:	$[Ag^+]$	$[NH_3]$	$[[Ag(NH_3)_2]^+]$
Initial concentration:	0.00	z	0.10
Change in concentration:	$+1.77 \times 10^{-8}$	$+2(1.77 \times 10^{-8})$	-1.77×10^{-8}
Equilibrium concentration:	1.77×10^{-8}	$\approx z$	≈ 0.10

Since 1.77×10^{-8} M is a very small concentration in comparison with 0.10 M, we can ignore its contribution in the equilibrium concentrations of NH_3 and $[Ag(NH_3)_2]^+$.

$$Ag^+(aq) + 2NH_3(aq) \rightleftharpoons [Ag(NH_3)_2]^+(aq)$$

$$K_f = \frac{\left[[Ag(NH_3)_2]^+\right]}{[Ag^+][NH_3]^2}$$

$$1.2 \times 10^{13} = \frac{0.10}{(1.77 \times 10^{-8})(z)^2}$$

$$z = [NH_3] = 0.59 \text{ M}$$

The concentration 0.59 M represents the *free* NH_3 in the solution. Since the equilibrium concentration of $[Ag(NH_3)_2]^+$ is 0.10 M, the concentration of NH_3 *within the complex ion* is 0.20 M. Therefore,

$$[NH_3]_{total} = 0.59 \text{ M} + 0.20 \text{ M} = \textbf{0.79 M}$$

We have seen that the formation of a complex ion increases the solubility of a sparingly soluble ionic compound. The next problem demonstrates how this solubility is calculated.

Sample Problem

The molar solubility of AgCl in pure water is 1.33×10^{-5} M. Calculate the molar solubility of AgCl in 0.200-M NH_3.

Solution

$$AgCl(s) \rightleftharpoons Ag^+(aq) + Cl^-(aq); K_{sp} = 1.77 \times 10^{-10}$$

$$Ag^+(aq) + 2NH_3(aq) \rightleftharpoons [Ag(NH_3)_2]^+(aq); K_f = 1.6 \times 10^7$$

We can combine these two simultaneous equilibria to produce a single reaction with a combined equilibrium constant (K):

$$AgCl(s) + 2NH_3(aq) \rightleftharpoons [Ag(NH_3)_2]^+(aq) + Cl^-(aq)$$

$$K = K_{sp} \cdot K_f = (1.77 \times 10^{-10}) \cdot (1.6 \times 10^7) = 2.8 \times 10^{-3}$$

Assume that x moles of AgCl dissolves per liter of solution:

$K_f = 2.8 \times 10^{-3}$	$AgCl(s) + 2NH_3(aq) \rightleftharpoons [Ag(NH_3)_2]^+(aq) + Cl^-(aq)$			
Substance:	[AgCl]	[NH_3]	[$[Ag(NH_3)_2]^+$]	[Cl^-]
Initial concentration:	constant	0.200	0.00	0.00
Change in concentration:	constant	$-2x$	$+x$	$+x$
Equilibrium concentration:	constant	$0.200 - 2x$	x	x

$$AgCl(s) + 2NH_3(aq) \rightleftharpoons [Ag(NH_3)_2]^+(aq) + Cl^-(aq)$$

$$K_f = \frac{\left[[Ag(NH_3)_2]^+\right][Cl^-]}{[NH_3]^2}$$

$$2.8 \times 10^{-3} = \frac{x^2}{(0.200 - 2x)^2} = \left(\frac{x}{0.200 - 2x}\right)^2$$

$$\sqrt{2.8 \times 10^{-3}} = \sqrt{\left(\frac{x}{0.200 - 2x}\right)^2}$$

$$5.3 \times 10^{-2} = \frac{x}{0.200 - 2x}$$

$$x = \text{molar solubility of AgCl} = \mathbf{9.6 \times 10^{-3}\ M}$$

Chapter 13: Review Questions

Multiple-Choice Questions

1. Which statements is (are) true?

   I. Chemical equilibrium requires reversible chemical reactions.
   II. Reversibility requires that all reactants and products be present.
   III. Reversibility requires that the rates of the forward and reverse reactions be equal.
   IV. Reversibility requires that all products and reactants are present in equal amounts.

   A. I only
   B. I and IV only
   C. I and II only
   D. II and IV only
   E. I, II, and IV only

2. Which is the equilibrium constant expression for this reaction?

   $Ni (s) + 4 CO (g) \rightleftharpoons Ni(CO)_4 (g)$

   A. $K_c = \dfrac{[Ni(CO)_4]}{[CO]}$

   B. $K_c = \dfrac{[Ni][CO]^4}{[Ni(CO_4)]}$

   C. $K_c = \dfrac{[CO]^4}{[Ni(CO_4)]}$

   D. $K_c = \dfrac{[Ni(CO_4)]}{CO^4}$

   E. $K_c = \dfrac{1}{[CO^4]}$

3. In the laboratory 1.0 mol of HBr is injected into a vessel, and 0.50 mol of H_2 and 0.50 mol Br_2 are injected into in an additional, but identical vessel. Both vessels are sealed and then maintained at 300°C. Equilibrium is then established in both systems.

   $2 HBr (g) \rightleftharpoons H_2 (g) + Br_2 (g)$

   Which is true?

   A. The contents of both vessels contain identical mixtures of HBr, H_2, and Br_2.

   B. Equilibrium is attained in each vessel when the reaction $2 HBr (g) \rightarrow H_2 (g) + Br_2 (g)$ goes to completion.

   C. Equilibrium can only be achieved between the contents of the two vessels if their components are mixed in one vessel.

   D. Equilibrium is attained in each vessel when the number of moles of HBr equals the number of moles of H_2 and Br_2.

   E. The contents of both vessels contain proportional mixtures of HBr, H_2, and Br_2, but the first container has twice as much of each.

4. What is the solubility product constant expression for copper(II) hydroxide?

A. $K_{sp} = \dfrac{[Cu^{2+}][OH^-]}{[Cu^{2+}(OH)_2]}$

B. $K_{sp} = [Cu^{2+}][OH^-]$

C. $K_{sp} = [Cu^{2+}(OH)_2]$

D. $K_{sp} = \dfrac{[Cu^{2+}(OH)_2]}{[Cu^{2+}][OH^-]}$

E. $K_{sp} = [Cu^{2+}][OH^-]^2$

5. At 1100 K, K_p for this reaction is 0.25. Which equation correctly calculates K_c for this reaction?

$2\ SO_2\ (g) + O_2\ (g) \rightarrow 2\ SO_3\ (g)$

A. $K_c = 0.25(0.0821 \times 1100)^{-2}$
B. $K_c = 0.25(0.0821 \times 1100)^{-1}$
C. $K_c = 0.25(0.0821 \times 1100)^1$
D. $K_c = 0.25(0.0821 \times 1100)^2$
E. $K_c = 0.25(0.0821 \times 1100)^5$

6. The Haber Process reaction at constant temperature has an equilibrium constant value $K_c = 1.3 \times 10^{-2}$.

$N_2\ (g) + 3\ H_2\ (g) \rightleftharpoons 2\ NH_3\ (g)$

Which variation of this reaction has a $K_c = 8.8$?

A. $N_2\ (g) + 3\ H_2\ (g) \rightleftharpoons 2\ NH_3\ (g)$

B. $2\ NH_3\ (g) \rightleftharpoons N_2\ (g) + 3\ H_2\ (g)$

C. $2\ N_2\ (g) + 6\ H_2\ (g) \rightleftharpoons 4\ NH_3(g)$

D. $\dfrac{1}{2} N_2\ (g) + \dfrac{3}{2} H_2\ (g) \rightleftharpoons NH_3\ (g)$

E. $NH_3\ (g) \rightleftharpoons \dfrac{1}{2} N_2\ (g) + \dfrac{3}{2} H_2\ (g)$

7. Which set of conditions maximizes the production of sulfur trioxide in this reaction?

$2\ SO_2\ (g) + O_2\ (g) \rightarrow 2\ SO_3\ (g);\ \Delta H = -198\ kJ$

A. Low pressure, large volume, low temperature

B. Low pressure, small volume, low temperature

C. High pressure, large volume, low temperature

D. High pressure, small volume, low temperature

E. Low pressure, large volume, high temperature

8. For an endothermic reaction at equilibrium, a shift towards the:

A. *reactants* occurs if the concentration of a product is *decreased* or if the temperature is *decreased*.

B. *reactants* occurs if the concentration of a product is *increased* or if the temperature is *decreased*

C. *reactants* occurs if the concentration of a product is *increased* or if the temperature is *increased*.

D. *products* occurs if the concentration of a product is *increased* or if the temperature is *decreased*.

E. *products* occurs if the concentration of a product is *decreased* or if the temperature is *decreased*.

9. Consider the following closed system at equilibrium and constant temperature:

$$CaCO_3 \text{ (s)} \rightleftharpoons CaO \text{ (s)} + CO_2 \text{ (g)}$$

What will happen if the number of moles of CaO in the vessel is doubled?

A. The equilibrium constant, K, is halved.

B. The equilibrium constant, K, is doubled.

C. The number of moles of $CaCO_3$ in the vessel is increased.

D. The number of moles of CO_2 present at equilibrium is halved.

E. The partial pressure of CO_2 in the vessel will remain unchanged.

10. At a constant temperature the equilibrium constant, K, equals 9.0 for the reaction:

$$H_2 \text{ (g)} + I_2 \text{ (g)} \rightleftharpoons 2 \text{ HI (g)} \quad K = 9.0$$

At this temperature, an equilibrium mixture of only these gases contains 0.60 moles of HI and 0.40 moles of H_2 in a 2.00 L flask. How many moles of I_2 are in the equilibrium mixture?

A. 0.050
B. 0.10
C. 0.17
D. 0.40
E. 0.085

Free-Response Questions

1. A 2.416 g sample of PCl_5 is placed in an evacuated 2.000 L flask. The flask is then heated and maintained at a constant temperature of 250.0°C. The PCl_5 completely vaporizes, and the pressure inside the flask is measured to be 358.7 mmHg. It is known that the phosphorus(V) chloride decomposes in an equilibrium reaction to form phosphorus(III) chloride and chlorine gas in a reversible reaction.

   (A) Write a balanced equation for the decomposition of PCl_5.
   (B) What is the initial pressure of PCl_5 in the flask?
   (C) Use a reaction table to determine the partial pressure of each of the three gases at equilibrium.
   (D) Calculate the numerical value of K_p for this reaction.

2. The decomposition of H_2S (g) is an endothermic reaction.

   $$H_2S \text{ (g)} \rightleftharpoons H_2 \text{ (g)} + S \text{ (s)} \quad \text{At 370 K, } K_c = 8 \times 10^{-6}$$

   (A) Write both K_p and K_c equilibrium constant expressions for the decomposition of H_2S.

   (B) What is the numerical value of K_p at 370 K?

   (C) A 1.00 L flask contains 2.00 moles of solid sulfur. When 0.200 mol of both H_2S and H_2 gas are injected into the flask, and constant temperature is maintained, will the reaction proceed in the direction of the products or the reactants to attain equilibrium? Briefly explain your reasoning.

   (D) Will the total amount of H_2S present at equilibrium increase, decrease, or remain the same, if the volume of the flask is doubled while the temperature is constant? Briefly explain your reasoning.

Multiple-Choice Answers

1. C   2. D   3. A   4. E   5. C   6. E   7. D   8. B   9. E   10. B

1. C

Reversibility requires that all reactants and products are present, and chemical equilibrium requires reversible chemical reactions. Neither the rates of reaction or the amounts of products and reactants involved in a reaction influence whether the reaction is reversible or not.

2. D

Compressed states of matter, solids, and liquids are left out of the equilibrium constant expression because their concentrations are constant. The concentration of products is always divided by the concentration of the reactants. The coefficient in front of the CO is accounted for in an exponent in the equilibrium expression.

3. A

Both vessels initially have the same number of bromine and hydrogen atoms, but in different molecules. The reaction proceeds in the forward direction in the first vessel, and in the reverse direction in the second vessel, until both vessels have identical mixtures of HBr, H_2, and Br_2.

4. E

The solubility product constant is based upon the net ionic equation of a compound. For copper (II) hydroxide, the reaction for this poorly soluble reactant is: $Cu(OH)_2 \rightleftharpoons Cu^{2+} + 2\ OH^-$

5. C

$K_p = K_c (RT)^{\Delta n}$. Rearranging the equation to solve for K_c gives $K_c = K_p (RT)^{-\Delta n}$. R is the molar gas constant of 0.0821. T is the temperature in Kelvins, 1100 K. Δn is the moles of product minus the moles of reactant, which is in this case is -1, and $-\Delta n$ is $+1$.

6. E

The equilibrium law expression for the Haber Process reaction is $K_c = \dfrac{[NH_3]^2}{[N_2][H_2]^3} = 1.3 \times 10^{-2}$. The

expression for answer (E) is $K_c = \dfrac{[N_2]^{1/2}[H_2]^{3/2}}{[NH_3]}$ which is the square root of the reciprocal of the original

reaction. So, $K_c = \sqrt{\dfrac{1}{1.3 \times 10^{-2}}} = 8.8$. The equilibrium expression for (A) is the same; (B) is the reciprocal;

(C) is squared; and (D) is the square root

7. D

This question uses Le Châtelier's Principle. Increasing the pressure will shift towards the products because the reactant side has more moles of gas than the product side (3 to 2). As the system returns to equilibrium the reaction will make more product (SO_3) to decrease the number of moles and compensate for the increased pressure. The effect of decreasing the temperature on this system would be to shift the reaction to the right, towards the products. This is an exothermic reaction, which makes heat a product. By decreasing the temperature, you are decreasing a product, and the system will compensate for the temperature decrease by making heat (and more SO_3).

8. B

For an *endothermic* reaction at equilibrium, a shift towards the reactants occurs if the concentration of a product is increased or if the temperature is decreased.

9. E

Adding solid calcium oxide to the vessel does not change anything. Because it is a solid, and there is already some CaO in the equilibrium system, there is no change in concentration. The partial pressure of CO_2 is unchanged if we can assume the CaO does not decrease the volume available for the CO_2 in the vessel.

10. B

The equilibrium law expression for this reaction is $K_c = \dfrac{[HI]^2}{[H_2][I_2]}$. The moles should be converted to molarities by dividing by the 2.00 L. This is a lucky student problem since the moles of reactant and moles of product are equal and the concentration term will cancel out. Substituting in the equation and solving for $[I_2]$ will result in 0.10 moles. If concentrations are used, it must be remembered to multiply by the 2.00 L since moles and not molarity is required.

Free-Response Answers

1. (A) PCl_5 (g) $\rightleftharpoons$ PCl_3 (g) + Cl_2 (g)

   (B) Use the ideal gas law equation and solve for pressure. $\dfrac{P = nRT}{V}$

$$P = \dfrac{\dfrac{2.416 \text{ g}}{208.24 \text{ g/mol}} \times 0.08206 \text{ L} \cdot \text{atm/mol} \cdot \text{K} \times 532.2 \text{ K}}{2.000 \text{ L}} = 0.2491 \text{ atm} \times 760. \text{ mmHg/atm} = 189.3 \text{ mmHg}$$

(C)

	PCl_5	PCl_3	Cl_2
Initial	189.3	0	0
Change	$-x$	$+x$	$+x$
Equilibrium	$189.3 - x$	x	x

$P_T = (189.3 - x) + x + x = 358.7$ mmHg $\qquad x = 169.4$ mmHg

$P_{PCl5} = 19.9$ mmHg and $P_{PCl_3} = P_{Cl_2} = 169.4$ mmHg

(D) $K_P = \dfrac{P_{PCl_3} \times P_{Cl_2}}{P_{PCl_5}} = \dfrac{(169.4)^2}{19.9} = 1440$ mmHg

2. (A) $K_P = \dfrac{P_{H_2}}{P_{H_2S}} \qquad K_c = \dfrac{[H_2]}{[H_2S]}$

(B) $K_P = K_c = 8 \times 10^{-6}$ since $\Delta n = 0$ for this reaction

(C) The reaction proceeds in the direction of the reactants as the system moves toward equilibrium. The equilibrium constant has a value < 1, so there must be significantly more reactants than products when the system is at equilibrium at this temperature.

(D) The amount of H_2S will remain the same. This system has an equal number of moles of gas on the product and the reactant side, and is not be affected by changes in volume or pressure.

Acids and Bases

CHAPTER FOURTEEN

The term "acid" is usually associated with substances that:

- Have the ability to function as strong or weak electrolytes in aqueous solution
- Have a sour taste
- Have the ability to change the color of substances known as *indicators*
- Have the ability to produce hydrogen gas when reacted with certain "active" metals
- May be corrosive

Similarly, the term "base" is associated with substances that

- Have the ability to function as strong or weak electrolytes in aqueous solution
- Have a bitter taste
- Are slippery to the touch
- Have the ability to change the color of substances known as *indicators*
- May be caustic

Moreover, acids and bases can react with each other in a way that can (at least partially) cancel their properties. Such a reaction is known as **neutralization**.

Some common acids and bases are given in the table below:

Name	Formula
Acids	
Acetic acid	$HC_2H_3O_2$
Hydrobromic acid	HBr
Hydrochloric acid	HCl
Hydrofluoric acid	HF
Nitric acid	HNO_3
Perchloric acid	$HClO_4$
Phosphoric acid	H_3PO_4
Sulfuric acid	H_2SO_4
Bases	
Ammonia	NH_3
Barium hydroxide	$Ba(OH)_2$
Calcium hydroxide	$Ca(OH)_2$
Potassium hydroxide	KOH
Sodium hydroxide	NaOH

While these properties define acids and bases operationally, they do not provide a chemical explanation for their behavior. During the past two centuries, various models have been developed to explain acid-base behavior.

THE ARRHENIUS MODEL

In 1887, the Swedish chemist Svante Arrhenius proposed a model of acid-base behavior that was an outgrowth of his theory of electrolytes. According to the Arrhenius model:

- An **Arrhenius acid** is any hydrogen-containing substance that is capable of releasing H^+ ions in water.

  For example, HCl is considered an acid in the Arrhenius sense because it ionizes into H^+ (and Cl^-) ions in water. The relevant reaction is: $HCl(aq) \rightarrow H^+(aq) + Cl^-(aq)$.

- An **Arrhenius base** is any –OH-containing substance that releases OH^- ions in water.

For example, the metallic hydroxide NaOH is considered a base in the Arrhenius sense because it dissociates into OH^- (and Na^+) ions in water. The relevant reaction is: $NaOH(aq) \rightarrow Na^+(aq) + OH^-(aq)$. Note that the organic compound known as ethanol, C_2H_5OH is *not* considered a base because it does not release OH^- ions when dissolved in water.

The Arrhenius model relates all acidic properties (sour taste, . . .) to the production of H^+ ions and all basic properties (bitter taste, . . .) to the production of OH^- ions.

In addition, the model explains the process of neutralization: When an acid such as HCl is reacted with a base such as NaOH, the overall reaction is:

$$HCl(aq) + NaOH(aq) \rightarrow NaCl(aq) + H_2O(\ell)$$

NaCl is known as a **salt**: the positive ion (Na^+) is derived from the base and the negative ion (Cl^-) from the acid. In order to view neutralization in the Arrhenius sense, we must recognize that both HCl and NaOH form ions in aqueous solution:

$$H^+(aq) + Cl^-(aq) + Na^+(aq) + OH^-(aq) \rightarrow Na^+(aq) + Cl^-(aq) + H_2O(\ell)$$

The $Na^+(aq) + Cl^-(aq)$ ions are **spectator ions**: they are unchanged by the reaction. Therefore, the *net ionic reaction* that describes neutralization is:

$$H^+(aq) + OH^-(aq) \rightarrow H_2O(\ell)$$

As we will see shortly, this reaction goes virtually to completion and it explains (at least in an Arrhenius sense) why the acidic and basic properties of the original solutions disappear.

THE BRØNSTED-LOWRY MODEL

The problem with the Arrhenius model was twofold: (1) it was limited to aqueous solutions and (2) it could not account for the basic properties of substances such as NH_3, which did not have an –OH within its structure. In 1923, the Danish chemist Johannes Brønsted and the English chemist Thomas Lowry independently proposed their own model of acid-base behavior. This model extended the definitions of acids and bases:

- A **Brønsted-Lowry acid** is a species (atom, molecule, or ion) that is capable of *donating a proton* to another species.

  In water, the following reaction occurs between HCl and H_2O:
  $$HCl(aq) + H_2O(\ell) \rightarrow H_3O^+(aq) + Cl^-(aq)$$

HCl is an acid in the Brønsted-Lowry sense because it donates a proton (H^+) to a water molecule. The H_3O^+ ion, known as a **hydronium ion**, is always formed when an acid such as HCl ionizes in water.

- A **Brønsted-Lowry base** is a species (atom, molecule, or ion) that is capable of *accepting a proton* from another species.

    Consider the following reaction that occurs in aqueous solution:

    $$H_3O^+(aq) + NH_3(aq) \rightarrow NH_4^+(aq) + H_2O(\ell)$$

    NH_3 is a base in the Brønsted-Lowry sense because it accepts a proton (H^+) from a water molecule.

One result of the Brønsted-Lowry definitions is the recognition that *any reaction that contains an acid must also contain a base.*

For example, in the reaction: $HCl(aq) + H_2O(\ell) \rightarrow H_3O^+(aq) + Cl^-(aq)$, HCl is the Brønsted-Lowry acid because it donates a proton; H_2O is the Brønsted-Lowry base because it accepts the proton from the HCl.

Sample Problem

Identify the Brønsted-Lowry acid and base in the reaction:
$HNO_3(sol) + NH_3(\ell) \rightarrow NH_4^+(sol) + NO_3^-(sol)$

(Note: The designation "(sol)" means that the solutes are dissolved in a solvent other than H_2O.)

Solution

HNO_3(sol) is the Brønsted-Lowry **acid**; **$NH_3(\ell)$** is the Brønsted-Lowry **base**.

Amphoterism

In the reaction $HCl(aq) + H_2O(\ell) \rightarrow H_3O^+(aq) + Cl^-(aq)$, H_2O is the Brønsted-Lowry base; however, in the reaction $NH_3(aq) + H_2O(\ell) \rightarrow NH_4^+(aq) + OH^-(aq)$, H_2O is the Brønsted-Lowry acid. Any species that is capable of either donating or accepting an H^+ ion is said to be **amphoteric**. Many ions are capable of amphoterism. For example, the HCO_3^- ion is capable of functioning as either a Brønsted-Lowry acid or base:

$$HCO_3^-(aq) + OH^-(aq) \rightarrow CO_3^{2-}(aq) + H_2O(\ell)$$

$$HCO_3^-(aq) + H_3O^+(aq) \rightarrow [H_2CO_3(aq)] \rightarrow CO_2(g) + H_2O(\ell)$$

By the way, this last reaction illustrates the principle behind Alka-Seltzer®–type medications: $NaHCO_3$ is combined with a solid acid, such as tartaric acid ($C_4H_6O_6$). When water is added, hydronium ions are released and react with the HCO_3^- ions, liberating CO_2 gas (the "fizz").

Reversible Acid-Base Reactions: Conjugate Acid-Base Pairs

Acid-base reactions are reversible reactions, although some go nearly to completion. We can write the reactions of acid HA and base B with water as:

$$HA(aq) + H_2O(\ell) \rightleftharpoons H_3O^+(aq) + A^-(aq)$$

$$B(aq) + H_2O(\ell) \rightleftharpoons BH^+(aq) + OH^-(aq)$$

We observe that in water, acids produce H_3O^+ ions, while bases produce OH^- ions. This is not too different from the definitions provided by the Arrhenius model.

We also observe that each reaction contains *two acids* and *two bases*: one pair for the forward reaction and one pair for the reverse reaction. For example, in the reverse reaction (between H_3O^+ and A^-), H_3O^+ serves as the Brønsted-Lowry acid and A^- serves as the Brønsted-Lowry base:

$$HA(aq) + H_2O(\ell) \rightleftharpoons H_3O^+(aq) + A^-(aq)$$
$$\text{acid} \qquad \text{base} \qquad\quad \text{acid} \qquad\ \text{base}$$

Sample Problem

Identify the Brønsted-Lowry acid and base in the reverse of the second reaction (between BH^+ and OH^-).

Solution

BH^+ is the Brønsted-Lowry **acid**; OH^- is the Brønsted-Lowry **base**:

$$B(aq) + H_2O(\ell) \rightleftharpoons BH^+(aq) + OH^-(aq)$$
$$\text{base} \qquad \text{acid} \qquad\quad \text{acid} \qquad\ \text{base}$$

In the reversible reaction between HA and H_2O:

$$HA(aq) + H_2O(\ell) \rightleftharpoons H_3O^+(aq) + A^-(aq)$$
$$\text{acid} \qquad \text{base} \qquad\quad \text{acid} \qquad\ \text{base}$$

the acid HA and the base A^- differ by a single H^+ ion (as do the base H_2O and the acid H_3O^+). We call all acid-base pairs related in this way **conjugate acid-base pairs**, and we say that HA is the **conjugate acid** of A^-, or, alternatively, that A^- is the **conjugate base** of HA.

Sample Problem

Determine the two conjugate acid-base pairs in the reversible reaction:

$$B(aq) + H_2O(\ell) \rightleftharpoons BH^+(aq) + OH^-(aq)$$

Solution

BH^+ **and B;** H_2O **and** OH^- (The first member of each pair is the conjugate acid.)

A conjugate acid always contains *exactly one more H^+* than its conjugate base. Therefore, we can determine the conjugate acid of a species by *adding* an H^+ to it; similarly, we can determine the conjugate base of a species by *subtracting* an H^+ from it.

Sample Problem

Determine the conjugate acid and the conjugate base of the $H_2PO_4^-$ ion.

Solution

Conjugate acid: $H_2PO_4^- + H^+ = \mathbf{H_3PO_4}$

Conjugate base: $H_2PO_4^- - H^+ = \mathbf{HPO_4^{2-}}$

Note that $H_2PO_4^-$ is an *amphoteric* ion because it can accept or donate H^+.

Neutralization

If every reaction that contains a Brønsted-Lowry acid also contains a Brønsted-Lowry base, then under what conditions does *neutralization* occur? According to the Arrhenius model, neutralization occurs when H_2O is formed by the reaction between an acid and a base. We also know that in an aqueous system, H_2O is the solvent. According to the Brønsted-Lowry model, neutralization occurs when the conjugate acid and base of the solvent react and regenerate the solvent itself. For example in an aqueous system, the conjugate acid of H_2O is H_3O^+ and the conjugate base is OH^-:

$$H_3O^+(aq) + OH^-(aq) \rightleftharpoons 2H_2O(\ell) \qquad \text{[neutralization]}$$

The "classic" neutralization reaction between HCl and NaOH obeys this definition because HCl produces H_3O^+ ions while NaOH produces OH^- ions when dissolved in water.

Sample Problem

(A) Determine the conjugate acid and base of $NH_3(\ell)$.

(B) Write the acid-base reaction that represents neutralization in liquid NH_3.

Solution

(A) Conjugate acid: $\mathbf{NH_4^+}$; conjugate base: $\mathbf{NH_2^-}$

(B) $NH_4^+(sol) + NH_2^-(sol) \rightleftharpoons 2NH_3(\ell)$

In the sections that follow, we will limit our discussions to acid-base reactions that occur in aqueous solution.

Acid-Base Strength

We can view acid-base reactions essentially as a competition for protons: A strong acid donates protons easily while a strong base accepts them easily. As a result, all acid-base reactions run "downhill" from a stronger acid and base to a weaker acid and base. Consider the following reaction, which goes *nearly* to completion:

$$HCl(aq) + NH_3(aq) \xrightarrow{\sim 100\%} NH_4^+(aq) + Cl^-(aq)$$

$$\text{stronger acid} \quad \text{stronger base} \qquad\qquad \text{weaker acid} \quad \text{weaker base}$$

Note that the stronger conjugate acid (HCl) is associated with the weaker conjugate base (Cl^-), and the stronger conjugate base (NH_3) is associated with the weaker conjugate acid (NH_4^+). This is not a coincidence, but a logical conclusion: an acid that donates protons readily *must* have a conjugate base that *accepts* protons only with difficulty, and a base that accepts protons readily *must* have a conjugate acid that *donates* protons only with difficulty. We can state these conclusions in one simple statement:

The strength of a conjugate acid is inversely related to the strength of its conjugate base.

The table below lists the relative strengths of some conjugate acid-base pairs:

		Relative Strengths of Conjugate Acid-Base Pairs		
	Formula	**Conjugate Acid** Name	Formula	**Conjugate Base** Name
	$HClO_4$	perchloric acid	ClO_4^-	perchlorate ion
	HI	hydriotic acid	I^-	iodide ion
	HBr	hydrobromic acid	Br^-	bromide ion
	HCl	hydrochloric acid	Cl^-	chloride ion
	H_2SO_4	sulfuric acid	HSO_4^-	hydrogen sulfate ion
	HNO_3	nitric acid	NO_3^-	nitrate ion
	H_3O^+	hydronium ion	H_2O	water
	HSO_4^-	hydrogen sulfate ion	SO_4^{2-}	sulfate ion
	H_3PO_4	phosphoric acid	$H_2PO_4^-$	dihydrogen phosphate ion
	HF	hydrofluoric acid	F^-	fluoride ion
	HNO_2	nitrous acid	NO_2^-	nitrite ion
	HCO_2H	formic (methanoic) acid	CO_2H^-	formate (methanoate) ion
	$HC_2H_3O_2$	acetic (ethanoic) acid	$C_2H_3O_2^-$	acetate (ethanoate) ion
	H_2CO_3	carbonic acid	HCO_3^-	hydrogen carbonate ion
	$H_2PO_4^-$	dihydrogen phosphate ion	HPO_4^{2-}	hydrogen phosphate ion
	NH_4^+	ammonium ion	NH_3	ammonia
	HCN	hydrocyanic acid	CN^-	cyanide ion
	HCO_3^-	hydrogen carbonate ion	CO_3^{2-}	carbonate ion
	HPO_4^{2-}	hydrogen phosphate ion	PO_4^{3-}	phosphate ion
	H_2O	water	OH^-	hydroxide ion
	NH_3	ammonia	NH_2^-	amide ion

(Left side, top to bottom: Strong acids → Weak acids; Acid strength increases ↑)
(Right side: Base strength increases ↓)

The terms *strong acid* and *strong base* are reserved for those substances that are essentially 100% ionized or dissociated in aqueous solution. For example, acids such as HCl and HNO_3 and bases such as $NaOH$ and KOH meet these criteria.

Factors That Affect Acid Strength

In this section, we consider the factors that affect the strength of acids. We shall be considering *binary acids* (such as HBr and H_2S) and *oxoacids* (such as HNO_3 and HOCl).

Binary Acids

The strength of a binary acid H–X depends on the ease with which X can release a proton. This, in turn, depends on (1) the enthalpy of the H–X bond, and (2) the polarity of the bond, as measured by the electronegativity difference between H and X. In summary:

- The lower the H–X bond enthalpy, the stronger the acid.
- The more polar the H–X bond, the stronger the acid.

Within a periodic group, bond enthalpy is the more important factor: As the size of atom X increases, there is a decrease in the bond enthalpy and an increase in acid strength. The table below lists the bond enthalpies for the binary acids of Group 17. :

Bond	ΔH /(kJ mol^{-1})
H–F	567
H–Cl	431
H–Br	366
H–I	299

The strength of these acids, in ascending order, is: HF << HCl < HBr < HI.

Within a single period, bond polarity is the more important factor: As the electronegativity of atom X increases, there is an increase in bond polarity and in the acid strength. The table below lists hydrogen compounds of Period 2 and the electronegativities of the constituent atoms:

Compound	Electronegativity
CH_3	2.55
NH_3	3.04
H_2O	3.44
HF	3.98

The strength of these acids, in ascending order, is: CH_4 < NH_3 << H_2O < HF.

Sample Problem

Arrange the following binary acids in order of increasing acid strength:

(A) H_2S, H_2O, H_2Se

(B) HCl, PH_3, H_2S

Solution

(A) These acids lie within a single group (16). As the size of X increases, the acid strength increases: $H_2O < H_2S < H_2Se$.

(B) These acids lie within a single period (3). As the electronegativity of X increases, the acid strength increases: $PH_3 < H_2S < HCl$.

Oxoacids

Oxoacids have the general formula H_mXO_n where X is a nonmetallic atom (such as Cl, N, S, C) that is always bonded to one or more –OH groups and, possibly, to additional oxygen atoms. In water an oxoacid undergoes the following reaction:

$$—X—O—H + H_2O \rightleftharpoons H_3O^+ + —X—O^-$$

Any factor that weakens the OH bond or increases its polarity, increases the strength of the acid. Two such factors are the electronegativity of X and the oxidation number of X.

For oxoacids that contain the same number of OH groups and the same number of oxygen atoms, as the electronegativity of X increases, the acid strength increases. This is illustrated in the table below:

Acid (H–O–X)	Electronegativity of X
H–O–I	2.66
H–O–Br	2.96
H–O–Cl	3.16

The strength of these acids in ascending order is: HOI < HOBr < HOCl

For oxoacids that contain the same atom X, but different numbers of oxygen atoms, as the number of oxygen atoms increases, the oxidation number of X also increases and this leads to an increase in the strength of the acid. This is illustrated in the table below:

Acid	Oxidation Number of X
H–O–Cl	+1
H–O–Cl–O	+3
H–O–Cl–O_2	+5
H–O–Cl–O_3	+7

The strength of these acids in ascending order is: $HOCl \ll HOClO < HOClO_2 \ll HOClO_3$

Sample Problem

Arrange the following oxoacids in order of increasing acid strength:

(A) $HBrO_3$, $HBrO$, $HBrO_2$

(B) $HClO_3$, $HBrO_3$, HIO_3

Solution

(A) $HBrO < HBrO_2 < HBrO_3$

(B) $HIO_3 < HBrO_3 < HClO_3$

ACID-BASE EQUILIBRIA

Ionization Constants

Since acid-base reactions are reversible, we can write equilibrium constants that can be used to assess the strength of an acid or a base:

Acid (HA):

$$HA(aq) + H_2O(\ell) \rightleftharpoons H_3O^+(aq) + A^-(aq)$$

$$K_a = \frac{[H_3O^+][A^-]}{[HA]}$$

The constant K_a is known as the **ionization constant of the acid**. The larger its value, the stronger the acid. The following table lists the K_a values for a number of acids:

Monoprotic Acids (Acids with one ionizable hydrogen atom)

Formula	Name	K_a
HIO_3	Iodic acid	1.6×10^{-1}
HNO_2	Nitrous acid	7.2×10^{-4}
HF	Hydrofluoric acid	6.6×10^{-4}
$HCHO_2$	Formic acid	1.8×10^{-4}
$HC_3H_5O_3$	Lactic acid	1.4×10^{-4}
$HC_7H_5O_2$	Benzoic acid	6.3×10^{-5}
$HC_4H_7O_2$	Butanoic acid	1.5×10^{-5}
HN_3	Hydrazoic acid	1.9×10^{-5}
$HC_2H_3O_2$	Acetic acid	1.8×10^{-5}
$HC_3H_5O_2$	Propanoic acid	1.3×10^{-5}
$HOCl$	Hypochlorous acid	2.9×10^{-8}
HCN	Hydrocyanic acid	6.2×10^{-10}
HC_6H_5O	Phenol	1.0×10^{-10}
H_2O_2	Hydrogen peroxide	2.2×10^{-12}

Diprotic Acids (Acids with two ionizable hydrogen atoms)

Formula	Name	K_{a1}	K_{a2}
H_2SO_4	Sulfuric acid	Large	1.1×10^{-2}
H_2CrO_4	Chromic acid	5.0	1.5×10^{-6}
$H_2C_2O_4$	Oxalic acid	5.4×10^{-2}	5.3×10^{-5}
H_2SO_3	Sulfurous acid	1.3×10^{-2}	6.2×10^{-8}
H_2SeO_3	Selenous acid	2.3×10^{-3}	5.4×10^{-9}
$H_2C_3H_2O_4$	Malonic acid	1.5×10^{-3}	2.0×10^{-6}
$H_2C_8H_4O_4$	Phthalic acid	1.1×10^{-3}	3.9×10^{-6}
$H_2C_4H_4O_6$	Tartaric acid	9.2×10^{-4}	4.3×10^{-5}
H_2CO_3	Carbonic acid	4.4×10^{-7}	4.7×10^{-11}

Triprotic Acids (Acids with three ionizable hydrogen atoms)

Formula	Name	K_{a1}	K_{a2}	K_{a3}
H_3PO_4	Phosphoric acid	7.1×10^{-3}	6.3×10^{-8}	4.2×10^{-13}
H_3AsO_4	Arsenic acid	6.0×10^{-3}	1.0×10^{-7}	3.2×10^{-12}
$H_3C_6H_5O_7$	Citric acid	7.4×10^{-4}	1.7×10^{-5}	4.0×10^{-7}

In the case of strong acids, the reaction goes nearly 100% to completion and it is not possible calculate a K_a with any accuracy. Therefore, the table lists only those values for *weak* acids.

Some acids are capable of donating more than one H^+ ion. In this case, there is a separate K_a value for each ionization (K_{a1}, K_{a2}, . . .). Since it becomes increasingly difficult to remove successive H^+ ions, the successive K_a values differ greatly in magnitude ($K_{a1} \gg K_{a2} \gg K_{a3}$).

Base (B):

$$B(aq) + H_2O(\ell) \rightleftharpoons BH^+(aq) + OH^-(aq)$$

$$K_b = \frac{[BH^+][OH^-]}{[B]}$$

The constant K_b is known as the **ionization constant of the base**. The larger its value, the stronger the base. The table below lists the K_b values for a number of bases:

Formula	Name	K_b
$(CH_3)_2NH$	Dimethylamine	6.9×10^{-4}
CH_3NH_2	Methylamine	4.2×10^{-4}
$CH_3CH_2NH_2$	Ethylamine	4.3×10^{-4}
$(CH_3)_3N$	Trimethylamine	6.3×10^{-5}
NH_3	Ammonia	1.8×10^{-5}
N_2H_4	Hydrazine	8.5×10^{-7}
C_5H_5N	Pyridine	1.5×10^{-9}
$C_6H_5NH_2$	Aniline	7.4×10^{-10}

In the case of strong bases, the reaction goes nearly 100% to completion and it is not possible calculate a K_b with any accuracy. Therefore, the table lists only those values for *weak* bases.

There is a third ionization constant that is associated with acid-base behavior, and is based on the **autoionization of water**:

$$2\ H_2O(\ell) \rightleftharpoons H_3O^+(aq) + OH^-(aq)$$

$$K_w = [H_3O^+][OH^-] = 1.00 \times 10^{-14} \quad (\text{at } 25^\circ C)$$

This constant is of great importance because it tells us that in every aqueous solution (1) both H_3O^+ and OH^- ions are always present, and (2) the product of the ion concentrations is constant. As a result, we can classify an aqueous solution as acidic, basic, or neutral depending on the relative concentrations of these ions:

- If $[H_3O^+] > [OH^-]$, the solution is acidic.
- If $[H_3O^+] < [OH^-]$, the solution is basic.
- If $[H_3O^+] = [OH^-]$, the solution is neutral.

We can express these relationships in terms of the molar concentrations of $[H_3O^+]$ and $[OH^-]$:

- In a *neutral* aqueous solution at 25°C, $[H_3O^+] = [OH^-] = 1.00 \times 10^{-7}$ M.
- In an *acidic* solution, $[H_3O^+] > 1.00 \times 10^{-7}$ M and $[OH^-] < 1.00 \times 10^{-7}$ M.
- In a *basic* solution $[H_3O^+] < 1.00 \times 10^{-7}$ M and $[OH^-] > 1.00 \times 10^{-7}$ M.

pH, pOH, and pK

Since $[H_3O^+]$, $[OH^-]$, and K values are generally less than 1—that is, powers of 10 with negative exponents—it is convenient to remove these exponents by expressing them in logarithmic form:

$$pH = -\log_{10}[H_3O^+]$$
$$pOH = -\log_{10}[OH^-]$$
$$pK = -\log_{10}K$$

Sample Problem

(A) The K_a for acetic acid is 1.8×10^{-5}. Calculate pK_a.

(B) If $[H_3O^+] = 4.6 \times 10^{-8}$, calculate the pH of the solution.

(C) If $[H_3O^+] = 4.6 \times 10^{-8}$, calculate the pOH of the solution.

Solution

(A) $pK_a = -\log_{10}(1.8 \times 10^{-5}) = \mathbf{4.7}$

(B) $pH = -\log_{10}(4.6 \times 10^{-8}) = \mathbf{7.3}$

(C) $[H_3O^+][OH^-] = K_w = 1.00 \times 10^{-14}$

  $(-\log[H_3O^+]) + (-\log[OH^-]) = -\log K_w = 14.0$

  $\mathbf{pH + pOH = 14.0}$

  $pOH = 14.0 - pH = 14.0 - 7.3 = \mathbf{6.7}$

In the solution to part (C), we derived a very important result that allows us to convert easily between pH and pOH:

$$pH + pOH = 14.0$$

In the next problem, we calculate $[H_3O^+]$, given the pH of the solution.

Sample Problem

The pH of a solution is 5.65. Calculate the $[H_3O^+]$ of this solution.

Solution

Since $pH = -\log_{10}[H_3O^+]$, it follows that $[H_3O^+] = 10^{-pH}$.

$$[H_3O^+] = 10^{-5.65} = 2.24 \times 10^{-6} \text{ M}$$

The relationships among $[H_3O^+]$, $[OH^-]$, and K_w allow us to construct a **pH scale** that describes the level of acidity or basicity in an aqueous solution. Note that a change of 1 pH unit is equivalent to a tenfold increase or decrease in $[H_3O^+]$.

pH Range	Description
0–2	Strongly acidic
2–5	Moderately acidic
5–(<)7	Weakly acidic
7	Neutral
(>)7–9	Weakly basic
9–12	Moderately basic
12–14	Strongly basic

The table below lists the pH values of some common substances:

Sample	pH value
Gastric juice (stomach)	1.0–2.0
Lemon juice	2.4
Vinegar	3.0
Grapefruit juice	3.2
Orange juice	3.5
Urine	4.8–7.5
Water (exposed to air)	5.5
Saliva	6.4–6.9
Milk	6.5
Water (pure)	7.0
Blood	7.35–7.45
Tears	7.4
Milk of magnesia	10.6
Ammonia (household)	11.5

Acid-Base Equilibrium Calculations

In this section, we solve a number of representative problems involving acid-base equilibria.

Sample Problem

(A) Calculate the pH of 0.023 M HCl, a strong acid.

(B) Calculate the pH of 0.071 M KOH, a strong base.

Solution

We defined the terms "strong" acid and "strong" base to mean nearly 100% ionization or dissociation.

(A) $[H_3O^+] = 0.023$ M; $pH = -\log_{10}[H_3O^+] = $ **1.6**

(B) $[OH^-] = 0.071$ M; $pOH = -\log_{10}[OH^-] = 1.1$; $pH = 14.0 - pOH = 14.0 - 1.1 = $ **12.9**

Sample Problem

Given the acetic acid equilibrium:

$$HC_2H_3O_2(aq) + H_2O(\ell) \rightleftharpoons H_3O^+(aq) + C_2H_3O_2{}^-(aq) \; [K_a = 1.8 \times 10^{-5}]$$

Calculate the pH, and the percent of $HC_2H_3O_2$ that is ionized in 0.10 M $HC_2H_3O_2$.

Solution

$K_a = 1.8 \times 10^{-5}$	$HC_2H_3O_2(aq) + H_2O(\ell) \rightleftharpoons H_3O^+(aq) + C_2H_3O_2{}^-(aq)$		
Substance:	$[HC_2H_3O_2]$	$[H_3O^+]$	$[C_2H_3O_2{}^-]$
Initial concentration:	0.10	0.00	0.00
Change in concentration:	$-x$	$+x$	$+x$
Equilibrium concentration:	$0.10 - x$	x	x

$$K_a = \frac{[H_3O^+][C_2H_3O_2{}^-]}{[HC_2H_3O_2]}$$

$$1.8 \times 10^{-5} = \frac{x^2}{0.10 - x} \approx \frac{x^2}{0.10}$$

$$x = [H_3O^+] = \sqrt{1.8 \times 10^{-6}} = 1.3 \times 10^{-3} \text{ M}$$

$$pH = -\log_{10}[H_3O^+] = -\log_{10}(1.3 \times 10^{-3}) = \textbf{2.9}$$

$$\% \text{ ionization} = \frac{[C_2H_3O_2{}^-]}{[HC_2H_3O_2]} = \frac{1.3 \times 10^{-3} \text{ M}}{0.10 \text{ M}} = \textbf{1.3\%}$$

Sample Problem

Repeat the last sample problem using 0.010 M $HC_2H_3O_2$.

Solution

$K_a = 1.8 \times 10^{-5}$	$HC_2H_3O_2(aq) + H_2O(\ell) \rightleftharpoons H_3O^+(aq) + C_2H_3O_2^-(aq)$		
Substance:	$[HC_2H_3O_2]$	$[H_3O^+]$	$[C_2H_3O_2^-]$
Initial concentration:	0.010	0.00	0.00
Change in concentration:	$-x$	$+x$	$+x$
Equilibrium concentration:	$0.010 - x$	x	x

$$K_a = \frac{[H_3O^+][C_2H_3O_2^-]}{[HC_2H_3O_2]}$$

$$1.8 \times 10^{-5} = \frac{x^2}{0.010 - x} \approx \frac{x^2}{0.010}$$

$$x = [H_3O^+] = \sqrt{1.8 \times 10^{-7}} = 4.2 \times 10^{-4} \text{ M}$$

$$pH = -\log_{10}[H_3O^+] = -\log_{10}(4.2 \times 10^{-4}) = \textbf{3.4}$$

$$\% \text{ ionization} = \frac{[C_2H_3O_2^-]}{[HC_2H_3O_2]} = \frac{4.2 \times 10^{-4} \text{ M}}{0.010 \text{ M}} \times 100 = \textbf{4.2\%}$$

When we compare the two problems, we should not be surprised that the $[H_3O^+]$ is smaller in the second problem, leading to a higher pH. However, the percent ionization in the second problem is larger than in the first by a factor of 4! This follows from LeChâtelier's principle: As the solution of $HC_2H_3O_2$ becomes more dilute, the equilibrium shifts toward the production of more dissolved particles (in this case, ions).

In the next problem, we calculate K_b given the pH of a basic solution.

Sample Problem

A 0.25-M solution of the weak base CH_3NH_2 has a pH of 12.0. Calculate the K_b of this base.

Solution

$$CH_3NH_2(aq) + H_2O(\ell) \rightleftharpoons CH_3NH_3^+(aq) + OH^-(aq)$$

$$pH + pOH = 14.0$$

$$pOH = 14.0 - 12.0 = 2.0$$

$$[OH^-] = 10^{-2.0} = 0.010 \text{ M}$$

$$\therefore [CH_3NH_3^+] = 0.010 \text{ M}; [CH_3NH_2] = 0.25 \text{ M} - 0.010 \text{ M} = 0.24 \text{ M}$$

$$K_b = \frac{[CH_3NH_3^+][OH^-]}{[CH_3NH_2]}$$

$$K_b = \frac{(0.010)(0.010)}{0.25 - 0.010} = \frac{(0.010)^2}{0.24} = \mathbf{4.2 \times 10^{-4}}$$

Dilute Solutions of Strong Acids

What is the pH of 1.00×10^{-8} M HCl, a strong acid? Generally, one has a "knee-jerk" reaction to this type of problem: After all, HCl is essentially 100% ionized; it follows that the $[H_3O^+]$ will be 1.00×10^{-8} M and the pH will be 8.00. This is an entirely unreasonable answer! A pH of 8.0 signifies that the solution is weakly basic and this is clearly impossible. In order to solve such problems, it is necessary to work out a different strategy:

- There are two sources of H_3O^+: the HCl and the H_2O itself. In this problem, H_2O actually provides more H_3O^+ ($\sim 10^{-7}$ M) than the HCl.

- The solution is electrically neutral and it follows that the total concentration of the positive ions $[H_3O^+]$ must equal the total concentration of the negative ions $[Cl^-]$ and $[OH^-]$:

$$[H_3O^+] = [Cl^-] + [OH^-]$$

- The $[Cl^-]$ must be 1.00×10^{-8} M since it comes from the complete ionization of HCl in water.

- Finally, we must remember that there is a constant relationship between $[H_3O^+]$ and $[OH^-]$:

$$K_w = [H_3O^+][OH^-]$$

The problem is solved as follows:

$$[H_3O^+] = [Cl^-] + [OH^-] = [Cl^-] + \frac{K_w}{[H_3O^+]} = (1.00 \times 10^{-8}) + \frac{1.00 \times 10^{-14}}{[H_3O^+]}$$

Multiply by $[H_3O^+]$, rearrange terms, and solve the quadratic equation:

$$[H_3O^+]^2 - (1.00 \times 10^{-8})[H_3O^+] - (1.00 \times 10^{-14}) = 0$$

$$[H_3O^+] = 1.05 \times 10^{-7} \text{ M}$$

$$pH = \textbf{6.98}$$

This answer makes perfect sense: the HCl solution is very, very slightly acidic.

Polyprotic Acids

Polyprotic acids have more than one acidic hydrogen atom and there are successive ionization reactions, each having its own K_a. The next two sample problems illustrate how equilibrium in a diprotic acid is treated.

Sample Problem

Given the following ionization reactions for carbonic acid H_2CO_3:

$$H_2CO_3(aq) + H_2O(\ell) \rightleftharpoons H_3O^+(aq) + HCO_3^-(aq) \quad K_{a1} = 4.4 \times 10^{-7}$$

$$HCO_3^-(aq) + H_2O(\ell) \rightleftharpoons H_3O^+(aq) + CO_3^{2-}(aq) \quad K_{a2} = 4.7 \times 10^{-11}$$

Calculate $[H_2CO_3]$, $[H_3O^+]$, $[HCO_3^-]$, $[CO_3^{2-}]$, and the pH of a 0.080-M solution of H_2CO_3.

Solution

We begin by doing the calculations for the first ionization:

$K_{a1} = 4.4 \times 10^{-7}$	$H_2CO_3(aq) + H_2O(\ell) \rightleftharpoons H_3O^+(aq) + HCO_3^-(aq)$		
Substance:	$[H_2CO_3]$	$[H_3O^+]$	$[HCO_3^-]$
Initial concentration:	0.080	0.000	0.000
Change in concentration:	$-x$	$+x$	$+x$
Equilibrium concentration:	$0.080 - x$	x	x

$$K_{a1} = \frac{[H_3O^+][HCO_3^-]}{[H_2CO_3]}$$

$$4.4 \times 10^{-7} = \frac{x^2}{0.080 - x} \approx \frac{x^2}{0.080}$$

$$x^2 = 3.5 \times 10^{-8}$$

$$x = [H_3O^+] = [HCO_3^-] = \sqrt{3.5 \times 10^{-8}} = \textbf{1.9} \times \textbf{10}^{-4} \textbf{ M}$$

$$[H_2CO_3] = 0.080 - 1.9 \times 10^{-4} = \textbf{0.080 M}$$

We continue with the calculations for the second ionization:

$K_{a2} = 4.7 \times 10^{-11}$	$HCO_3^-(aq) + H_2O(\ell) \rightleftharpoons H_3O^+(aq) + CO_3^{2-}(aq)$		
Substance:	$[HCO_3^-]$	$[H_3O^+]$	$[CO_3^{2-}]$
Initial concentration:	1.9×10^{-4}	1.9×10^{-4}	0.000
Change in concentration:	$-y$	$+y$	$+y$
Equilibrium concentration:	$1.9 \times 10^{-4} - y$	$1.9 \times 10^{-4} + y$	y

$$K_{a2} = \frac{[H_3O^+][CO_3^{2-}]}{[HCO_3^-]}$$

$$4.7 \times 10^{-11} = \frac{(1.9 \times 10^{-4} + y)y}{(1.9 \times 10^{-4} - y)} \approx \frac{(1.9 \times 10^{-4})y}{(1.9 \times 10^{-4})}$$

$$y = [CO_3^{2-}] = 4.7 \times 10^{-11} \text{ M}$$

Since y is negligible when compared with 1.9×10^{-4}:

$$[H_3O^+] \approx 1.9 \times 10^{-4} \text{ M}$$

$$pH = 3.7$$

The next problem involves H_2SO_4, a rather special diprotic acid because the first ionization goes nearly to completion.

Sample Problem

Given the following ionization reactions for sulfuric acid H_2SO_4:

$$H_2SO_4(aq) + H_2O(\ell) \xrightarrow{\sim 100\%} H_3O^+(aq) + HSO_4^-(aq) \ [K_{a1} = \text{very large}]$$

$$HSO_4^-(aq) + H_2O(\ell) \rightleftharpoons H_3O^+(aq) + SO_4^{2-}(aq) \ [K_{a2} = 1.1 \times 10^{-2}]$$

Calculate $[H_3O^+]$, $[HSO_4^-]$, $[SO_4^{2-}]$, and the pH of a 0.021-M solution of H_2SO_4.

Solution

We begin by doing the calculations for the first ionization, which we assume is 100% complete:

$K_{a1} = \text{very large}$	$H_2SO_4(aq) + H_2O(\ell) \xrightarrow{\sim 100\%} H_3O^+(aq) + HSO_4^-(aq)$		
Substance:	$[H_2SO_4]$	$[H_3O^+]$	$[HSO_4^-]$
Initial concentration:	0.021	0.000	0.000
Change in concentration:	−0.021	+0.021	+0.021
Equilibrium concentration:	0.000	0.021	0.021

We continue with the calculations for the second ionization:

$K_{a2} = 1.1 \times 10^{-2}$	$HSO_4^-(aq) + H_2O(\ell) \rightleftharpoons H_3O^+(aq) + SO_4^{2-}(aq)$		
Substance:	$[HSO_4^-]$	$[H_3O^+]$	$[SO_4^{2-}]$
Initial concentration:	0.021	0.021	0.000
Change in concentration:	$-y$	$+y$	$+y$
Equilibrium concentration:	$0.021 - y$	$0.021 + y$	y

$$K_{a2} = \frac{[H_3O^+][SO_4^{2-}]}{[HSO_4^-]}$$

$$1.1 \times 10^{-2} = \frac{(0.021+y)y}{(0.021-y)}$$

We must expand and solve this equation directly because of the large value of K_{a2}:

$$y^2 + 0.032y - 0.00023 = 0$$
$$y = [SO_4^{2-}] = \textbf{0.0061 M}$$
$$[HSO_4^-] = 0.021 - 0.0061 = \textbf{0.015 M}$$
$$[H_3O^+] = 0.021 + 0.0061 = \textbf{0.027 M}$$
$$pH = \textbf{1.6}$$

pH and Solubility

Recall that the solubility equilibrium of $Mg(OH)_2$ is: $Mg(OH)_2(s) \rightleftharpoons Mg^{2+}(aq) + 2OH^-(aq)$. Since the equilibrium system contains $OH^-(aq)$ ions, the solubility of this compound will be pH-dependent. As the pH of the solution increases, the $[OH^-]$ also increases and the equilibrium is shifted to the left. As a result, the solubility of $Mg(OH)_2$ is lowered when the pH is raised. Conversely, as the pH is lowered, the solubility of the compound is increased.

Sample Problem

For the equilibrium system: $Mg(OH)_2(s) \rightleftharpoons Mg^{2+}(aq) + 2OH^-(aq)$, $K_{sp} =$ is 5.61×10^{-12} at 25°C.

(A) Calculate the molar solubility of $Mg(OH)_2$ in water, and calculate the pH of this solution.

(B) Calculate the molar solubility of $Mg(OH)_2$ in a solution whose pH is maintained at 9.00.

(C) Calculate the molar solubility of $Mg(OH)_2$ in a solution whose pH is maintained at 12.0.

Solution

In each case, the solubility of the compound is determined by the $[Mg^{2+}]$.

(A) $K_{sp} = [Mg^{2+}][OH^-]^2$. If $[Mg^{2+}] = x$, then $[OH^-] = 2x$.

$$5.61 \times 10^{-12} = (x)(2x)^2 = 4x^3$$

$$x = [Mg^{2+}] = \textbf{Solubility} = \textbf{1.12} \times \textbf{10}^{-4} \textbf{ M}$$

$$[OH^-] = 2x = 2.24 \times 10^{-4}; \text{pOH} = 3.65; \textbf{pH} = \textbf{10.4}$$

(B) If pH = 9.00, pOH = 5.00 and $[OH^-] = 1.00 \times 10^{-5}$ M.

$$K_{sp} = [Mg^{2+}][OH^-]^2$$

$$5.61 \times 10^{-12} = [Mg^{2+}] \cdot (1.00 \times 10^{-5})^2$$

$$[Mg^{2+}] = \textbf{Solubility} = \textbf{5.61} \times \textbf{10}^{-2} \textbf{ M}$$

(C) If pH = 12.0, pOH = 2.00 and $[OH^-] = 1.00 \times 10^{-2}$ M.

$$K_{sp} = [Mg^{2+}][OH^-]^2$$

$$5.61 \times 10^{-12} = [Mg^{2+}] \cdot (1.00 \times 10^{-2})^2$$

$$[Mg^{2+}] = \textbf{Solubility} = \textbf{5.61} \times \textbf{10}^{-8} \textbf{ M}$$

As we predicted, the solubility of $Mg(OH)_2$ decreases with increasing pH.

Ionization Constants of Conjugate Acid-Base Pairs

Consider the acid *HA* and its conjugate base A^-. In water, the conjugate acid undergoes the reaction:

$$HA(aq) + H_2O(\ell) \rightleftharpoons H_3O^+(aq) + A^-(aq) \qquad K_a = \frac{[H_3O^+][A^-]}{[HA]}$$

while the conjugate base undergoes the reaction:

$$A^-(aq) + H_2O(\ell) \rightleftharpoons HA(aq) + OH^-(aq) \qquad K_b = \frac{[HA][OH^-]}{[A^-]}$$

If we multiply the two ionization constants together, we obtain the important result:

$$K_a K_b = \frac{[H_3O^+][A^-]}{[HA]} \cdot \frac{[HA][OH^-]}{[A^-]} = [H_3O^+][OH^-]$$

$$\boxed{K_a K_b = K_w}$$

Sample Problem

The K_a for acetic acid ($HC_2H_3O_2$) is 1.8×10^{-5}. What is the K_b for its conjugate base ($C_2H_3O_2^-$)?

Solution

$$K_aK_b = K_w$$

$$K_b = \frac{K_w}{K_a} = \frac{1.0 \times 10^{-14}}{1.8 \times 10^{-5}} = 5.6 \times 10^{-10}$$

Acid-Base Properties of Salts

When an acid neutralizes a base, the ionic compound that is formed is known as a **salt**. Aqueous solutions of salts can be neutral, acidic, or basic, depending on the properties of the positive and negative ions present in the salt.

Salts Formed from Strong Acids and Strong Bases

A salt such as KCl is formed by the neutralization of the strong acid HCl and the strong base KOH. The K^+(aq) ion has virtually no acid-base properties. We can analyze the behavior of the Cl^-(aq) ion by considering the relative strengths of the conjugate acid-base pairs present in the salt solution:

$$HCl \gg H_3O^+ \Rightarrow Cl^- \ll H_2O$$

Therefore, Cl^- will not display any basic properties. As a result, the solution will be neutral.

In summary: *Salts formed from strong acids and strong bases will produce neutral aqueous solutions.*

Salts Formed from Weak Acids and Strong Bases

A salt such as NaCN is formed from the neutralization of the weak acid HCN and the strong base NaOH. We will focus our attention on the relative strengths of HCN and CN^- in aqueous solution:

$$HCN < H_3O^+ \Rightarrow CN^- > H_2O$$

Since the CN^- ion is a stronger base than H_2O, it will accept a proton from the H_2O, forming OH^- ions:

$$CN^-(aq) + H_2O(\ell) \rightleftharpoons HCN(aq) + OH^-(aq)$$

This reaction is commonly known as the **hydrolysis** of the CN^- ion and, as a result, the solution will be basic.

In summary: *Salts formed from weak acids and strong bases will produce basic aqueous solutions.*

Sample Problem

Calculate the pH of 0.2M NaCN. $[K_a \text{ (HCN)} = 6.2 \times 10^{-10}]$

Solution

$$CN^-(aq) + H_2O(\ell) \rightleftharpoons HCN(aq) + OH^-(aq); K_b(CN^-) = \frac{K_w}{K_a(HCN)} = \frac{1.0 \times 10^{-14}}{6.2 \times 10^{-10}} = 1.6 \times 10^{-5}$$

$K_b = 1.6 \times 10^{-5}$	$CN^-(aq) + H_2O(\ell) \rightleftharpoons HCN(aq) + OH^-(aq)$		
Substance:	$[CN^-]$	$[HCN]$	$[OH^-]$
Initial concentration:	0.20	0.00	0.00
Change in concentration:	$-x$	$+x$	$+x$
Equilibrium concentration:	$0.20 - x$	x	x

$$K_b = \frac{[HCN][OH^-]}{[CN^-]}$$

$$1.6 \times 10^{-5} = \frac{x^2}{0.20 - x} \approx \frac{x^2}{0.20}$$

$$x = [OH^-] = \sqrt{3.4 \times 10^{-6}} = 1.8 \times 10^{-3} \text{ M}$$

$$pOH = -\log_{10}[OH^-] = -\log_{10}[1.8 \times 10^{-3}] = 2.7$$

$$pH = 14.0 - pOH = 14.0 - 2.7 = \mathbf{11.3}$$

Salts Formed from Strong Acids and Weak Bases

A salt such as NH_4Br is formed from the neutralization of the strong acid HBr and the weak base NH_3. Since HBr >> H_3O^+ ⇒ Br^- << H_2O and Br^- ions will not accept a proton from H_2O. Instead, we will focus our attention on the relative strengths of NH_3 and NH_4^+ in aqueous solution:

$$NH_3 < OH^- \Rightarrow NH_4^+ > H_2O$$

Since the NH_4^+ ion is a stronger acid than H_2O, it will donate a proton to the H_2O, forming H_3O^+ ions:

$$NH_4^+ (aq) + H_2O(\ell) \rightleftharpoons H_3O^+(aq) + NH_3(aq)$$

The hydrolysis of the NH_4^+ ion produces a solution that is acidic.

In summary: *Salts formed from strong acids and weak bases will produce acidic aqueous solutions.*

Sample Problem

Calculate the pH of 0.10 M NH_4Cl. $[K_b (NH_3) = 1.8 \times 10^{-5}]$

$$NH_4^+ (aq) + H_2O(\ell) \rightleftharpoons H_3O^+(aq) + NH_3(aq); K_a(NH_4^+) = \frac{K_w}{K_b(NH_3)} = \frac{1.0 \times 10^{-14}}{1.8 \times 10^{-5}} = 5.6 \times 10^{-10}$$

$K_a = 5.6 \times 10^{-10}$	$NH_4^+ (aq) + H_2O(\ell) \rightleftharpoons H_3O^+(aq) + NH_3(aq)$		
Substance:	$[NH_4^+]$	$[H_3O^+]$	$[NH_3]$
Initial concentration:	0.10	0.00	0.00
Change in concentration:	$-x$	$+x$	$+x$
Equilibrium concentration:	$0.10 - x$	x	x

$$K_a = \frac{[H_3O^+][NH_3]}{[NH_4]}$$

$$5.6 \times 10^{-10} = = \frac{x^2}{0.10 - x} \approx \frac{x^2}{0.10}$$

$$x = [H_3O^+] = \sqrt{5.6 \times 10^{-11}} = 7.5 \times 10^{-5} \text{ M}$$

$$pH = -\log_{10}[H_3O^+] = -\log_{10}(7.5 \times 10^{-5}) = \mathbf{5.1}$$

Sample Problem

Calculate the pH of 0.10 M $NaHSO_4$.

Solution

HSO_4^- is the conjugate base of H_2SO_4, an acid that is much stronger than H_3O^+. Therefore, there is virtually no tendency for HSO_4^- to receive a proton from H_2O. Instead, we must consider the acidic nature of the HSO_4^- ion ($K_{a2} = 1.1 \times 10^{-2}$):

$K_{a2} = 1.1 \times 10^{-2}$	$HSO_4^-(aq) + H_2O(\ell) \rightleftharpoons H_3O^+(aq) + SO_4^{2-}(aq)$		
Substance:	$[HSO_4^-]$	$[H_3O^+]$	$[SO_4^{2-}]$
Initial concentration:	0.10	0.00	0.00
Change in concentration:	$-y$	$+y$	$+y$
Equilibrium concentration:	$0.10 - y$	y	y

$$K_a = \frac{[H_3O^+][SO_4^{2-}]}{[HSO_4^-]}$$

$$1.1 \times 10^{-2} = \frac{y^2}{(0.10 - y)}$$

We must expand and solve this equation directly because of the large value of K_{a2}.

$$y^2 + 0.011y - 0.0011 = 0$$
$$y = [H_3O^+] = 2.8 \times 10^{-2}\,M$$
$$pH = \mathbf{1.6}$$

Salts Containing Hydrated Metallic Ions

Another type of salt that produces an acidic solution is one that contains a small, highly charged metallic ion such as Al^{3+}. The relevant reaction for this hydrated ion is:

$$Al(H_2O)_6^{3+}(aq) + H_2O(\ell) \rightleftharpoons H_3O^+(aq) + Al(H_2O)_5(OH)^{2+}(aq)$$

The high charge of the Al^{3+} ion draws the electrons in the bound water molecules toward it, resulting in weakened O–H bonds. As a result, a proton is transferred to a solvent water molecule, as is shown in the reaction immediately above. Other hydrated metallic ions that produce acidic solutions are $Fe^{3+}(aq)$, $Cr^{3+}(aq)$, and $Zn^{2+}(aq)$. Each of these hydrated metal ions has a measurable K_a associated with it. For example, the K_a for $Al(H_2O)_6^{3+}(aq)$ is 1.4×10^{-5}, which makes it a much stronger acid than NH_4^+ ($K_a = 5.6 \times 10^{-10}$).

Salts Formed from Weak Acids and Weak Bases

In order to predict the acid-base properties of salts formed from both weak acids and bases, one must compare the K_a of the conjugate acid to the K_b of the conjugate base. For example, in the salt $(NH_4)_2CO_3$, the K_a (NH_4^+) is 5.6×10^{-10}, and the K_b (CO_3^{2-}) is 1.8×10^{-4}. This salt would produce a basic aqueous solution.

The Common-Ion Effect in Acid-Base Equilibria

Earlier, we examined the acetic acid equilibrium:

$$HC_2H_3O_2(aq) + H_2O(\ell) \rightleftharpoons H_3O^+(aq) + C_2H_3O_2^-(aq) \; [K_a = 1.8 \times 10^{-5}]$$

and we found that the pH of 0.10-molar $HC_2H_3O_2$ was 2.9. Suppose $C_2H_3O_2^-$ was added to this solution. What should we expect? The addition of the common ion would shift the equilibrium to the left. As a result, the $[H_3O^+]$ would decrease and the pH would rise. This is verified by the next sample problem.

Sample Problem

Calculate the pH of a solution that contains 0.10 M $HC_2H_3O_2$ and 0.050 M $NaC_2H_3O_2$.

Solution

$K_a = 1.8 \times 10^{-5}$	$HC_2H_3O_2(aq) + H_2O(\ell) \rightleftharpoons H_3O^+(aq) + C_2H_3O_2^-(aq)$		
Substance:	$[HC_2H_3O_2]$	$[H_3O^+]$	$[C_2H_3O_2^-]$
Initial concentration:	0.10	0.00	0.050
Change in concentration:	$-x$	$+x$	$+x$
Equilibrium concentration:	$0.10 - x$	x	$0.050 + x$

$$K_a = \frac{[H_3O^+][C_2H_3O_2^-]}{[HC_2H_3O_2]}$$

$$1.8 \times 10^{-5} = \frac{x(0.050 + x)}{0.10 - x} \approx \frac{0.050x}{0.10}$$

$$x = [H_3O^+] = 3.6 \times 10^{-5} \text{ M}$$

$$pH = -\log_{10}[H_3O^+] = -\log_{10}(3.6 \times 10^{-5}) = \mathbf{4.4}$$

As we predicted, the pH rose from 2.9 to 4.4.

Buffer Systems

A **buffer system** is one that maintains the pH of an aqueous solution within certain prescribed limits. Buffer systems contain either:

- a weak acid and its conjugate base, such as $HC_2H_3O_2$ and $C_2H_3O_2^-$, or

- a weak base and its conjugate acid, such as NH_3 and NH_4^+.

The choice of a particular buffer system depends primarily on the *value* of the pH that is to be maintained.

The Mechanism of Buffer Action

In essence, a buffer system is a chemical "sponge" for H_3O^+ or OH^- ions that may be produced within a solution. As an example, let us consider the $HC_2H_3O_2$–$C_2H_3O_2^-$ buffer system.

- When additional H_3O^+ ions are added, the following reaction takes place:

$$H_3O^+(aq) + C_2H_3O_2^-(aq) \xrightarrow{\sim 100\%} = HC_2H_3O_2(aq) + H_2O(\ell)$$

That is, the acidic H_3O^+ ions react with the *conjugate base* component of the buffer system to produce H_2O and $HC_2H_3O_2$, a weak acid that is only slightly ionized. As a result, the pH drops only minimally.

- When additional OH^- ions are added, the following reaction takes place:

$$OH^-(aq) + HC_2H_3O_2(aq) \xrightarrow{\sim 100\%} C_2H_3O_2^-(aq) + H_2O(\ell)$$

That is, the basic OH^- ions react with the *acid* component of the buffer system to produce H_2O and $C_2H_3O_2^-$, a weak base. As a result, the pH rises only minimally.

Equilibrium Considerations

Since the K_a of $HC_2H_3O_2$ is small, the equilibrium concentrations of $HC_2H_3O_2$ and $C_2H_3O_2^-$ are very nearly their *initial* concentrations. As a result, we can write:

$$K_a = \frac{[H_3O^+][C_2H_3O_2^-]}{[HC_2H_3O_2]} \approx \frac{[H_3O^+][C_2H_3O_2^-]_0}{[HC_2H_3O_2]_0}$$

Rearranging the terms and solving for $[H_3O^+]$, we obtain:

$$[H_3O^+] = K_a \frac{[HC_2H_3O_2]_0}{[C_2H_3O_2^-]_0}$$

We can generalize this expression, so that it is applicable for all weak acid–conjugate base buffer systems:

$$[H_3O^+] = K_a \frac{[\text{Weak acid}]_0}{[\text{Conjugate base}]_0}$$

This expression tells us that in a buffer system consisting of a weak acid and its conjugate base, the $[H_3O^+]$ is determined by:

- The K_a of the weak acid
- The ratio of the concentration of the weak acid to the concentration its conjugate base

At times, it is useful to convert the buffer relationship shown above into its logarithmic form:

$$pH = pK_a + \log_{10} \frac{[\text{Conjugate base}]_0}{[\text{Weak acid}]_0}$$

In this form, the relationship is known as the **Henderson-Hasselbalch equation.**

If a buffer contains a weak base such as NH_3 and its conjugate acid NH_4^+, the expression becomes:

$$K_b = \frac{[NH_4^+][OH^-]}{[NH_3]} \approx \frac{[NH_4^+][OH^-]_0}{[NH_3]_0}$$

Solving for $[OH^-]$, we obtain:

$$[OH^-] = K_b \frac{[NH_3]_0}{[NH_4^+]_0}$$

and the generalized form of this expression is:

$$[OH^-] = K_b \frac{[\text{Weak base}]_0}{[\text{Conjugate acid}]_0}$$

In a buffer system consisting of a weak base and its conjugate acid, the $[OH^-]$ is determined by:

- The K_b of the weak base
- The ratio of the concentration of the weak base to the concentration of its conjugate acid

The Henderson-Hasselbalch equation for a basic buffer is:

$$pOH = pK_b + \log_{10} \frac{[\text{Conjugate acid}]_0}{[\text{Weak base}]_0}$$

A buffer system works most effectively when its concentration ratio is equal to 1. Under these conditions:

- The $[H_3O^+]$ of a weak acid/conjugate base buffer equals the K_a of the weak acid
- The $[OH^-]$ of a weak base/conjugate acid buffer equals the K_b of the weak base

Sample Problem

Calculate the pH at which the following buffer systems are most effective:

(A) $HC_7H_5O_2 / C_7H_5O_2^-$ ($K_a = 6.3 \times 10^{-5}$)

(B) $CH_3CH_2NH_2 / CH_3CH_2NH_3^+$ ($K_b = 4.3 \times 10^{-4}$)

Solution

(A) $[H_3O^+] = K_a = 6.3 \times 10^{-5}$; pH $= -\log_{10}(6.3 \times 10^{-5}) = \mathbf{4.2}$

(B) $[OH^-] = K_b = 4.3 \times 10^{-4}$; pOH $= -\log_{10}(4.3 \times 10^{-4}) = 3.4$; pH $= 14.0 - 3.4 = \mathbf{10.6}$

In general, the effective pH *range* of a buffer system is: (optimum pH $\pm$ 1.0 pH unit). In the last sample problem, the effective pH range for the first system is 3.2–5.2, and 9.6–11.6 for the second system.

Sample Problem

In part (B) of the last sample problem, we indicated that the effective range of pH was 9.6–11.6. Suppose that we wanted to use this system to create a buffer that maintained the pH close to 10.2. How should we proceed?

Solution

$$pH = 10.2 \Rightarrow pOH = 14.0 - 10.2 = 3.8 \Rightarrow [OH^-] = 1.6 \times 10^{-4} \text{ M}$$

$$[OH^-] = K_b \frac{[\text{Weak base}]_0}{[\text{Conjugate acid}]_0}$$

$$1.6 \times 10^{-4} \text{ M} = (4.3 \times 10^{-4}) \frac{[\text{Weak base}]_0}{[\text{Conjugate acid}]_0}$$

$$\frac{[\text{Weak base}]_0}{[\text{Conjugate acid}]_0} = 0.37$$

As long as the concentration ratio is maintained at 0.37:1, the system will be buffered at a pH of 10.2. For example, we could prepare a solution in which $[CH_3CH_2NH_2] = 0.37$ M and $[CH_3CH_2NH_3^+] = 1.0$ M.

We return to the $HC_2H_3O_2$–$C_2H_3O_2^-$ system in order to illustrate how this buffer "sponges" up additional H_3O^+ or OH^- ions. Suppose we prepare a buffer by dissolving 0.20 mole of $HC_2H_3O_2$ and 0.30 mole of $C_2H_3O_2^-$ in 1.0 liter of solution, that is, $[HC_2H_3O_2]_0 = 0.20$ M and $[C_2H_3O_2^-]_0 = 0.30$ M. As a first step, let us calculate the pH of this buffer system:

$$[H_3O^+] = K_a \frac{[HC_2H_3O_2]_0}{[C_2H_3O_2^-]_0}$$

$$[H_3O^+] = (1.8 \times 10^{-5}) \frac{0.20}{0.30} = 1.2 \times 10^{-5} \text{ M}$$

$$pH = -\log_{10}(1.2 \times 10^{-5}) = 4.9$$

Now suppose that 0.050 mole of H_3O^+ is added to this system in a way that does not change the volume significantly. As we saw earlier, the following reaction will occur:

$$H_3O^+(aq) + C_2H_3O_2^-(aq) \xrightarrow{\sim 100\%} HC_2H_3O_2(aq) + H_2O(\ell)$$

As a result, $[C_2H_3O_2^-]$ will *decrease* by 0.050 M to 0.25 M, and $[HC_2H_3O_2]$ will *increase* by 0.050 M to 0.25 M. The pH under these circumstances is:

$$[H_3O^+] = K_a \frac{[HC_2H_3O_2]}{[C_2H_3O_2^-]}$$

$$[H_3O^+] = (1.8 \times 10^{-5}) \frac{0.20 + 0.050}{0.30 - 0.050} = (1.8 \times 10^{-5}) \frac{0.25}{0.25} = 1.8 \times 10^{-5} \text{ M}$$

$$pH = -\log_{10}(1.8 \times 10^{-5}) = 4.7$$

The pH has fallen, but only by 0.2 pH unit. If 0.050 mole of H_3O^+ had been added to 1 liter of a solution whose *unbuffered* pH was 4.9, the pH would have fallen to 1.3, a change of 3.6 pH units! (These calculations are left to the reader.)

Sample Problem

(A) Calculate the pH of a buffer system consisting of $[CH_3CH_2NH_2]_0 = 0.23$ M and $[CH_3CH_2NH_3^+]_0 = 0.46$ M. ($K_b = 4.3 \times 10^{-4}$)

(B) Calculate the new pH if 0.021 M $[OH^-]$ is added to the system. (Assume that the volume change is negligible.)

Solution

(A)

$$[OH^-] = K_b \frac{[CH_3CH_2NH_2]_0}{[CH_3CH_2NH_3^+]_0}$$

$$[OH^-] = (4.3 \times 10^{-4}) \frac{0.46}{0.23} = 8.6 \times 10^{-4} \text{ M}$$

$$pOH = -\log_{10}(2.2 \times 10^{-4}) = 3.1$$

$$pH = 14.0 - 3.1 = \textbf{10.9}$$

(B) When OH^- is added to this system, the following reaction occurs:

$$OH^-(aq) + CH_3CH_2NH_3^+(aq) \xrightarrow{\sim 100\%} CH_3CH_2NH_2(aq) + H_2O(\ell)$$

$$[OH^-] = K_b \frac{[CH_3CH_2NH_2]}{[CH_3CH_2NH_3^+]}$$

$$[OH^-] = (4.3 \times 10^{-4}) \frac{0.46 + 0.021}{0.23 - 0.021} = (4.3 \times 10^{-4}) \frac{0.48}{0.21} = 9.8 \times 10^{-4} \text{ M}$$

$$pOH = -\log_{10}(9.8 \times 10^{-4}) = 3.0$$

$$pH = 14.0 - 3.0 = \textbf{11.0}$$

There is an increase of 0.1 pH unit.

Buffer Capacity

Earlier, we learned that the $[H_3O^+]$ for the $HC_2H_3O_2/C_2H_3O_2^-$ buffer ($K_a = 1.8 \times 10^{-5}$) can be calculated from the relationship:

$$[H_3O^+] = K_a \frac{[HC_2H_3O_2]_0}{[C_2H_3O_2^-]_0}$$

Suppose that we prepare three $HC_2H_3O_2/C_2H_3O_2^-$ buffer solutions as follows:

Solution	$[HC_2H_3O_2]_0$ /M	$[C_2H_3O_2^-]_0$ /M	Final Volume /L
A	0.200	0.200	1.00
B	0.400	0.400	1.00
C	0.800	0.800	1.00

Since the concentration ratio of each buffer is equal to 1.00, all three solutions will have the same initial pH (4.74). Now suppose that we add 0.0500 mole of OH^- ions to each solution *in a way that does not alter the total volume*. The table below shows the changes that occur within each solution:

Solution	$[HC_2H_3O_2]$ /M	$[C_2H_3O_2^-]$ /M	Final pH	ΔpH
A	0.15	0.25	4.97	+0.23
B	0.35	0.45	4.85	+0.11
C	0.75	0.85	4.80	+0.059

Buffer "C," made with the highest concentrations of acid and conjugate base, shows the smallest change in pH when OH^- ions are added. Moreover, this solution will be capable of absorbing more H_3O^+ *or* OH^- ions than the other two buffer solutions. As we make a buffer solution more concentrated, we say that we have increased its **buffer capacity**.

As a final note to this section, we add the following caveat : All buffers depend on the presence of *both* a conjugate and its conjugate base. When one of these substances is exhausted, the solution loses *all* buffering capacity.

ACID-BASE TITRATION

The molar concentration of an unknown acid or base can be determined by employing a laboratory procedure known as **acid-base titration**. In this procedure, a quantity of a *standardized solution* (base or acid) is added to a premeasured volume of the unknown until complete neutralization occurs. This is known as the **equivalence point** of the titration. (Note: We use the term "complete neutralization" to mean that *stoichiometrically equivalent* quantities of acid and base are used. The pH of the final solution need not be 7.0.)

The standard solution, known as the **titrant**, is generally delivered with a long tube known as a **burette**. The burette is calibrated so that the volume of the standard solution can be read directly. Generally, an acid-base indicator or a pH meter is also present and signals the **end**

point of the titration, that is, the visual cue that the titration is complete. Obviously, the end point must not be significantly different than the equivalence point.

When a monoprotic acid such as HCl is completely neutralized by a monohydroxide base such as NaOH, *equal numbers of moles* of H_3O^+ and OH^- ions react according to the equation: $H_3O^+(aq) + OH^-(aq) \rightarrow 2H_2O(\ell)$. For this simple acid-base neutralization, we can derive a mathematical relationship as follows:

$$H_3O^+(aq) + OH^-(aq) \rightarrow 2H_2O(\ell)$$

At neutralization: $n_{H_3O^+} = n_{OH^-}$

$$M = \frac{n}{V} \Rightarrow n = (M \cdot V)$$

$$\boxed{(M \cdot V)_{acid} = (M \cdot V)_{base}}$$

Sample Problem

When 25.0 mL of HCl is neutralized by 0.100-M NaOH, 35.0 mL of the base are consumed. Calculate the molarity of the acid.

Solution

$$(M \cdot V)_{acid} = (M \cdot V)_{base}$$

$$M_{acid} \cdot (25.0 \text{ mL}) = (0.100 \text{ M}) \cdot (35.0 \text{ M})$$

$$M_{acid} = \textbf{0.140 M}$$

If the acid contains more than one ionizable proton, and/or the base contains more than one dissociable hydroxide ion, then the equation given above must be modified in order to include these numbers:

$$\boxed{(M \cdot V \cdot \#_{H_3O^+})_{acid} = (M \cdot V \cdot \#_{OH^-})_{base}}$$

where the "#" sign indicates the number of acidic hydrogens or dissociable hydroxide ions in the chemical formula.

Sample Problem

What volume of 0.200-M H_2SO_4 is needed to be completely neutralized by 300. mL of 0.450-M KOH?

Solution

H_2SO_4 contains 2 ionizable hydrogen atoms and KOH contains 1 dissociable OH ion.

$$(M \cdot V \cdot \#_{H_3O^+})_{acid} = (M \cdot V \cdot \#_{OH^-})_{base}$$

$$(0.200 \text{ M}) \cdot (V_{acid}) \cdot (2) = (0.450 \text{ M}) \cdot (300\text{mL}) \cdot (1)$$

$$V_{acid} = \textbf{338 mL}$$

It should be recognized that these problems are fundamentally stoichiometry problems and, as such, can be solved by the factor-label method, as shown below.

Sample Problem

What volume of 0.200-M H_2SO_4 is needed to be completely neutralized by 300. mL of 0.450-M KOH?

Solution

$$H_2SO_4(aq) + 2KOH(aq) \rightarrow K_2SO_4(aq) + 2H_2O(\ell)$$

$$V_{H_2SO_4} = (300 \text{ mL KOH}) \cdot \left(\frac{0.450 \text{ mmol KOH}}{1.00 \text{ mL KOH}}\right) \cdot \left(\frac{1 \text{ mmol } H_2SO_4}{2 \text{ mmol KOH}}\right) \cdot$$

$$\left(\frac{1 \text{ mL } H_2SO_4}{0.200 \text{ mmol } H_2SO_4}\right) = \textbf{338 mL}$$

Acid-Base Indicators

An **acid-base indicator** is a substance that changes color within a specified pH range. Indicators are, themselves, weak acids or bases and, as such, undergo acid-base equilibria:

$$HIn(aq) + H_2O(\ell) \rightleftharpoons H_3O^+(aq) + In^-(aq) \qquad K_{in} = \frac{[H_3O^+][In^-]}{[HIn]}$$

The species HIn is known as the *acid form* of the indicator, and In^- is known as the *basic form* of the indicator. Each form has its own distinct color. We can rearrange the equilibrium constant expression to yield:

$$[H_3O^+] = K_{in}\frac{[HIn]}{[In^-]}$$

When $[HIn] > [In^-]$, the indicator has the color associated with its acid form; when $[In^-] > [HIn]$, the indicator has the color associated with its basic form. When $[HIn] = [In^-]$, the color of the indicator changes; at this point, $[H_3O^+] = K_{in}$. Ideally, this condition should represent the end point of the titration. In order to select a suitable indicator for a titration, it is best if the pK_{in} lies within one pH unit of the equivalence point:

$$pK_{in} = pH_{equivalence\ point} \pm 1$$

As we examine various titration curves in the next section, we will modify this rule somewhat.

The table below lists a number of common acid-base indicators and the colors present within their pH ranges:

pH RANGES OF SELECTED ACID-BASE INDICATORS

Indicator	pH range	Color at low end of pH range	Color at high end of pH range	pK_{in}
Methyl violet	0.0–1.6	yellow	blue	
Thymol blue (I)*	1.2–2.8	red	yellow	1.7
2,4 Dinitrophenol	2.0–4.7	colorless	yellow	4.0
Methyl orange	3.2–4.4	red	yellow	3.5
Bromcresol green	3.8–5.4	yellow	blue	4.9
Methyl red	4.2–6.2	red	yellow	5.0
Litmus	4.5–8.3	red	blue	6.5
Alizarin (I)*	5.6–7.2	yellow	red	
Bromthymol blue	6.0–7.6	yellow	blue	7.3
Phenol red	6.8–8.4	yellow	red	8.0
Metacresol purple	7.4–9.0	yellow	purple	8.3
Thymol blue (II)*	8.0–9.6	yellow	blue	9.0
Phenolphthalein	8.3–10.0	colorless	pink	9.5
Alizarin yellow R	10.1–12.0	yellow	violet	11.2
Alizarin (II)*	11.0–12.4	red	purple	11.7

*These indicators have *two* pH ranges

TITRATION CURVES

A **titration curve** is one that monitors the pH of a solution as an acid-base titration proceeds from beginning to well beyond the equivalence point. In general, the pH is plotted on the *y*-axis and the volume of acid (or base) delivered is plotted on the *x*-axis. The shape of a titration curve makes it possible to identify the equivalence point and is an aid in selecting an appropriate indicator for the titration. In addition, a titration curve can be used to calculate the K_a or K_b of a weak acid or base.

In this section, we will discuss four types of titrations and the curves they produce: (1) strong acid-strong base, (2) weak acid-strong base, (3) weak base-strong acid, and (4) polyprotic

Strong Acid-Strong Base Titrations

We begin with 30.0 milliliters of 0.100-M HCl, to which we add 0.100-M NaOH in small volumes, and calculate the pH after each addition. We will not be sticklers about precision: we assume that all concentrations and volumes are exact. The table below tracks the titration of HCl with NaOH:

NaOH Added /mL	OH⁻ Added /mmol	H_3O^+ (OH^-) Remaining /mmol	Total Volume /mL	$[H_3O^+]$ ($[OH^-]$) /M	pH
0	0	3	30	0.1000	1.000
5	0.5	2.5	35	0.07143	1.146
10	1	2	40	0.05000	1.301
15	1.5	1.5	45	0.03333	1.477
20	2	1	50	0.02000	1.699
25	2.5	0.5	55	0.009091	2.041
29	2.9	0.1	59	0.001695	2.771
29.9	2.99	0.01	59.9	0.0001669	3.777
29.99	2.999	0.001	59.99	0.00001667	4.778
29.999	2.9999	0.0001	59.999	0.000001667	5.778
30	**3**	**"0"**	**60**	**0.00000010000**	**7.000**
30.001	3.0001	(0.0001)	60.001	(0.000001667)	8.222
30.01	3.001	(0.001)	60.01	(0.00001667)	9.222
30.1	3.01	(0.01)	60.1	(0.0001664)	10.221
31	3.1	(0.1)	61	(0.001639)	11.215
35	3.5	(0.5)	65	(0.007692)	11.886
40	4	(1)	70	(0.01429)	12.155
45	4.5	(1.5)	75	(0.02000	12.301
50	5	(2)	80	(0.02500)	12.398
55	5.5	(2.5)	85	(0.02941)	12.469
60	6	(3)	90	(0.03333)	12.523

In constructing the table, we determine the number of millimoles of excess H_3O^+ or OH^- ions present after each addition of NaOH. We divided this excess by the total volume of the solution in order to obtain the molar concentration, and then we calculated the pH of the solution.

The completed titration curve appears below:

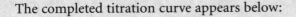

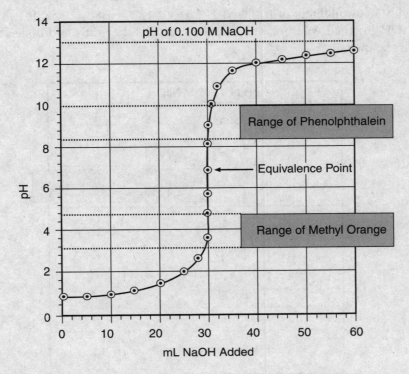

We can divide the titration curve into three segments:

1. Prior to the equivalence point (0–29.999 mL NaOH):

    • The pH rises slowly at first because the solution contains a sizeable excess of H_3O^+ ions. We begin to see a significant rise in pH after 29.0 mL of NaOH has been added.

2. At the equivalence point (30.00 mL NaOH):

    • All excess H_3O^+ ions have been removed by the NaOH. The $[H_3O^+]$ is now 1.000×10^{-7} M, which corresponds to a pH of 7.000.

3. Beyond the equivalence point (30.001–60.000 mL NaOH):

    • The solution now contains an excess of OH^- ions and as the pH rises, it approaches that of the titrant (0.1 M NaOH; pH = 13.00).

Selecting an Appropriate Indicator

If we examine the shape of the titration curve, we note that the region around the equivalence point is nearly vertical. This represents a drastic change in pH that is brought about by the addition of less than 0.100 milliliter of titrant. Any indicator that changes color within this vertical portion of the curve is suitable. As we can see from the curve, either methyl orange (pH range: 3.2–4.4) or phenolphthalein (pH range: 8.3–10.0) falls within the vertical region and is suitable as an indicator.

Weak Acid–Strong Base Titrations

We begin with 30.0 milliliters of 0.100-M $HC_2H_3O_2$ ($K_a = 1.8 \times 10^{-5}$), to which we add 0.100-M NaOH in small volumes, and calculate the pH after each addition. The titration curve appears below:

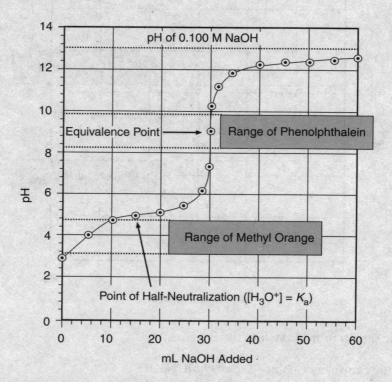

Once again, we divide the titration curve into segments:

1. Before NaOH has been added:

   - The pH of the solution (0.100 M $HC_2H_3O_2$) is 2.87.

2. Prior to the equivalence point (0–29.9 mL NaOH):

   - As NaOH is added below the equivalence point, the following reaction occurs:
   $$HC_2H_3O_2(aq) + OH^-(aq) \rightarrow C_2H_3O_2^-(aq) + H_2O(\ell)$$

   - The solution is now a buffer system that contains both the weak acid ($HC_2H_3O_2$) and its conjugate base ($C_2H_3O_2^-$).

   - At the point of half-neutralization (15.00 mL of NaOH added), $[HC_2H_3O_2] = [C_2H_3O_2^-]$ and $[H_3O^+] = K_a$; (pH = 4.74).

3. At the equivalence point (30.00 mL NaOH):

    •   All of the $HC_2H_3O_2$ has been exhausted and the total volume (60.00 mL) is now double the original volume of the acid. The solution is composed of 0.0500-M $C_2H_3O_2^-$. The equivalence point corresponds to a pH of 8.72 (rather than 7.00).

4. Beyond the equivalence point (30.1–60.000 mL NaOH):

    •   The solution now contains an excess of OH^- ions and as the pH rises, it approaches that of the titrant (0.1 M NaOH; pH = 13.00).

Selecting an Appropriate Indicator

If we examine the titration curve, we see that phenolphthalein falls within the vertical region and is suitable for use as an indicator. Methyl orange, however, falls within the "buffer region" of the curve and is unsuitable in this case.

Weak Base–Strong Acid Titrations

We begin with 30.0 milliliters of 0.100-M NH_3 ($K_b = 1.8 \times 10^{-5}$), to which we add 0.100-M HCl in small volumes, and calculate the pH after each addition. The titration curve appears below:

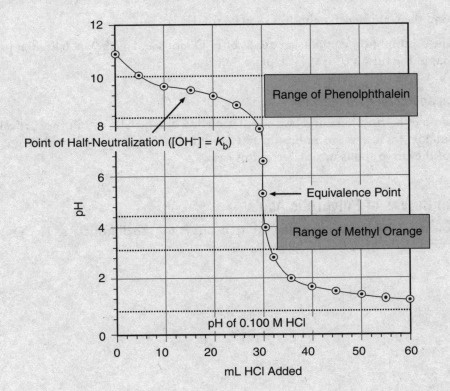

Once again, we divide the titration curve into segments:

1. Before HCl has been added:

   • The pH of the solution (0.100-M NH_3) is 11.13.

2. Prior to the equivalence point (0–29.9 mL HCl):

   • As HCl is added below the equivalence point, the following reaction occurs:

$$NH_3(aq) + H_3O^+(aq) \rightarrow NH_4^+(aq) + H_2O(\ell)$$

   • The solution is now a buffer system that contains both the weak base (NH_3) and its conjugate acid (NH_4^+).

   • At the point of half-neutralization (15.00 mL of HCl added), $[NH_3] = [NH_4^+]$ and $[OH^-] = K_b$; (pH = 9.26).

3. At the equivalence point (30.00 mL HCl):

   • All of the NH_3 has been exhausted and the total volume (60.00 mL) is now double the original volume of the acid. The solution is composed of 0.0500-M NH_4^+. The equivalence point corresponds to a pH of 5.28 (rather than 7.00).

4. Beyond the equivalence point (30.1–60.00 mL HCl):

   • The solution now contains an excess of H_3O^+ ions and as the pH falls, it approaches that of the titrant (0.1 M HCl; pH = 1.00).

Selecting an Appropriate Indicator

If we examine the titration curve, we see that methyl orange falls within the vertical region and is suitable for use as an indicator. Phenolphthalein, however, falls within the "buffer region" of the curve and is unsuitable in this case.

Titration Curves of Polyprotic Acids

The following diagrams represent the titration curves for the diprotic acid H_2CO_3 and the triprotic acid H_3PO_4:

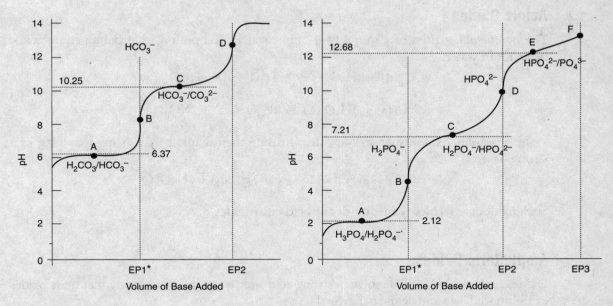

* EP = Equivalence Point

We can draw the following conclusions by examining these curves:

- The titration curve of a polyprotic acid has as many *equivalence points* as there are acidic hydrogen atoms. The curve for H_2CO_3 has two (points *B* and *D*), and the curve for H_3PO_4 has three (points *B*, *D*, and *F*).

- Points *A*, *C*, and *E* represent the optimum buffering pH values for each conjugate acid-base pair. These pH values correspond to the successive K_a values of each acid.

- With each successive ionization, the characteristic shape of the titration curve becomes less distinct.

ACID-BASE PROPERTIES OF OXIDES

Basic Oxides

Water-soluble *metallic* oxides, such as Na_2O, are ionic compounds that react with water to form hydroxide bases:

$$Na_2O(s) + H_2O(\ell) \rightarrow 2Na^+(aq) + 2OH^-(aq)$$

Such oxides are termed **basic oxides** or **base anhydrides**. Water-insoluble oxides, such as MgO, will dissolve in a strong acid as the H_3O^+ ion reacts with the O^{2-} ion:

$$MgO(s) + 2H_3O^+(aq) \rightarrow Mg^{2+}(aq) + 3H_2O(\ell)$$

The oxides of Groups 1 and 2 of the Periodic Table demonstrate basic properties in water.

Acidic Oxides

The *nonmetallic* oxides of Groups 14–17 are covalent compounds that demonstrate acidic behavior in water:

$$CO_2(g) + 2H_2O(\ell) \rightleftharpoons H_3O^+(aq) + HCO_3^-(aq)$$

$$N_2O_5(s) + 3H_2O\ (\ell) \rightarrow 2H_3O^+(aq) + 2NO_3^-(aq)$$

Water-insoluble oxides, such as SiO_2 demonstrate their acidic behavior by dissolving in a strong base:

$$SiO_2(s) + 2OH^-(aq) \rightarrow SiO_3^{2-}(aq) + 2H_2O(\ell)$$

Such oxides are termed **acidic oxides** or **acid anhydrides**.

Amphoteric Oxides

Oxides such as Al_2O_3 will dissolve in strong acid and strong base, indicating that these oxides exhibit both basic and acidic behavior:

Basic behavior: $Al_2O_3(s) + 6H_3O^+(aq) \rightarrow 2Al^{3+}(aq) + 9H_2O(\ell)$

Acidic behavior: $Al_2O_3(s) + 2OH^-(aq) + 3H_2O(\ell) \rightarrow 2Al(OH)_4^-(aq)$

Oxides that exhibit both acidic and basic behavior are known as **amphoteric oxides**.

LEWIS ACIDS AND BASES

In 1923, the American chemist G. N. Lewis proposed a model of acid-base behavior that was even more general than the Brønsted-Lowry model. In this model, acids and basis are classified according to their behavior with regard to electron-pairs:

Lewis acid: Any species that *accepts* a pair of electrons in a chemical reaction.

Lewis base: Any species that *donates* a pair of electrons in a chemical reaction.

When a Lewis acid reacts with a Lewis base, a covalent bond is formed, producing a superstructure known as a **Lewis acid-base complex**.

The following are examples of Lewis acid-base reactions:

- $H^+ + :NH_3 \rightarrow [NH_4^+]$
  In this reaction, the Lewis acid-base complex (NH_4^+) is formed when the proton accepts a pair of electrons from NH_3, creating a fourth N–H bond. The H^+ is the Lewis acid and the NH_3 is the Lewis base.

- $Cu^{2+} + 4 :NH_3 \rightarrow [Cu(NH_3)_4]^{2+}$
  In the formation of this complex ion (the Lewis acid-base complex), Cu^{2+} accepts four pairs of electrons from four NH_3 molecules; Cu^{2+} is the Lewis acid and NH_3 is the Lewis base.

- $BF_3 + :NH_3 \rightarrow F_3B–NH_3$
  In BF_3, boron has only six electrons in its valence shell. NH_3 donates a pair of electrons that completes the valence shell of boron and forms a covalent bond between B and N. BF_3, which has accepted a pair of electrons from NH_3 is the Lewis acid, and NH_3 is the Lewis base. The Lewis acid-base complex is the compound F_3BNH_3.

Chapter 14: Practice Questions

Multiple-Choice Questions

1. Which statement is true for the reaction between boron trifluoride and ammonia?

   $$BF_3 \text{ (aq)} + NH_3 \text{ (aq)} \rightarrow BF_3NH_3 \text{ (aq)}$$

   A. BF_3 is a Lewis base.
   B. BF_3 is an Arrhenius base.
   C. NH_3 is an Arrhenius acid.
   D. NH_3 is a Brønsted–Lowry base.
   E. BF_3 accepts a pair of electrons from ammonia.

2. The K_b of the weak base, methylamine, CH_3NH_2 is 4.2×10^{-4}. What is the K_a of methyl ammine?

   A. 4.2×10^{-18}
   B. 2.4×10^{-11}
   C. -4.2×10^{-4}
   D. 4.2×10^{-4}
   E. 4.2×10^{10}

3. Rank boric acid, formic acid, and hydrocyanic acid in order of increasing acidity.

Acid	Formula	K_a
Boric acid	H_3BO_3	5.9×10^{-10}
Formic acid	HCO_2H	1.7×10^{-4}
Hydrocyanic acid	HCN	4.9×10^{-10}

   A. Boric acid < formic acid < hydrocyanic acid
   B. Hydrocyanic acid < formic acid < boric acid
   C. Formic acid < boric acid < hydrocyanic acid
   D. Hydrocyanic acid < boric acid < formic acid
   E. Boric acid < hydrocyanic acid < formic acid

4. Which mixture should be chosen to prepare a buffer with a pH close to 3.4?

Acid	Formula	K_a
Acetic acid	CH_3COOH	1.7×10^{-5}
Ammonium ion	NH_4^+	5.6×10^{-10}
Hydrogen carbonate ion	HCO_3^-	4.7×10^{-11}
Hypochlorous acid	$HOCl$	3.5×10^{-8}
Nitrous acid	HNO_2	4.5×10^{-4}

   A. NH_4NO_3 and NH_3
   B. $HOCl$ and $NaOCl$
   C. CH_3COOH and $NaCH_3COO$
   D. HNO_2 and $NaNO_2$
   E. $NaHCO_3$ and Na_2CO_3

5. Which is *not* an acid-base conjugate pair?

   A. HS^-, S^{2-}
   B. H_3O^+, OH^-
   C. HNO_2, NO_2^-
   D. $CH_3NH_4^+$, CH_3NH_3
   E. C_6H_5COOH, $C_6H_5COO^-$

6. Which gives a basic solution when dissolved in water?

   I. $Na\ CH_3COO$

   II. $NaNO_3$

   III. NH_4NO_3

   A. I only
   B. II only
   C. I and II only
   D. I, II, and III
   E. None of the above

7. Which is true about the solution prepared mixing 50.00 mL of 1.00 M HCl and 25.00 mL of 1.00 M CH_3COOH ($K_a = 1.8 \times 10^{-5}$)?

    A. It is a solution with a pH greater than 7 that is not a buffer.

    B. It is a buffer solution with a pH between 7 and 10.

    C. It is a solution with a pH of 7.

    D. It is a buffer solution with a pH between 4 and 7.

    E. It is a solution with a pH less than 7 that is not a buffer.

8. Acid base indicators are complex molecules which are either acids or base and have different colored conjugate acid-base pairs. Which is true about indicators strength as acids or bases and where the color changes occur?

	Made up of:	Color changes at:
A.	strong acid–strong base	stoichiometric point
B.	strong acid–weak base	neutral pH
C.	weak acid–weak base	equivalence point
D.	weak acid–weak base	pH 7
E.	weak acid–strong base	low pH

9. The titration of a 0.100 M solution of the weak base cyanide ion, CN^- (HCN: $K_a = 4.9 \times 10^{-10}$), with a 0.100 M solution of the strong acid HNO_3 results in a solution at the equivalence point having a pH of:

    A. between 10 and 13.
    B. between 7 and 10.
    C. 7.
    D. between 4 and 7.
    E. between 1 and 3.

10. A 25.00 mL sample of a monoprotic weak acid ($K_a = 4.0 \times 10^{-6}$) is titrated with 0.1000 M NaOH. The equivalence point is reached when 28.54 mL of the base has been added. Which is a suitable indicator to detect the equivalence point?

	Indicator	Acid Color	Base Color	pH range
A.	Bromocresol green	Yellow	Blue	3.8–5.4
B.	Methyl red	Red	Yellow	4.2–6.1
C.	Bromthymol blue	Yellow	Blue	6.0–7.6
D.	Phenolphthalein	Colorless	Pink	8.0–9.8
E.	Thymolphthalein	Colorless	Blue	9.4–10.6

Free-Response Questions

1. A solution is made by dissolving 1.00 mole of propionic acid ($C_3H_5O_2H$; $K_a = 1.3 \times 10^{-5}$) and 0.40 mol of NaOH in enough water to make 1.00 L of solution.

   (A) Write a balanced equation to depict the reaction that occurs.
   (B) How many moles of propionic acid and its conjugate base, propionate ion are present after the reaction?
   (C) Calculate the pH of the solution.
   (D) What is the pH after the addition of 0.40 g of NaOH?

2. A 50.0 mL sample of 0.10 M acetic acid ($K_a = 1.7 \times 10^{-5}$) is titrated with 0.10 M NaOH.

   (A) What is the pH at half-equivalence?
   (B) What is the pH at the equivalence point?
   (C) Describe how a weak base–strong acid titration curve differs from a weak acid–strong base titration curve.

Multiple-Choice Answers

1. E   2. B   3. D   4. D   5. B   6. A   7. D   8. C   9. D   10. D

1. E

The correct answer is: boron trifluoride accepts a pair of electrons from ammonia. Boron trifluoride accepts a pair of electrons from ammonia, making it a Lewis acid. Since it neither produces nor donates a hydrogen ion, it is neither an Arrhenius acid nor a Brønsted–Lowry acid.

2. B

The equation used for the question is $K_b \times K_a = K_w$ and it is known that $K_w = 1 \times 10^{-14}$. With $K_b = 4.2 \times 10^{-4}$, simply plug this into the equation above to determine the value of K_a.

$$K_a = \frac{1 \times 10^{-14}}{4.2 \times 10^{-4}} = 2.4 \times 10^{-11}$$

3. D

The larger the value of the K_a, the stronger the acid. The strongest acid has the highest K_a, and the weakest acid has the lowest K_a. Formic acid is the strongest; hydrocyanic acid is the weakest; and boric acid falls in between. Note that 5.9×10^{-10} is bigger than 4.9×10^{-10}.

4. D

The buffer prepared by mixing equal quantities of nitrous acid and nitrite ion will have a $[H^+] = K_a$. The pH of this buffer solution is $pK_a = -\log K_a = -\log [4.5 \times 10^{-4}] = 3.35$, which is very close to the desired pH.

5. B

A conjugate acid-base pair is separated by a single proton which is donated by one and accepted by the other. The H_3O^+, OH^- pair is separated by *two* protons.

6. A

Sodium ions in aqueous solution are not acid or bases. They do not have protons to donate, and their positive charge does not allow them to accept protons. Nitrate ions are not acids (no H^+), and NO_3^- is the conjugate base of the strong acid HNO_3 (strong acid $\rightarrow$ no base), which makes them neutral also. NH_4^+ is the conjugate acid of a weak base (NH_3), and CH_3COO^- is the conjugate base of a weak acid (CH_3COOH). The only base in the group is acetate ion, CH_3COO^-.

7. D

The strong acid HCl reacts with the weak base, CH_3COO^-, forming the acetic acid–acetate ion buffer. $pK_a = -\log K_a = -\log [1.8 \times 10^{-5}] = 4.74$, a pH between 4 and 7.

8. C

Acid-base indicators are complex molecules that are either weak acids or weak bases. Their protonated (acid) and unprotonated (base) forms have different colors. The color changes at the equivalence point.

9. D

The equation for the reaction is HCN (aq) + H_2O (ℓ) $\rightleftharpoons$ H_3O^+ (aq) + CN^- (aq) and at the equivalence point $[H_3O^+] = [CN^-]$. Also at the equivalence point, the [HCN] is about 0.05 M, since the roughly equal amounts of weak base and strong acid will have been diluted by about half.

$\dfrac{x^2}{0.05} = 4.9 \times 10^{-10}$, with $x = [H_3O^+] = 5 \times 10^{-6}$. This leads to a pH of about 5.3.

10. D

The equation for the reaction is A^- + H_2O (ℓ) $\rightleftharpoons$ HA (aq) + OH^- (aq) and at the equivalence point $[HA] = [OH^-]$. The titration of a weak acid with a strong base will result in a basic solution containing A^-. The initial concentration of HA is by calculation, 0.114 M. At the equivalence point, the $[A^-]$ can be approximated at about 0.05 M, since the roughly equal amounts of weak base and strong acid have been diluted by about half. The basic A^- reacts with water, and its $K_b = \dfrac{K_w}{K_a} = 2.5 \times 10^{-9}$. $K_b = \dfrac{x^2}{0.05} = 2.5 \times 10^{-9}$, with $x = [OH^-] = 1.2 \times 10^{-5}$. This leads to a pH of about 9.0. The phenolphthalein with a range including this will be the indicator of choice.

Free-Response Answers

1. (A) The reaction between the base and acid is: $C_3H_5O_2^- + H_2O \rightarrow C_3H_5O_2H + OH^-$. This reaction can be either written the way shown, which as the K_b equation, or as the reverse equation.

    (B) The initial concentration of propionic acid is 1.00 M.

	$C_3H_5O_2^-$	$C_3H_5O_2H$	OH^-
Initial	0	1.00	0.400
Change	+0.400	−0.400	−0.400
Equilibrium	0.400	0.6000	

   (C) To determine the pH, the Henderson–Hasselbalch equation is used. The final concentrations of the acid and the base are 0.600 M and 0.400 M, respectively.

   $pH = pK_a + \log([base]/[acid])$

   $pH = 4.89 + \log(0.400/0.600) = 4.71$

   (D) The 0.40 g of NaOH is 0.010 mole added to the 1.00 L of solution.

	$C_3H_5O_2^-$	$C_3H_5O_2H$	OH^-
Initial	0.400	0.600	0.010
Change	+0.010	−0.010	−0.010
Equilibrium	0.410	0.590	0

   This is a buffer solution, so the pH should not change much.

   To determine the pH, the Henderson–Hasselbalch equation is again used. The final concentrations of the acid and the base are 0.590 M and 0.410 M, respectively.

   $pH = pK_a + \log([base]/[acid])$

   $pH = 4.89 + \log(0.410/0.590) = 4.73$

2. (A) The pH at the half-equivalence is $pK_a = -\log(1.7 \times 10^{-5}) = 4.77$.

(B) The equivalence point is reached when the amount of added base reacts with all of the initial H^+. The initial concentration of $[H^+]$ is $(0.10 \times 50.0) / 75.0 = 0.0013$ M, so at the equivalence point there will be 0.0013 M CH_3COO^- in solution, which is 6.5×10^{-5} mol. This is a weak base that accepts a proton from water to form OH^-. The K_b of this reaction is 5.9×10^{-10}. The volume of NaOH needed to react with the H^+ is 0.65 L.

$$CH_3COO^- \text{ (aq)} + H_2O \text{ (}\ell\text{)} \rightleftharpoons CH_3COOH \text{ (aq)} + OH^- \text{ (aq)}$$

	CH_3COO^-	CH_3COOH	OH^-
Initial	0.0013	0	0
Change	$-x$	$+x$	$+x$
Equilibrium	$0.0013 - x$	x	x

The $[CH_3COO^-]$ at equilibrium is 0.0013 M, since x is small.

$$5.9 \times 10^{-10} = \frac{x^2}{1.3 \times 10^{-3}}$$

$$x = [OH^-] = 8.8 \times 10^{-7}$$

$$pOH = -\log [OH^-] = 6.06$$

$$pH = 14.00 - 6.06 = 7.94$$

(C) The titration curve for a weak acid–strong base titration has the same shape as the elongated S-shaped titration curve for a weak base–strong acid titration, except that the curve is inverted. For a weak acid–strong base titration, the initial pH is low and it increases as base is added. For a weak base–strong acid titration, the initial pH is high and it decreases as acid is added.

Thermodynamics II: Spontaneous Reactions and Electrochemistry

In chapter 7, we learned that the First Law of Thermodynamics described energy transfers within chemical systems. We begin this chapter by studying the **Second Law of Thermodynamics**, a scientific principle that governs the *direction of spontaneous processes* within chemical systems.

SPONTANEOUS REACTIONS

A **spontaneous process** is a physical or chemical event that can occur without the intervention of some continuous external influence, such as work. As an example, consider a light bulb with a tungsten filament. As it operates, the temperature of the filament is sufficiently high to cause it to emit visible light. When the current to the bulb is turned off, the filament begins to cool immediately—a spontaneous process that does not require any intervention on our part. In contrast, consider the reverse process: the filament will remain hot and glowing only if we supply a continuous source of electric current to it. This is an example of a **nonspontaneous process**. We can generalize this example in the following two statements:

- A spontaneous process occurs without any external interference.

- The reverse of a spontaneous process is always nonspontaneous.

As another example, consider the glass bulb arrangements shown in the diagram below:

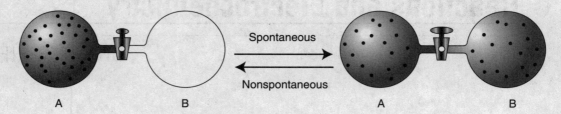

Before the stopcock is opened, gas molecules are present only in bulb *A* on the left. When the stopcock is opened, the gas flows spontaneously into bulb *B* on the right until the gas concentrations in both bulbs are equal. No external agent was needed to accomplish this process.

Now let us consider the reverse process: returning the bulbs to the initial condition shown on the left side of the diagram is virtually impossible unless external work (such as employing a pump) is continuously applied. We see, once again, that the reverse of any spontaneous process is nonspontaneous.

The word "spontaneous" is associated with the *natural direction* of a process and *not* with the speed at which it occurs: Spontaneous processes can be very slow, such as the rusting of iron, or very rapid, such as the formation of a precipitate when solutions of two ionic compounds (such as $AgNO_3(aq)$ and $NaCl(aq)$) are mixed.

Under what conditions can we expect a chemical system to react spontaneously? There is a very simple answer to this question:

- A spontaneous change will occur when both energy and matter become more disordered.

In the example of the cooling bulb filament, the thermal energy stored in the atoms of the hot filament spread to its surroundings, that is, the energy has become more disordered. The reverse process is not observed (spontaneously) because it is highly improbable that the energy in the much larger surroundings will reconcentrate itself in the small filament. In the example of the expanding gas, the molecules have assumed a more disordered arrangement when they fill both containers. Again it is highly unlikely that the reverse process will occur spontaneously.

Entropy: The Measure of Disorder

The property that measures disorder is a state function known as **entropy**, *S*. Just as internal energy measures the *quantity* of the energy stored within a system, entropy measures the *disorder* of that stored energy as well as the disorder of the matter present in the system. If the entropy of a system is low, so is the degree of disorder. For example, the entropy of $H_2O(s)$ is lower than that of $H_2O(g)$ because an ice crystal is a much more orderly arrangement of H_2O molecules than is present in water vapor.

THERMODYNAMICS II: SPONTANEOUS REACTIONS AND ELECTROCHEMISTRY

Entropy is also dependent on temperature. Since the absolute temperature is a measure of the average molecular kinetic energy, increasing the temperature of a substance increases its disorder and, therefore, its entropy. The following diagram illustrates how the entropy of a substance might change with temperature:

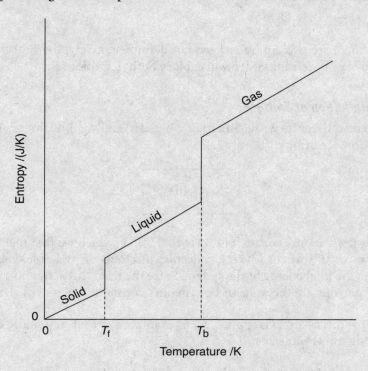

Note that *the entropy of the solid approaches zero as the absolute temperature approaches zero.* This statement is known as the **Third Law of Thermodynamics**.

Also note that while phase changes (melting and boiling) occur at constant temperature, the entropy of the substance continues to rise. Recall that heat is absorbed during phase changes, and this absorbed energy increases the disorder of the atoms.

How can we predict the change in the entropy of a chemical system? In general, the entropy *increases* when:

- An "upward" phase change occurs (solid → liquid; liquid → gas; solid → gas):
  $CO_2(s) \rightarrow CO_2(g)$

- There is an increase in the number of moles of products, compared with reactants:
  $4NH_3(g) + 5O_2(g) \rightarrow 4NO(g) + 6H_2Og)$

- The products have a molecular structure that is simpler than that of the reactants:
  $PCl_5(g) \rightarrow PCl_3(g) + Cl_2(g)$

- A solid produces a gas in a chemical reaction:

  $CaCO_3(s) \rightarrow CaO(s) + CO_2(g)$

- A solute dissolves in a solvent:

  $C_6H_{12}O_6(s) \rightarrow C_6H_{12}O_6(aq)$

Mathematically, there are two equivalent ways of defining entropy: the statistical definition and the thermodynamic definition. We will explore both definitions.

The Statistical Definition of Entropy

In the mid-nineteenth century, the Austrian physicist Ludwig Boltzmann proposed the following definition of entropy:

$$S = k \ln W$$

The quantity k is Boltzmann's constant $(1.38 \times 10^{-23}$ J/K) (which we first met in chapter 6), and W is the number of ways in which a collection of atoms or molecules can be arranged without changing the total energy of the system. Since the term "$\ln W$" has no units, it follows that the units of entropy are the units of Boltzmann's constant: joules per kelvin (J/K).

In order to see how this formula is applied, let us examine a crystal containing 100 molecules of NO, a simple diatomic molecule.

If the crystal is perfect at 0 K, there is only one possible arrangement of the molecules, which is shown in the diagram below:

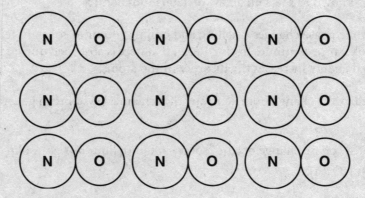

The entropy of this system is:

$$S = k \ln W = (1.38 \times 10^{-23} \text{ J/K}) \cdot \ln 1 = 0 \text{ J/K}$$

THERMODYNAMICS II: SPONTANEOUS REACTIONS
AND ELECTROCHEMISTRY

If the crystal is imperfect, there is more than one possible arrangement. Let us assume that each molecule can assume only one of two possible orientations, as shown in the diagram below:

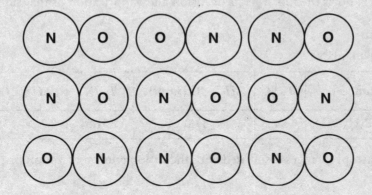

In this case, the number of possible arrangements is 2^{100}, and the entropy of the system is:

$S = k \ln W = (1.38 \times 10^{-23} \text{ J/K}) \cdot \ln(2^{100}) = (100) \cdot (1.38 \times 10^{-23} \text{ J/K}) \cdot \ln 2 = 9.57 \times 10^{-22} \text{ J/K}$

Clearly, the entropy of the imperfect crystal is larger than that of the perfect crystal.

Finally, suppose that the imperfect contained one mole of molecules, a situation with possible arrangements:

$S = k \ln W = (1.38 \times 10^{-23} \text{ J/K}) \cdot \ln 2^{6.02 \times 10^{23}} = (6.02 \times 10^{23}) \cdot (1.38 \times 10^{-23} \text{ J/K}) \cdot \ln 2 = 5.76 \text{ J/K}$

Since the number of molecules in this case is measured in terms of moles, S is usually called the **molar entropy** and its units are reported as J/mol·K.

The Thermodynamic Definition of Entropy

The thermodynamic definition of entropy is one that is based on heat and temperature and was proposed as the result of work done by the French engineer Sadi Carnot and the German physicist Rudolph Clausius. In the thermodynamic sense, the change in entropy, ΔS, is defined as:

$$\Delta S = \frac{q_{\text{rev}}}{T}$$

in which q_{rev} represents a *reversible* transfer of heat occurring at a constant absolute temperature T. By "reversible" we mean that the transfer takes place at an *infinitesimal rate*, an ideal situation that allows the transfer to be reversed at any moment. If heat is absorbed, ΔS is positive; if heat is released, ΔS is negative.

Phase changes are good examples of heat transfers that occur at constant temperature. For example, the entropy changes associated with the melting and boiling of a substance can be calculated if one knows the molar heats of fusion and vaporization of that substance.

Sample Problem

Given the following data:

Substance	T_f/K	ΔH_{fus}^{o}/(kJ/mol)	T_b/K	ΔH_{vap}^{o}/(kJ/mol)
Ethanol (C_2H_5OH)	158.7	4.60	351.5	43.5

Calculate the entropy changes associated with the fusion and vaporization of ethanol.

Solution

$$\Delta S_{fus} = \frac{\Delta H_{fus}^{o}}{T_f} = \frac{+4{,}600 \text{ J/mol}}{158.7 \text{ K}} = \textbf{+29.0 J/mol} \cdot \textbf{K}$$

$$\Delta S_{vap} = \frac{\Delta H_{vap}^{o}}{T_b} = \frac{+43{,}500 \text{ J/mol}}{351.5 \text{ K}} = \textbf{+124 J/mol} \cdot \textbf{K}$$

According to the thermodynamic definition, the change in entropy is proportional to the quantity of heat that is transferred. We should expect this result: For example, as more heat is absorbed by an object, the thermal energy of the atoms increases and leads to a more disordered state.

This definition also indicates that the change in entropy is inversely proportional to the absolute temperature. At a high temperature, a substance already has considerable entropy, and any change in this quantity will be proportionately smaller than the change encountered at a lower temperature.

Sample Problem

Calculate the change in entropy when 16.0 joules of heat is transferred reversibly out of a system:

(A) at 25°C   (298 K)
(B) at 100°C   (373 K)

Solution

$$\Delta S = \frac{q_{rev}}{T}$$

(A) $\Delta S = \dfrac{-16.0 \text{ J}}{298 \text{ K}} = \textbf{-0.0537 J/K}$

(B) $\Delta S = \dfrac{-16.0 \text{ J}}{373 \text{ K}} = \textbf{-0.0429 J/K}$

As we can see, the change in entropy is smaller at the higher temperature.

The Second Law of Thermodynamics

As we indicated at the beginning of this chapter, the Second Law of Thermodynamics is a statement that governs the direction of spontaneous processes. In terms of entropy, the Second Law is embodied in the following statement:

> In any spontaneous process, the entropy of the universe increases: $\Delta S_{univ} > 0$

What exactly is ΔS_{univ}? Recall that we can divide the universe into (1) a *system* that we study and (2) all the rest, which we call the *surroundings*:

$$\Delta S_{univ} = \Delta S_{system} + \Delta S_{surroundings}$$

In a chemical system, the surroundings need not be the rest of the *entire* universe: for practical purposes, it could be the immediate environment in which the system is placed.

As a corollary to the Second Law, we can state that $\Delta S_{univ} < 0$ means that a process is nonspontaneous and that $\Delta S_{univ} = 0$ means that the system and its surroundings are in equilibrium.

Sample Problem

Suppose that 10.0 joules of heat is reversibly transferred from a system maintained 200 K, to its surroundings maintained at 400 K. Show that this process must be nonspontaneous.

Solution

$$\Delta S_{univ} = \Delta S_{system} + \Delta S_{surroundings}$$

$$\Delta S_{univ} = \frac{-10.0 \text{ J}}{200 \text{ K}} + \frac{+10.0 \text{ J}}{400 \text{ K}} = -0.0500 \text{ J/K} + 0.0250 \text{ J/K} = -0.0250 \text{ J/K}$$

Since, $\Delta S_{univ} < 0$, the process is nonspontaneous

It might seem a bit weird that we define the entropy change in terms of a *reversible process* and yet use it to describe *spontaneous processes*, which are, by their very nature, irreversible. Remember, however, that entropy is a *state function*: its change depends on the initial and final states of the system and not *how* the change is made.

Standard Molar Entropies

Using the Third Law of Thermodynamics (see page 373) and a number of advanced techniques, it is possible to calculate the absolute entropy of a substance at a given temperature. It is customary to calculate the entropy associated with one mole of substance under standard conditions, a quantity known as the **standard molar entropy**, $S°$. The usual reference temperature is 298.15 K, and the units of $S°$ are joules per mole·kelvin (J/mol·K). A table of standard molar entropies is given in Appendix B.

The Standard Entropy Change of a Chemical Reaction

We calculate the standard entropy change of a chemical reaction, $\Delta S°$, in the same manner as we calculate $\Delta H°$ (See chapter 7):

$$\Delta S° = \left(\Sigma n \cdot S°\right)_{products} - \left(\Sigma m \cdot S°\right)_{reactants}$$

The symbols n and m represent the respective coefficients of the products and reactants.

Sample Problem

Using the appropriate standard molar entropies in Appendix B, calculate the standard entropy change for the reaction: $Fe_2O_3(s) + 6HCl(g) \rightarrow 2FeCl_3(s) + 3H_2O(g)$.

Solution

$$\Delta S° = (2 \cdot S°[FeCl_3(s)] + 3 \cdot S°[H_2O(g)]) - (1 \cdot S°[Fe_2O_3(s)] + 6 \cdot S°[HCl(g)])$$

$$\Delta S° = (2 \text{ mol} \cdot (142.3 \text{ J/mol·K}) + 3 \text{ mol} \cdot (188.8 \text{ J/mol·K})) - (1 \text{ mol} \cdot (87.4 \text{ J/mol·K}) + 6 \text{ mol} \cdot (186.9 \text{ J/mol·K}))$$

$$\Delta S° = -357.8 \text{ J/K}$$

This is a result we should have expected: There are fewer moles of product (5) than reactant (7), and an additional mole of solid product has been produced by the reaction.

Predicting the Spontaneity of a Chemical Reaction: Free Energy

Let us consider the following spontaneous reaction:

$$CH_4(g) + 2O_2(g) \rightarrow CO_2(g) + 2H_2O(\ell) \ [\Delta H° = -890.3 \text{ kJ}]$$

Clearly, the system has lost potential energy, which appears as heat in the surroundings. Since the reaction is spontaneous, we can conclude that the expelled heat increased the entropy of the surroundings in a way that caused the entropy of the universe to increase.

Now let us consider another spontaneous reaction, the dissociation of N_2O_4 into NO_2:

$$N_2O_4(g) \rightarrow 2NO_2(g) \ [\Delta H° = +57.1 \text{ kJ}]$$

Evidently, this reaction, which absorbs heat from its surroundings, must also be capable of increasing the entropy of the universe.

In order to predict whether a reaction will be spontaneous, we must be able to calculate the entropy changes of a system and its surroundings. It would be far more helpful if we could focus *exclusively* on the reaction and avoid examining its surroundings.

THERMODYNAMICS II: SPONTANEOUS REACTIONS
AND ELECTROCHEMISTRY

Let us examine a hypothetical spontaneous reaction that occurs at constant temperature and pressure and follow the various changes, step by step:

- We know that the entropy of the universe must increase:
  $$\Delta S_{univ} = \Delta S_{system} + \Delta S_{surroundings} > 0$$

- Since the reaction occurs at constant pressure, $q_p = \Delta H$

- The definition of entropy change states: $\Delta S = \dfrac{q}{T} = \dfrac{\Delta H}{T}$

- Combining terms, we obtain:
  $$\Delta S_{univ} = \Delta S_{system} + \Delta S_{surroundings} = \Delta S_{system} + \left(\dfrac{\Delta H}{T}\right)_{surroundings} > 0$$

- Since energy is always conserved: $\Delta H_{surroundings} = -\Delta H_{system}$

- Since we can express all of the quantities in terms of the *system*, there is no need to use the subscript "system" any longer: $\Delta S - \dfrac{\Delta H}{T} > 0$

- Finally, we multiply both sides of the inequality by $(-T)$ to obtain a very important criterion that must be met if a reaction is to be spontaneous:

$$\boxed{\Delta H - T\Delta S < 0}$$

Both terms in this expression have energy units (such as kJ/mol). Since a negative energy change indicates that energy is released, it follows that *some form of energy must be released by the system into the environment if the reaction is to be spontaneous*. This form of energy is known as the **Gibbs Free Energy**, whose change is defined as:

$$\boxed{\Delta G = \Delta H - T\Delta S}$$

- If $\Delta G < 0$, the reaction is spontaneous.
- If $\Delta G > 0$, the reaction is nonspontaneous.
- If $\Delta G = 0$, the system is at equilibrium.

When a reaction is spontaneous, the system loses free energy to the environment. It can be shown that the free energy change is the maximum amount of energy that is available for *useful work*. This implies that a portion of the energy of the system is unavailable for such work and is frequently termed "waste heat." (If you studied heat engines in physics, you learned that a heat engine can *never* convert all of its heat input into work because a part of the input is always sent to the "cold reservoir" as waste heat.)

According to the definition of free energy change, we need to measure three quantities in order to determine whether a reaction will be spontaneous: enthalpy change, entropy change, and the absolute temperature at which the reaction occurs. The following table summarizes this information:

Enthalpy Change	Entropy Change	Example	Spontaneous?
Exothermic ($\Delta H < 0$)	Increase ($\Delta S > 0$)	$2NO_2(g) \rightarrow 2N_2(g) + O_2(g)$	Yes, always
Exothermic ($\Delta H < 0$)	Decrease ($\Delta S < 0$)	$H_2O(\ell) \rightarrow H_2O(s)$	Only at lower temperatures
Endothermic ($\Delta H > 0$)	Increase ($\Delta S > 0$)	$2NH_3(g) \rightarrow N_2(g) + 3H_2(g)$	Only at higher temperatures
Endothermic ($\Delta H > 0$)	Decrease ($\Delta S < 0$)	$3O_2(g) \rightarrow 2O_3(g)$	No, never

Standard Free Energies of Formation

The free energy change associated with a formation reaction is known as the **standard free energy of formation** (ΔG_f°). The unit of this quantity is kilojoule per mole of compound (kJ/mol). The thermochemical equation for the formation of $Al_2O_3(s)$ is shown below:

$$Al(s) + \frac{3}{2}O_2(s) \rightarrow Al_2O_3(s) \qquad\qquad \Delta G_f^\circ = -1{,}582.3 \text{ kJ/mol}$$

This means that under standard conditions and at 298.15 K the formation reaction is spontaneous and the reverse reaction, the decomposition of $Al_2O_3(s)$ is nonspontaneous. Compounds that have negative free energies of formation are said to be *thermodynamically stable compounds*. Such compounds have no tendency to decompose spontaneously into their constituent elements. By definition, all elements in their stable states at 298.15 K are assigned a standard free energy of formation of zero. A listing of standard free energies of formation is given in Appendix B.

Standard Free Energies of Reactions

The standard free energy of a reaction, ΔG°, can be calculated by: (1) using the definition of free energy change directly ($\Delta G^\circ = \Delta H^\circ - T\Delta S^\circ$), and (2) by using the standard free energies of formation of each component in the reaction. The first method is more tedious since ΔH° and ΔS° for the reaction must be calculated first. However, since we calculated these values for the reaction $Fe_2O_3(s) + 6HCl(g) \rightarrow 2FeCl_3(s) + 3H_2O(g)$ previously (in chapter 7 and in this chapter), we will use these results to calculate ΔG° at 298.15 K:

Sample Problem

Calculate ΔG° for the reaction: $Fe_2O_3(s) + 6HCl(g) \rightarrow 2FeCl_3(s) + 3H_2O(g)$, given that $\Delta H^\circ = -146.4$ kJ and $\Delta S^\circ = -357.8$ J/K at 298.15 K.

Solution

Note: It is necessary to convert $\Delta S°$ to kJ/mol · K

$$\Delta G° = \Delta H° - T\Delta S° = (-146.4 \text{ kJ}) - (298.15 \text{ K}) \cdot (-0.3578 \text{ kJ/K}) = \textbf{−39.72 kJ}$$

∴ The reaction is spontaneous.

We can use the definition of free energy change to calculate the minimum or maximum temperature at which a given reaction will be spontaneous. This is shown in the next sample problem.

Sample Problem

Calculate the maximum temperature at which the forward reaction given in the last sample problem will be spontaneous.

Solution

When both $\Delta H°$ and $\Delta S°$ are negative, a reaction will be spontaneous (that is, $\Delta G° < 0$) at *lower temperatures*. If we determine the temperature at which $\Delta G° = 0$, we will be calculating the *maximum* temperature at which the reaction will be spontaneous.

$$\Delta G° = \Delta H° - T\Delta S°$$

$$0 \text{ kJ} = (-146.4 \text{ kJ}) - T(-0.3578 \text{ kJ/K})$$

$$T = \frac{(146.4 \text{ kJ})}{(0.3578 \text{ kJ/K})} = \textbf{409.2 K}$$

∴ At any temperature less than 409.2 K, the forward reaction will be spontaneous.

In calculating the standard free energy of a reaction from standard free energies of formation, one uses the same technique that was developed for enthalpies of formation in chapter 7. The relevant equation is:

$$\Delta G° = \left(\sum n \cdot \Delta G_f°\right)_{products} - \left(\sum m \cdot \Delta G_f°\right)_{reactants}$$

The symbols n and m represent the respective coefficients of the products and reactants, respectively.

Sample Problem

Using the appropriate standard free energies of formation found in Appendix B, calculate the standard free energy change for the reaction: $Fe_2O_3(s) + 6HCl(g) \rightarrow 2FeCl_3(s) + 3H_2O(g)$.

Solution

$$\Delta G° = (2 \cdot \Delta G_f° [FeCl_3(s)] + 3 \cdot \Delta G_f° [H_2O(g)]) - (1 \cdot \Delta G_f° [Fe_2O_3(s)] + 6 \cdot \Delta G_f° [HCl(g)])$$

$$\Delta G° = (2 \text{ mol} \cdot (-334.0 \text{ kJ/mol}) + 3 \text{ mol} \cdot (-228.6 \text{ kJ/mol})) - (1 \text{ mol} \cdot (-742.2 \text{ kJ/mol})$$
$$+ 6 \text{ mol} \cdot (-95.3 \text{ kJ/mol}))$$

$$\Delta G° = \textbf{−39.80 kJ}$$

The two methods for calculating the standard free energy of this reaction yield results that are in very close agreement: the two values are within 0.2% of one another.

Free Energy and Composition

The free energy of a reaction varies with the composition of its components. The diagram below tracks the change in free energy with the progress of a reaction.

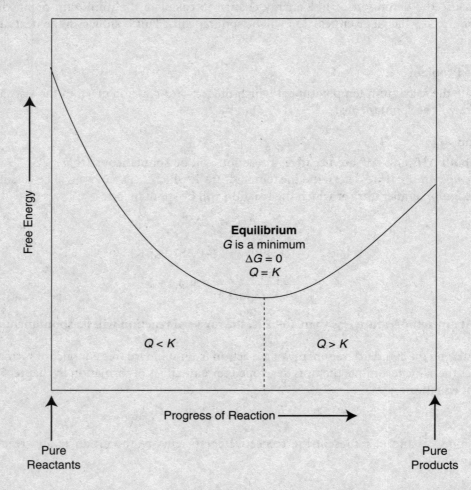

At the lowest point on the curve, the free energy is at a minimum. Since the system is unable to lose any free energy at this point, $\Delta G = 0$ and the system is at equilibrium. The reaction quotient (Q) at this point equals the value of the equilibrium constant.

At the left side of the diagram, only the reactants are present. Since the free energy of the reactants are greater than the free energy of the equilibrium mix, free energy will be lost ($\Delta G < 0$) and the reaction will proceed spontaneously in the forward direction toward equilibrium. In this region, $Q < K$.

THERMODYNAMICS II: SPONTANEOUS REACTIONS AND ELECTROCHEMISTRY

At the right side of the diagram, only the products are present. Since the free energy of the products are also greater than the free energy of the equilibrium mix, free energy will be lost ($\Delta G < 0$) and the reaction will proceed spontaneously in the reverse direction toward equilibrium. In this region, $Q > K$.

In the diagram, the position of the equilibrium point is nearly midway between the reactants and products. In this case we expect to have comparable quantities of reactants and products in the equilibrium mix. If the equilibrium point were positioned very near the pure products, we would classify this as a reaction that "goes nearly to completion." Conversely, if the equilibrium point were positioned very near the pure reactants, we would classify this as a reaction that "hardly goes forward."

The free energy of a reaction (ΔG) is dependent on temperature as well as on composition, and is given by the relation:

$$\Delta G = \Delta G^\circ + RT\ln Q$$

Under standard conditions, all solids and liquids are in their pure states and assigned a composition of 1. In addition, the partial pressures of all gases are set at 1 atmosphere, and the concentrations of all dissolved substances are set at 1 M. Under these conditions, $Q = 1$ ($\ln Q = 0$) and $\Delta G = \Delta G^\circ$, as we would expect. Under nonstandard conditions, we need to use the expression given above.

Sample Problem

For the reaction $N_2O_4(g) \rightleftharpoons 2NO_2(g)$, $\Delta G^\circ = +2.8$ kJ at 298.15 K. Calculate the free energy change of this reaction if the partial pressure of both gases are 100. atm.

Solution

$$\Delta G = \Delta G^\circ + RT\ln Q = \Delta G^\circ + RT\ln \frac{P_{NO_2}^2}{P_{N_2O_4}}$$

Note: We need to convert kilojoules to joules.

$$\Delta G = +2.8 \text{ kJ} \cdot \left(\frac{1{,}000 \text{ J}}{1 \text{ kJ}}\right) + (8.315 \text{ J/mol} \cdot \text{K})(298.15 \text{ K})\ln\frac{(100.)^2}{(100.)} = +14.2 \text{ kJ}$$

At this temperature and composition, the forward reaction is not spontaneous.

Recall that at equilibrium, $\Delta G = 0$ and $Q = K$. Substituting these conditions into the free energy expression and rearranging terms, we obtain:

$$\Delta G^\circ = -RT\ln K$$

We can draw the following conclusions from this expression:

- If $\Delta G° < 0$, $K > 1$
- If $\Delta G° > 0$, $K < 1$
- If $\Delta G° = 0$, $K = 1$

Sample Problem

Calculate the equilibrium constant, K_P, for the reaction given in the last sample problem.

Solution

$$\Delta G° = -RT\ln K$$

$$+2800 \text{ J} = -(8.315 \text{ J/mol} \cdot \text{K})(298.15 \text{ K})\ln K$$

$$\ln K = \frac{(+2,800)}{-(8.315)(298.15)} = -1.13$$

$$K = e^{-1.13} = \mathbf{0.32}$$

OXIDATION AND REDUCTION

Oxidation and reduction involve the transfer or the apparent transfer of electrons. When the process of oxidation occurs, one or more electrons are lost by a species; when reduction occurs, one or more electrons are gained.

Oxidation: A loss of electrons

Reduction: A gain of electrons

Not only are oxidation and reduction opposite processes, but they must be linked together: If one species loses electrons, another species must be able to gain them. We call all chemical reactions involving oxidation and reduction **redox** reactions. In any redox reaction, the species that is oxidized is known as the **reducing agent**, and the species that is reduced is known as the **oxidizing agent**.

In order to determine whether a chemical reaction is a redox reaction, we need to examine the oxidation number of each element in the reaction. An **oxidation number** is a real or apparent charge on an atom or ion. The rules for assigning oxidation numbers are as follows:

- *A free element* (such as He, Al, N_2, or P_4) *is assigned an oxidation number of 0.*

- *The oxidation number of an ion is its charge.* For example, the oxidation number of Na^+ is 1+ and the oxidation number of O^{2-} is 2−.

THERMODYNAMICS II: SPONTANEOUS REACTIONS AND ELECTROCHEMISTRY

- *In molecular compounds, oxygen nearly always has an oxidation number of 2–.* (A notable exception is the class of compounds known as *peroxides*, such as H_2O_2, in which the oxidation number of oxygen is 1–.)

- *In molecular compounds, hydrogen nearly always has an oxidation number of 1+.* (A notable exception is the class of compounds known as *hydrides*, such as NaH, in which the oxidation number is 1–.)

- *The sum of all the oxidation numbers in a compound must total 0.* For example, in the compound P_2O_5, the total oxidation number of the five oxygen atoms is 10–. Therefore, the phosphorus atoms must have a total oxidation number of 10+. Since there are two phosphorus atoms, each atom must have an oxidation number of 5+.

- *The sum of all the oxidation numbers in a polyatomic ion must equal the charge on the ion.* For example, in the ClO_4^- ion, the total oxidation number of the four oxygen atoms is 8–. Therefore, the chlorine atom must have an oxidation number of 7+.

Since redox reactions involve real or apparent electron transfers, the species involved will undergo changes in oxidation number: *the species that is oxidized will exhibit an increase in oxidation number, and the species that is reduced will exhibit a decrease in oxidation number.* If there is no change in oxidation number, the reaction is not a redox reaction.

Sample Problem

Determine whether the following equations (which may or may not be balanced) represent redox reactions. For those reactions that are redox reactions, identify the species that is oxidized and the species that is reduced.

(A) $N_2 + H_2 \rightarrow NH_3$

(B) $H^+ + OH^- \rightarrow H_2O$

(C) $Cu + HNO_3 \rightarrow Cu(NO_3)_2 + NO_2 + H_2O$

(D) $KOH + HBr \rightarrow KBr + H_2O$

Solution

In each case we rewrite the equations with all of the oxidation numbers included:

(A) $\overset{0}{N_2} + \overset{0}{H_2} \rightarrow \overset{3-}{N}\overset{+}{H_3}$

This is a redox reaction. The oxidation number of "N" decreases from 0 to 3–: it is reduced; the oxidation number of "H" increases from 0 to 1+: it is oxidized.

(B) $\overset{+}{H} + \overset{2-}{O}\overset{+}{H} \rightarrow \overset{+}{H_2}\overset{2-}{O}$

This is *not* a redox reaction: there are no changes in oxidation numbers.

(C) $\overset{0}{C}u + \overset{+}{H}\overset{+}{N}\overset{2-}{O}_3 \rightarrow \overset{2+}{C}u(\overset{5+}{N}\overset{2-}{O}_3)_2 + \overset{4+}{N}\overset{2-}{O}_2 + \overset{+}{H}_2\overset{2-}{O}$

This is a redox reaction. The oxidation number of "Cu" increases from 0 to 2+: it is oxidized; the oxidation number of (some of the) "N" decreases from 5+ to 4+: it is reduced.

(D) $\overset{+}{K}\overset{2-}{O}\overset{+}{H} + \overset{+}{H}\overset{-}{B}r \rightarrow \overset{+}{K}\overset{-}{B}r + \overset{+}{H}_2\overset{2-}{O}$

This is *not* a redox reaction: there are no changes in oxidation numbers.

Half-Reactions

It is possible to split redox reactions into two **half-reactions**: one for oxidation and one for reduction. In writing half-reactions, we include the numbers of electrons that are lost or gained.

For example, in part (C) of the last sample problem, the half reactions are:

$\overset{0}{C}u \rightarrow \overset{2+}{C}u + 2\overset{-}{e}$ (oxidation half-reaction)

$\overset{5+}{N} + \overset{-}{e} \rightarrow \overset{4+}{N}$ (reduction half-reaction)

While the oxidation half-reaction involves a *real* loss of two electrons, the gain of one electron in the reduction half-reaction is only *apparent* since $\overset{5+}{N}$ and $\overset{4+}{N}$ do *not* exist as separate species.

Therefore, a half-reaction is only a conceptual way of illustrating oxidation or reduction.

Sample Problem

Write the half-reactions for the redox reaction given in part (A) of the last sample problem.

Solution

$\overset{0}{H} \rightarrow \overset{+}{H} + \overset{-}{e}$ (oxidation half-reaction)

$\overset{0}{N} + 3\overset{-}{e} \rightarrow \overset{3}{N}$ (reduction half-reaction)

BALANCING REDOX REACTIONS

Balancing Redox Reactions Using Half-Reactions

Let us begin by balancing the simple redox reaction $N_2 + H_2 \rightarrow NH_3$. In the two last sample problems we identified the oxidation numbers and the oxidation and reduction half-reactions:

$$\overset{0}{N_2} + \overset{0}{H_2} \rightarrow \overset{3-}{N}\overset{+}{H_3}$$

$$\overset{0}{H} \rightarrow \overset{+}{H} + \overset{-}{e} \qquad \text{(oxidation half-reaction)}$$

$$\overset{0}{N} + 3\overset{-}{e} \rightarrow \overset{3-}{N} \qquad \text{(reduction half-reaction)}$$

Observe that while three electrons are gained by the nitrogen, only one electron is lost by the hydrogen. In order to ensure that electric charge is conserved, we must modify the half-reactions so that the number of electrons lost equals the number of electrons gained. This is accomplished by multiplying the first half-reaction by 3:

$$3\cdot(\overset{0}{H} \rightarrow \overset{+}{H} + \overset{-}{e})$$

$$\overset{0}{N} + 3\overset{-}{e} \rightarrow \overset{3-}{N}$$

We can now add these half-reactions together to obtain the **balanced redox couple:**

$$\overset{0}{N} + 3\overset{0}{H} \rightarrow \overset{3-}{N} + 3\overset{+}{H} + \overset{-}{e}$$

We can now insert the appropriate coefficients back into the original equation, remembering that molecular nitrogen and hydrogen are both *diatomic*:

$$\frac{1}{2}N_2 + \frac{3}{2}H_2 \rightarrow NH_3$$

or, using simplest whole-number coefficients:

$$N_2 + 3H_2 \rightarrow 2NH_3$$

Many redox reactions can be very difficult to balance simply by inspection. A good case in point is the reaction between copper and nitric acid:

$$Cu(s) + HNO_3(aq) \rightarrow Cu(NO_3)_2(aq) + NO(g) + H_2O(\ell)$$

Equations involving redox reactions can be balanced, however, by employing the same technique that was used to balance the N_2–H_2–NH_3 equation shown above.

The steps in this technique are outlined as follows:

(1) Rewrite the equation with its oxidation numbers.

$$\overset{0}{Cu} + \overset{+}{H}\overset{5+}{N}\overset{2-}{O_3} \rightarrow \overset{2+}{Cu}(\overset{5+}{N}\overset{2-}{O_3})_2 + \overset{2+}{N}\overset{2-}{O} + \overset{+}{H_2}\overset{2-}{O}$$

(2) Identify the species that are oxidized and reduced.

Cu is oxidized; its oxidation number increases from 0 to 2+.

N is reduced; its oxidation number decreases from 5+ to 2+.

(3) Write the oxidation and reduction half-reactions.

$$\overset{0}{Cu} \rightarrow \overset{2+}{Cu} + 2\overset{-}{e} \quad \text{(oxidation half-reaction)}$$

$$\overset{5+}{N} + 3\overset{-}{e} \rightarrow \overset{2+}{N} \quad \text{(reduction half-reaction)}$$

(4) Multiply each half-reaction so that the number of electrons lost by the oxidized species is equal to the number of electrons gained by the reduced species.

$$3 \cdot (\overset{0}{Cu} \rightarrow \overset{2+}{Cu} + 2\overset{-}{e})$$

$$2 \cdot (\overset{5+}{N} + 3\overset{-}{e} \rightarrow \overset{2+}{N})$$

(5) Add the two "balanced" half-reactions together to obtain a redox couple.

$$3\overset{0}{Cu} + 2\overset{5+}{N} \rightarrow 3\overset{2+}{Cu} + 2\overset{2+}{N}$$

(6) Insert the coefficients from the skeleton back into the original equation by matching each species with its oxidation number. Do not insert the coefficient of any species that appears in more than one place in the equation.

$$3\overset{0}{Cu} + \overset{+}{H}\overset{5+}{N}\overset{2-}{O_3} \rightarrow 3\overset{2+}{Cu}(\overset{5+}{N}\overset{2-}{O_3})_2 + 2\overset{2+}{N}\overset{2-}{O} + \overset{+}{H_2}\overset{2-}{O}$$

Since N^{5+} appears twice, we do not insert the coefficient for it at this time.

$$3Cu + \ldots HNO_3 \rightarrow 3Cu(NO_3)_2 + 2NO + \ldots H_2O$$

(7) Balance the rest of the equation (the nonredox part) by inspection.

$$3Cu + 8HNO_3 \rightarrow 3Cu(NO_3)_2 + 2NO + 4H_2O$$

Sample Problem

Balance the following redox reactions using half-reactions:

(A) $Cu + HNO_3 \rightarrow Cu(NO_3)_2 + NO_2 + H_2O$

(B) $HNO_3 + I_2 \rightarrow HIO_3 + NO_2 + H_2O$

Solution

(A) $Cu + 4HNO_3 \rightarrow Cu(NO_3)_2 + 2NO_2 + 2H_2O$

(B) $10HNO_3 + I_2 \rightarrow 2HIO_3 + 10NO_2 + 4H_2O$

Balancing Redox Reactions by the Ion-Electron Method

The redox equation between iodine and nitric acid: $I_2 + HNO_3 \rightarrow HIO_3 + NO_2 + H_2O$ can be written as an ionic equation, stripped of all the species that do not actually participate in oxidation-reduction:

$$I_2 + NO_3^- \rightarrow HIO_3 + NO_2 \text{ (acidic solution)}$$

We need to indicate that the solution is *acidic*, because HNO_3 is not written out.

This ionic redox reaction can be balanced by a technique known as the **ion-electron method**. It is simpler than balancing by using half-reactions because oxidation numbers do not have to be assigned. Instead, we balance for both mass and charge in a step-by-step fashion:

(1) Separate the equation into half-reactions by selecting similar species on both sides of the arrow.

$$NO_3^- \rightarrow NO_2$$
$$I_2 \rightarrow HIO_3 + NO_2$$

(2) Balance the half-reactions for all atoms that are neither hydrogen nor oxygen.

$$NO_3^- \rightarrow NO_2$$
$$I_2 \rightarrow \mathbf{2}HIO_3$$

(3) Balance the half-reactions for oxygen by adding one H_2O molecule for each oxygen atom needed. Add the H_2O molecules to the appropriate side of the half-reaction.

$$NO_3^- \rightarrow NO_2 + \mathbf{H_2O}$$
$$\mathbf{6H_2O} + I_2 \rightarrow 2HIO_3$$

(4) Balance the half-reactions for hydrogen by adding one H^+ ion for each hydrogen atom needed. Add the H^+ ions to the appropriate side of the half-reaction. (Remember: acid is present.)

$$\mathbf{2H^+} + NO_3^- \rightarrow NO_2 + H_2O$$
$$6H_2O + I_2 \rightarrow 2HIO_3 + \mathbf{10H^+}$$

(5) Balance the half-reactions for electric charge by adding electrons (e^-) to the appropriate side of the half-reaction.

$$\mathbf{e^-} + 2H^+ + NO_3^- \rightarrow NO_2 + H_2O$$
$$6H_2O + I_2 \rightarrow 2HIO_3 + 10H^+ + \mathbf{10e^-}$$

(6) Multiply the half-reactions in order to conserve electric charge.

$$\mathbf{10} \cdot (e^- + 2H^+ + NO_3^- \rightarrow NO_2 + H_2O)$$
$$\mathbf{1} \cdot (6H_2O + I_2 \rightarrow 2HIO_3 + 10H^+ + 10e^-)$$

(7) Add the half-reactions and eliminate excess H_2O and/or H^+. The resulting equation is now balanced.

$$20H^+ + 10NO_3^- + 6H_2O + I_2 \rightarrow 2HIO_3 + 10H^+ + 10NO_2 + 10H_2O$$

This reduces to:

$$10H^+ + 10NO_3^- + I_2 \rightarrow 2HIO_3 + 10NO_2 + 4H_2O$$

Sample Problem

Use the ion-electron method to balance the following redox reaction that occurs in *acid* solution:

$$Cu + NO_3^- \rightarrow Cu^{2+} + NO$$

Solution

$$3Cu + 8H^+ + 2NO_3^- \rightarrow 3Cu^{2+} + 2NO + 4H_2O$$

The ion-electron method can also be used to balance redox reactions that occur in *basic* solution. The approach we use in this book is to balance the reaction as though it occurred in an acid solution. Then, at the last step, *we add sufficient OH^- ions to remove all of the H^+ ions present.* (Remember that the combination of H^+ and OH^- produces H_2O.)

Let us balance the following redox equation, which occurs in basic solution:

$$NO_2^- + Al \rightarrow AlO_2^- + NH_3$$

If this reaction were balanced in acid solution, it would look like this:

$$2H_2O + NO_2^- + 2Al \rightarrow 2AlO_2^- + NH_3 + H^+$$

To remove the one H^+, we add one OH^- to each side of the equation:

$$OH^- + 2H_2O + NO_2^- + 2Al \rightarrow 2AlO_2^- + NH_3 + H^+ + OH^-$$

This becomes:

$$OH^- + 2H_2O + NO_2^- + 2Al \rightarrow 2AlO_2^- + NH_3 + H_2O$$

which reduces to:

$$OH^- + H_2O + NO_2^- + 2Al \rightarrow 2AlO_2^- + NH_3$$

Sample Problem

Use the ion-electron method to balance the following redox reaction that occurs in *basic* solution:

$$IO_3^- + H_2S + \rightarrow 3I_2 + SO_3^{2-}$$

Solution

$$6IO_3^- + 5H_2S + 4OH^- \rightarrow 3I_2 + 5SO_3^{2-} + 7H_2O$$

THERMODYNAMICS II: SPONTANEOUS REACTIONS
AND ELECTROCHEMISTRY

SPONTANEOUS REDOX REACTIONS

The diagram below illustrates what happens when a strip of Zn(s) is placed in a beaker containing $Cu^{2+}(aq)$ ions:

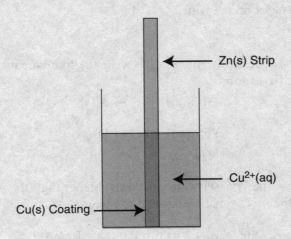

A spontaneous reaction is immediately evident. The metallic strip becomes coated with copper metal and the characteristic blue color of the $Cu^{2+}(aq)$ is lost as it is replaced by $Zn^{2+}(aq)$. The reaction that has occurred is:

$$Zn(s) + Cu^{2+}(aq) \rightarrow Zn^{2+}(aq) + Cu(s)$$

Like all redox reactions, this reaction involves a competition for electrons. Here, Zn(s) loses electrons more easily than Cu(s), and Cu^{2+} (aq) gains electrons more easily than Zn^{2+}(aq).

How can we predict whether another redox reaction will be spontaneous? In Appendix B, there is a table entitled "Standard Electrode Potentials," which lists a series of reduction half-reactions. This list is in descending order; that is, a reduction half-reaction that is higher on the list has a greater tendency to occur than a reduction half-reaction that lies below it. For example, the half-reaction:

$$Ag^+ + e^- \rightarrow Ag(s)$$

has a greater tendency to occur than the half-reaction:

$$Mg^{2+}(aq) + 2e^- \rightarrow Mg(s)$$

As a result of the relative placement of these half-reactions, we can conclude that:

- $Ag^+(aq)$ is more easily reduced than $Mg^{2+}(aq)$
- Mg(s) is more easily oxidized than Ag(s)

The overall spontaneous reaction is:

$$Ag^+(aq) + Mg(s) \rightarrow Ag(s) + Mg^{2+}(aq)$$

We can use the table of Standard Electrode Potentials to write spontaneous redox reactions by employing the following rule:

- Combine the half-reaction that is *higher* on the table with the *reverse* of the half-reaction that is *lower* on the table.

Sample Problem

Use the table of Standard Electrode Potentials to write the spontaneous reaction that occurs, given the following half-reactions:

$$Al^{3+}(aq) + 3e^- \rightarrow Al(s)$$

$$Pb^{2+}(aq) + 2e^- \rightarrow Pb(s)$$

Solution

We see that on the table the Pb^{2+}/Pb half-reaction is higher than the Al^{3+}/Al half-reaction. Therefore, we *reverse* the Al^{3+}/Al half-reaction and obtain:

$$Pb^{2+}(aq) + Al(s) \rightarrow Pb(s) + Al^{3+}(aq)$$

This redox equation is unbalanced because the number of electrons lost by $Al(s)$ does not equal the number of electrons gained by $Pb^{2+}(aq)$. Balancing is accomplished by multiplying the Al^{3+}/Al half-reaction by 2 and the Pb^{2+}/Pb half-reaction by 3 to obtain:

$$3Pb^{2+}(aq) + 2Al(s) \rightarrow 3Pb(s) + 2Al^{3+}(aq)$$

ELECTROCHEMICAL CELLS

An **electrochemical cell** is a device that is associated with a redox reaction in one of two ways:

- A **galvanic** (or **voltaic**) **cell** is an electrochemical cell that utilizes a *spontaneous* redox reaction to generate electrical energy by facilitating the passage of electrons through an *external circuit*.
- An **electrolytic cell** is an electrochemical cell that utilizes an *external source* of electrical energy to *force* a nonspontaneous redox reaction to take place.

Galvanic Cells

Let us construct a simple design for a galvanic cell involving the spontaneous redox reaction:

$$Zn(s) + Cu^{2+}(aq) \rightarrow Zn^{2+}(aq) + Cu(s)$$

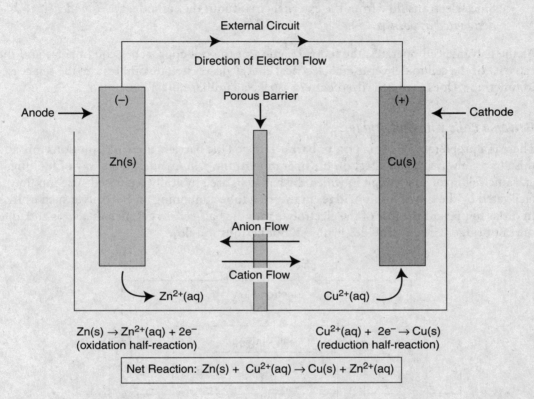

The cell is divided into two compartments, known as **half-cells**, that are separated by a porous barrier. Oxidation occurs in the left half-cell. The electrons that are released as a result of the oxidation half-reaction travel through the strip of $Zn(s)$ and across the external circuit (in this case, a wire), and enter the strip of $Cu(s)$. Reduction occurs in the right half-cell. The metal strips are called **electrodes**. The electrode at which oxidation occurs is called the **anode**; the electrode at which reduction occurs, the **cathode**.

> The direction of the electron flow in a galvanic cell is always from the anode to the cathode.

The function of the porous barrier is to allow the migration of positive and negative ions between the half-cells, thus completing the electrical circuit.

The flow of electrons provides a usable source of electrical energy. If it were feasible to use this galvanic cell commercially, we would call it a **battery**.

The signs of the electrodes are determined as follows:

- Since electrons flow out of the anode and into the external circuit, *the anode is designated as negative* (−).

- Since electrons flow from the external circuit into the cathode, *the cathode is designated as positive* (+).

As the galvanic cell operates, the redox reaction of the cell approaches equilibrium and the capacity of the cell to deliver useful electrical energy decreases. At equilibrium, the cell ceases to function. (Does the term "dead battery" strike a familiar note?)

Galvanic Cells with Salt Bridges

There is a problem with using porous barriers. Inside the barriers the ionic solutions mix and this almost always has an effect on the operation of the cell. An alternative way of building a galvanic cell involves a design in which the half-cells are physically separated. The ion flow is facilitated by the use of a **salt bridge**, an inverted tube containing an electrolyte such as KCl. In order to prevent the loss of the electrolyte into the half-cells, a gel such as *agar* is added to the salt bridge. The gel provides firmness, but permits ion flow.

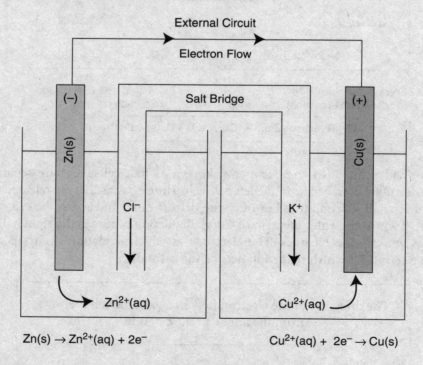

$$Zn(s) \rightarrow Zn^{2+}(aq) + 2e^-$$
$$Cu^{2+}(aq) + 2e^- \rightarrow Cu(s)$$

Since $Zn^{2+}(aq)$ ions are produced in the half-cell on the left, Cl^- ions flow in from the salt bridge in order to maintain electrical neutrality. In the half-cell on the right, $Cu^{2+}(aq)$ ions are being consumed. Electrical neutrality is maintained by the influx of K^+ ions from the salt bridge.

THERMODYNAMICS II: SPONTANEOUS REACTIONS
AND ELECTROCHEMISTRY

A number of galvanic cells do not use metal strips as electrodes. For example, consider the following half-cell reactions:

$$H_2(g) \rightarrow 2H^+(aq) + 2e^-$$

$$Co^{3+}(aq) + e^- \rightarrow Co^{2+}(aq)$$

In such cases, a chemically inert electrode such as platinum (Pt) or graphite (C(gr)) is inserted and the half-cell reaction occurs at the site of the inert electrode. The diagram below illustrates the operation of a galvanic cell employing both of these half-cell reactions.

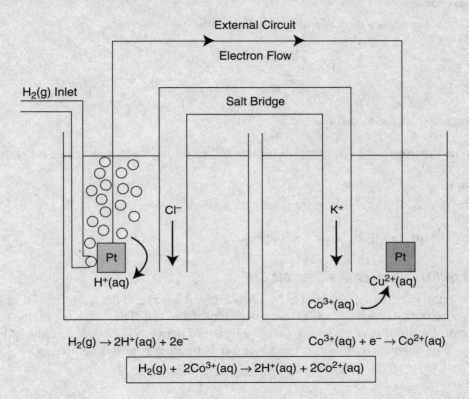

$$H_2(g) \rightarrow 2H^+(aq) + 2e^- \qquad\qquad Co^{3+}(aq) + e^- \rightarrow Co^{2+}(aq)$$

$$H_2(g) + 2Co^{3+}(aq) \rightarrow 2H^+(aq) + 2Co^{2+}(aq)$$

Galvanic Cell Notation

There is a short-hand notation that is used to describe the reacting components of a galvanic cell. For example, the notation for the cell with the porous barrier (on page 393) is:

$$Zn(s)|Zn^{2+}(aq)|Cu^{2+}(aq)|Cu(s)$$

The order, from left to right, is from anode to cathode, and from reactant to product. In the example shown above, Zn(s) is the anode and forms $Zn^{2+}(aq)$; $Cu^{2+}(aq)$ forms Cu(s), which is the cathode. The vertical bar represents the boundary between two phases. When a salt bridge is present, a double vertical bar is used as shown below:

$$Zn(s)|Zn^{2+}(aq)||Cu^{2+}(aq)|Cu(s)$$

When a platinum electrode is present, it is placed at the left and/or right end of the cell notation. For example, the cell whose reaction is $H_2(g) + Co^{3+}(aq) \rightarrow 2H^+(aq) + Co^{2+}(aq)$ has the notation:

$$Pt(s)|H_2(g)|H^+(aq)||Co^{3+}(aq), Co^{2+}(aq)|Pt(s)$$

Since $Co^{3+}(aq)$ and $Co^{2+}(aq)$ are present in the same phase, they are separated by a comma rather than by a vertical bar.

Sample Problem

A galvanic cell consists of the following half-reactions:

$$Mn(s) \rightarrow Mn^{2+}(aq) + 2e^-$$

$$Cu^{2+}(aq) + e^- \rightarrow Cu^+(aq)$$

(A) Write the net reaction for this cell.

(B) Write the shorthand cell notation for this cell assuming that a salt bridge is present.

Solution

(A) $Mn(s) + 2Cu^{2+}(aq) \rightarrow Mn^{2+}(aq) + 2Cu^+(aq)$

(B) $Mn(s)|Mn^{2+}(aq)||Cu^{2+}(aq), Cu^+(aq)|Pt$

The Potential Difference of a Galvanic Cell

The electrical energy that is delivered by a galvanic cell is equal to the quantity of useful work that can be obtained as a result of the operation of the cell. This work, w, is measured in relation to the amount of charge, q, that is transferred between the anode and the cathode of the cell. This quantity is known as the **potential difference**, E, and is defined as:

$$E = \frac{w}{q}$$

The SI unit of potential difference is the joule per coulomb, which is also known as the **volt** (V).

Sample Problem

A galvanic cell delivers 1.10 joules of useful work for every 2.00 coulombs of charge transferred between the anode and the cathode. Calculate the potential difference of the cell.

Solution

$$E = \frac{w}{q} = \frac{1.10 \text{ J}}{2.00 \text{ C}} = \textbf{0.550 V}$$

THERMODYNAMICS II: SPONTANEOUS REACTIONS
AND ELECTROCHEMISTRY

The potential difference of a galvanic cell is also known by a number of other names: **cell potential**, **cell voltage**, and **cell EMF** (for *electromotive force*). For the remainder of this chapter, we will use the term *cell potential* exclusively.

The cell potential depends on a number of factors, including:

- The nature of the reactions taking place at the anode and cathode
- The molar concentration of any reacting substance that is dissolved
- The partial pressure of any reacting gas
- The temperature at which the cell operates

In order to compare cell potentials for different cells, it is useful to define a cell that operates under *standard conditions*:

- All solids and liquids are in their pure states.
- The concentration of each dissolved substance is set at 1 M.
- The partial pressures of each gas is set at 1 atm (actually, at 100 kPa).
- The usual reference temperature is set at 298.15 K.

Under these conditions, the cell potential is known as the **standard cell potential**, $E°$. For example, for the copper-zinc cell described above, we can write:

$$Zn(s)|Zn^{2+}(aq)||Cu^{2+}(aq)|Cu(s) \qquad\qquad E° = 1.10 \text{ V}$$

Standard Electrode (Reduction) Potentials

In the table of Standard Electrode Potentials found in Appendix B, note that each half-reaction is associated with a signed numerical value, known as a **standard electrode potential** (or a **standard reduction potential**), $E°$. The more positive a standard electrode potential is, the greater is the oxidizing power of the accompanying redox half-reaction. Conversely, the more negative a standard electrode potential is, the greater is the reducing power of the reverse redox half-reaction. For example, the half-reaction $F_2(g) \rightarrow 2F^-(aq) + 2e^-$ is associated with the most positive standard electrode potential listed on the table. Therefore, we can expect F_2 to be the best oxidizing agent on the table. In contrast, the half-reaction $Li^+(aq) + e^- \rightarrow Li(s)$ is associated with the most negative standard electrode potential on the table. Therefore, we can expect Li to be the best reducing agent on the table.

In general, we identify a particular standard electrode potential by appending the notation "(reduced species/oxidized species)" after the $E°$ notation. For example, $E°(Ni^{2+}/Ni)$ represents the standard electrode potential for the half reaction $Ni^{2+}(aq) + 2e^- \rightarrow Ni(s)$.

The value of a standard electrode potential can be determined as follows:

- A galvanic cell, operating under standard conditions, is constructed using a special reference half-cell known as a **standard hydrogen electrode** and a test half-cell in which a different half-reaction will occur. The standard hydrogen electrode consists of a platinum electrode in the presence of $H_2(g)$ at 1 atm, and $H^+(aq)$ at 1 M. *It is assigned a standard electrode potential of exactly* 0.00 V.

- The standard cell potential, $E°_{cell}$, is then measured and the standard electrode potential of the test half-cell is then determined by using the following expression:

$$E°_{cell} = E°_{cathode} - E°_{anode}$$

For example, consider the following cell:

$$Zn(s)|Zn^{2+}(aq,\ 1\ M)||H^+(aq,\ 1\ M)|H_2(g,\ 1\ atm)|Pt \qquad E°_{cell} = +0.76\ V$$

In this cell, the zinc electrode is the anode and the hydrogen electrode is the cathode. Applying the expression given above, we obtain:

$$E°_{cell} = E°_{cathode} - E°_{anode}$$

$$+0.76\ V = E°(H^+/H_2) - E°(Zn^{2+}/Zn) = 0.00\ V - E°(Zn^{2+}/Zn)$$

$$E°(Zn^{2+}/Zn) = -0.76\ V$$

Note that $E°$ for the zinc electrode refers to the *reduction* half-reaction, even though zinc served as the anode in this cell.

Sample Problem

Determine the standard electrode potential for the copper electrode in the cell:

$$Pt|H_2(g,\ 1\ atm)|H^+(aq,\ 1\ M)||\ Cu^{2+}(aq,\ 1\ M)|Cu(s) \qquad E°_{cell} = +0.34\ V$$

Solution

In this cell, the hydrogen electrode is the anode and the copper electrode is the cathode.

$$E°_{cell} = E°_{cathode} - E°_{anode}$$

$$+0.34\ V = E°(Cu^{2+}/Cu) - E°(H^+/H_2) = E°(Cu^{2+}/Cu) - 0.00\ V$$

$$E°(Cu^{2+}/Cu) = \textbf{+0.34 V}$$

The expression given above is perfectly general: We can use it to determine $E°_{cell}$ for *any* two half-cells listed in the table of Standard Electrode Potentials. The electrode associated with

THERMODYNAMICS II: SPONTANEOUS REACTIONS
AND ELECTROCHEMISTRY

the half-reaction that has the more positive $E°$ will be the cathode and the other electrode will be the anode.

Sample Problem

Determine $E°_{cell}$ for the cell: $Zn(s)|Zn^{2+}(aq)||Cu^{2+}(aq)|Cu(s)$

Solution

$E°_{cell} = E°_{cathode} - E°_{anode} = E°(Cu^{2+}/Cu) - E°(Zn^{2+}/Zn)$ $(+0.34\ V) - (-0.76\ V) = $ **+1.10 V**

Note that the calculated cell potential is a *positive* number. Whenever a cell potential is positive, it signifies that the accompanying redox reaction is *spontaneous*.

Sample Problem

Determine $E°_{cell}$ for the cell in which the following redox reaction occurs:

$$Cr(s) + 3Ag^{+}(aq) \rightarrow Cr^{3+}(aq) + Ag(s)$$

Solution

In this cell, $Cr(s)$ is the anode and Ag is the cathode.

$E°_{cell} = E°_{cathode} - E°_{anode} = E°(Ag^{+}/Ag) - E°(Cr^{3+}/Cr)$ $(+0.80\ V) - (-0.73\ V) = $ **+1.53 V**

Sample Problem

A galvanic cell has the following notation:

$$Mn|Mn^{2+}||X^{3+}|X \qquad\qquad E°_{cell} = +2.68\ V$$

Determine $E°(X^{3+}/X)$ and identify the components of the half-cell.

Solution

$$E°_{cell} = E°_{cathode} - E°_{anode}$$

$$+2.68\ V = E°(X^{3+}/X) - E°(Mn^{2+}/Mn) = E°(X^{3+}/X) - (-1.18\ V)$$

$$E°(X^{3+}/X) = \textbf{+1.50 V}$$

Based on the table of Standard Electrode Potentials, the components of the half-cell are $Au^{3+}(aq)$ and $Au(s)$.

Standard Oxidation Potentials

Just as we developed a table for standard reduction potentials, we could have developed a table base on *oxidation* half-reactions. The standard potential associated with an oxidation half-reaction is known as a **standard oxidation potential**. Since an oxidation half-reaction is the *reverse* of a reduction half-reaction, it follows that both standard oxidation and reduction potentials will have the same numerical value, but they will have *opposite signs*: If a standard reduction potential is –0.76 V, then the corresponding standard oxidation potential is +0.76 V. In calculating cell potentials, we could have included both types of half-cell potentials by employing the relationship:

$$E^\circ_{cell} = E^\circ_{red}(\text{cathode}) + E^\circ_{ox}(\text{anode})$$

For example, in the reaction: $Zn(s) + Cu^{2+}(aq) \rightarrow Zn^{2+}(aq) + Cu(s)$, $E^\circ_{red}(Cu^{2+}/Cu) = +0.34$ V and $E^\circ_{ox}(Zn/Zn^{2+}) = +0.76$ V. Therefore, $E^\circ_{cell} = (+0.34 \text{ V}) + (+0.76 \text{ V}) = +1.10$ V. We will use this relationship later in the chapter when we calculate K_{sp} from cell potential data.

Free Energy and Cell Potential

Since useful work can be measured by the change in the free energy of a spontaneous reaction, and by the potential of a galvanic cell, it stands to reason that the two quantities must be related. This relationship is given by the following expression:

$$\Delta G = -nFE$$

The variable n represents the number of moles of electrons transferred in the reaction. The variable F is known as the **Faraday constant** and represents the charge of one mole of electrons. The value of F is 96,485 coulombs per mole. Since a volt is equal to a joule per coulomb, both sides of this expression reduce to joules. The negative sign is introduced because of sign conventions: In a spontaneous reaction, the free energy change is *negative*, yet the cell potential is *positive*. Under standard conditions, we can write:

$$\Delta G^\circ = -nFE^\circ$$

Sample Problem

The standard cell potential for the balanced redox reaction $Zn(s) + Cu^{2+}(aq) \rightarrow Zn^{2+}(aq) + Cu(s)$ is +1.10 V. Calculate the standard free energy change for this reaction.

Solution

$$\Delta G^\circ = -nFE^\circ = -(2\text{mol}) \cdot (96{,}485 \text{ C/mol}) \cdot (+1.10 \text{ J/C}) = -212{,}000 \text{ J} = -212 \text{ kJ}$$

Nonstandard Cell Potentials: The Nernst Equation

Earlier in this chapter, we learned that a change in free energy was dependent on temperature and composition and could be expressed by the relationship:

$$\Delta G = \Delta G^\circ + RT\ln Q$$

THERMODYNAMICS II: SPONTANEOUS REACTIONS
AND ELECTROCHEMISTRY

We can express this relationship in terms of cell potentials by incorporating the relationship: $\Delta G = -nFE$.

We then obtain an expression that is known as the **Nernst equation:**

$$E = E° - \frac{RT}{nF}\ln Q$$

Since cell potentials are commonly evaluated at 298.15 K (25°C), we can substitute values for the RT/F term, and the Nernst equation can be expressed in either of two forms:

$$E = E° - \frac{0.02569\text{ V}}{n}\ln Q$$

$$E = E° - \frac{0.05916\text{ V}}{n}\log_{10}Q$$

In using these equations, *no units* are assigned either to the variable n or to the values that appear in Q. The second equation, which is expressed in terms of common logarithms, is particularly useful in relating cell potential to pH.

The next few sample problems illustrate how the Nernst equation is used in electrochemistry.

Sample Problem

Calculate the cell potential for the following cell:

$$Cr(s)|Cr^{3+}(aq, 0.25\text{ M})||\ Ni^{2+}(aq, 0.10\text{ M})\ M)|Ni(s) \qquad E°_{cell} = +0.50\text{ V}$$

Solution

The balanced redox reaction for this cell is: $2Cr(s) + 3Ni^{2+}(aq) \rightarrow 2Cr^{3+}(aq) + 3Ni(s)$, and $n = 6$.

The reaction quotient expression for this reaction is: $Q = \dfrac{[Cr^{3+}]^2}{[Ni^{2+}]^3}$

$$E = E° - \frac{0.02569\text{ V}}{n}\ln Q = 0.50\text{ V} - \frac{0.02569\text{ V}}{6}\ \ln\frac{(0.25)^2}{(0.10)^3} = \textbf{+0.48 V}$$

The reduction in cell potential means that the reaction was driven to the *left*, an indication that, under these conditions, Q is greater than the equilibrium constant, K.

Sample Problem

Calculate the cell potential for the following cell:

$$Pt|H_2(g, 0.20\text{ atm})|H^+(aq, 0.010\text{ M})||\ Cu^{2+}(aq, 0.10\text{ M})\ M)|Cu(s) \qquad E°_{cell} = +0.34\text{ V}$$

Solution

The net reaction for this cell is: $H_2(g) + Cu^{2+}(aq) \rightarrow 2H^+(aq) + Cu(s)$, and $n = 2$.

The reaction quotient expression for this reaction is: $Q = \dfrac{[H^+]^2}{P_{H_2}[Cu^{2+}]}$

$$E = E^\circ - \frac{0.02569 \text{ V}}{n} \ln Q = 0.34 \text{ V} - \frac{0.02569 \text{ V}}{2} \ln \frac{(0.010)^2}{(0.20)(0.10)} = \textbf{+0.41 V}$$

The increase in cell potential means that the reaction was driven to the *right*, an indication that, under these conditions, Q is less than the equilibrium constant, K.

Concentration Cells

Consider the following cell:

$$Cu(s)|Cu^{2+}(aq, 0.0010 \text{ M})||Cu^{2+}(aq, 1.0 \text{ M})|Cu(s)$$

This is a cell with the same anode and cathode, only the *concentrations* of the aqueous solutions of Cu^{2+} are different. This cell is known as a **concentration cell**. Since $E^\circ_{anode} = E^\circ_{cathode}$, it follows that $E^\circ_{cell} = 0$. The half-reactions for this cell are:

$$Cu(s) \rightarrow Cu^{2+}(aq, 0.0010 \text{ M}) + 2e^-$$

$$Cu^{2+}(aq, 1.0 \text{ M}) + 2e^- \rightarrow Cu(s)$$

The net reaction for this cell is:

$$Cu^{2+}(aq, 1.0 \text{ M}) \rightarrow Cu^{2+}(aq, 0.0010 \text{ M})$$

As this cell operates spontaneously, the concentrated solution becomes more dilute and the dilute solution becomes more concentrated. When the concentrations in both half-cells are equal, the cell will be in an equilibrium state and E_{cell} will equal 0.00 volt. The next sample problem verifies that this concentration cell operates spontaneously.

Sample Problem

Calculate E_{cell} for the concentration cell described above.

Solution

$$E^\circ = 0.00 \text{ V and } n = 2$$

Applying the Nernst equation, we obtain:

$$E = E^\circ - \frac{0.02569 \text{ V}}{n} \ln Q = 0.00 \text{ V} - \frac{0.02569 \text{ V}}{2} \ln \frac{(0.0010)}{(1.0)} = \textbf{+0.089 V}$$

Since $E > 0$, we can conclude that the reaction is spontaneous.

THERMODYNAMICS II: SPONTANEOUS REACTIONS
AND ELECTROCHEMISTRY

In theory (though not in practice), the pH of a solution could be measured using a hydrogen concentration cell. Consider the following cell:

$$Pt|H_2(g, 1 \text{ atm})|H^+(aq, x \text{ M})||H^+(aq, 1.0 \text{ M})|H_2(g, 1 \text{ atm})|Pt$$

The balanced redox reaction for this cell is: $H_2(g, 1 \text{ atm}) + 2H^+(aq, 1.0 \text{ M}) \rightarrow 2H^+(aq, x \text{ M}) + H_2(g, 1 \text{ atm})$. There are 2 moles of electrons transferred ($n = 2$) and the reaction quotient expression for this reaction is:

$$Q = \frac{P_{H_2(g, 1 \text{ atm})}[H^+(aq, x \text{ M})^2}{P_{H_2(g, 1 \text{ atm})}[H^+(aq, 1 \text{ M})]^2}$$

We assume that the partial pressure of the H_2 in both half-cells is maintained at 1 atm, and the H^+ ion concentration at the cathode is maintained at 1 M. We now apply the second form of the Nernst equation:

$$E = E^o - \frac{0.05916 \text{ V}}{n}\log_{10}Q = 0.00 \text{ V} - \frac{0.05916 \text{ V}}{2}\log_{10}\frac{P_{H_2(g, 1 \text{ atm})}[H^+(aq, x \text{ M})]^2}{P_{H_2(g, 1 \text{ atm})}[H^+(aq, 1 \text{ M})]^2}$$

$$E = -\frac{0.05916 \text{ V}}{2}\log_{10}\frac{(1)(x)^2}{(1)(1)^2} = -\frac{0.05916 \text{ V}}{2}\log_{10}x^2$$

Since $\log_{10}x^2 = 2\log_{10}x$, and $-\log x = \text{pH}$, the expression can be simplified further:

$$E = 0.05916 \text{ V} \cdot \text{pH}$$

We have shown that, in this cell, the pH is proportional to the cell potential. (In practice, electronic pH meters use other cathode half-reactions.)

Sample Problem

A hydrogen concentration cell produces a cell potential of 0.211 V. Calculate the pH of the solution at the anode of the cell.

Solution

$$E = 0.05916 \text{ V} \cdot \text{pH}$$

$$\text{pH} = \frac{E}{0.05916 \text{ V}} = \frac{0.211 \text{ V}}{0.05916 \text{ V}} = \textbf{3.57}$$

Cell Potential and Equilibrium

The cell potential of a galvanic cell will be 0 when its redox reaction reaches equilibrium. At this point, the reaction quotient is equal to the equilibrium constant. The Nernst equation can be used to determine the equilibrium constant, K, of a redox reaction:

$$E = E^o - \frac{0.02569 \text{ V}}{n}\ln Q$$

At equilibrium $E° = 0$:

$$E° = \frac{0.02569 \text{ V}}{n} \ln K$$

Rearranging terms, we obtain:

$$\ln K = \frac{nE°}{0.02569 \text{ V}}$$

If $E°$ is a positive number, it follows that $K > 1$; if $E°$ is negative, $K < 1$.

Sample Problem

If $E° = +1.10$ V for the reaction $Zn(s) + Cu^{2+}(aq) \rightarrow Zn^{2+}(aq) + Cu(s)$ (at 25°C), calculate the equilibrium constant of the reaction.

Solution

For this reaction, $n = 2$.

$$\ln K = \frac{nE°}{0.02569 \text{ V}} = \frac{(2) \cdot (+1.10 \text{ V})}{0.02569 \text{ V}} = 85.6$$

$$K = e^{85.6} = 1.55 \times 10^{37}$$

Is it any wonder that we write this reaction with a single, rather than a double arrow?

Equilibrium Constants of Nonspontaneous Reactions: Determination of K_{sp}

An interesting application of the Nernst Equation is the determination of the K_{sp} of a solubility equilibrium such as: $AgCl(s) \rightleftharpoons Ag^+(aq) + Cl^-(aq)$. The relevant half-reactions and their standard half-cell potentials (which can be found in the table of Standard Electrode Potentials) are:

$$AgCl(s) + e^- \rightarrow Ag(s) + Cl^- \qquad E°_{red} = +0.22 \text{ V}$$

$$Ag(s) \rightarrow Ag^+(aq) + e^- \qquad E°_{ox} = -0.80 \text{ V}$$

Note that we used the standard *oxidation* potential of the second half-reaction. The net overall reaction is the solubility equilibrium described above: $AgCl(s) \rightleftharpoons Ag^+(aq) + Cl^-(aq)$. The standard cell potential can be determined by *adding* the standard reduction and oxidation potentials of the half-reactions:

$$E°_{cell} = E°_{red} + E°_{ox} = (+0.22 \text{ V}) + (-0.80 \text{ V}) = -0.58 \text{ V}$$

Note that this is a *nonspontaneous* reaction.

THERMODYNAMICS II: SPONTANEOUS REACTIONS
AND ELECTROCHEMISTRY

Now we employ the Nernst equation relationship, remembering that $n = 1$ for this reaction:

$$\ln K_{sp} = \frac{nE^\circ}{0.02569 \text{ V}} = \frac{(1) \cdot (-0.58 \text{ V})}{0.02569 \text{ V}} = -22.6$$

$$K_{sp} = e^{-22.6} = 1.6 \times 10^{-10}$$

Sample Problem

Determine the K_{sp} of $PbSO_4$ using the data found in the table of Standard Electrode Potentials (Appendix B).

Solution

$$PbSO_4(s) + 2e^- \rightarrow Pb(s) + SO_4^{2-} \qquad\qquad E^\circ_{red} = -0.35 \text{ V}$$

$$Pb(s) \rightarrow Pb^{2+}(aq) + 2e^- \qquad\qquad E^\circ_{ox} = +0.13 \text{ V}$$

The net overall reaction is: $PbSO_4(s) \rightleftharpoons Pb^{2+}(aq) + SO_4^{2-}(aq)$.

$$E^\circ_{cell} = (-0.35 \text{ V}) + (+0.13 \text{ V}) = -0.22 \text{ V}$$

For this reaction, $n = 2$.

$$\ln K_{sp} = \frac{nE^\circ}{0.02569 \text{ V}} = \frac{(2) \cdot (-0.22 \text{ V})}{0.02569 \text{ V}} = -17.1$$

$$K_{sp} = e^{-17.1} = \mathbf{3.6 \times 10^{-8}}$$

Electrolysis and Electrolytic Cells

Consider the following balanced redox reaction, which takes place at 25°C under standard conditions:

$$2Na^+(aq) + 2Cl^-(aq) \rightarrow 2Na(s) + Cl_2(g)$$

The relevant half-reactions are:

$$2Na^+(aq) + 2e^- \rightarrow 2Na(s)$$

$$2Cl^-(aq) \rightarrow Cl_2(g) + 2e^-$$

$$E^\circ_{cell} = E^\circ_{red}(Na^+/Na) + E^\circ_{ox}(Cl_2/Cl^-) = (-2.71 \text{ V}) + (-1.36 \text{ V}) = -4.07 \text{ V}$$

Evidently, this redox reaction is *not* spontaneous and we could not construct a galvanic cell based it. Nonspontaneous reactions must be forced to occur by supplying work from an external source: In this case, 4.07 volts represents the minimum potential difference that needs to be supplied by an external source of electrical energy in order to effect this reaction under standard conditions.

When electrical energy is used in this fashion, the process is called **electrolysis** (literally, "breaking up with electricity"), and any electrochemical cell employed for this purpose is known as an **electrolytic cell**. In practice, electrolysis always requires more than the minimum potential difference to drive the nonspontaneous reaction. The additional potential difference that must be supplied is known as **overpotential** and is dependent on the nature of the half-reactions that occur at the electrodes and the types of electrodes that are used in the electrolytic process.

Electrolysis of Molten Salts

The simplest type of electrolysis utilizes a molten salt such as $NaCl(\ell)$. In the liquid phase, NaCl is a good conductor because the Na^+ and Cl^- ions are now free to move. The electrolytic cell that accomplishes the electrolysis of molten NaCl is shown below:

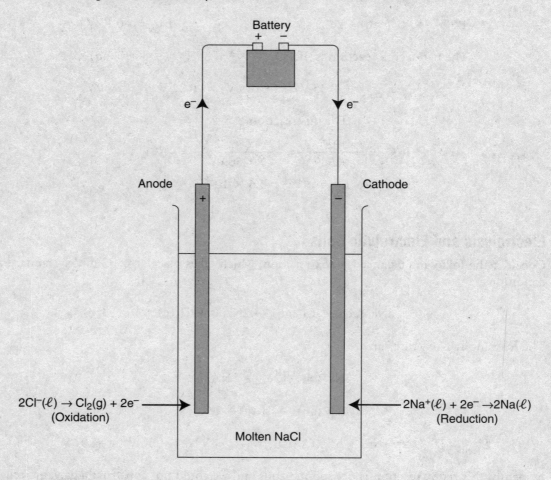

The positive and negative electrodes are connected, respectively, to the positive and negative terminals of the battery. Ions of *opposite charge* migrate to these electrodes and react. *The anode is the positive electrode* because oxidation occurs at this site. Similarly, *the cathode is the negative electrode* because reduction occurs at this site. Note that the anode and cathode of an electrolytic cell have signs opposite to those in a galvanic cell.

THERMODYNAMICS II: SPONTANEOUS REACTIONS AND ELECTROCHEMISTRY

Electrolysis of Water

If a very small amount of a substance such as Na_2SO_4 is added to water, the dilute solution will conduct electricity readily. The half-reactions that occur at the (inert) electrodes are:

Anode: $2H_2O(\ell) \rightarrow O_2(g) + 4H^+(aq) + 4e^-$ $E^{\circ}_{ox} = -1.23$ V

Cathode: $2H_2O(\ell) + 2e^- \rightarrow H_2(g) + 2OH^-(aq)$ $E^{\circ}_{red} = -0.83$ V

The overall reaction that occurs is:

$6H_2O(\ell) \rightarrow 2H_2(g) + O_2(g) + 4H^+(aq) + 4OH^-(aq)$ $E^{\circ}_{cell} = -2.06$ V

If the electrolysis is carried out in a single container, the $H^+(aq)$ and the $OH^-(aq)$ ions combine to yield 4 moles of additional water molecules. The net reaction is then:

$2H_2O(\ell) \rightarrow 2H_2(g) + O_2(g)$

Electrolysis of Aqueous Solutions

When more concentrated aqueous solutions are electrolyzed, we must consider the possibility that unexpected half-reactions will occur. When molten NaCl is electrolyzed, $Na(\ell)$ and $Cl_2(g)$ are produced, as expected. If an aqueous solution of NaCl (known as brine) is electrolyzed, however, we have to consider all of the possibilities by comparing standard half-cell potentials of the substance that is dissolved with the half-cell potentials of water:

At the cathode, the possible half-reactions are:

$2Na^+(aq) + 2e^- \rightarrow 2Na(s)$ $E^{\circ}_{red} = -2.71$ V

$2H_2O(\ell) + 2e^- \rightarrow H_2(g) + 2OH^-(aq)$ $E^{\circ}_{red} = -0.83$ V

The standard reduction potentials indicate that H_2O has a greater ability to be reduced, and the half-reaction occurring at the cathode is: $2H_2O(\ell) + 2e^- \rightarrow H_2(g) + 2OH^-(aq)$.

At the anode, the possible half-reactions are:

$2Cl^-(aq) \rightarrow Cl_2(g) + 2e^-$ $E^{\circ}_{ox} = -1.36$ V

$2H_2O(\ell) \rightarrow O_2(g) + 4H^+(aq) + 4e^-$ $E^{\circ}_{ox} = -1.23$ V

Based on the standard oxidation potentials, we would expect H_2O to be oxidized at the anode. This does *not* occur, however. The overpotential for the formation of O_2 is greater than that for the formation of Cl_2, and the reaction that occurs at the anode is: $2Cl^-(aq) \rightarrow Cl_2(g) + 2e^-$.

In summary, when brine is electrolyzed, the half-reactions are:

Cathode: $2H_2O(\ell) + 2e^- \rightarrow H_2(g) + 2OH^-(aq)$

Anode: $2Cl^-(aq) \rightarrow Cl_2(g) + 2e^-$

The net reaction that occurs is: $2Cl^-(aq) + 2H_2O(\ell) \rightarrow Cl_2(g) + H_2(g) + 2OH^-(aq)$.

The $Na^+(aq)$ serve as spectator ions.

Electrolysis with Active Electrodes

There are practical applications of electrolysis that require the presence of an **active electrode**, that is, an electrode that takes part in the electrolysis reaction. One important application is known as **electroplating**, a process in which one metal is coated with another. This diagram illustrates how an electroplating cell is used to plate a steel strip with nickel:

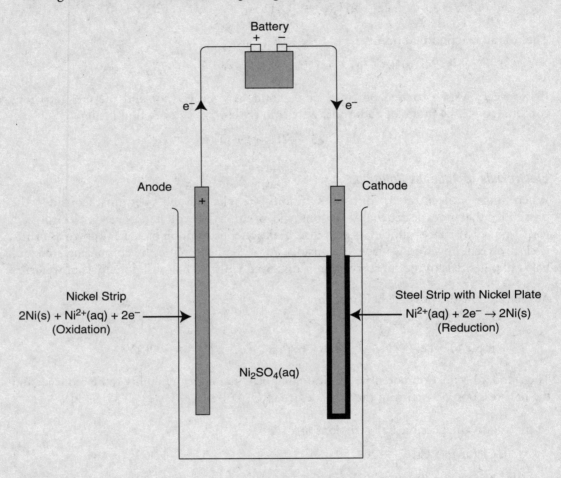

The plating solution is $NiSO_4$, chosen because the SO_4^{2-} ion does not participate in the plating reaction. A strip of nickel (the active electrode) is the anode and is oxidized to Ni^{2+} ions. A steel strip, which is to be plated, is the cathode (negative electrode) and is the site at which metallic nickel is deposited—the result of reduction of Ni^{2+} ions. The net effect of the electroplating process is to move a mass of metallic nickel from the anode to the cathode: $Ni(s, anode) \rightarrow Ni(s, cathode)$.

THERMODYNAMICS II: SPONTANEOUS REACTIONS AND ELECTROCHEMISTRY

Quantitative Aspects of Electrolysis

It is often important to know the *quantity* of a substance produced in an electrolytic reaction. For example, we may want to know the mass of sodium metal that is produced when molten NaCl is electrolyzed. Or, we may want to know the mass of nickel that has been deposited in a nickel-plating operation.

In order to answer these questions, we need to know:

- The relevant half-reaction that takes place
- The electric current that is passed through the electrolytic cell
- The time that the electrolytic cell operates

Electric current is defined as the rate at which electric charge passes through a circuit. The SI unit of electric current is the **ampere**, **A**, which is equivalent to a *coulomb per second* (C/s).

Once we have this information, we can solve the problem using the factor-label method. For example: Suppose that a nickel-plating cell operates for 2.00 hours and draws an electric current of 12.0 amperes. What mass of metallic nickel will be deposited at the cathode of the cell? The relevant half-reaction is:

$$Ni^{2+}(aq) + 2e^- \rightarrow Ni(s)$$

The half-reaction tells us that 1.00 mole of Ni(s) will be deposited for every 2.00 moles of e^- that pass through the circuit. In solving the problem, we will employ the following "road map":

$$\text{Time (s)} \xrightarrow{\text{current (A)}} \text{Charge (C)} \xrightarrow{F} \text{Moles of electrons}$$

$$\xrightarrow[\text{half-reaction}]{\text{coefficients of}} \text{Moles of substance} \xrightarrow{M} \text{Mass of substance (g)}$$

The text over the arrows represents the quantities or constants we need to employ in order to carry the conversion to the next step.

The solution is:

$$\text{Mass of nickel} = (2.00 \text{ h})\left(\frac{3{,}600 \text{ s}}{1 \text{ h}}\right)\left(\frac{12.0 \text{ C}}{1 \text{ s}}\right)\left(\frac{1 \text{ mol } e^-}{96{,}485 \text{ C}}\right)\left(\frac{1 \text{ mol Ni}}{2 \text{ mol } e^-}\right)\left(\frac{58.69 \text{ g Ni}}{1 \text{ mol Ni}}\right) = \textbf{26.3 g Ni}$$

Sample Problem

$Al^{3+}(aq)$ is reduced to Al(s) in an electrolytic cell. If the electrolysis takes 0.50 hour and a 15.0-ampere electric current is present, what mass of Al(s) is produced at the cathode?

Solution

$$\text{Mass of Al(s)} = (0.50 \text{ h})\left(\frac{3{,}600 \text{ s}}{1 \text{ h}}\right)\left(\frac{15.0 \text{ C}}{1 \text{ s}}\right)\left(\frac{1 \text{ mol } e^-}{96{,}485 \text{ C}}\right)\left(\frac{1 \text{ mol Al}}{3 \text{ mol } e^-}\right)\left(\frac{26.98 \text{ g Al}}{1 \text{ mol Al}}\right) = \textbf{2.52 g Al}$$

Chapter 15: Practice Questions

Multiple-Choice Questions

1. A reaction with a negative ΔG is:

   A. spontaneous
   B. nonspontaneous
   C. exothermic
   D. endothermic
   E. non-thermic

2. Which processes are spontaneous?

   I. A scoop of ice cream melting on a hot summer day.

   II. A snowball rolling up a hill.

   III. NaCl (s) dissolving into Na^+ (aq) and Cl^- (aq) ions when added to water.

   IV. A book falling off of a shelf.

   V. Water freezing at 1 atm and $-5°C$.

   A. I and II only
   B. I and IV only
   C. I, III, and V
   D. I, III, IV, and V
   E. I, III, and IV

3. For the reaction $2\ C(s) + 3\ H_2(g) \rightarrow C_2H_6(g)$ the entropy change at 25°C (J / mol · K) is:

	S° (J/mol · K)
C (s)	5.74
C_2H_6 (g)	229.5
H_2 (g)	130.6

   A. -173.8
   B. -162.3
   C. -150.9
   D. 93.2
   E. 229.5

4. What is the standard enthalpy change for the reaction of methane with oxygen to form carbon dioxide, carbon monoxide, and steam?

   $4\ CH_4\ (g) + 7\ O_2\ (g) \rightarrow 2\ CO_2\ (g) + 2\ CO\ (g) + 8\ H_2O\ (g)$

	$\Delta H°$ (kJ/mol)
CH_4 (g)	-74.9
O_2 (g)	0
CO_2 (g)	-393.5
CO (g)	-110.5
H_2O (g)	-241.8

   A. $-2,642.8$ kJ
   B. $-3,242.0$ kJ
   C. $2,200.8$ kJ
   D. 9.82 kJ
   E. 1.02×10^{-1} kJ

5. For a reaction, ΔS is positive and ΔH is positive. In what temperature range is the reaction spontaneous?

   A. The reaction is always spontaneous.
   B. The reaction is never spontaneous.
   C. The reaction is spontaneous at $T > \Delta H/\Delta S$.
   D. The reaction is spontaneous at $T < \Delta H/\Delta S$.
   E. The reaction is spontaneous at $T = \Delta H/\Delta S$.

6. For the reaction $HBr\ (g) \rightarrow \frac{1}{2}\ H_2\ (g) + \frac{1}{2}\ Br_2\ (g)$, calculate the free-energy change, $\Delta G°$, in kilojoules. The free energy of formation of HBr is -53.43 kJ / mol and Br_2 (g) is 3.144 kJ / mol.

   A. -51.86
   B. -50.29
   C. $+53.43$
   D. $+55.00$
   E. $+56.57$

THERMODYNAMICS II: SPONTANEOUS REACTIONS AND ELECTROCHEMISTRY

7. Sulfur dioxide reacts with oxygen to produce sulfur trioxide. At what temperature is the reaction spontaneous?

   $2 SO_2 (g) + O_2 (g) \rightarrow 2 SO_3 (g)$

   $\Delta H° = -198$ kJ/mol

   $\Delta S° = -0.187$ kJ/mol $\cdot$ K

   A. The reaction is always spontaneous.
   B. The reaction is never spontaneous.
   C. The reaction is spontaneous at temperatures above 1060 K.
   D. The reaction is spontaneous at temperatures below 1060 K.
   E. The reaction is spontaneous at temperatures below room temperature.

8. Which describes the equilibrium constant for a spontaneous reaction?

   A. $K = 0$
   B. $K < 0$
   C. $K = 1$
   D. $K < 1$
   E. $K > 1$

9. Which of is true at equilibrium?

   A. $Q = 0$
   B. $K = 0$
   C. $Q = K$
   D. $\Delta G > 0$
   E. $\Delta G° = 1$

10. For which reaction is $\Delta S° < 0$?

    A. $C_6H_6 (s) \rightarrow C_6H_6 (\ell)$

    B. $2 IBr (g) \rightarrow I_2 (s) + Br_2 (\ell)$

    C. $2 NO_2 (g) \rightarrow N_2 (g) + 2 O_2 (g)$

    D. $NH_4Cl (s) \rightarrow NH_3 (g) + HCl (g)$

    E. $(NH_4)_2CO_3 (s) \rightarrow 2 NH_3 (g) + H_2O (g) + CO_2 (g)$

11. Calculate the equilibrium constant at 25°C for the reaction:

    $3 C (s) + 4 H_2 (g) \rightarrow C_3H_8 (g)$

    The free energy of formation of the C_3H_8 is $\Delta G_f° = -23.49$ kJ/mol $\cdot$ K.

    A. 3.0×10^9
    B. 1.3×10^4
    C. 1.0
    D. 7.6×10^{-1}
    E. 3.3×10^{-10}

12. Consider the cell Cd (s) | Cd^{2+} (1.0 M) || Cu^{2+} (1.0 M) | Cu. In order to make this cell with a more positive voltage using the same substance, you will need to:

    A. increase both the $[Cd^{2+}]$ and $[Cu^{2+}]$ to 2.00 M.
    B. increase only the $[Cd^{2+}]$ to 2.00 M.
    C. decrease both the $[Cd^{2+}]$ and $[Cu^{2+}]$ to 0.100 M.
    D. decrease only the $[Cd^{2+}]$ to 0.100 M.
    E. decrease only the $[Cu^{2+}]$ to 0.100 M.

13. The oxidation number of the nitrogen in the nitrate ion, NO_3^-, is:

    A. $+1$
    B. $+2$
    C. $+3$
    D. $+4$
    E. $+5$

14. Which set must be true for a reaction to be spontaneous?

    A. $\Delta G° < 0$ and $E° > 0$
    B. $\Delta G° < 0$ and $E° < 0$
    C. $K > 1$ and $E° < 0$
    D. $K > 1$ and $E° > 0$
    E. $K > 1$ and $E° < 0$

15. The voltage of the cell Zn (s) | Zn^{2+} (1.0 M) || Pb^{2+} (1.0 M) | Pb (s) is:

Half-Reaction	Reduction Potential, $E°$
Zn^{2+} (aq) + 2 e^- → Zn (s)	−0.763 V
Pb^{2+} (aq) + 2 e^- → Pb (s)	−0.126 V

    A. 0.889 volts
    B. 0.637 volts
    C. 0.511 volts
    D. −0.637 volts
    E. −0.889 volts

16. Which will reduce Ag^+ to Ag but not Ni^{2+} to Ni?

Half-Reaction	Reduction Potential, $E°$, volts
Mg^{2+} (aq) + 2 e^- → Mg (s)	−2.363
Al^{3+} (aq) + 3 e^- → Al (s)	−1.662
Zn^{2+} (aq) + 2 e^- → Zn (s)	−0.763
Cd^{2+} (aq) + 2 e^- → Cd (s)	−0.410
Ni^{2+} (aq) + 2 e^- → Ni (s)	−0.230
Pb^{2+} (aq) + 2 e^- → Pb (s)	−0.126
Ag^+ (aq) + e^- → Ag (s)	+0.799

    A. Al
    B. Cd
    C. Mg
    D. Pb
    E. Zn

17. If all species are in their standard states, which is the strongest oxidizing agent?

Half-Reaction	Reduction Potential, $E°$, volts
Mn^{2+} (aq) + 2 e^- → Mn (s)	−1.180
Zn^{2+} (aq) + 2 e^- → Zn (s)	−0.763
Fe^{2+} (aq) + 2 e^- → Fe (s)	−0.440
Co^{2+} (aq) + 2 e^- → Co (s)	−0.280
Br_2 (aq) + 2 e^- → 2 Br^- (s)	+1.087

    A. Br^-
    B. Co^{2+}
    C. Fe^{2+}
    D. Mn^{2+}
    E. Zn^{2+}

18. For the net cell potential of the cell Zn (s) | Zn^{2+} || Cd^{2+}| Cd (s) is 0.360 volts. What is the standard free-energy change, $\Delta G°$, for this reaction in kilojoules at 25°C?

    A. −225
    B. −34.7
    C. −69.5
    D. 69.5
    E. 113

19. For the reaction between permanganate ion and oxalate ion in basic solution, the unbalanced equation is:

    $2 MnO_4^- + 3 C_2O_4^{2-} \rightarrow 2 MnO_2 + CO_3^{2-}$

    Which is true regarding the number of OH⁻ ions when the equation is balanced with smallest, whole-number coefficients?

    A. 0
    B. 2 on the right
    C. 2 on the left
    D. 4 on the right
    E. 4 on the left

20. The pair of vertical lines in the standard cell notation represents which connection between the anode and the cathode?

    A. Wire
    B. Voltmeter
    C. Salt bridge
    D. Space between the beakers
    E. Separation between solution and electrodes

Free-Response Questions

1. Consider the reaction between nitrogen(II) oxide and oxygen to produce nitrogen(IV) oxide at 25°C.

   $2 NO (g) + O_2 (g) \rightarrow 2 NO_2 (g)$

	ΔH_f° (kJ/mol)	S_f° (J/mol · K)
NO (g)	+90.25	210.7
O_2 (g)	0	205.0
NO_2 (g)	+33.18	240.0

   (A) What is the standard enthalpy change of the reaction at 25°C?
   (B) What is the standard entropy change of the reaction at 25°C?
   (C) What is the standard free-energy change of the reaction at 25°C?
   (D) Is this reaction spontaneous at 25°C? State the reason for your answer.
   (E) What is the equilibrium constant for the reaction at 25°C?

2. Reactions may or may not be spontaneous, depending sometimes upon temperature.

   (A) List at least two conditions of enthalpy change and entropy change where temperature is the determining factor on whether or not a reaction is spontaneous.

   (B) At what temperatures will this reaction be spontaneous if $\Delta S^\circ = -232.5$ J/mol · K and $\Delta H^\circ = +140.0$ kJ/mol?
   $C_2H_2 (g) + 2 H_2 (g) \rightarrow C_2H_6 (g)$

   (C) At what temperatures will this reaction be spontaneous if $\Delta S^\circ = +134$ J/mol · K and $\Delta H^\circ = +131$ kJ/mol?
   $C (s) + H_2O (g) \rightarrow CO (g) + H_2 (g)$

   (D) Account for why ΔS° decreases in Reaction (B) and increases in Reaction (C).

3. (A) Write the standard cell notation for this redox reaction:

   $Zn\ (s) + Cu^{2+}\ (aq) \rightarrow Zn^{2+}\ (aq) + Cu\ (s)$

   (B) What is the overall redox reaction for the standard cell notation shown below?
   $Cr\ (s)\ |\ Cr^{3+}\ (aq)\ ||\ Ag^+\ (aq)\ |\ Ag\ (s)$

   (C) What is the difference between "cell potential" and "standard reduction potential"?

   (D) The free energy of a redox reaction involving the transfer of two electrons is –44.4 kJ. What is the voltage of the electrochemical cell in which this reaction takes place?

4. The Nernst equation is $E_{cell} = E^\circ - \dfrac{RT}{nF}\ \ln Q$.

   (A) When is there a need to take concentration into account when working with electrochemical cells?
   (B) Write the Nernst equation. What are the units on each term in the Nernst equation?
   (C) What is the standard potential for the following electrochemical cell?
   $Cr\ (s) + 3\ Ag^+\ (aq) \rightarrow Cr^{3+}\ (aq) + 3\ Ag\ (s)$

Reaction	E°, volts
$Cr^{3+}\ (aq) + 3\ e^- \rightarrow Cr\ (s)$	–0.74
$Ag^+\ (aq) + e^- \rightarrow Ag\ (s)$	+0.80

   (D) A voltaic cell consists of a solid chromium electrode in a 0.500 M solution of Cr^{3+} (aq) and a silver electrode in a 0.250 M solution of Ag + (aq). What is the voltage of this cell at 25°C?

Multiple-Choice Answers

1. A   2. D   3. A   4. A   5. C   6. C   7. D   8. E   9. C   10. B

11. B   12. D   13. E   14. A   15. B   16. D   17. B   18. D   19. E   20. C

1. A

ΔG, or Gibbs free energy, is equal to $\Delta H - T\Delta S$. If ΔG is negative, the process is said to be spontaneous. If ΔG is positive, the process is said to be nonspontaneous. Given this information, we can determine that the correct answer is (A).

2. D

A snowball will not spontaneously roll up a hill because this would require a continuous input of energy. The melting of a scoop of ice cream on a hot summer day occurs spontaneously because it doesn't require a continuous input of energy. Because Na^+ and Cl^- ions have a greater affinity for water than they do for each other, a continuous input of energy is not required for the dissolution of NaCl (s) in water, and this process occurs spontaneously. An initial input of energy, such as a push or an earthquake, is required for a book to fall off a shelf. However, once the book is airborne, no more energy is required, so it will fall spontaneously. The transition of liquid water into solid ice at 1 atm and −5°C is spontaneous because again it doesn't require a continuous input of energy.

3. A

The change in entropy for a reaction is equal to: $\Delta S = \Sigma S°$ (Products) $- \Sigma S°$ (Reactants)

The entropies of elements are not zero unless they are in the standard state at absolute zero. The coefficients in the balanced reaction come into play because the entropies listed in the table are given a per mole.

Change in entropy = $((229.5) - ((2)(5.74) + (3)(130.6))) = -173.8$ J

4. A

The change in enthalpy for a reaction is equal to: $\Delta H = \Sigma \Delta H_f°(\text{Products}) - \Sigma \Delta H_f°(\text{Reactants})$

The coefficients in the balanced reaction come into play because the enthalpies listed in the table are given per mole.

Change in enthalpy = $((2)(-393.5) + (8)(-241.8)) + (2)(-110.5)) - ((4)(-74.9) + (7)(0)) = -2642.8$ kJ

5. C

$\Delta G = \Delta H - T\Delta S$; if $\Delta G < 0$ it is spontaneous. At $\Delta G = 0$, $T = \Delta H/\Delta S$.

If ΔH is positive and nonspontaneous, and ΔS is positive but spontaneous, the reaction will be spontaneous at temperatures larger than $\Delta H/\Delta S$. The reaction becomes spontaneous, since $T\Delta S > \Delta H$, and ΔG becomes negative and spontaneous.

6. C

$\Delta G° = \Sigma \Delta G_f°(\text{Products}) - \Sigma \Delta G_f°(\text{Products})$. The free energies of formation of the elements in their standard states is zero. Bromine as a gas is not in its standard state, which is the liquid state.

$\Delta G° = ((1/2 \times (3.144) + (0)) - (-53.43) = 55.00$ kJ

7. D

The reaction is spontaneous if ΔG is less than zero (ΔG is negative). Solving for T to find the temperature on the cusp of making it spontaneous gives:

$T = \Delta H \div \Delta S = 198 \div .187 = 1060$ K

At this temperature, the reaction would be at equilibrium. ΔH is spontaneous; ΔS is not. Above the temperature of 1060 K the reaction is nonspontaneous, and below 1060 K it is spontaneous.

8. E

The equilibrium constant, K, is a measure of the tendency of a reaction to go to completion and is calculated by dividing the concentration of the products by the concentration of the reactants. In a spontaneous reaction, more products will be produced than reactants—the reaction spontaneously makes more products than reactants. For spontaneous reactions K > 1, there is more product than reactant. When K = 0, the reaction is not occurring at all—there are no products. When K < 0 is impossible (negative products?). A value of K = 1 indicates the product concentrations is equal to the reactants.

9. C

By definition, when a reaction is at equilibrium, the value of Q will equal the value of K.

10. (B)

Entropy change is a measure of chaos. $\Delta S°$ of Gases > Liquids > Solids. In answer (B) the chaotic gaseous IBr is forming a less chaotic solid and a liquid. This is not spontaneous. Entropy changes which proceed in the direction of increases chaos are spontaneous.

11. B

$\Delta G° = -RT \ln K$ and $\ln K = \dfrac{\Delta G°}{-RT}$. $R = 8.314$ J / K $\cdot$ mol

The reaction is for the formation of a product from its elements, so $\Delta G° = \Delta G_f° = -23.49$. Watch out! ΔG is in kJ, and R is in J. ΔG must be multiplied or R must be divided by 1,000.

$$\ln K = \frac{-(-23,490)}{(8.314 \times 298)} = 9.481$$

$$\therefore K = e^{9.481} = 1.3 \times 10^4$$

12. D

The overall reaction is $Cd(s) + Cu^{2+} \rightarrow Cd^{2+} + Cu$. The Nernst equation calculates the cell voltage when the concentrations are not the standard 1.0 M and 1 atm. $E_{cell} = E° - (0.059/n) \log Q$. Q, the reaction quotient takes the same form as K, the equilibrium constant, [Products] / [Reactants]. If the [Reactants] > [Products] Q is a fraction whose log is a negative number, and the log term is additive to the standard cell potential. This will cause $E_{cell} > E°$, the desired result.

13. E

Oxygen's oxidation number is generally −2, unless in the peroxide form when it is −1. In an ion, the sum of the oxidation numbers equals the charge on the ion. For NO_3^-, the oxygen atoms contribute −6, which means for the ion to have a −1 total, the oxidation number of nitrogen must be +5.

14. A

For a spontaneous reaction the ΔG must be a negative number. The equilibrium constant, K, must show the production of more product than reactant and be greater than 1. The clincher is that all spontaneous reactions will have a positive cell voltage, $E°$.

15. B

The zinc is oxidized, and the lead ion is reduced in order to provide a positive cell potential. Oxidation is the reverse of reduction, so the voltage for the Zn being converted to zinc ions is +0.763. The potential (voltage) for this cell is (+0.763) + (−0.126) or 0.637 volts.

16. D

The reduction reaction for lead ions is just above that of silver ions and below that of nickel ions. In a reaction between Pb and Ag^+, the cell potential is positive. In the reaction between Pb and Ni^{2+}, the cell potential is negative, and will not occur. The Pb − Ni^{2+} relationship is true for all the other ions above Pb in the table.

17. B

A substance that is an oxidizing agent is itself reduced in the redox reaction. The best oxidizing agent is the substance in the list that is most easily reduced. The cobalt ion with a reduction potential of −0.280 volts, the lowest of the group, is most easily reduced. The bromide ion is oxidized, not reduced, and is not a candidate in this question.

18. D

$\Delta G = -nFE° = $ (2 moles)(96,500 J / mole • volt)(0.360 volts) = 69,500 J = 69.5 kJ

19. E

To balance this redox reaction in basic solution follow these steps.

Write the two half-reactions and balance the number of atoms of elements other than O. Balance the oxygen by adding 2 OH^- ions for each oxygen you need, and a 1 H_2O on the other side of the reaction for each 2 OH^- you have added. Then balance the charge by adding electrons to the side deficient in them. Then balance the gain and loss of electrons by multiplying each half-reaction by the factor needed so the gain or loss equals a lowest common denominator number of electrons. Add the two half-reactions and simplify. If all goes well, the atoms and the charges will be balanced.

$2 \ [MnO_4^- + 2 \ H_2O + 3 \ e^- \rightarrow MnO_2 + 4OH^-]$

$3 \ [C_2O_4^{2-} + 4OH^- \rightarrow 2 \ CO_3^{2-} + 2 \ H_2O + 2 \ e^-]$

$2 \ MnO_4^- + 3 \ C_2O_4^{2-} + 4OH^- \rightarrow 2 \ MnO_2 + 6 \ CO_3^{2-} + 2 \ H_2O$

20. C

The pair of vertical lines represents the salt bridge which allows ions to migrate between the anode and cathode to maintain the electrical neutrality in each half-cell.

Free-Response Answers

1. (A) $\Delta H° = \Sigma \Delta H_f°(\text{Products}) - \Sigma \Delta H_f°(\text{Reactants})$

   $\Delta H° = (2 \times +33.18) - ((2 \times +90.25) + 0) = -114.1$ kJ

   (B) $S° = \Sigma S_f°(\text{Products}) - \Sigma S_f°(\text{Reactants})$
   $\Delta S° = (2 \times +240.0) - ((2 \times +210.7) + (1 \times +205.0)) = -146.4$ J

   (C) $\Delta G° = \Delta H° - \dfrac{T\Delta S°}{1000}$. The $T\Delta S$ must be divided by 1000 since ΔH is in kJ and ΔS in Joules.

   $\Delta G° = (-114.1) - \dfrac{(298)(-146.4)}{1000} = -70.52$ kJ

   (D) The reaction is spontaneous at this temperature since the $\Delta G° < 0$.

   (E) $\Delta G° = -RT \ln K$ and $\ln K = -\dfrac{\Delta G°}{RT}$

   $\ln K = \dfrac{-(-70.52 \text{ kJ})}{(0.008314 \text{ kJ / mol} \cdot \text{K}) \times 298 \text{ K}} = 28.46$

   $K = 2.299 \times 10^{12}$

2. (A) When ΔH and ΔS both have the same sign, either positive or negative, one is favorable for spontaneous reaction and the other is unfavorable. When ΔH is positive (unfavorable), high temperature must be achieved to make the $T\Delta S$ term larger so that it exceeds ΔH, making ΔG negative and the reaction spontaneous. Low temperatures will lead to spontaneous reaction when both ΔH and ΔS are negative.

   (B) At no temperature will this reaction be spontaneous since ΔH is positive (unfavorable) and ΔS is negative (unfavorable).

   (C) When $\Delta G = 0$, $T = \dfrac{\Delta H}{\Delta S} = \dfrac{(+131 \text{ kJ / mol})}{(+0.134 \text{ kJ / mol} \cdot \text{K})} = 978$ K

   At temperatures greater than 978 K, the reaction is spontaneous.

   (D) In Reaction (B) the chaos decreases since 3 molecules are joining to make 1, and chaos decreases. In Reaction (C) a solid and a gas are combining to make 2 gases. Gases are more chaotic than solids.

3. (A) $\text{Zn (s)} \mid \text{Zn}^{2+} \text{(aq)} \parallel \text{Cu}^{2+} \text{(aq)} \mid \text{Cu (s)}$

   Zn and Cu are the anode and cathode, respectively. The phase boundaries are shown by $\mid$, and the salt bridge by $\parallel$. The anode half-cell is routinely on the left, and the cathode half-cell on the right.

   (B) $\text{Cr (s)} + 3 \text{ Ag}^+ \text{(aq)} \rightarrow \text{Cr}^{3+} \text{(aq)} + 3 \text{ Ag (s)}$

   (C) The cell potential (also known as the electromotive force or cell voltage) is the difference in electrical potentials of two electrodes in an electrochemical cell, which is the driving force for the movement of electrons. The standard reduction potential is the cell potential for a reduction half-reaction in which reactants and products are in their standard states, and no current is flowing.

The difference between these two terms is that the cell potential is the total voltage for the oxidation and reduction half-reactions, whereas the standard reduction potential is the voltage of just the reduction half-reaction.

(D) $\Delta G° = -nFE°$

-44.4 kJ $= -4.44 \times 10^4$ J $= (-2$ mol e$^-$)(96,500 °C/mol e$^-$)($E°$)

$$E° = \frac{(-4.44 \times 10^4 \text{ J})}{[(-2 \text{ mol e}^-)(96,500 \text{ °C/mol e}^-)}] = 0.230 \text{ J/°C} = 0.230 \text{ V}$$

4. (A) Free energy depends on concentration, and since free energy and cell potential are related, it follows that cell potential is dependent on concentration. The cell potential is calculated directly from standard reduction potentials only when all species in the redox reaction are present in their standard states. For aqueous ions, this means 1 M. If concentration is anything other than 1 M, we must include the concentration values in the calculation of E_{cell}.

(B) The terms and their units are:
E_{cell} = cell potential, in volts
$E°$ = standard cell potential, in volts
R = gas constant in units of J/mol · K (or volt-coulomb/mol · K)
T = absolute temperature, in kelvins
n = number of moles of electrons transferred, in units of mol e$^-$
F = Faraday constant = 96,500 coulombs/mol e$^-$
Q = reaction quotient (unit-less)

(C) Oxidation half-reaction: Cr (s) → Cr^{3+} (aq) + 3 e$^-$; $E°$ = -(-0.74) V = + 0.74 V
Reduction half-reaction: 3 Ag + (aq) + 3 e$^-$ → 3 Ag (s); $E°$ = + 0.80 V
Overall reaction: $E°$ = 0.74 V + 0.80 V = 1.54 V

(D) The overall reaction is: Cr (s) + 3 Ag$^+$ (aq) → Cr^{3+} (aq) + 3 Ag (s). Three electrons are transferred in this reaction, so n = 3. Use the condensed form of the Nernst equation, since T = 25°C (298 K):

$E = E° - (0.0592/n)\log Q$

Obtain $E°$ from part (C): $E°$ = 1.54 V

$$E = 1.54 \text{ V} - \left(\frac{0.0592}{3}\right) \log \left(\frac{[\text{Cr}^{3+}]}{[\text{Ag}^+]^3}\right) = 1.54 \text{ V} - \left(\frac{0.0592}{3}\right) \log (32) = 1.51 \text{ V}$$

Organic Chemistry

Organic chemistry is the study of carbon and its compounds (excluding such compounds as CO_2, carbonates, and hydrogen carbonates). In organic chemistry, the *structure* of a molecule and the *functional groups* that it contains is important in determining its properties and reactivity. A **functional group** is an entity containing one or more atoms that determines an organic molecule's reactivity. We will examine the main functional groups of organic molecules later in this chapter. Ultimately, the source of many natural and synthetic organic compounds is petroleum and its derivatives.

COMPARISON OF ORGANIC AND INORGANIC COMPOUNDS

While there are always exceptions to every rule, a number of general comparisons can be made between organic and inorganic compounds:

- Inorganic compounds are generally soluble in polar solvents such as water, while organic compounds are generally soluble in nonpolar solvents such as benzene.
- When dissolved in solution, many inorganic compounds conduct an electric current because of ionization or dissociation of the solute. Organic compounds are generally nonelectrolytes.
- Generally, inorganic compounds have higher melting and boiling points than organic compounds because of their stronger intermolecular forces.
- In general, inorganic compounds react more rapidly than organic compounds, because reacting inorganic compounds require fewer bond rearrangements than those of organic compounds.

HYDROCARBONS AND HOMOLOGOUS SERIES

The simplest organic compounds contain only carbon and hydrogen and are called *hydrocarbons*, and they form the basis of all other organic compounds. Hydrocarbons are obtained from the refining of petroleum, a complex mixture of natural hydrocarbons, by fractional distillation. The simplest hydrocarbon is *methane* (CH_4), which is the chief component of natural gas. Methane is a nonpolar molecule that has a tetrahedral structure due to the sp^3 hybridization of the carbon atom (See chapter 9).

Hydrocarbons with related structures and properties are usually separated into "families" known as *homologous series*. Each member of a series differs from the next member by a single carbon atom. With each successive member, melting and boiling points usually increase steadily.

The Alkanes

Alkanes are hydrocarbons that contain only *single* bonds between carbon atoms. The first member of the alkane series is methane (CH_4). In chapter 8, we discovered that methane is a *tetrahedral* molecule because carbon forms four sp^3 hybrid orbitals—as it does in every alkane molecule. The table below lists the first ten members of the alkane series:

Name of Alkane	Molecular Formula	Phase at STP
Methane	CH_4	Gas
Ethane	C_2H_6	Gas
Propane	C_3H_8	Gas
Butane	C_4H_{10}	Gas
Pentane	C_5H_{12}	Liquid
Hexane	C_6H_{14}	Liquid
Heptane	C_7H_{16}	Liquid
Octane	C_8H_{18}	Liquid
Nonane	C_9H_{20}	Liquid
Decane	$C_{10}H_{22}$	Liquid
Eicosane	$C_{20}H_{42}$	Solid

As the number of carbon atoms increases, the London forces between the molecules become stronger and there is a progression from gas to liquid to solid. The general formula for an alkane with n carbon atoms is C_nH_{2n+2}.

Note that the names of all alkanes end with the suffix *–ane*. The prefixes that precede the *–ane* ending (highlighted in *italics* in the list of alkanes) indicate the number of carbon atoms, not only in alkane molecules, but in *all* organic molecules: thus, *meth–* stands for one carbon atom, *but–* for four carbon atoms, *oct–* for eight carbon atoms, and so on.

Chemists write the formulas of organic compounds in several ways:

- A *molecular* formula, such as C_3H_8, indicates how many atoms of each element are present, but it fails to provide information about how the atoms are connected.

- A *dashed structural* formula uses straight lines to represent chemical bonds. For example, ethane (C_2H_6) would be written as:

$$\begin{matrix} & H & & H & \\ & | & & | & \\ H - & C & - & C & - H \\ & | & & | & \\ & H & & H & \end{matrix}$$

The advantage of a dashed structural formula lies in its ability to display the connections among the atoms. However, more complex formulas can be cumbersome to write in this style.

- A *condensed structural* formula represents a compromise between the simplicity of a molecular formula and the information provided by a dashed structural formula. In this style, atoms are usually written *without* dashes, and hydrogen atoms are written to the *right* of the atom to which they are connected. For example, the condensed structural formula of alkane known as butane is:

$$CH_3CH_2CH_2CH_3$$

In this book, we will combine condensed and dashed structural formulas when it makes the structure of the molecule more apparent. For example, the alkane known as isobutane would be written as:

$$\begin{matrix} & CH_3 & \\ & | & \\ CH_3 & CH & CH_3 \end{matrix}$$

There are other ways of writing formulas for organic compounds, but they will not be used in this book.

Isomers

The compounds known as *n*-butane and isobutane (shown above) have the same molecular formula (C_4H_{10}), but different structural formulas. Substances that have the same molecular formula but have different structural formulas are called **isomers**. Isomers differ from one another in both their physical and chemical properties.

Constitutional Isomers

Two or more compounds are **constitutional isomers** if they have the same molecular formula but *their atoms are connected in a different order*. The two butane molecules shown above are constitutional isomers.

Sample Problem

The group of alkanes collectively called the hexanes (C_6H_{14}) consists of five constitutional isomers.

Write the two-dimensional structural formula for any two of these isomers.

Solution

All five structural formulas are shown below:

I $CH_3CH_2CH_2CH_2CH_2CH_3$

$$CH_3$$
$$|$$
II $CH_3CHCH_2CH_2CH_3$

$$CH_3$$
$$|$$
III $CH_3CH_2CHCH_2CH_3$

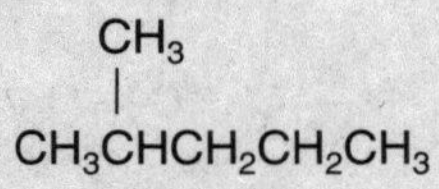

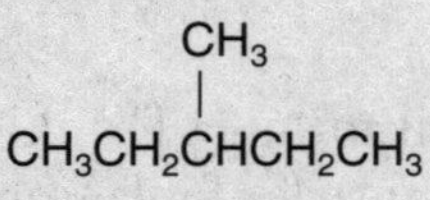

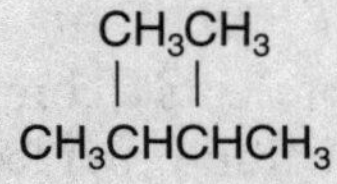

$$CH_3 \ CH_3$$
$$| \ \ \ |$$
IV $CH_3CHCHCH_3$

$$CH_3$$
$$|$$
V $CH_3CCH_2CH_3$
$$|$$
$$CH_3$$

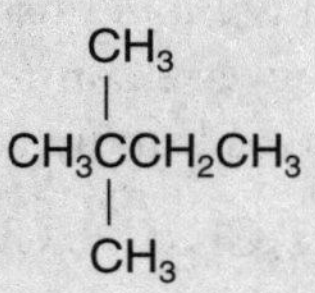

It must be emphasized that each constitutional isomer is a *different* compound, possessing unique physical and chemical properties. The table below lists the melting point, boiling point, and density of the five constitutional isomers of C_6H_{14}.

Isomer Number	Melting Point /°C	Boiling Point /°C	Density (20°C) /(g/cm^3)
I	−95	69	0.659
II	−154	60	0.653
III	−163	63	0.664
IV	−129	58	0.661
V	−99	50	0.649

Enantiomers: Mirror-Image Isomerism

Suppose we imagine a molecule in which a carbon atom forms four single bonds with four different groups, *W*, *X*, *Y*, and *Z*. We know that the shape of the molecule around this carbon atom will be tetrahedral as shown in the diagram below.

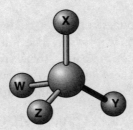

Surprisingly, this molecule actually this exists in *two* forms that are *mirror images* of each other. The diagram below illustrates both forms of the molecule:

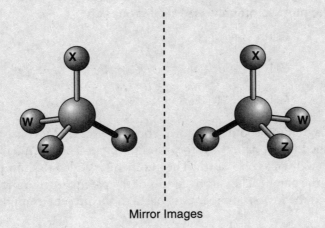

Mirror Images

The molecules cannot be superimposed on one another, just as a left hand cannot be superimposed on its mirror image: a right hand. Such mirror-image isomers are called **enantiomers**. Atoms that can form enantiomers are said to exhibit the property of *chirality* (from the Greek word for "hand"). Chiral molecules are especially important in biological systems because certain molecules can be metabolized only if they have a "left-handed" or a "right-handed" configuration.

The IUPAC System of Nomenclature

As the number of carbon atoms in an organic compound increases, the number of isomers rises sharply. For example, the molecular formula $C_{10}H_{22}$ corresponds to 75 isomers, and the molecular formula $C_{20}H_{42}$ corresponds to 366,319 isomers. Providing a unique name for each and every isomer would be impossible unless a simple and logical way to name organic compounds were developed. The *International Union of Pure and Applied Chemistry* (IUPAC) is responsible for overseeing such a system of nomenclature. In the IUPAC system, organic compounds are considered derivatives of a *parent hydrocarbon*: the longest, unbroken chain of carbon atoms. For example, in the compound known as isobutane:

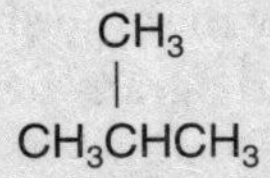

$$CH_3$$
$$|$$
$$CH_3CHCH_3$$

there are *three* carbon atoms in the longest, unbroken chain. Since only single bonds are present, the name of the parent hydrocarbon is *propane*. The full name of an organic compound is determined by modifying the name of the parent structure according to the functional *groups* that are attached to the parent. These organic compounds are known as **straight-chain** hydrocarbons. For example, in the diagram below, the hydrocarbon is named *pentane*:

$$CH_3CH_2CH_2CH_2CH_3$$

When groups of carbon atoms—known as **alkyl groups**—are attached to the parent hydrocarbon, the compound is known as a **branched-chain** hydrocarbon. Alkyl groups are named by attaching the suffix *–yl* to the prefix that indicates how many carbon atoms the group has. The three most commonly used alkyl groups are:

CH_3- *methyl*

CH_3CH_2- or C_2H_5- *ethyl*

$CH_3CH_2CH_2-$ or C_3H_7- *propyl*

If we refer to the diagram of isobutane, shown above, we see that it consists of a one-carbon group (methyl) joined to the parent hydrocarbon propane. The IUPAC name for isobutane is *2-methylpropane* (written as one word). The number 2 is added to show precisely *where* the group is located on the parent. The carbon atoms of the parent are numbered in order so that the group is assigned the lowest number possible: 2 in the present example.

The five constitutional isomers of C_6H_{14} (illustrated on page 424) are named according to the IUPAC system in the table below:

Isomer Number	IUPAC Name
I	hexane
II	2-methylpentane
III	3-methylpentane
IV	2,3-dimethylbutane
V	2,2-dimethylbutane

In structure V note that each methyl group is entitled to its own number, even if the number is repeated.

Reactions of Alkanes

Combustion Alkanes are widely used as fuels. In the presence of excess oxygen, the alkane combusts completely to form carbon dioxide and water as its only products:

$$C_3H_8(g) + 7O_2(g) \rightarrow 3CO_2(g) + 4H_2O(g)$$

If the supply of oxygen is limited, carbon monoxide (CO) or particles of carbon (C), known as *soot*, may form.

Substitution In a **substitution** reaction one type of atom or group is replaced with another type. In an alkane, a variety of atoms may replace one or more hydrogen atoms in the formula. Substitution reactions using halogens such as chlorine, bromine, and iodine as the substituting atoms are common. Halogen substitution is a stepwise process, in which the halogen molecules are split into single atoms (*free radicals*); the halogenated molecules that are formed are known as **haloalkanes**. The chlorination of methane is shown below (the IUPAC names of the haloalkanes are given in parentheses):

$$CH_4 + Cl_2 \rightarrow HCl + CH_3Cl \qquad \text{(chloromethane)}$$

$$CH_3Cl + Cl_2 \rightarrow HCl + CH_2Cl_2 \qquad \text{(\textit{di}chloromethane)}$$

$$CH_2Cl_2 + Cl_2 \rightarrow HCl + CHCl_3 \qquad \text{(\textit{tri}chloromethane)}$$

$$CHCl_3 + Cl_2 \rightarrow HCl + CCl_4 \qquad \text{(\textit{tetra}chloromethane)}$$

According to the IUPAC system, a haloalkane is named by placing the combining form of the halogen name (*fluoro–, chloro–, bromo–, iodo–*) before the name of the alkane. If necessary, a prefix (*di, tri*, etc.) and numbers are assigned to indicate the number of halogen atoms present and their locations.

Sample Problem

Name the following haloalkane according to the IUPAC system:

$$\begin{array}{cc} Br & Br \\ | & | \\ \end{array}$$
$$CH_2CH_2CHCH_2CH_3$$

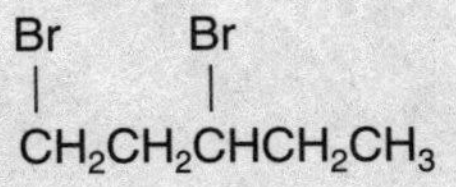

Solution

The parent hydrocarbon contains five carbon atoms: Its name is pentane. Counting from the left, bromine atoms are located on the first and third carbon atoms. The compound's name is **1,3-dibromopentane**.

The Alkenes

Alkenes are hydrocarbons that contain one carbon-carbon double bond. The first member of the alkene series is *ethene*, C_2H_4 (commonly called *ethylene*):

$$\begin{array}{ccc} H & & H \\ \diagdown & & \diagup \\ & C = C & \\ \diagup & & \diagdown \\ H & & H \end{array}$$

If we examine the structural formula of ethene (C_2H_4) given above we find that it is a planar (flat) molecule and the bond angles around each carbon atom are close to 120°. In chapter 9, we learned that a planar arrangement of three 120° bond angles was associated with the formation of three sp^2 hybrid orbitals.

The two carbon atoms in ethene form a total of six sp^2 hybrid orbitals. Four of them overlap with 1*s* orbitals of the hydrogen atoms and the remaining two join the carbon atoms. These bonds are sigma (σ) bonds because they are symmetrical about the internuclear axes, and all lie in the same plane. In addition, each carbon atom has a remaining 2*p* orbital containing a single electron. These overlap *sideways* to produce a pi (π) bond that lies above and below the plane of the sigma bonds. The orbital diagram of ethene is shown below:

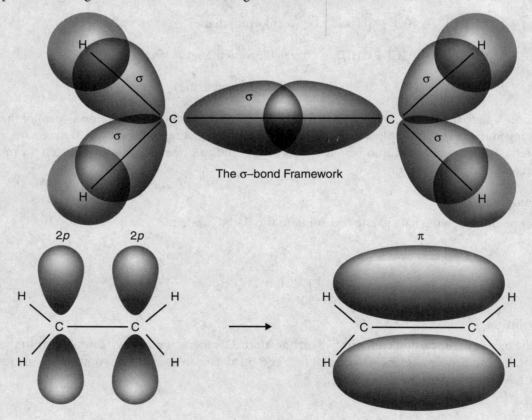

The Formation of a π-bond by the Overlap of the Half-Filled 2*p* Orbitals

The names of all alkenes names end with the suffix *–ene*. The general formula for an alkene with *n* carbon atoms is $C_n H_{2n}$.

Sample Problem

What are the molecular and structural formulas of *butene*?

Solution

There are three constitutional isomers of butene. All have the molecular formula C_4H_8. Two of the constitutional isomers differ in the position of the double bond:

$$CH_2 = CHCH_2CH_3 \qquad CH_3CH = CHCH_3$$
1-butene   2-butene

The third isomer has the structure:

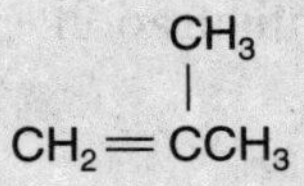

$$\begin{array}{c} CH_3 \\ | \\ CH_2 = CCH_3 \end{array}$$

2-methyl-1-propene

Stereoisomerism

Stereoisomers are isomers whose atoms have different arrangements in space. Enantiomers are one example of stereoisomers. Another example is the placement of atoms or groups about a double bond. There are two stereoisomers of 2-butene. While there is *free rotation* about a C–C single bond, the way in which C=C double bonds are formed causes the atoms around each carbon to be held rigidly in position. These two stereoisomers are shown below:

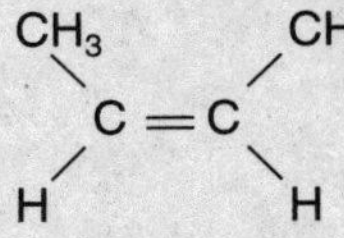

cis-2-butene 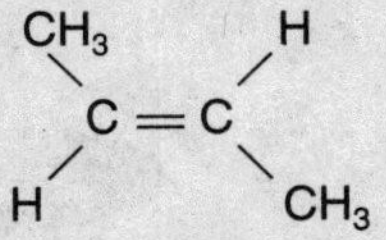   trans-2-butene

In the first stereoisomer, known as the *cis* form, the methyl groups are closer together than in the second stereoisomer, which is known as the *trans* form. Therefore, we name the isomer on the left *cis*-2-butene and the one on the right *trans*-2-butene.

Sample Problem

Draw the structural formulas of the two stereoisomers of 1,2-diiodoethene.

Solution

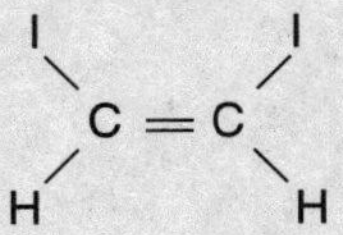

cis-1,2-diiodoethene 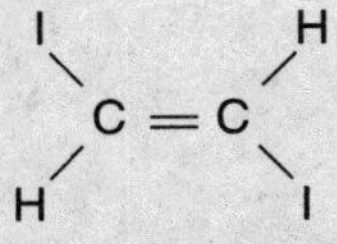   trans-1,2-diiodoethene

The Alkynes

Alkynes are hydrocarbons that contain one carbon-carbon *triple* bond. The first member of the alkyne series is ethyne, C_2H_2 (commonly called acetylene):

$$CH \equiv CH$$

Ethyne
(Acetylene)

The names of all alkynes end with the suffix *–yne*, and the general formula for an alkyne with *n* carbon atoms is C_nH_{2n-2}.

Ethyne is a linear molecule and the bond angles around each carbon atom are 180°. In chapter 9, we learned that this type of configuration was associated with the formation of two *sp* hybrid orbitals.

The two carbon atoms in ethyne form a total of four *sp* hybrid orbitals. Two of them overlap with hydrogen atoms and the remaining two join the carbon atoms. Each of these bonds is a sigma (σ) bond. In addition, each carbon atom has two 2*p* orbitals, each containing a single electron. These overlap sideways to produce two pi (π) bonds. Therefore, *the triple bond is composed of one sigma and two pi bonds* as shown in the following illustration.

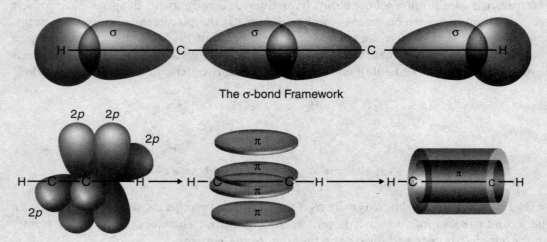

The σ-bond Framework

Formation of π-bonds by the Overlap of Half-filled 2*p* Orbitals

Organic compounds that contain double or triple bonds between carbon atoms are said to be **unsaturated**, while organic compounds containing only single bonds between carbon atoms are said to be **saturated**.

Reactions of Unsaturated Compounds

Addition In an **addition** reaction, a molecule breaks a double or triple bond, which is more reactive than a single bond. As a result, saturated compounds can be produced from unsaturated compounds. A number of addition reactions are illustrated below:

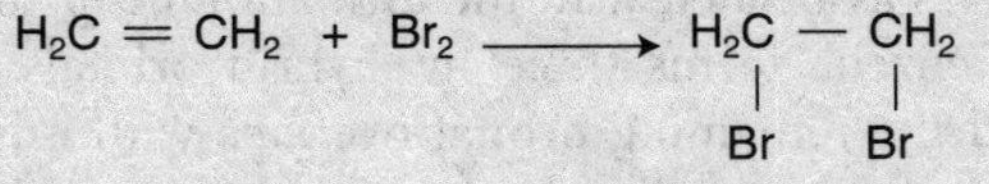

$$H_2C = CH_2 + Br_2 \longrightarrow \begin{array}{cc} H_2C - CH_2 \\ | \quad\quad | \\ Br \quad Br \end{array}$$

$$CH_3CH = CHCH_3 + H_2 \xrightarrow{Ni, 500°C} CH_3CH_2CH_2CH_3$$

$$CH_2 = CH_2 + HBr \longrightarrow CH_3CH_2Br$$

$$CH_2 = CH_2 + H_2O \xrightarrow{H_2SO_4} CH_3CH_2OH$$

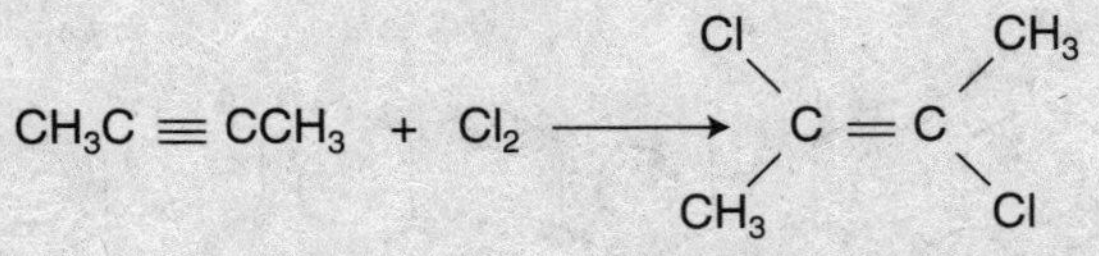

$$CH_3C \equiv CCH_3 + Cl_2 \longrightarrow \begin{array}{c} Cl \quad\quad CH_3 \\ \diagdown \quad\quad \diagup \\ C = C \\ \diagup \quad\quad \diagdown \\ CH_3 \quad\quad Cl \end{array}$$

2-butyne   *trans*-2,3-dichloro-2-butene

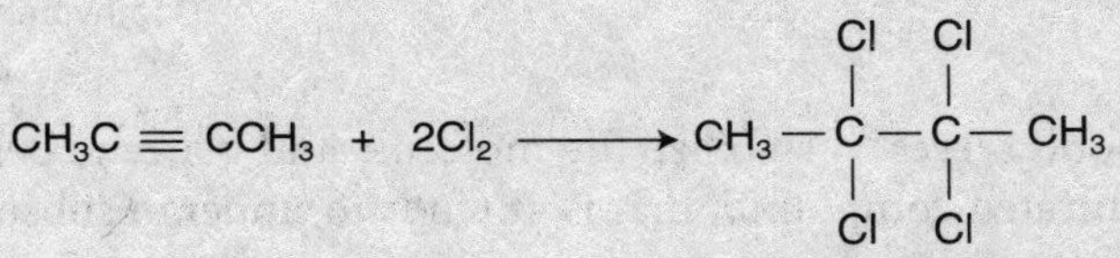

$$CH_3C \equiv CCH_3 + 2Cl_2 \longrightarrow \begin{array}{c} Cl \quad Cl \\ | \quad\quad | \\ CH_3 - C - C - CH_3 \\ | \quad\quad | \\ Cl \quad Cl \end{array}$$

2-butyne   2,2,3,3-tetrachlorobutane

Addition reactions take place more rapidly than substitution reactions because the pi bonds in alkenes and alkynes are more reactive than the sigma bonds. Addition is characteristic of unsaturated compounds. The addition of halogen atoms such as chlorine and bromine is a common test for unsaturation in hydrocarbons. The addition of hydrogen to an unsaturated compound is usually called *hydrogenation*; it is used commercially to solidify liquid vegetable oils. Although halogen addition is a relatively rapid reaction, hydrogenation is slow and finely divided metal catalysts must be employed to insure a reasonable reaction rate.

Cyclic Compounds

In addition to straight-chain and branched-chain compounds, organic molecules can be formed by linking the first and last carbon atom. Such compounds are known as **cyclic** or **ring compounds**. An example of a cyclic compound is *cyclopentane*, C_5H_{10}, whose structure is shown below:

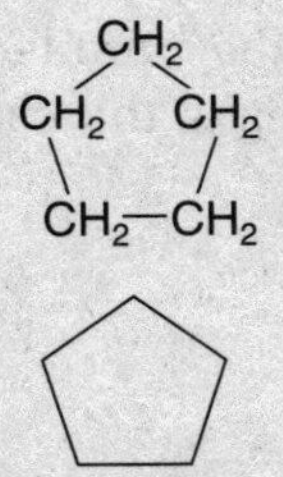

$$\begin{array}{c} CH_2 \\ CH_2 \quad CH_2 \\ CH_2 - CH_2 \end{array}$$

Cyclopentane

Aromatic Compounds

Aromatic compounds are organic compounds that are related to *benzene*, which has the molecular formula C_6H_6. The structure of benzene is quite unique: it is a flat, hexagonal molecule and the bond angles around each carbon atom are 120°. In chapter 9, we saw that this type of configuration was associated with the formation of sp^2 hybrid orbitals. Each of the six carbon atoms in benzene forms three sp^2 hybrid orbitals: two overlap with the adjacent carbon atoms and the remaining orbital overlaps with a hydrogen atom. All are sigma (σ) bonds. In addition, each carbon atom has one $2p$ orbital containing a single electron. The overlap of these p orbitals creates delocalized pi (π) bond that resides above and below the plane of the ring of carbon atoms. Usually, the benzene ring is drawn as a hexagon and the delocalized pi bond is usually represented by a solid circle. A number of aromatic hydrocarbons are illustrated below:

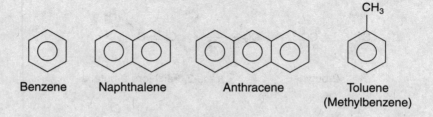

Benzene   Naphthalene   Anthracene   Toluene
(Methylbenzene)

The delocalized pi bond serves to stabilize the molecule and benzene behaves more like an alkane than an unsaturated compound, that is, it tends to undergo substitution rather than addition:

Functional Groups

Earlier, we defined a **functional group** as an entity containing one or more atoms that determines an organic molecule's reactivity. We will now investigate various functional groups and the organic compounds they produce. The following table describes the functional groups that we will be studying in this chapter.

Functional Group	Type of Compound	Suffix	Example	Systematic Name (Common Name)
\C = C/ (alkene structure)	Alkene	*-ene*	H₂C = CH₂ (ethene structure)	Ethene (Ethylene)
—C ≡ C—	Alkyne	*-yne*	H — C ≡ C — H	Ethyne (Acetylene)
—C—Ö—H	Alcohol	*-ol*	H — C — Ö — H (methanol structure)	Methanol (Methyl Alcohol)
—C—Ö—C—	Ether	*-oxy*	H — C — Ö — C — H (dimethyl ether structure)	Methoxymethane (Dimethyl Ether)
—C—Ẍ: (X = Halogen)	Haloalkane	*halo-*	H — C — Cl: (chloromethane structure)	Chloromethane (Methyl Chloride)
—C—N̈—	Amine	*-amine*	H — C — C — N̈ — H (ethylamine structure)	Ethylamine
:O: ‖ —C—H	Aldehyde	*-al*	H — C — C(=O) — H (ethanal structure)	Ethanal (Acetaldehyde)
:O: ‖ —C—C—C—	Ketone	*-one*	H — C — C(=O) — C — H (2-propanone structure)	2-Propanone (Acetone)
:O: ‖ —C—Ö—H	Carboxylic Acid	*-oic acid*	H — C — C(=O) — Ö — H (ethanoic acid structure)	Ethanoic Acid (Acetic Acid)
:O: ‖ —C—Ö—C—	Ester	*-oate*	H — C — C(=O) — Ö — C — H (methyl ethanoate structure)	Methyl Ethanoate (Methyl Acetate)
:O: ‖ —C—N̈—	Amide	*-amide*	H — C — C(=O) — N̈ — H (ethanamide structure)	Ethanamide (Acetamide)

Alcohols

All alcohols contain the *hydroxyl* functional group (–OH). Alcohols can be classified in two ways: (1) according to the number of hydroxyl groups present in a molecule, and (2) according to the way in which the hydroxyl groups are attached to the parent hydrocarbon.

An alcohol is named according to the IUPAC system by adding the suffix –*ol* to the name of the parent hydrocarbon. The simplest *monohydroxy* alcohol contains one carbon atom and is named *methanol* (commonly known as methyl alcohol or wood alcohol). Methanol is a popular solvent. The next member of this series, *ethanol* (commonly known as ethyl alcohol or grain alcohol), contains two carbon atoms. Ethanol is used as an antiseptic and is also a component of alcoholic beverages. The structural formulas of methanol and ethanol are given below:

CH_3OH

Methanol

CH_3CH_2OH

Ethanol

Methanol is produced commercially by reacting carbon monoxide and hydrogen gases under pressure and in the presence of a catalyst:

$$CO(g) + 2H_2(g) \xrightarrow[400°C]{200–300 \text{ atm}} CH_3OH(g)$$

Ethanol is produced by **fermentation**, a process in which yeast enzymes catalyze the conversion of glucose into ethanol and carbon dioxide anaerobically:

$$C_6H_{12}O_6(aq) \xrightarrow{yeast} 2C_2H_5OH(aq) + 2CO_2(g)$$

Alcohols (as well as certain other types of organic compounds) can be classified structurally as *primary*, *secondary*, or *tertiary* because of the differences in the chemical behavior among these types of alcohols. The classification is based on the number of carbon atoms attached to the carbon atom that contains the hydroxyl group:

- An alcohol in which *zero* or *one* carbon atom is bonded to the hydroxyl-containing carbon atom is known as a **primary alcohol**. The general formula and an example (ethanol) are given below.

```
        H                          H
        |                          |
  R — C — OH               CH3 — C — OH
        |                          |
        H                          H

  Primary Alcohol              Ethanol
```

Note that the letter R *in the general formula stands for any unspecified alkyl group.*

- An alcohol in which *two* carbon atoms are bonded to the hydroxyl-containing carbon atom is known as a **secondary alcohol**. The general formula and an example (2-propanol, commonly called *isopropyl alcohol* or *rubbing alcohol*) are given below.

```
        H                          H
        |                          |
  R1 — C — OH               CH3 — C — OH
        |                          |
        R2                         CH3

  Secondary Alcohol            2-Propanol
```

- An alcohol in which *three* carbon atoms are bonded to the hydroxyl-containing carbon atom is known as a **tertiary alcohol**. The general formula and an example (2-methyl-2-propanol) are given below.

```
        R3                        CH3
        |                          |
  R1 — C — OH               CH3 — C — OH
        |                          |
        R2                        CH3

  Tertiary Alcohol        2-Methyl-2-Propanol
```

Alcohols can contain more than one hydroxyl group, though not on the same carbon atom. The dihydroxy alcohol *1,2-ethanediol* (commonly called *ethylene glycol*) is one component of automobile antifreeze. The trihydroxy alcohol *1,2,3-propanetriol* (commonly called *glycerol* or *glycerine*) is important, both commercially and in a number of biological processes. The structures of these alcohols are given below.

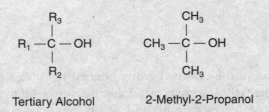

```
    OH  OH                   OH  OH  OH
    |   |                    |   |   |
  H—C — C—H                H—C — C — C —H
    |   |                    |   |   |
    H   H                    H   H   H

  1,2,-Ethanediol           1,2,3,-Propanetriol
  (Ethylene Glycol)            (Glycerol)
```

Ethers

An **ether** is an organic compound in which an oxygen atom is singly bonded to two carbon atoms The structure of dimethyl ether (IUPAC name: methoxymethane) is shown below:

$$CH_3 - O - CH_3$$

An ether can be prepared by *dehydrating* a primary alcohol, such as ethanol, in the presence of H_2SO_4:

$$2C_2H_5OH \xrightarrow{H_2SO_4} C_2H_5OC_2H_5 + H_2O$$

Ethanol   Ethoxyethane
(Diethyl Ether)

Dehydration reactions are also known as *condensation reactions*.

Diethyl ether was widely used at one time as a general anesthetic. In general, ethers are excellent solvents in which organic reactions are frequently carried out.

The Carbonyl Group

One of the most important functional groups in organic chemistry is the *carbonyl group*, which consists of a carbon atom connected to an oxygen atom by a double bond:

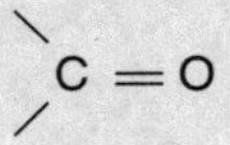

$$C = O$$

Aldehydes

In **aldehydes**, the carbon of the carbonyl group bonded to at least one hydrogen atom. An aldehyde is named in the IUPAC system by adding the suffix *–al* to the name of the parent hydrocarbon. Here are the general formula and two examples:

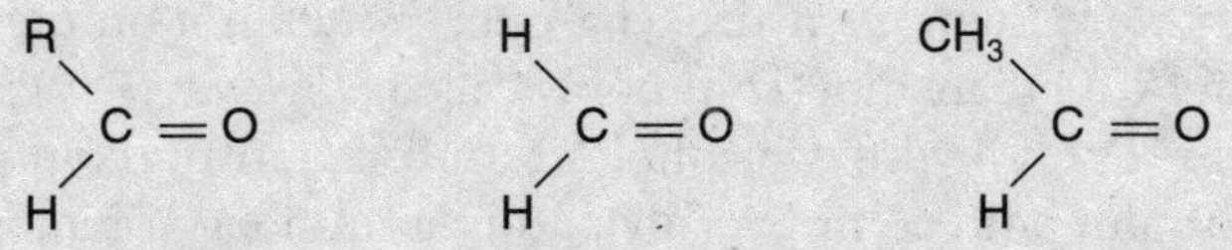

R   H CH_3
C = O   C = O   C = O
H   H   H

Aldehyde   Methanal   Ethanal

Methanal (commonly known as *formaldehyde*) is used as a biological preservative. Other aldehydes, such as benzaldehyde, cinnamaldehyde, and vanillin are the basis of familiar food flavorings. Their structures are shown below:

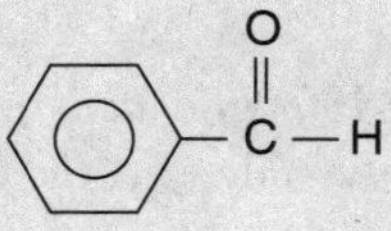

Benzaldehyde
(Almond Flavor)

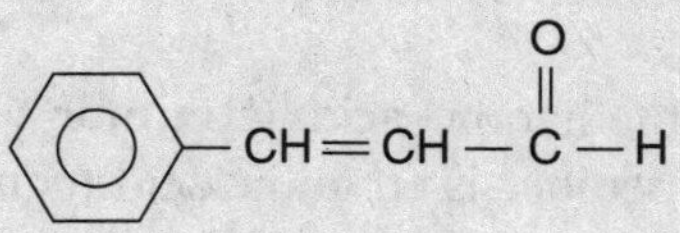

Cinnamaldehyde
(Cinnamon Flavor)

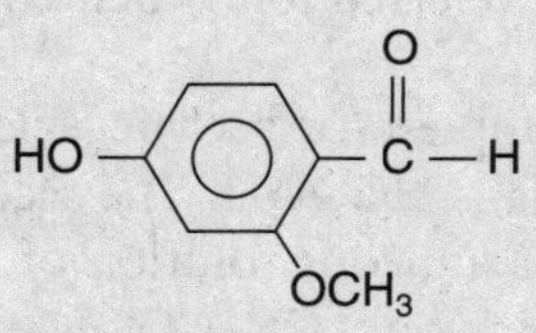

Vanillin
(Vanilla Flavor)

Ketones

Ketones have *no* hydrogen atoms directly attached to the carbonyl group. In the IUPAC system, a ketone is named by attaching the suffix *–one* to the name of the parent hydrocarbon. The general formula and an example are given below:

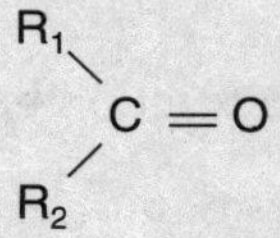

Ketone

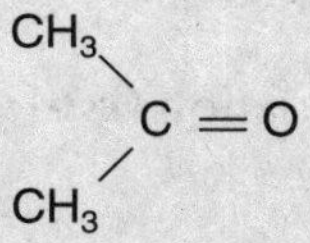

2-Propanone

2-Propanone (commonly known as *acetone*) is the principal solvent in nail polish remover. There are other, more exotic ketones, as shown below:

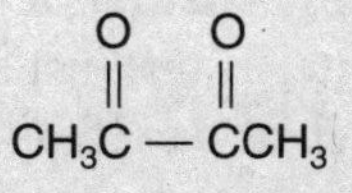

2,3-Butanedione
(Butter Flavoring)

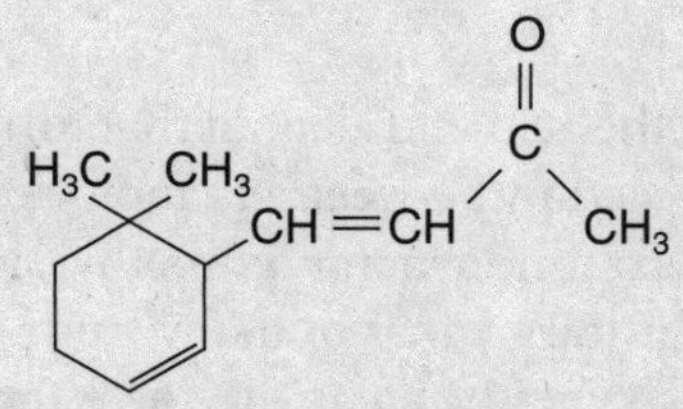

Irone
(Odor of Violets)

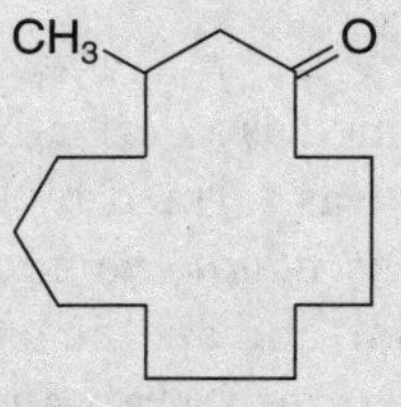

Muscone
(Musk Oil, an
Ingredient in Perfumes)

The Carboxyl Group: Carboxylic Acids

The name *carboxyl* is a contraction of *carbonyl* and *hydroxyl*. The carboxyl functional group has this formula:

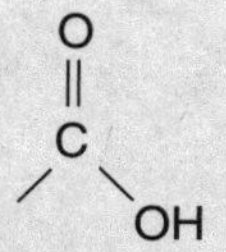

which is usually abbreviated as *COOH*. The presence of the carboxyl group confers acidic properties on the organic compound. In the IUPAC system, an organic acid is named by adding the suffix *–oic* and the word *acid* to the name of the parent hydrocarbon. They are usually referred to by their common names, however.

The simplest carboxylic acid is methanoic acid (commonly called *formic acid*), whose formula is HCOOH. The second member is ethanoic acid (commonly called *acetic acid*). A 5%–6% aqueous solution of ethanoic acid is popularly known as vinegar. The formula for ethanoic acid is CH_3COOH or $HC_2H_3O_2$. Note that the formulas of carboxylic acids separate the acidic hydrogen atom (that is the hydrogen on the carboxyl group) from the other hydrogen atoms in the formula.

Esters

Esters have the general formula:

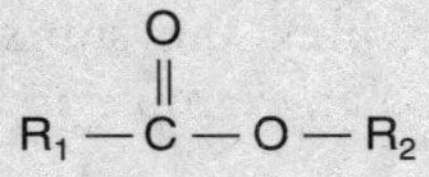

$$R_1 - \overset{\overset{\textstyle O}{\|}}{C} - O - R_2$$

They are prepared by a condensation (*esterification*) reaction between an organic acid and an alcohol, as shown below:

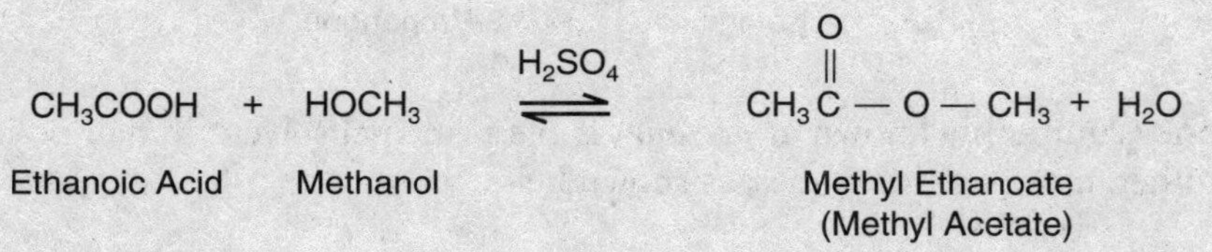

$$CH_3COOH \;+\; HOCH_3 \;\underset{}{\overset{H_2SO_4}{\rightleftharpoons}}\; CH_3\overset{\overset{\textstyle O}{\|}}{C} - O - CH_3 \;+\; H_2O$$

Ethanoic Acid   Methanol   Methyl Ethanoate (Methyl Acetate)

An ester is named as a derivative of the acid and alcohol that formed it. For example, the ester shown above was formed by the reaction between ethanoic acid and methanol. Its IUPAC name is *methyl ethanoate* and its common name is methyl acetate. Esters usually have a distinctive odor, and are the basis for the aromas of many fruits (such as bananas) and plants (such as wintergreen). Fats and oils, types of *lipids*, are the esters of long-chained carboxylic acids (known as fatty acids) and glycerol.

The decomposition of an ester into an acid and an alcohol is known as *hydrolysis*. In animals, special enzymes are used catalyze lipid formation by esterification and lipid breakdown by hydrolysis. Another way of decomposing an ester is to use a base such as NaOH: the alcohol is formed along the sodium salt of the carboxylic acid:

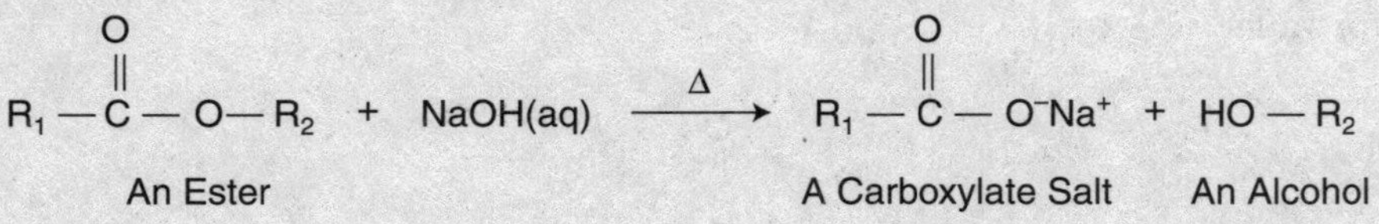

$$R_1 - \overset{\overset{\textstyle O}{\|}}{C} - O - R_2 \;+\; NaOH(aq) \;\overset{\Delta}{\longrightarrow}\; R_1 - \overset{\overset{\textstyle O}{\|}}{C} - O^-Na^+ \;+\; HO - R_2$$

An Ester   A Carboxylate Salt   An Alcohol

This reaction, known as *saponification*, is important in the soap-making industry.

Amines and Amides

Amines

An **amine** is an organic compound that contains one or more **amino groups**: a nitrogen atom bonded to two hydrogen atoms as illustrated below:

$$-NH_2$$

The simplest amine is aminomethane (common name: *methyl amine*), CH_3NH_2.

Amino Acids

An **amino acid** is a carboxylic acid that contains one or more amino groups. The simplest amino acids are aminoethanoic acid and aminopropanoic acid, commonly known as *glycine* and *alanine*. Their structures are given below:

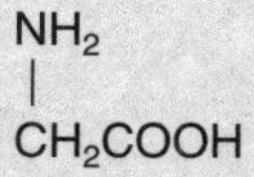

$$\begin{array}{cc} NH_2 & \quad\quad NH_2 \\ | & \quad\quad | \\ CH_2COOH & \quad CH_3CHCOOH \\[4pt] \text{Glycine} & \quad\quad \text{Alanine} \end{array}$$

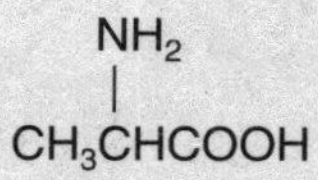

Note that alanine consists of two enantiomers since the central carbon contains four distinct groups bonded to it. Only one enantiomer (the "left-handed" form) of this compound is biologically active in mammals.

Amides

An **amide** is a derivative of an organic acid, in which an amino group ($-NH_2$) replaces the $-OH$ on the carboxyl group. The general formula for an amide is:

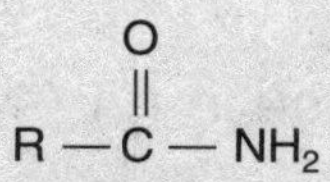

$$\begin{array}{c} O \\ \| \\ R - C - NH_2 \end{array}$$

In the IUPAC system, an amide is named by dropping the "–oic" from the name of the acid, and adding "–amide" in its place. For example, the amide whose formula is:

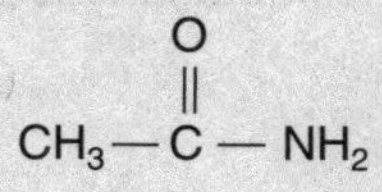

$$\begin{array}{c} O \\ \| \\ CH_3 - C - NH_2 \end{array}$$

is derived from *ethanoic acid*. Its IUPAC name is *ethanamide* and its common name is *acetamide*.

Polymers and Polymerization

A **polymer** is a large molecule composed of many repeating units called *monomers*. In **polymerization**, the monomer units are joined to produce the polymer:

$$n \text{ monomer} \rightarrow (\text{monomer})_n$$

Natural polymers include proteins (monomer: amino acids), cellulose and starch (monomer: glucose). Synthetic polymers include the plastic polyethylene (monomer: ethene), nylon, and polyester.

Addition Polymerization

Monomers that are unsaturated may undergo polymerization by undergoing a series of internal addition reactions. In this process, the double or triple bonds are reduced to single or double bonds. Polyethylene is an example of an addition polymer. The polymerization process is illustrated below:

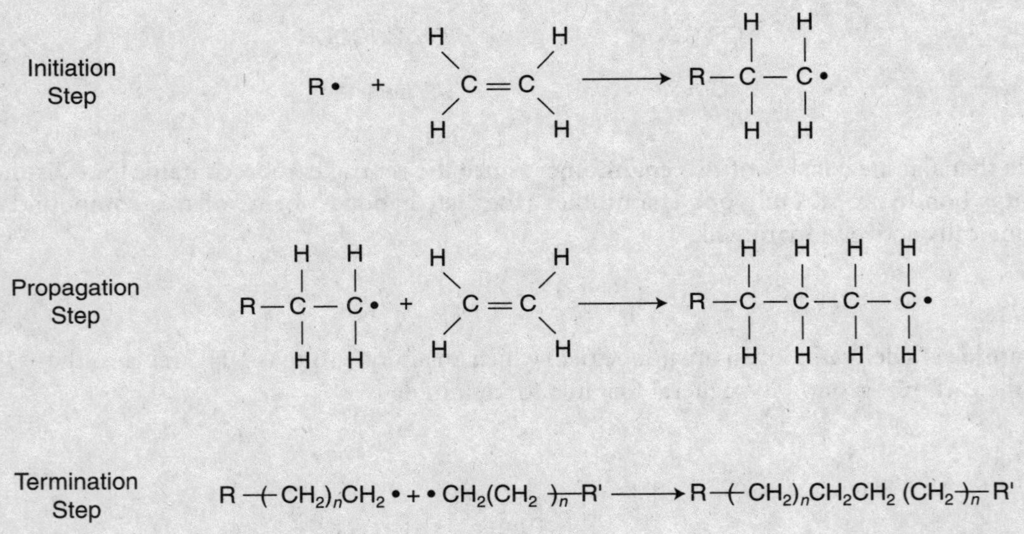

Initiation Step

Propagation Step

Termination Step

$$R \text{---} (CH_2)_n CH_2 \bullet + \bullet CH_2 (CH_2 \text{---})_{\overline{n}} R' \longrightarrow R \text{---} (CH_2)_n CH_2 CH_2 (CH_2 \text{---})_{\overline{n}} R'$$

Note that this type of polymerization reaction involves *free radicals* (i.e., atoms with unpaired electrons) and consists of a chain initiating step, followed by a chain propagation step, and a chain terminating step. As a result, the small alkene molecules are converted into a series of large alkane molecules.

Condensation Polymerization

In **condensation polymerization**, the monomers are joined by a dehydration reaction. The first step in this process is illustrated on the next page, as amino acid molecules condense to form a *polypeptide* in which the amino acid molecules are joined by amide bonds:

$$n\text{HO}-\overset{\overset{\text{O}}{\|}}{\text{C}}-\text{CH}_2-\overset{\overset{\text{H}}{|}}{\text{N}}-\text{H} \quad + \quad n\text{HO}-\overset{\overset{\text{O}}{\|}}{\text{C}}-\text{CH}_2-\overset{\overset{\text{H}}{|}}{\text{N}}-\text{H}$$

$$\longrightarrow \left[\overset{\overset{\text{O}}{\|}}{\text{C}}-\text{CH}_2-\underset{\underset{\text{Amide Bond}}{|}}{\text{N}}-\overset{\overset{\text{O}}{\|}}{\text{C}}-\text{CH}_2-\overset{\overset{\text{H}}{|}}{\text{N}}\right]_n \quad + \quad 2n\,\text{H}_2\text{O}$$

In this way, proteins are synthesized within living systems.

Examples of synthetic condensation polymers include silicones, nylons, and polyesters.

Chapter 16: Practice Questions

Multiple-Choice Questions

Questions 1–5 refer to five organic compounds listed. A choice may be used once, more than once, or not at all.

    A.   propane
    B.   1-propanol
    C.   propanal
    D.   2-propanone
    E.   propoxypropane

1. An alcohol

2. An ether

3. A constitutional isomer of 1-hexanol

4. Has the general formula C_nH_{2n+2}

5. Does *not* contain oxygen

6. A molecule of ethene, $CH_2=CH_2$, contains:
    A.   one sigma (σ) bond and one pi (π) bond.
    B.   one sigma (σ) bond and four pi (π) bonds.
    C.   four sigma (σ) bonds and one pi (π) bond.
    D.   one sigma (σ) bond and five pi (π) bonds.
    E.   five sigma (σ) bonds and one pi (π) bond.

7. The dehydration (condensation) product of two molecules of ethanol produces:
    A.   an ether.
    B.   an alkene.
    C.   an aldehyde.
    D.   a ketone.
    E.   an aromatic compound.

8. Which formula represents an alkyne?
    A. C_2H_4
    B. CH_3CHO
    C. C_3H_4
    D. $CH_3CH_2NH_2$
    E. C_4H_{10}

9. The complete combustion of butane produces:
    A.   CO and H_2O.
    B.   C and H_2O.
    C. CO_2 and H_2O.
    D.   CO and H_2O_2.
    E. CO_2 and H_2O_2.

10. An aqueous solution of which organic compound has a pH near 5.0?
    A. CH_3OCH_3
    B. $CH_3CH_2NH_2$
    C. $CH_3CH_2CH_3$
    D. CH_3COOH
    E. CH_3COCH_3

Free-Response Question

1. The organic compound hexane has the molecular formula C_6H_{14}.

    (A) Write the Lewis structures for the five structural isomers of hexane. (Condensed structural formulas are acceptable.)

    (B) Supply the proper IUPAC names for each of the isomers you have drawn.

Multiple-Choice Answers

1. B   2. E   3. E   4. A   5. A   6. E   7. A   8. C   9. C   10. D

1. B

The name for an alcohol tends to end in –*ol*.

2. E

The name for an ether contains what is known as an alkoxy group. The only answer choice that contains an alkoxy group is (E), propoxypropane.

3. E

Constitutional isomers are three-dimensional rearrangements of compounds possessing the same molecular formula. These different rearrangements can give the compounds drastically different properties. You are looking for an isomer of 1-hexanol. The root of this alcohol is hexane, which means that it is a six-carbon compound. This means the correct answer must also possess six carbons. The only compound of the five answer choices to have six carbons is (E), propoxypropane. This compound has a 3-carbon propane base that is attached to a 3-carbon propoxy side group.

4. A

Propane is a saturated hydrocarbon with all single bonds that has the general formula C_nH_{2n+2}.

5. A

Propane is a hydrocarbon and only contains carbon and hydrogen.

6. E

A hydrocarbon that has –*ene* at the end of its name contains a double bond. Ethene's formula is C_2H_4. It contains one double bond and four single bonds. A single bond is a sigma bond; a double bond consists of one pi bond and one sigma bond. This means that there will be one pi bond and five sigma bonds.

7. A

When a dehydration reaction occurs between two molecules of ethanol, CH_3CH_2OH, it results in the formation of $CH_3CH_2OCH_2CH_3$ and H_2O. $CH_3CH_2OCH_2CH_3$ is an ether. The generic formula for an ether is R-O-R′ and the product from this dehydration reaction matches that description.

8. C

An alkyne is a hydrocarbon that contains a triple bond. Answer choice (A) is an alkene; answer choice (B) is an aldehyde; answer choice (C) is an alkyne; answer choice (D) is an amine; answer choice (E) is a fully saturated alkane. Answer choice (C) is correct.

9. C

Combustion of hydrocarbons always results in the production of H_2O and CO_2.

10. D

CH_3COOH is a carboxylic acid and will have a pH significantly lower than 7.0. Answer choice (A) is an ether; answer choice (B) is an amine and is basic; answer choice (C) is a hydrocarbon; answer choice (E) is a ketone.

Free-Response Answer

1. (A)

$$CH_3\ CH_2CH_2CH_2CH_2CH_3 \quad (I)$$

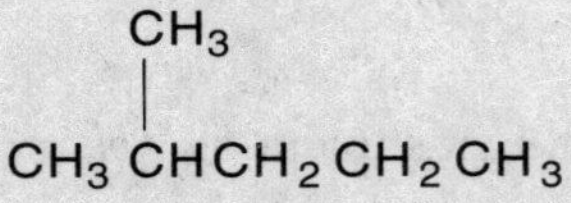

$$\begin{array}{c} CH_3 \\ | \\ CH_3\ CHCH_2\ CH_2\ CH_3 \quad (II) \end{array}$$

$$\begin{array}{c} CH_3 \\ | \\ CH_3\ CH_2\ CHCH_2\ CH_3 \quad (III) \end{array}$$

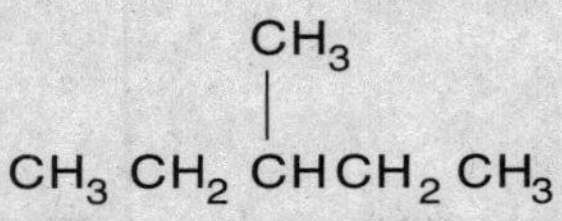

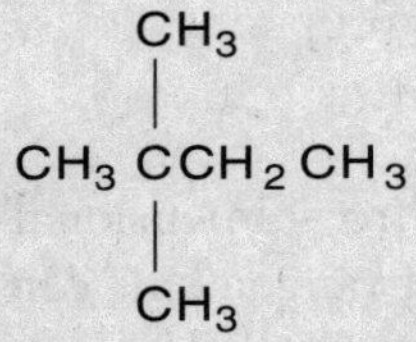

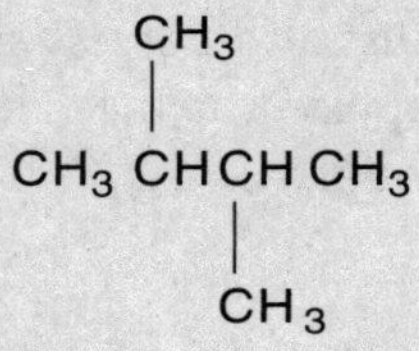

$$\begin{array}{c} CH_3 \\ | \\ CH_3\ CCH_2\ CH_3 \quad (IV) \\ | \\ CH_3 \end{array}$$

$$\begin{array}{c} CH_3 \\ | \\ CH_3\ CHCH\ CH_3 \quad (V) \\ | \\ CH_3 \end{array}$$

(B) (I)   hexane
    (II)   2-methylpentane
    (III) 3-methylpentane
    (IV) 2,2-dimethylbutane
    (V)   2,3-dimethylbutane

Nuclear Chemistry

In the preceding chapters we have studied what amounts to the chemistry of the *electron*. In this chapter we focus on the nucleus of the atom and the changes that it can undergo. The phenomenon of **radioactivity**, the spontaneous decay of certain atoms, was discovered in the last decade of the nineteenth century. Since that time, the study nuclear chemistry has produced a host of practical applications.

THE NUCLEUS AND NUCLEAR PARTICLES

We already know that the nucleus contains **nucleons**: protons and neutrons. A nucleus with a *specified* number of protons and neutrons is called a **nuclide** and is symbolized as follows:

$$^{A}_{Z}X$$

The letter X is the symbol of the element; the letter Z is the **atomic number** (the number of protons in the nucleus); and the letter A is the **mass number** (the number of protons and neutrons in the nucleus). Nuclides with the same atomic number but different mass numbers are called **isotopes** of an element. For example, and are two isotopes of nitrogen. Ways of writing these nuclides include ^{14}N, ^{15}N; nitrogen-14, nitrogen-15; and N-14, N-15.

In addition, other nuclear particles appear in nuclear reactions. The principal ones are listed in the table below:

Particle	Symbol(s)	Remarks
proton	$^{1}_{1}H$	nucleus of hydrogen–1 atom
electron (beta minus)	$^{0}_{-1}e$ (β^{-})	
positron (beta plus)	$^{0}_{+1}e$ (β^{+})	antiparticle of electron
alpha	$^{4}_{2}He$ (α)	nucleus of helium–4 atom
gamma photon	γ	particle of radiant energy

NUCLEAR EQUATIONS

A **nuclear equation** is a representation of a change that occurs within or among atomic nuclei:

$$^{14}_{7}N + ^{1}_{0}n \rightarrow ^{14}_{6}C + ^{1}_{1}H$$

In order to insure that both charge and mass are conserved in the reaction, a nuclear equation must be *balanced*. A nuclear equation is balanced if:

- The sum of the atomic numbers on the left side (7 + 0) equals the sum of the atomic numbers on the right side (6 + 1)

- The sum of the mass numbers on the left side of the equation (14 + 1) equals the sum of the mass numbers on the right side (14 + 1)

In the reaction shown above, one element (nitrogen-14) is changed to another (carbon-14). This change is known as **transmutation**.

RADIOACTIVE DECAY

The spontaneous radioactive decay of a nuclide is accompanied by the emission of subatomic particles and/or photons such as X-rays. Many nuclides are radioactive, including all of the nuclides whose atomic numbers are 84 or greater.

The five types of decay we will study are alpha (α) particle emission, beta (β) particle emission, gamma photon emission, positron emission, and electron capture. These processes are summarized in the following table:

Mode of decay	Particle emitted	Nuclear changes	
		Atomic number	Mass number
Alpha particle emission	$^{4}_{2}H$	−2	−4
Beta (β) particle emission	$^{0}_{-1}e$	+1	0
Gamma photon emission	$^{0}_{0}\gamma$	0	0
Positron emission	$^{0}_{+1}e$	−1	0
Electron capture	X-ray photon	−1	0

Alpha Particle Emission

The following nuclear reaction is an example of alpha particle emission:

$$^{238}_{92}U \rightarrow \,^{234}_{90}Th + \,^{4}_{2}He$$

In this reaction, the uranium-238 nucleus (the *parent nucleus*) breaks down to produce a thorium-234 nucleus (the *daughter nucleus*) and a helium-4 nucleus (the *alpha particle*).

Alpha particles are relatively massive, but they have little *penetrating power*: they can be stopped by a sheet of paper. Many heavier radioactive nuclei, especially those with atomic numbers greater than 83, undergo alpha decay as a way of reducing the number of protons and neutrons present.

Sample Problem

The nuclide $^{222}_{86}Rn$ undergoes alpha decay. Write a balanced equation that illustrates this process.

Solution

$$^{222}_{86}Rn \rightarrow \,^{218}_{84}Po + \,^{4}_{2}He$$

Beta (β) Particle Emission

Certain nuclei undergo radioactive decay and produce an electron, a beta (β) particle, in the reaction. The following equation is an example of beta (β) decay:

$$^{234}_{90}Th \rightarrow \,^{234}_{91}Pa + \,^{0}_{-1}e$$

In beta (β) decay, a neutron is converted into a proton and an electron, which is expelled from the nucleus. Beta (β) particles have greater penetrating power than alpha particles: they can pass through aluminum foil 2 to 3 millimeters thick.

Sample Problem

The nuclide $_1^3$H undergoes beta (β) decay. Write a balanced nuclear equation that illustrates this process.

Solution

$$_1^3H \rightarrow _2^3He + _{-1}^0e$$

Gamma Photon Emission

Gamma-ray photons are a highly penetrating form of electromagnetic radiation. Like the electrons outside of the nucleus, the nucleus itself contains energy levels and, occasionally, a nucleus will enter an excited state, known as a **nuclear isomer**. A nuclear isomer is symbolized by placing the letter m next to its mass number. When the isomer returns to its ground state, a gamma-ray photon is emitted. The following equation is an example of gamma decay:

$$_{90}^{230m}Th \rightarrow _{90}^{230}Th + \gamma$$

Positron Emission

The name *positron* is a contraction of the terms *posi*tive and ele*ctron*. It is called the *antiparticle* of the electron. The following equation is an example of positron decay:

$$_{13}^{26}Al \rightarrow _{12}^{26}Mg + _{+1}^0e$$

In positron emission, a proton is converted into a neutron and a positron, which is expelled from the nucleus. Positrons have very little penetrating power because they are annihilated when they encounter an electron within a sample of matter.

Electron Capture

In this type of reaction, the nucleus absorbs one of the inner-shell electrons of the atom. In the nucleus, the captured electron combines with a proton to form a neutron. The following equation is an example of electron capture:

$$_{53}^{125}I \rightarrow _{-1}^0e + _{52}^{125}Te$$

Radioactive Decay Series

The purpose of all radioactive decay is to produce more stable nuclei. Sometimes, this is a multistep process, as with the decay of $_{92}^{238}U$ decays to the stable nuclide $_{82}^{206}Pb$ in a series of steps involving both alpha and beta (β) particle emissions.

The graph below summarizes the uranium-238 decay series:

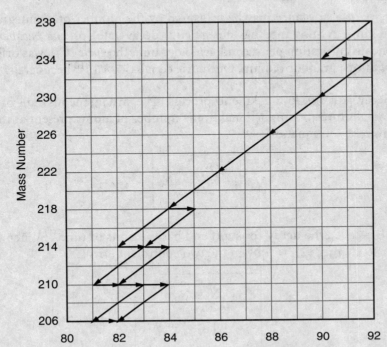

The diagonal arrows represent alpha particle emissions, and the horizontal arrows represent beta (β) particle emissions. Note also that beginning with the formation of polonium (Po)-218, several alternative paths are possible in the series.

Sample Problem

Write equations for the alternative paths by which bismuth (Bi)-210 can be converted to lead (Pb)-206.

Solution

$$^{210}_{83}\text{Bi} \rightarrow ^{210}_{84}\text{Po} + ^{0}_{-1}\text{e}$$

$$^{210}_{84}\text{Po} \rightarrow ^{206}_{82}\text{Pb} + ^{4}_{2}\text{He}$$

$$^{210}_{83}\text{Bi} \rightarrow ^{206}_{81}\text{Tl} + ^{4}_{2}\text{He}$$

$$^{206}_{81}\text{Tl} \rightarrow ^{206}_{82}\text{Pb} + ^{0}_{-1}\text{e}$$

RATES OF RADIOACTIVE DECAY

The radioactivity of a substance may be measured by the number of disintegrations taking place per unit time. A substance has an *activity* (decay rate) of *one becquerel* (Bq) if it experiences one disintegration per second. An older unit, the *curie* (Ci), was originally based on the activity of one gram or radium. One curie equals 3.7×10^{10} becquerels.

Radioactive decay is a first-order kinetic process. The instantaneous rate of decay of N radioactive atoms is directly proportional to the number of atoms present at that instant in time, and is given by the equation:

$$-\frac{dN}{dt} = \lambda N$$

The constant λ is called the **decay constant** and has the units of time^{-1}. When the equation is solved (refer to chapter 12), we obtain the two equivalent forms:

$$(1)\ N_t = N_0 e^{-\lambda t}$$
$$(2)\ \ln\left(\frac{N_t}{N_0}\right) = -\lambda t$$

We can substitute quantities such as mass and activity for N since they are also proportional to the number of atoms present at any given time.

The graphs below represent the radioactive decay of a hypothetical substance, exponentially and logarithmically.

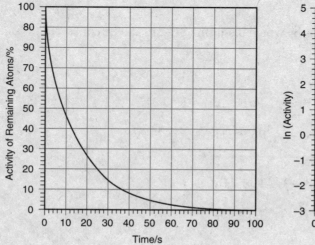
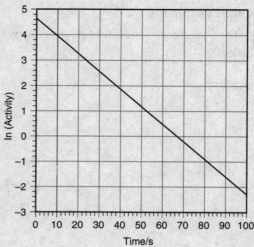

In the logarithmic graph, the slope of the straight line equals the *negative* of the decay constant.

Half-Life

An interesting feature of radioactive decay (and first-order kinetics in general) is the constancy of the quantity known as *half-life*. As we have seen in chapter 12, the **half-life** is defined as the time required for a substance to decay to one-half of its initial value. As an example, consider the decay of the radioactive isotope iodine-131, whose half-life is approximately 8 days. If we have an initial sample of 4.0 milligrams, the decay over a period of time occurs as follows:

$$4.0 \text{ mg} \xrightarrow{8d} 2.0 \text{ mg} \xrightarrow{8d} 1.0 \text{ mg} \xrightarrow{8d} 0.50 \text{ mg} \xrightarrow{8d} 0.25 \text{ mg} \ldots$$

After 32 days of decay, 0.25 milligram of iodine-131 remains unchanged. The half-life is constant: it cannot be altered by changes in pressure, temperature, or chemical combination. We can solve the logarithmic equation given above to obtain the relationship:

$$t_{0.5} = \frac{\ln 2}{\lambda} = \frac{0.693}{\lambda}$$

where $t_{0.5}$ represents the half-life of the substance.

Sample Problem

The half-life of cobalt-60 is 5.27 years; the molar mass of cobalt-60 is 59.93 grams per mole.

(A) Calculate the decay constant of cobalt-60.

(B) Calculate the instantaneous decay rate of 5.00×10^{-4} gram of cobalt-60 in *atoms per year*.

(C) How much of the original sample in (B) remains after 22.0 years of decay?

Solution

(A) $t_{0.5} = \dfrac{0.693}{\lambda}$

$\lambda = \dfrac{0.693}{t_{0.5}} = \dfrac{0.693}{5.27 \text{y}} = \textbf{0.131 y}^{-1}$

(B) $-\dfrac{dN}{dt} = \lambda N = (0.131 \text{ y}^{-1}) \cdot (5.00 \times 10^{-4} \text{ g}) \cdot \left(\dfrac{1 \text{ mol}}{59.93 \text{ g}}\right) \cdot \left(\dfrac{6.02 \times 10^{23} \text{ atoms}}{1 \text{ mol}}\right)$
$= \textbf{6.58} \times \textbf{10}^{17} \textbf{ atoms y}^{-1}$

(C) $N_t = N_o e^{-\lambda t} = (5.00 \times 10^{-4} \text{ g}) \cdot e^{-(0.131 \text{ y}^{-1}) \cdot (22.0 \text{ y})} = \textbf{2.80} \times \textbf{10}^{-5} \textbf{ g}$

USES OF RADIOACTIVE ISOTOPES

Radioactive isotopes can be used in a variety of applications that depend on the properties of the particular isotope. Because radioisotopes (such as hydrogen-3) are chemically similar to their stable counterparts (such as hydrogen-1), they can be used as *tracers* to follow the course of a reaction. This application is particularly useful in uncovering how certain chemical reactions occur.

Radioisotopes also are used in a variety of medical applications that depend on the isotope's radioactivity and short half-life. These applications include medical diagnoses, such as the use of iodine-131 in uncovering thyroid disorders. A short half-life is necessary to ensure rapid decay and elimination of the radioisotope from the body. Radioactive isotopes such as cobalt-60 are used in treating certain malignant tumors.

Certain radioactive isotopes are used to determine the age of various artifacts, such as wooden bowls. For example, carbon-14 has a half-life of 5,730 years. This radioactive isotope is produced in the upper atmosphere by the bombardment of nitrogen-14 with neutrons from cosmic radiation:

$$\ce{^{14}_{7}N + ^{1}_{0}n \rightarrow ^{14}_{6}C + ^{1}_{1}H}$$

Carbon-14 undergoes beta (β) particle emission:

$$\ce{^{14}_{6}C \rightarrow ^{14}_{7}N + ^{0}_{-1}e}$$

The carbon-14 is incorporated into atmospheric CO_2. As a result of a dynamic equilibrium between carbon-14 production and decay, the ratio this isotope to carbon-12, its nonradioactive counterpart, is essentially constant. When a plant incorporates carbon into its structure, this ratio is maintained *as long as the plant is alive. The activity (that is, decay rate) of carbon-14 in living tissue is approximately 15 becquerels.* When a plant such as a tree dies or is cut down, the activity of the carbon-14 decreases with time. By comparing the activity of carbon-14 in a wooden bowl to the activity of carbon-14 in living tissue, the age of the bowl can be determined with considerable accuracy if the age of the sample lies between 300 and 50,000 years.

Sample Problem

When an ancient wooden artifact is subjected to carbon-14 dating, the observed decay rate is 8.0 becquerels. Calculate the age of the artifact.

Solution

$$t_{0.5} = \frac{0.693}{\lambda}$$

$$\lambda = \frac{0.693}{t_{0.5}} = \frac{0.693}{5730y} = 1.21 \times 10^{-4}\, y^{-1}$$

$$\ln\left(\frac{N_t}{N_0}\right) = -\lambda t$$

$$t = \frac{\ln\left(\frac{N_t}{N_0}\right)}{-\lambda} = \frac{\ln\left(\frac{8.0\ \text{Bq}}{15\ \text{Bq}}\right)}{-1.21 \times 10^{-4}\, y^{-1}} = \mathbf{5.2 \times 10^3\, y}$$

INDUCED NUCLEAR REACTIONS

In addition to natural processes, nuclear changes can be effected by bombardment with a variety of particles, a process known as **induced transmutation**. Neutrons are effective bombarding particles because they lack charges and are not repelled by the positive nuclei. Positively charged particles must be given considerable kinetic energy in order to overcome nuclear repulsion, and this is accomplished by means of a **particle accelerator**, a device that uses electric and magnetic fields to provide the necessary energy.

- In 1919, Rutherford bombarded nitrogen-14 nuclei with alpha particles and produced the first example of induced transmutation:

$$\,^{14}_{7}\text{N} + \,^{4}_{2}\text{He} \rightarrow \,^{17}_{8}\text{O} + \,^{1}_{1}\text{H}$$

- In 1932, the English physicist Sir James Chadwick bombarded beryllium-9 and identified a stream of uncharged particles that we now call *neutrons*. The reaction is shown below:

$$\,^{9}_{4}\text{Be} + \,^{4}_{2}\text{He} \rightarrow \,^{1}_{0}\text{n} + \,^{12}_{6}\text{C}$$

- In 1934, French physicists Frédéric Joliot-Curie and Irène Joliot-Curie bombarded aluminum-27 and produced the first artificially radioactive isotope, phosphorus-30:

$$\,^{27}_{13}\text{Al} + \,^{4}_{2}\text{He} \rightarrow \,^{1}_{0}\text{n} + \,^{30}_{15}\text{P}$$

The phosphorus-30 undergoes positron emission and forms silicon-30.

- In 1940, the first of the **transuranium elements** (elements with atomic numbers greater than 92) were discovered. The bombardment of uranium-238 with a neutron produces uranium-239, which then decays by beta (β) particle emissions to neptunium-239 and plutonium-239:

$$\,^{238}_{92}\text{U} + \,^{1}_{0}\text{n} \rightarrow \,^{239}_{92}\text{U}$$

$$\,^{239}_{92}\text{U} \rightarrow \,^{239}_{93}\text{Np} + \,^{0}_{-1}\text{e}$$

$$\,^{239}_{93}\text{Np} \rightarrow \,^{239}_{94}\text{Pu} + \,^{0}_{-1}\text{e}$$

NUCLEAR STABILITY

One important factor in determining nuclear stability is the *number* of protons and neutrons present in the nucleus. Most naturally occurring stable isotopes (approximately 160) have *even* numbers of protons and neutrons (for example, carbon-12). Only four stable nuclides have *odd* numbers of protons and neutrons (for example, nitrogen-14). The remaining stable nuclides (approximately 100) have even-odd combinations. Another important factor is the *ratio* of the number of neutrons (N) to the number of protons (Z) present. The graph below illustrates that all of the naturally occurring stable nuclides fall within a belt of stability:

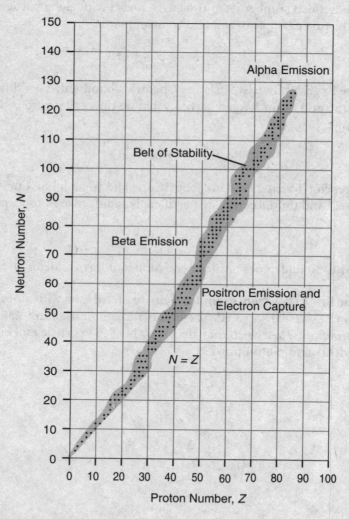

The N/Z ratio is approximately 1.0 for the smaller atomic numbers, and it rises gradually to 1.5 for the higher atomic numbers. The graph terminates at $Z = 83$ since no stable nuclides exist above this atomic number. Unstable nuclides undergo the decay modes shown in the graph.

NUCLEAR ENERGY

Nuclear energy finds its origin in the relationship between mass and energy, developed by Einstein in 1905:

$$E = mc^2$$

This relationship illustrates the essential equivalence of mass and energy; the constant in the equation (c^2) is the square of the speed of light.

Nuclear energies are generally expressed in terms of the **electronvolt** (eV), and its multiple, the **megaelectronvolt** (MeV = 10^6 eV):

$$1 \text{ eV} = 1.602 \times 10^{-19} \text{ J}; 1 \text{ MeV} = 1.602 \times 10^{-13} \text{ J}$$

Nuclear masses are frequently expressed in terms of their *energy equivalents*. Using Einstein's relationship, it can be shown that 1 atomic mass unit of mass (1 u = 1.661×10^{-27} kg) is equivalent to 931.5 MeV of energy.

Sample Problem

Calculate the energy lost in the decay of uranium-238 by alpha particle emission:
$$^{238}_{92}\text{U} \rightarrow {}^{234}_{92}\text{Th} + {}^{4}_{2}\text{He}$$

The relevant masses are: uranium-238 = 238.0508 u; thorium-234 = 234.0436 u; helium-4 = 4.0026 u.

Solution

The net change in mass is: (234.0436 u + 4.0026 u) − (238.0508 u) = −0.0046 u

The energy equivalent is: $(-0.0046) \cdot \left(\dfrac{931.5 \text{ MeV}}{1 \text{ u}} \right) = $ **−4.3 MeV**

The negative sign indicates that energy is *released*: it is converted into the kinetic energy of the expelled alpha particle.

Mass Defect and Binding Energy

The stability of a nucleus can also be understood by comparing the mass of the nucleus to the total mass of the nucleons found in that nucleus. The following sample problem illustrates how this comparison is made:

Sample Problem

(A) Compare the mass of a $_{26}^{56}$Fe nucleus with the total mass of its nucleons, given the following:

Masses: iron-56 = 55.9206 u; proton = 1.007276 u; neutron = 1.008665 u

(B) Calculate the energy equivalent of the difference in mass.

Solution

This nuclide contains 26 protons and 30 neutrons:

(A) We assume that the nucleus is formed from its nucleons: $26\,_1^1\text{H} + 30\,_0^1\text{n} \rightarrow\,_{26}^{56}\text{Fe}$

Mass of 26 protons =	$(26)\cdot(1.007276 \text{ u}) = 26.1892 \text{ u}$
Mass of 30 neutrons =	$(30)\cdot(1.008665 \text{ u}) = 30.2600 \text{ u}$
Total mass of nucleons	$= 56.4492 \text{ u}$
Difference in mass	$= 55.9206 \text{ u} - 56.4492 \text{ u} = \textbf{-0.5286 u}$

(B) Energy equivalent $= (-0.5286 \text{ u}) \cdot \left(\dfrac{931.5 \text{ MeV}}{1 \text{ u}}\right) = \textbf{-492.4 MeV}$

The difference in mass is known as the **mass defect** of the nucleus. Its energy equivalent is known as the **binding energy** of the nucleus. The negative value of the binding energy in the last problem is an indication that the nucleus is more stable than its component nucleons. Generally, binding energy is expressed as a *positive* value: 492.4 MeV is the energy needed to separate an iron-56 nucleus into its component nucleons.

Average Binding Energy Per Nucleon

One way of comparing nuclear stabilities is by calculating a quantity known as the average **binding energy per nucleon**: the binding energy of the nucleus divided by the mass number (the number of nucleons present). In the problem above, the binding energy of 492.4 MeV is divided by 56 nucleons to yield 8.79 MeV per nucleon. The larger the binding energy per nucleon, the more stable is the nucleus.

The graph below illustrates how the binding energy per nucleon varies with the number of nucleons:

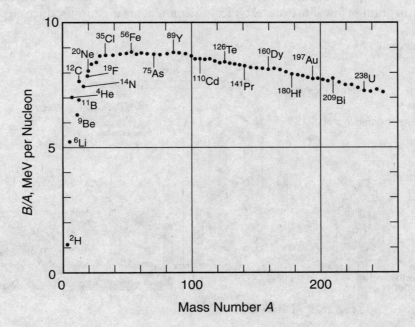

Note that the binding energy per nucleon rises to a maximum at ^{56}Fe and then decreases steadily. We can draw the conclusion that light nuclei (such as ^{2}H) can become more stable if they combine to form heavier nuclei, a process known as **nuclear fusion**. Similarly, the heavier elements can become more stable if a split into lighter nuclei, a process known as **nuclear fission**.

NUCLEAR FISSION

In the 1934, Enrico Fermi, an Italian-American physicist, suggested that neutrons be used in bombardment reactions because they would not be repelled by target nuclei. Fermi and Emilio Segrè bombarded uranium-238 with slow neutrons (known as *thermal neutrons*), but they could not establish the source of the beta (β) particles that were emitted. In 1938, the German chemists Otto Hahn and Fritz Strassman showed that the source of the radiation came from radioactive elements much lighter than uranium. A subsequent discovery showed that the nuclear fission occurred, not in uranium-238, but in a minor isotope, uranium-235.

One possible path of the fission of uranium 235 is shown below:

$$^{235}_{92}U + {}^{1}_{0}n \rightarrow \left[{}^{236}_{92}U \right] \rightarrow {}^{141}_{56}Ba + {}^{92}_{36}Kr + 3{}^{1}_{0}n + \text{Energy}$$

The energy released (about 200 MeV per fission event) is the result of mass-energy conversion. Each fission event releases additional neutrons (three in the path shown above), which can be used to initiate more fission events, known as a **chain reaction**. If chain reaction is not controlled, and a sufficient quantity of uranium (known as the **critical mass**) is present, an explosion will occur.

Fission Reactors

A **fission reactor** employs a *controlled* chain reaction to provide a continuous source of useful energy. The diagram below illustrates a typical fission reactor used to produce electricity.

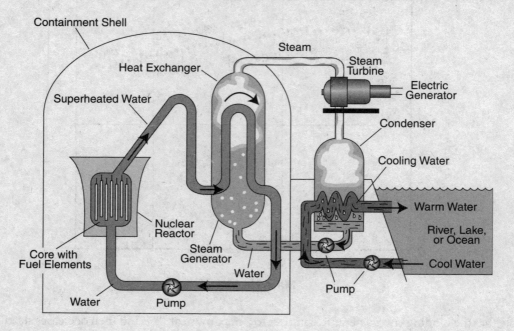

The primary system of a fission reactor has the following components:

• The **containment shell** (concrete and steel) provides shielding for the reactor.

• The **fuel rods**, located in the **core**, serve as sources of energy. Enriched uranium-235 is used as a fuel.

• The **moderator**, also located in the core, creates thermal neutrons by lowering their kinetic energies to approximately the average kinetic energy of air molecules at room temperature. For this reason, they are known as *thermal neutrons*. Moderators are usually composed of water (containing either hydrogen-1 or hydrogen-2), graphite, or beryllium.

• The **control rods** in the core regulate the rate of fission by absorbing neutrons. Control rods are usually made of cadmium.

• The **coolant** (water or liquid sodium) removes thermal energy from the core.

• The **heat exchanger** receives the thermal energy and produces steam for the generation of electrical energy by the turbine connected to the reactor.

Uranium-238 does not undergo fission. When it absorbs a neutron, however, it is converted into plutonium-239 in three steps:

$$_{92}^{238}U + _{0}^{1}n \rightarrow _{92}^{239}U$$

$$_{92}^{239}U \rightarrow _{93}^{239}Np + _{-1}^{0}e$$

$$_{93}^{239}Np \rightarrow _{94}^{239}Pu + _{-1}^{0}e$$

The plutonium-239 *does* fission, and *breeder reactors* that produce their own fuel have been constructed. No breeder reactors have been operated commercially in the United States, however, because of the possible health and environmental hazards attributable to plutonium.

A number of problems are associated with *any* fission reactor. For example, the heat energy produced by the reactor contributes to *thermal pollution*. There is also the serious problem of *radioactive waste disposal*.

Fission reactors are also used to produce radioactive isotopes for a variety of applications. For example, when the stable isotope is bombarded with neutrons from a reactor, the radioisotope $_{27}^{60}Co$, used in cancer therapy, is produced.

NUCLEAR FUSION

Nuclear fusion is the process by which stars produce their energy. In stars, fusion is a series of reactions that depend on the temperature of the particular star. The *net* fusion reaction in our sun is as follows:

$$4_{1}^{1}H \rightarrow _{2}^{4}H + 2_{+1}^{0}e + 2\gamma + 2\nu + Energy$$

where γ represents a gamma photon and ν represents another subatomic particle called a *neutrino*.

For a self-sustaining fusion reaction to occur, temperatures on the order of 40,000,000 K are needed. At these temperatures, the gases are completely ionized into a mixture of positive nuclei and electrons, known as a **plasma**. The plasma must be confined at an extremely high density in order for the fusion to occur. In some types of fusion reactors, the plasma is confined within a magnetic field. For a given mass of fuel, nuclear fusion yields more energy than nuclear fission and there are fewer pollution problems with the products of fusion reactions. For this reason, work is being done to develop fusion reactors as a means of producing power. Four fusion reactions that are under investigation are shown below:

$$_{1}^{2}H + _{1}^{2}H \rightarrow _{1}^{3}H \rightarrow _{1}^{1}H + Energy$$

$$_{1}^{2}H + _{1}^{2}H \rightarrow _{2}^{4}He + Energy$$

$$_{1}^{2}H + _{1}^{2}H \rightarrow _{2}^{3}He + _{0}^{1}n + Energy$$

$$_{1}^{2}H + _{1}^{3}H \rightarrow _{2}^{4}He + _{0}^{1}n + Energy$$

At the present time, no successful fusion reactor has been constructed. The problems to be overcome include the production of a high ignition temperature, the packing of nuclei into a space small enough to allow a sufficient number of collisions, and the control of the fusion reaction once it has begun.

Chapter 17: Practice Questions

Multiple-Choice Questions

1. How many nuclear particles are in an atom of lead-210, ^{210}Pb?

   A.   0
   B.   82
   C.   128
   D.   210
   E.   292

2. Which of these nuclides is (are) radioactive?

   I. $^{146}_{30}Nd$

   II. $^{20}_{8}O$

   III. $^{236}_{90}Th$

   IV. $^{32}_{16}S$

   A.   II only
   B.   III only
   C.   I, II, and III
   D.   I and IV only
   E.   II and III only

3. A nuclide that lies above the band of stability is unstable. Which decay process will move it towards the band of stability?

   A.   α–emission
   B.   β–emission
   C.   γ–emission
   D.   Electron capture
   E.   Positron emission

4. Which of the following equations or pairs of equations properly describes the decay of P to Cl?

   A. $^{35}_{15}P = {}^{35}_{17}Cl + {}^{0}_{1}\beta$

   B. $^{35}_{15}P = {}^{35}_{14}Si + {}^{0}_{-1}\beta$
        $^{35}_{14}Si = {}^{35}_{17}Cl + {}^{0}_{-1}\beta$

   C. $^{35}_{15}P = {}^{35}_{17}Cl + {}^{0}_{-1}\beta$

   D. $^{35}_{15}P + e^- = {}^{35}_{16}S$

   E. $^{5}_{16}S + e^- = {}^{35}_{17}Cl$
        $^{35}_{15}P = {}^{35}_{16}S + {}^{0}_{-1}\beta$
        $^{35}_{16}S = {}^{35}_{17}Cl + {}^{0}_{-1}\beta$

5. When $^{214}_{83}Bi$ decays by the emission of one α-particle and two β-particles, what is the product nuclide?

   A. $^{210}_{82}Pb$

   B. $^{208}_{83}Bi$

   C. $^{210}_{83}Bi$

   D. $^{214}_{83}Bi$

   E. $^{210}_{84}Po$

6. Which of the following statements is (are) correct?

  I.   To be stable, the ratio of neutrons to protons must be less than 1:1 for nuclides whose atomic number is between 20 and 83.
  II. Electron capture transforms a neutron into a proton in the nucleus, and a γ-ray photon is emitted.
  III. Proton-rich nuclides can undergo either positron emission or electron capture to reduce the number of protons in the nucleus.
  IV. Nuclides with even numbers of protons and neutrons are more likely to be stable than those with odd numbers.

  A. I and II only
  B. I and IV only
  C. II only
  D. III and IV only
  E. I, III, and IV

7. The transuranium elements lawrencium and californium were first synthesized in the cyclotron at the Lawrence Berkeley Laboratory in California. How many neutrons are emitted in the conversion of californium-262, $_{98}^{262}\text{Cf}$, to lawrencium-256, $_{103}^{256}\text{Lr}$?

  A. 12
  B. 10
  C. 8
  D. 6
  E. 4

8. The half-life of ^{99m}Tc is 6.0 h. The delivery of a sample of ^{99m}Tc from the reactor to the nuclear medicine lab of a hospital takes 3.0 h. What is the minimum amount of ^{99m}Tc that must be shipped in order for the lab to receive 10.0 mg?

  A. 14.1 mg
  B. 15.0 mg
  C. 18.6 mg
  D. 20.0 mg
  E. 40.0 mg

9. What is the missing fission product in the reaction?

$$_{92}^{235}\text{U} + _{0}^{1}n \rightarrow _{57}^{146}\text{La} + \underline{\quad\quad} + 3\ _{0}^{1}n$$

  A. $_{32}^{87}\text{Ge}$

  B. $_{32}^{89}\text{Ge}$

  C. $_{35}^{86}\text{Br}$

  D. $_{35}^{87}\text{Br}$

  E. $_{35}^{89}\text{Br}$

10. Which type of radioactive decay will nuclides with *high* neutron to proton ratios generally undergo?

  A. α-decay
  B. β-decay
  C. γ-decay
  D. Positron emission
  E. Electron capture

Free-Response Questions

1. Lanthanum-127 is an unstable isotope with a half-life of 3.5 minutes. Radon-222 is a natural decay product of uranium, is less widespread, and has a much shorter half-life (38 seconds) than the more abundant Rn-226. There are also a number of manmade radioactive isotopes of iron, including iron-61.

   (A) Explain why lanthanum-127 is unstable.
   (B) Write the complete nuclear equation(s) for lanthanum-127 decay.
   (C) Explain why radon-222 is unstable.
   (D) Write the complete nuclear equation(s) for iron-61 decay.
   (E) Explain why iron-61 is unstable.
   (F) Write the complete nuclear equation for radon-222 decay.

2. Uranium-238 decays to lead-206 and the half-life for this decay is 4.51×10^9 years. The ^{206}Pb / ^{238}U ratio (corrected for any lead present) of a mineral is 0.297.

   (A) Write a balanced nuclear equation for the decay of uranium-238 all the way to lead-206.

   (B) How old is the mineral sample?

   (C) Lead has 4 naturally occurring stable isotopes. What is the atomic weight of lead to the proper number of significant figures, using this data?

Isotope	% Natural Abundance	Atomic Weight
Pb-204	1.4	203.973
Pb-206	24.1	205.9745
Pb-207	22.1	206.9759
Pb-208	52.4	207.9766

Multiple-Choice Answers

1. D   2. E   3. B   4. D   5. C   6. D   7. D   8. A   9. E   10. B

1. D

An atom of lead-210 has 210 nuclear particles. The "210" is the mass number, which gives the number of protons + neutrons since both protons and neutrons each carry a mass number of 1.

2. E

O is radioactive because its atomic number is less than 20 and its n/p (neutron : proton) ratio is greater than 1. Thorium, Th, is radioactive because its atomic number is greater than 83. All isotopes with atomic numbers greater than 83 are unstable.

3. B

The band of stability is an area on an isotope plot that contains all the "stable" nuclei. The axes for the plot are neutrons versus protons. As the atomic number for an atom rises, the ratio of neutrons to protons tends to increase. When an isotope falls to the left, or sits above the band of stability on the plot, it tends to be a beta-emitter. When an isotope falls to the right or sits below the plot, it tends to be a positron-emitter. Therefore, this nuclide sitting above the band of stability would undergo beta emission.

4. D

$^{35}_{15}P$ lies to the left of the band of stability and therefore will undergo β-decay to $^{35}_{16}S$. $^{35}_{16}S$ also lies to the left of the band of stability and therefore will undergo β-decay to $^{35}_{17}Cl$.

5. C

The correct answer is: bismuth-210, $^{210}_{83}Bi$. The emission of one α-particle and two β-particles would lead to no net change in protons, so the product nuclide would still be bismuth; however, it would lead to a loss of four neutrons, changing the mass number to 210.

6. D

The correct answer is: III and IV only. I is incorrect because to be stable, the ratio of neutrons to protons must be *greater* than 1:1 for nuclides whose atomic number is between 20 and 83. II is incorrect because electron capture transforms a proton into a neutron in the nucleus and an X-ray photon is emitted.

7. D

$^{256}_{98}Cf \rightarrow \ ^{256}_{103}Lr + 6\ ^{1}_{0}n + 5\ ^{0}_{-1}\beta$ The sum of the mass numbers on the left side of the equation is 262, while the sum of the mass number on the right side of the equation is 256. Therefore, $262 - 256 = 6$ neutrons must be emitted to balance the equation.

8. A

The half-life equation is $\ln \dfrac{N_o}{N_t} = 0.693 \dfrac{t}{t_{1/2}}$. Values can be substituted and the equation solved for N_o: $\ln (N_o / N_t) = 0.347$. $N_o / N_t = 1.414$ and $N_o = 1.414 \times 10.0 \text{ mg} = 14.1 \text{ mg}$

9. E

To get the mass number from the total mass of the parent side of the reaction, deduct the total mass of the daughter side of the reaction. Mass number of missing particle $= (235 + 1) - (146 + 3) = 89$. To get the atomic number: atomic number $= 92 - 57 = 35$. The missing particle is identified by its atomic number as being bromine.

10. B

Nuclides with too *many* neutrons lie to the left of the band of stability and undergo β-decay to lower their neutron : proton ratio. Nuclides with too *few* neutrons lie to the right of the band of stability and undergo positron (β^-) decay or electron capture to raise their neutron : proton ratio.

Free-Response Answers

1. (A) Lanthanum-127 is *below* the band of stability. It decays to increase the neutron : proton ratio by making more neutrons. There are two ways to increase the ratio—positron emission and electron capture—since both of these transforms a proton into a neutron.

   (B) $^{127}_{57}\text{La} + ^{\ 0}_{-1}\text{e} \rightarrow ^{127}_{56}\text{Ba} + \text{photon}$

   or

   $^{127}_{57}\text{La} \rightarrow ^{127}_{56}\text{Ba} + ^{\ 0}_{-1}\text{e} + \text{photon}$

   (C) The atomic number of radon is 86, and all atoms with more than 83 protons are unstable. They generally undergo α-decay in order to produce a smaller nucleus.

   (D) $^{61}_{26}\text{Fe} \rightarrow ^{61}_{27}\text{Co} + ^{\ 0}_{-1}\text{e}$

   (E) The neutron : proton ratio of iron-61 is too *high* for the nuclide to be stable. Fe-61 undergoes β-decay to lower the ratio. β-decay transforms a neutron into a proton.

   (F) $^{222}_{86}\text{Rn} \rightarrow ^{218}_{84}\text{Rn} + ^{4}_{2}\text{He}$

2. (A) $^{238}_{92}\text{U} \rightarrow\, ^{206}_{82}\text{Pb} + 8\,^{4}_{2}\text{He} + 6\,^{0}_{-1}\text{e}$

(B) The half-life equation with the daughter nuclide after decay in the numerator is $\text{In}\,\dfrac{N}{N_\circ} = -kt$, with k being the decay constant for the reaction. It also known that $k = 0.693/\,t_{1/2}$. Substituting for k in the equation gives: $\text{In}\,\dfrac{N}{N_\circ} = \dfrac{0.693}{t_{1/2}}$. Now its time for "plug and chug":

$$t = -\frac{t_{1/2}\text{In}(N/N_o)}{0.693} = -\frac{4.51 10^9 \text{ yrs} \times \text{In}(0.297)}{0.693} = 7.90 \times 10^9 \text{ yrs}$$

(C) Atomic weight = $(0.014 \times 203.973) + (0.241 \times 205.9745) + (0.221 \times 206.9759) + (0.524 \times 207.9766) = 207.2$. The significant figures are determined by the addition rule. Each multiplication gives a valid answer to the tenths place. When the 4s are added, the last complete column is the tenths column, which accounts for the answer as given.

PRACTICE TESTS AND EXPLANATIONS

Information in this table may be useful in answering the questions in the Practice Tests.

Periodic Table of the Elements

Atomic Number — 6
Symbol — C
Atomic Mass — 12.01

* Numbers in parentheses are the *mass numbers* of the most stable isotope of the element.

1	2	3	4	5	6	7	8	9	10	11	12	13	14	15	16	17	18
1 **H** 1.008																	2 **He** 4.003
3 **Li** 6.941	4 **Be** 9.012											5 **B** 10.81	6 **C** 12.01	7 **N** 14.01	8 **O** 16.00	9 **F** 19.00	10 **Ne** 20.18
11 **Na** 22.99	12 **Mg** 24.31											13 **Al** 26.98	14 **Si** 28.09	15 **P** 30.97	16 **S** 32.07	17 **Cl** 35.45	18 **Ar** 39.95
19 **K** 39.10	20 **Ca** 40.08	21 **Sc** 44.96	22 **Ti** 47.88	23 **V** 50.94	24 **Cr** 52.00	25 **Mn** 54.94	26 **Fe** 55.85	27 **Co** 58.93	28 **Ni** 58.69	29 **Cu** 63.55	30 **Zn** 65.39	31 **Ga** 69.72	32 **Ge** 72.61	33 **As** 74.92	34 **Se** 78.96	35 **Br** 79.90	36 **Kr** 83.80
37 **Rb** 85.47	38 **Sr** 87.62	39 **Y** 88.91	40 **Zr** 91.22	41 **Nb** 92.91	42 **Mo** 95.94	43 **Tc** (98)	44 **Ru** 101.1	45 **Rh** 102.9	46 **Pd** 106.4	47 **Ag** 107.9	48 **Cd** 112.4	49 **In** 114.8	50 **Sn** 118.7	51 **Sb** 121.8	52 **Te** 127.6	53 **I** 126.9	54 **Xe** 131.3
55 **Cs** 132.9	56 **Ba** 137.3	57–70	72 **Hf** 178.5	73 **Ta** 181.0	74 **W** 183.8	75 **Re** 186.2	76 **Os** 190.2	77 **Ir** 192.2	78 **Pt** 195.1	79 **Au** 197.0	80 **Hg** 200.6	81 **Tl** 204.4	82 **Pb** 207.2	83 **Bi** 209.0	84 **Po** (209)	85 **At** (210)	86 **Rn** (222)
87 **Fr** (223)	88 **Ra** 226.0	89–102	104 **Rf** (261)	105 **Db** (262)	106 **Sg** (263)	107 **Bh** (262)	108 **Hs** (265)	109 **Mt** (268)	110 **Uun** (269)	111 **Uuu** (272)	112 **Uub** (277)	113 **Uut**	114 **Uuq** (289)	115 **Uup**	116 **Uuh**	117 **Uus**	118 **Uuo**

57 **La** 138.9	58 **Ce** 140.1	59 **Or** 140.9	60 **Nd** 144.2	61 **Pm** (145)	62 **Sm** 150.4	63 **Eu** 152.0	64 **Gd** 157.3	65 **Tb** 158.9	66 **Dy** 162.5	67 **Ho** 164.9	68 **Er** 167.3	69 **Tm** 168.9	70 **Yb** 173.0	71 **Lu** 175.0
89 **Ac** 227.0	90 **Th** 232.0	91 **Pa** 231.0	92 **U** 238.0	93 **Np** 237.0	94 **Pu** (244)	95 **Am** (243)	96 **Cm** (247)	97 **Bk** (247)	98 **Cf** (251)	99 **Es** (252)	100 **Fm** (257)	101 **Md** (258)	102 **No** (259)	103 **Lr** (260)

Information in this table may be useful in answering the questions in Section II of the Practice Tests.

STANDARD REDUCTION POTENTIALS IN AQUEOUS SOLUTION AT 25°C

Half-reaction			$E°(V)$
$F_2(g) + 2\,e^-$	$\rightarrow$	$2F^-$	2.87
$Co^{3+} + e^-$	$\rightarrow$	Co^{2+}	1.82
$Au^{3+} + 3\,e^-$	$\rightarrow$	$Au(s)$	1.50
$Cl_2(g) + 2\,e^-$	$\rightarrow$	$2Cl^-$	1.36
$O_2(g) + 4\,H^+ + 4\,e^-$	$\rightarrow$	$2H_2O(\ell)$	1.23
$Br_2(\ell) + 2\,e^-$	$\rightarrow$	$2Br^-$	1.07
$2\,Hg^{2+} + 2\,e^-$	$\rightarrow$	Hg_2^{2+}	0.92
$Ag^+ + e^-$	$\rightarrow$	$Ag(s)$	0.80
$Hg_2^{2+} + 2\,e^-$	$\rightarrow$	$2Hg(\ell)^-$	0.79
$Fe^{3+} + e^-$	$\rightarrow$	Fe^{2+}	0.77
$I_2(s) + 2\,e^-$	$\rightarrow$	$2I^-$	0.53
$Cu^+ + e^-$	$\rightarrow$	$Cu(s)$	0.52
$Cu^{2+} + 2\,e^-$	$\rightarrow$	$Cu(s)$	0.34
$Cu^{2+} + e^-$	$\rightarrow$	Cu^+	0.15
$Sn^{4+} + 2\,e^-$	$\rightarrow$	Sn^{2+}	0.15
$S(s) + 2\,H^+ + 2\,e^-$	$\rightarrow$	$H_2S(g)$	0.14
$2\,H^+ + 2\,e^-$	$\rightarrow$	$H_2(g)$	0.00
$Pb^{2+} + 2\,e^-$	$\rightarrow$	$Pb(s)^-$	−0.13
$Sn^{2+} + 2\,e^-$	$\rightarrow$	$Sn(s)^-$	−0.14
$Ni^{2+} + 2\,e^-$	$\rightarrow$	$Ni(s)^-$	−0.25
$Co^{2+} + 2\,e^-$	$\rightarrow$	$Co(s)^-$	−0.28
$Tl^+ + e^-$	$\rightarrow$	$Tl(s)^-$	−0.34
$Cd^{2+} + 2\,e^-$	$\rightarrow$	$Cd(s)^-$	−0.40
$Cr^{3+} + e^-$	$\rightarrow$	Cr^{2+}	−0.41
$Fe^2 + 2\,e^-$	$\rightarrow$	$Fe(s)^-$	−0.44
$Cr^{3+} + 3\,e^-$	$\rightarrow$	$Cr(s)$	−0.74
$Zn^{2+} + 2\,e^-$	$\rightarrow$	$Zn(s)^-$	−0.76
$Mn^{2+} + 2\,e^-$	$\rightarrow$	$Mn(s)^-$	−1.18
$Al^{3+} + 3\,e^-$	$\rightarrow$	$Al(s)$	−1.66
$Be^{2+} + 2\,e^-$	$\rightarrow$	$Be(s)^-$	−1.70
$Mg^{2+} + 2\,e^-$	$\rightarrow$	$Mg(s)^-$	−2.37
$Na^+ + e^-$	$\rightarrow$	$Na(s)^-$	−2.71
$Ca^{2+} + 2\,e^-$	$\rightarrow$	$Ca(s)^-$	−2.87
$Sr^{2+} + 2\,e^-$	$\rightarrow$	$Sr(s)^-$	−2.89
$Ba^{2+} + 2\,e^-$	$\rightarrow$	$Ba(s)^-$	−2.90
$Rb^+ + e^-$	$\rightarrow$	$Rb(s)^-$	−2.92
$K^+ + e^-$	$\rightarrow$	$K(s)^-$	−2.92
$Cs^+ + e^-$	$\rightarrow$	$Cs(s)^-$	−2.92
$Li^+ + e^-$	$\rightarrow$	$Li(s)^-$	−3.05

Information in this table may be useful in answering the questions in Section II of the Practice Tests.

AP CHEMISTRY EQUATIONS AND CONSTANTS

Atomic Structure

$\Delta E = h\nu$

$c = \lambda\nu$

$\lambda = \dfrac{h}{mv}$

$p = mv$

$E_n = \dfrac{-2.178 \times 10^{-18}}{n^2} \text{ joule}$

Equilibrium

$K_a = \dfrac{[H^+][A^-]}{[HA]}$

$K_b = \dfrac{[OH^-][HB^+]}{[B]}$

$K_w = [OH^-][H^+] = 1.0 \times 10^{-14} \text{ at } 25°C$
$\quad = K_a \times K_b$

$pH = -\log[H^+]$

$pOH = -\log[OH^-]$

$14 = pH + pOH$

$pH = pK_a + \log\dfrac{[A^-]}{[HA]}$

$pOH = pK_b + \log\dfrac{[HB^+]}{[B]}$

$pK_a = -\log K_a$

$pK_b = -\log K_b$

$K_p = K_c(RT)^{\Delta n}, \text{ where}$
$\quad \Delta n = \text{moles product gas} - \text{moles reactant gas}$

Thermochemistry

$\Delta S° = \Sigma S° \text{ products} - \Sigma S° \text{ reactants}$

$\Delta H° = \Sigma H_f° \text{ products} - \Sigma H_f° \text{ reactants}$

$\Delta G° = \Sigma G_f° \text{ products} - \Sigma G_f° \text{ reactants}$

$\Delta G° = \Delta H° - T\Delta S°$

$\quad = -RT \ln K = -2.303\, RT \log K$

$\quad = -n\,\mathfrak{F}\, E°$

$\Delta G° = \Delta G° + RT \ln Q = \Delta G° + 2.303\, RT \log Q$

$q = mc\Delta T$

$G_p = \dfrac{\Delta H}{\Delta T}$

$m = \text{mass}$
$E = \text{energy}$
$\nu = \text{frequency}$
$\lambda = \text{wavelength}$
$p = \text{momentum}$
$v = \text{velocity}$
$n = \text{principal quantum number}$

Speed of light, $c = 3.0 \times 10^8 \text{ m s}^{-1}$
Electron charge, $e = -1.602 \times 10^{-19}$ coulomb
1 electron volt per atom $= 96.5 \text{ kJ mol}^{-1}$
Planck's constant, $h = 6.63 \times 10^{-34}$ J s
Boltzmann's constant, $k = 1.38 \times 10^{-23}$ J K^{-1}
Avogadro's number $= 6.022 \times 10^{23}$
molecules mol^{-1}

Equilibrium Constants

K_a (weak acid)

K_b (weak base)

K_c (molar concentrations)

K_p (gas pressure)

K_w (water)

$S° =$ standard entropy

$H° =$ standard enthalpy

$G° =$ standard free energy

$E° =$ standard reduction potential

$n =$ moles

$m =$ mass

$T =$ temperature

$q =$ heat

$c =$ specific heat capacity

$C_p =$ molar heat capacity at constant pressure

1 faraday $\mathscr{F} =$ 96,500 coulombs

Gases, Liquids, and Solutions

$$PV = nRT$$

$$\left(P + \frac{n^2 a}{V^2}\right)(V - nb) = nRT$$

$$P_A = P_{total} \times X_A, \text{ where } X_A = \frac{\text{moles } A}{\text{total moles}}$$

$$P_{total}P_A = P_A + P_B + P_C + \ldots$$

$$n = \frac{m}{\mathcal{M}}$$

$$K = °C + 273$$

$$\frac{P_1 V_1}{T_1} = \frac{P_2 V_2}{T_2}$$

$$D = \frac{m}{V}$$

$$u_{rms} = \sqrt{\frac{3kT}{m}} = \sqrt{\frac{3RT}{\mathcal{M}}}$$

$$KE \text{ per molecule} = \frac{1}{2} mv^2$$

$$KE \text{ per mole} = \frac{3}{2} RT$$

$$\frac{r_1}{r_2} = \sqrt{\frac{\mathcal{M}_2}{\mathcal{M}_1}}$$

molarity, $M =$ moles solute per liter solution

molality $=$ moles solute per kilogram solvent

$$\Delta T_f = iK_f \times \text{molality}$$

$$\Delta T_b = iK_b \times \text{molality}$$

$$\pi = \frac{nRT}{V} i$$

Oxidation-Reduction; Electrochemistry

$$Q = \frac{[C]^c\,[D]^d}{[A]^a\,[B]^b}, \text{ where } a\,A + b\,B \rightarrow c\,C + d\,D$$

$$I = \frac{q}{t}$$

$$E_{cell} = E^\circ{}_{cell}\frac{RT}{n\Im}\ln Q = E^\circ{}_{cell} - \frac{0.0592}{n}\log Q \text{ at } 25°C$$

$$\log K = \frac{nE^\circ}{0.0592}$$

$P =$ pressure

$V =$ volume

$T =$ temperature

$n =$ number of moles

$D =$ density

$v =$ velocity

$E^\circ =$ standard reduction potential

$K =$ equilibrium constant

$KE =$ kinetic energy

$r =$ rate of effusion

$\mathcal{M} =$ molar mass

$t =$ time (seconds)

$\pi =$ osmotic pressure

$i =$ van't Hoff factor

$K_f =$ molal freezing-point depression constant

$K_b =$ molal boiling-point elevation constant

$Q =$ reaction quotient

$I =$ current (ampreres)

$q =$ charge (coulombs)

$u_{rms} =$ root-mean-square speed

Gas constant, $R = $ 8.31 J mol^{-1} K^{-1}

$= $ 0.0821 L atm mol^{-1} K^{-1}

$= $ 8.31 volt coulomb mol^{-1} K^{-1}

Boltzmann's constant, $k = 1.38 \times 10^{-23}$ J K^{-1}

K_f for $H_2O = $ 1.86 K kg mol^{-1}

K_b for $H_2O = $ 0.512 K kg mol^{-1}

1 atm $= $ 760 mm Hg

$= $ 760 torr

STP $= $ 0.000°C and 1.000 atm

Faraday's constant, $\Im = $ 96,500 coulombs per mole of electrons

Practice Test One

On the following pages you'll find a complete practice examination in a format similar to that found on the AP Chemistry exam.

Section I consists of 75 multiple-choice questions and constitutes 45% of your final grade. Only a "bare bones" Periodic Table may be used; you may *not* use a calculator for this section. Use the answer sheet provided to record your multiple-choice answers. The time allotted for this section is 90 minutes.

Section II consists of eight free-response questions, each of which will have multiple parts. Section II makes up 55% of your total score. In addition to the Periodic Table, you may use the table of standard reduction potentials and the list of AP Chemistry equations and constants provided at the beginning of this section. Use the space provided to record your free-response answers. (On the actual AP Chemistry exam, you will be asked to record your answers to Section II on lined pages following each question.)

Be sure to pay careful attention to the instructions for Section II:

- You are allotted 40 minutes to complete Part A. You may use a calculator on this part only.

- There are three questions in Part A. You must answer Question 1, and you will be asked to choose between Question 2 and Question 3.

- On Part B, calculators are *not* permitted. You are allotted 50 minutes for this part.

- There are five questions in Part B. You must answer Questions 5 and 6, and choose between Questions 7 and 8. Question 4 has explicit instructions to write five of eight net reactions.

At the end of the practice test you will find an answer key, along with a discussion of the answers to each multiple-choice and free-response question. Scoring information is located on page 564.

Good luck!

PRACTICE TEST ONE ANSWER SHEET

1 Ⓐ Ⓑ Ⓒ Ⓓ Ⓔ	20 Ⓐ Ⓑ Ⓒ Ⓓ Ⓔ	39 Ⓐ Ⓑ Ⓒ Ⓓ Ⓔ	58 Ⓐ Ⓑ Ⓒ Ⓓ Ⓔ
2 Ⓐ Ⓑ Ⓒ Ⓓ Ⓔ	21 Ⓐ Ⓑ Ⓒ Ⓓ Ⓔ	40 Ⓐ Ⓑ Ⓒ Ⓓ Ⓔ	59 Ⓐ Ⓑ Ⓒ Ⓓ Ⓔ
3 Ⓐ Ⓑ Ⓒ Ⓓ Ⓔ	22 Ⓐ Ⓑ Ⓒ Ⓓ Ⓔ	41 Ⓐ Ⓑ Ⓒ Ⓓ Ⓔ	60 Ⓐ Ⓑ Ⓒ Ⓓ Ⓔ
4 Ⓐ Ⓑ Ⓒ Ⓓ Ⓔ	23 Ⓐ Ⓑ Ⓒ Ⓓ Ⓔ	42 Ⓐ Ⓑ Ⓒ Ⓓ Ⓔ	61 Ⓐ Ⓑ Ⓒ Ⓓ Ⓔ
5 Ⓐ Ⓑ Ⓒ Ⓓ Ⓔ	24 Ⓐ Ⓑ Ⓒ Ⓓ Ⓔ	43 Ⓐ Ⓑ Ⓒ Ⓓ Ⓔ	62 Ⓐ Ⓑ Ⓒ Ⓓ Ⓔ
6 Ⓐ Ⓑ Ⓒ Ⓓ Ⓔ	25 Ⓐ Ⓑ Ⓒ Ⓓ Ⓔ	44 Ⓐ Ⓑ Ⓒ Ⓓ Ⓔ	63 Ⓐ Ⓑ Ⓒ Ⓓ Ⓔ
7 Ⓐ Ⓑ Ⓒ Ⓓ Ⓔ	26 Ⓐ Ⓑ Ⓒ Ⓓ Ⓔ	45 Ⓐ Ⓑ Ⓒ Ⓓ Ⓔ	64 Ⓐ Ⓑ Ⓒ Ⓓ Ⓔ
8 Ⓐ Ⓑ Ⓒ Ⓓ Ⓔ	27 Ⓐ Ⓑ Ⓒ Ⓓ Ⓔ	46 Ⓐ Ⓑ Ⓒ Ⓓ Ⓔ	65 Ⓐ Ⓑ Ⓒ Ⓓ Ⓔ
9 Ⓐ Ⓑ Ⓒ Ⓓ Ⓔ	28 Ⓐ Ⓑ Ⓒ Ⓓ Ⓔ	47 Ⓐ Ⓑ Ⓒ Ⓓ Ⓔ	66 Ⓐ Ⓑ Ⓒ Ⓓ Ⓔ
10 Ⓐ Ⓑ Ⓒ Ⓓ Ⓔ	29 Ⓐ Ⓑ Ⓒ Ⓓ Ⓔ	48 Ⓐ Ⓑ Ⓒ Ⓓ Ⓔ	67 Ⓐ Ⓑ Ⓒ Ⓓ Ⓔ
11 Ⓐ Ⓑ Ⓒ Ⓓ Ⓔ	30 Ⓐ Ⓑ Ⓒ Ⓓ Ⓔ	49 Ⓐ Ⓑ Ⓒ Ⓓ Ⓔ	68 Ⓐ Ⓑ Ⓒ Ⓓ Ⓔ
12 Ⓐ Ⓑ Ⓒ Ⓓ Ⓔ	31 Ⓐ Ⓑ Ⓒ Ⓓ Ⓔ	50 Ⓐ Ⓑ Ⓒ Ⓓ Ⓔ	69 Ⓐ Ⓑ Ⓒ Ⓓ Ⓔ
13 Ⓐ Ⓑ Ⓒ Ⓓ Ⓔ	32 Ⓐ Ⓑ Ⓒ Ⓓ Ⓔ	51 Ⓐ Ⓑ Ⓒ Ⓓ Ⓔ	70 Ⓐ Ⓑ Ⓒ Ⓓ Ⓔ
14 Ⓐ Ⓑ Ⓒ Ⓓ Ⓔ	33 Ⓐ Ⓑ Ⓒ Ⓓ Ⓔ	52 Ⓐ Ⓑ Ⓒ Ⓓ Ⓔ	71 Ⓐ Ⓑ Ⓒ Ⓓ Ⓔ
15 Ⓐ Ⓑ Ⓒ Ⓓ Ⓔ	34 Ⓐ Ⓑ Ⓒ Ⓓ Ⓔ	53 Ⓐ Ⓑ Ⓒ Ⓓ Ⓔ	72 Ⓐ Ⓑ Ⓒ Ⓓ Ⓔ
16 Ⓐ Ⓑ Ⓒ Ⓓ Ⓔ	35 Ⓐ Ⓑ Ⓒ Ⓓ Ⓔ	54 Ⓐ Ⓑ Ⓒ Ⓓ Ⓔ	73 Ⓐ Ⓑ Ⓒ Ⓓ Ⓔ
17 Ⓐ Ⓑ Ⓒ Ⓓ Ⓔ	36 Ⓐ Ⓑ Ⓒ Ⓓ Ⓔ	55 Ⓐ Ⓑ Ⓒ Ⓓ Ⓔ	74 Ⓐ Ⓑ Ⓒ Ⓓ Ⓔ
18 Ⓐ Ⓑ Ⓒ Ⓓ Ⓔ	37 Ⓐ Ⓑ Ⓒ Ⓓ Ⓔ	56 Ⓐ Ⓑ Ⓒ Ⓓ Ⓔ	75 Ⓐ Ⓑ Ⓒ Ⓓ Ⓔ
19 Ⓐ Ⓑ Ⓒ Ⓓ Ⓔ	38 Ⓐ Ⓑ Ⓒ Ⓓ Ⓔ	57 Ⓐ Ⓑ Ⓒ Ⓓ Ⓔ	

SECTION I

Time—90 minutes

<u>Directions</u>: Select the best answer choice and fill in the corresponding oval on the answer sheet. **You may NOT use a calculator for this section. You may use the Periodic Table provided on page 469 of this book.**

1. Solid sulfur is burned producing sulfur dioxide gas.

   $$S\ (s) + O_2\ (g) \rightarrow SO_2\ (g) \qquad \Delta H = -296\ \text{kJ/mol}$$

   What is the enthalpy change for the burning of 1.00 kilogram of solid sulfur?

   A. $-9{,}230$ kJ
   B. -296 kJ
   C. -9.49 kJ
   D. -9.23 kJ
   E. $+9230$ kJ

2. A 250.0 mL sample of ethanol is heated from 25.0°C to 47.0°C at constant pressure. How much energy is added to the sample of ethanol? The specific heat capacity of ethanol is 2.4 J/g·K and its density is 0.789 g/mL.

   A. 1.0×10^4 J
   B. 1.3×10^4 J
   C. 2.2×10^4 J
   D. 2.8×10^4 J
   E. 4.3×10^3 J

3. What is the frequency of a photon of blue light that has a wavelength of 500 nm? The speed of light, c, is 3.00×10^8 m/s.

   A. 1.67×10^{-15} s^{-1}
   B. 1.5×10^2 s^{-1}
   C. 3×10^8 s^{-1}
   D. 1.5×10^{11} s^{-1}
   E. 6×10^{14} s^{-1}

4. Which have the highest frequency and the highest energy per photon?

   A. Ultraviolet rays
   B. Infrared rays
   C. Gamma rays
   D. Microwaves
   E. Radio waves

5. An electron probability density is the picture that emerges when we superimpose lots of instantaneous positions of electrons, revealing the atomic orbitals shape as a cloud of points surrounding the nucleus. Why is this the most reliable image of an atomic orbital?

   A. An electron is in all the positions in an orbital simultaneously.

   B. The electron is moving a defined pathway in an orbital.

   C. The electrons are frozen in positions in an orbital around the atom.

   D. The film that is most often used for atoms takes multiple exposures.

   E. The location of an electron is best described as the probability of finding an electron in a given region of space.

6. What is the shape of an orbital in a p sublevel?

   A. 8 lobes
   B. Dumbbell
   C. Hyperbolic
   D. Parabolic
   E. Spherical

7. In a krypton atom, which atomic orbitals contain the valence electrons?

   A. $3d$
   B. $4p$
   C. $3s$ and $3p$
   D. $4s$ and $4p$
   E. $4s$, $3d$, and $4p$

GO ON TO THE NEXT PAGE. ➡

8. How many valence electrons are there in the Lewis structure of the sulfate ion, SO_4^{2-}?

   A.   32
   B.   30
   C.   26
   D.   24
   E.   12

9. For a given pair of atoms, which of the following statements is (are) true?

   I. As bond length increases, bond energy increases.

   II. As bond order increases, bond energy increases.

   III. As bond order increases, bond length decreases.

   A.   I only
   B.   II only
   C.   III only
   D.   II and III only
   E.   I, II, and III

10. How do geometric isomers differ?

   I. Bonds

   II. Formula

   III. Physical properties

   IV. Arrangement of atoms in space

   A.   III only
   B.   III and IV only
   C.   II, III, and IV only
   D.   I, III, and IV only
   E.   I, II, III, and IV

11. A 0.7500 g sample of sucrose, $C_{12}H_{22}O_{11}$, is completely burned with excess oxygen in a 3.40 L vessel maintained at a constant temperature of 25.00°C. What is the partial pressure, in mmHg, of the CO_2 gas in the vessel after the combustion?

   $C_{12}H_{22}O_{11}$ (s) + 12 O_2 (g) → 12 CO_2 (g) + 11 H_2O (ℓ)

   A.   0.190
   B.   6.21
   C.   12.0
   D.   72.0
   E.   144

12. Which is nonpolar, but contains polar bonds?

   A. H_2O
   B.   HCl
   C. NO_2
   D. SO_2
   E. SO_3

For questions 13–19, select the molecular formula that fits the description given. Answers may be used more than once.

   A. CS_2
   B. C_2Cl_4
   C. Br_2
   D.   HCN
   E. $BaCl_2$

13. Has only one pi (π) bond

14. Has two double bonds

15. Has one triple bond

16. Exhibits ionic bonding

17. Polar molecules at room temperature

18. Is a solid at room temperature

19. Has a nonpolar single bond

GO ON TO THE NEXT PAGE. ➡

20. Which of the following statements is (are) true for isotopes of an element?

    I. They are atoms of the same mass with different atomic numbers.

    II. The only difference in composition between isotopes of an element is the number of neutrons.

    III. The atomic weight of an element is an average of the weights of its isotopes, in the proportions in which they naturally occur in nature.

    A. I only
    B. II only
    C. III only
    D. II and III only
    E. I, II, and III

21. An atom with the electron configuration s^2p^4 in the outer shell is in the same family as:

    A. beryllium.
    B. carbon.
    C. fluorine.
    D. manganese.
    E. oxygen.

22. Which is true regarding the bond order and magnetism of oxygen, O_2?

	Bond Order	Magnetism
A.	1	paramagnetic
B.	1	diamagnetic
C.	2	paramagnetic
D.	2	diamagnetic
E.	3	paramagnetic

23. How many grams of $K_2Cr_2O_7$ (molar mass = 294 g/mol) are required to prepare 200. mL of a 0.100 M solution?

    A. 2.94
    B. 4.82
    C. 5.88
    D. 2.94
    E. 58.8

24. Water has a vapor pressure of 23.76 Torr at 25°C. What is the vapor pressure of a solution of sucrose whose molality is 23.8 mol / kg H_2O?

    A. 7.13 Torr
    B. 15.2 Torr
    C. 16.6 Torr
    D. 23.8 Torr
    E. 30.9 Torr

25. Which reaction produces chlorine gas when they are heated in a test tube placed in a laboratory hood?

    A. $HCl + Br_2$
    B. $HCl + KMnO_4$
    C. $NaCl + MnO_2$
    D. $NaCl + HNO_3$
    E. $NaCl + H_2SO_4$

26. Which statement best accounts for the fact that gases can be greatly compressed?

    A. Molecules occupy space.

    B. The collisions of molecules are elastic.

    C. Molecules of gases are in constant motion.

    D. The molecules of a given gas are identical.

    E. Molecules of gases are relatively far from each other.

27. Which measurement technique is most suitable for the determination of the molar mass of oxyhemoglobin, a molecule with a molar mass of many thousands?

   A. Osmotic pressure
   B. Boiling point elevation
   C. Vapor pressure lowering
   D. Freezing point depression
   E. Henry's Law and the partial pressure of the gas above the solution

28. What will happen if 50.0 mL of 0.050 M $BaCl_2$ is mixed with 50.0 mL of 0.050 M K_2SO_4? The K_{sp} of barium sulfate $= 1.1 \times 10^{-10}$.

   A. KCl precipitates

   B. $BaSO_4$ precipitates

   C. No reaction occurs

   D. The concentration of Ba^{2+} in the resulting solution is 0.025 M

   E. The concentration of SO_4^{2-} in the resulting solution is 0.050 M

29. Which is true regarding the reaction $H_2O\ (\ell) \rightleftharpoons H_2O\ (g)$ at 100°C and 1.0 atm pressure?

   A. $\Delta S = 0$
   B. $\Delta H = 0$
   C. $\Delta H = \Delta G$
   D. $\Delta H = \Delta E$
   E. $\Delta H = T\Delta S$

30. Which atom has the same number of neutrons as ^{85}Rb?

   A. ^{85}Sr
   B. ^{86}Sr
   C. ^{86}Kr
   D. ^{85}Kr
   E. ^{87}Rb

31. The titration curve shows HCl being titrated with:

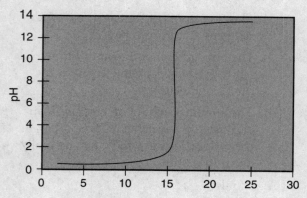

   A. Acetic acid
   B. Ammonia
   C. Methylamine
   D. Sodium hydroxide
   E. Sulfuric acid

32. In the reaction $HCl + H_2O \rightarrow H_3O^+ + Cl^-$, the water acts as:

   A. a Lewis acid.
   B. a Brønsted acid.
   C. a Brønsted base.
   D. an oxidizing agent.
   E. an inert substance.

33. Which of these substances has no covalent bonds?

   A. H_2
   B. Br_2
   C. LiH
   D. SiF_4
   E. $Ca(OH)_2$

34. What is the percent of water of crystallization in $MgSO_4 \cdot 7H_2O$?

   A. 4.8%
   B. 6.6%
   C. 25.6%
   D. 48.8%
   E. 51.2%

GO ON TO THE NEXT PAGE. ➡

35. The graphs below represent the behavior of various gases under different conditions. Which gas exhibits behavior that differs most from that of an ideal gas?

A. Gas A at constant P and n

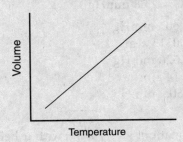

B. Gas B at constant T and n

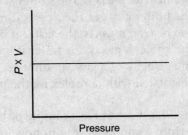

C. Gas C at constant V and n

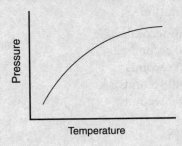

D. Gas D at constant n

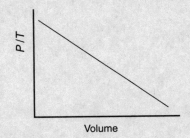

E. Gas E at constant T and n

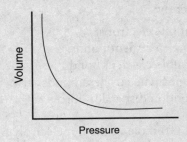

36. What are the formal charges on C, O, and Cl, respectively, in the $COCl_2$ molecule?

A. 0, 0, 0
B. 0, −1, +1
C. 0, +1, −1
D. +1, −1, 0
E. −1, 0, +1

37. Which is true about iodine-131, ^{131}I?

A. Isotopes with atomic number greater than 83 generally undergo alpha decay.

B. Isotopes that fall to the right of the band of stability generally undergo beta decay.

C. Isotopes that fall to the left of the band of stability generally undergo positron decay.

D. Lighter, stable nuclei tend to have equal numbers of protons and neutrons.

E. Lighter, radioactive nuclei tend to have unequal numbers of protons and neutrons.

38. Given the half-reaction reduction potentials:

Al^{3+} (aq) + 3e$^-$ → Al (s) $E° = -1.66$ V

Cl_2 (g) + 2e$^-$ → 2Cl$^-$ (aq) $E° = +1.36$ V

What is the net voltage, $E°$, for the reaction:

2 Al (s) + 3 Cl_2 (g) → 2 Al^{3+} (aq) + 6 Cl$^-$ (aq)

A. −7.40 V
B. −3.02 V
C. 0.36 V
D. 3.02 V
E. 7.40 V

GO ON TO THE NEXT PAGE. ➡

39. Layers of carbon atoms in graphite are held together by:

    A. mobile electrons.
    B. single covalent bonds.
    C. double covalent bonds.
    D. coordinate covalent bonds.
    E. London dispersion forces.

40. One (1.00) mole of the nonelectrolyte ethylene glycol is dissolved in 500 g of water. What is the freezing point of the solution? The freezing point depression constant, K_f, for water is 1.86°C / m.

    A. −0.93 °C
    B. −1.86 °C
    C. −2.79 °C
    D. −3.72 °C
    E. −5.58 °C

41. This molecule has the formula AsF_5. What is the hybridization of the central arsenic atom?

    A. sp
    B. sp^2
    C. sp^3
    D. sp^3d
    E. sp^3d^2

42. A 100.0 mL sample of oxygen was collected at 27.0°C and 836 Torr. What is the volume at 127°C and standard pressure?

    A. 68.2 mL
    B. 82.5 mL
    C. 121mL
    D. 147 mL
    E. 411 mL

43. A 342.8 mL sample of the gas above a solution of water and dissolved carbon dioxide was collected at 25°C. The sample contained 0.030 moles. What is the partial pressure of the carbon dioxide? (The vapor pressure of water at 25°C = 23.8 mmHg.)

    A. 1,603 mmHg
    B. 0.0313 atm
    C. 1,626 mmHg
    D. 2.14 atm
    E. 160.3 mmHg

44. The equation: $v = \sqrt{\dfrac{3RT}{\mu}}$, with μ being the molar mass, shows us that:

    A. the heavier a gas is, the slower it moves.
    B. the hotter a gas is, the slower it moves.
    C. the heavier a gas is, the faster it moves.
    D. the cooler a gas is, the faster it moves.
    E. the velocity of a gas can only be calculated with complex mathematics.

45. It takes 20.0 seconds for a sample of helium to effuse from a container. About how long will it take for a similar sample of nitrogen to effuse from the same container at the same temperature?

    A. 2.6 seconds
    B. 20 seconds
    C. 53 seconds
    D. 140 seconds
    E. 980 seconds

GO ON TO THE NEXT PAGE. ➡

46. Consider the reaction at equilibrium:

    $4 NO_2 (g) + 6 H_2O (g) \rightarrow 4 NH_3 (g) + 7 O_2 (g)$

    If the partial pressure of NH_3 (g) increases, which best describes what happens to the partial pressures of the other substances?

    A. The partial pressures of NO_2 (g), H_2O (g), and O_2 (g) all increase.

    B. The partial pressures of NO_2 (g), H_2O (g), and O_2 (g) all decrease.

    C. The partial pressures of NO_2 (g) and H_2O (g) increase and the amount of O_2 (g) decreases.

    D. The partial pressures of NO_2 (g) and O_2 (g) increase and the amount of H_2O (g) decreases.

    E. The partial pressures of NO_2 (g) and H_2O (g) decrease and the amount O_2 (g) increases.

47. The equilibrium concentrations in which reaction are *not* affected by a change in pressure?

    A. $2 NO_2 (g) \rightarrow N_2 (g) + 2 O_2 (g)$

    B. $CO (g) + \frac{1}{2} O_2 (g) \rightarrow CO_2 (g)$

    C. $PCl_3 (g) + Cl_2 (g) \rightarrow PCl_5 (g)$

    D. $CO (g) + H_2O (g) \rightarrow CO_2 (g) + H_2 (g)$

    E. $N_2 (g) + 3 H_2 (g) \rightarrow 2 NH_3 (g)$

48. Only 4.17% of acetic acid dissociates in solution. What is the pH of a solution where 0.30 mol of acetic acid is dissolved in enough water to make 500.0 mL of solution?

    A. 0.097
    B. 0.22
    C. 1.60
    D. 3.20
    E. 4.38

49. When the Arrhenius equation is algebraically arranged in the form of a straight line, the slope of the line can be used to determine the:

    A. activation energy.
    B. frequency factor.
    C. rate constant.
    D. reactant concentration.
    E. temperature.

50. The first law of thermodynamics:

    I. states that the total entropy of the system of interest and its surroundings is constant.

    II. accounts for the energy transfer only in the direction that increases the total energy of the system and its surroundings.

    III. implies that processes that occur within a system in which the total entropy is constant must necessarily be nonspontaneous.

    IV. states that the total energy of the system of interest and its surroundings is constant.

    A. I only
    B. IV only
    C. I and II only
    D. II, III, and IV only
    E. None of these statements is correct.

GO ON TO THE NEXT PAGE.

51. Which is true regarding the solubility product of an insoluble ionic compound?

   I. A solubility product is an equilibrium constant.

   II. The solubility product of an ionic salt has a fixed numerical value.

   III. The solubility product of an insoluble ionic salt is always a number smaller than 1.

   IV. The units of solubility product are M^2 or mol^2 / L^2.

   A. I and II only
   B. II and III only
   C. III and IV only
   D. I and III only
   E. II and IV only

52. What is the pH of a mixture of acetic acid and sodium acetate where the ratio of $[CH_3COO^-] / [CH_3COOH]$ is 4.00? The K_a of CH_3COOH is 1.8×10^{-5}.

   A. $-\log(1.8 \times 10^{-5}) - \log(4.00/1.00)$
   B. $-\log(1.8 \times 10^{-5}) + \log(1.00/4.00)$
   C. $-\log(1.8 \times 10^{-5})$
   D. $\log(1.8 \times 10^{-5}) + \log(4.00/1.00)$
   E. $-\log(1.8 \times 10^{-5}) + \log(1.00/4.00)$

53. For which of the substances does $\Delta H_f^\circ = 0$?

   A. Br_2 (g)
   B. C (g)
   C. CO (g)
   D. N (g)
   E. Ne (g)

54. Bronze is an alloy of which metals?

   A. Copper and tin
   B. Copper and zinc
   C. Aluminum and nickel
   D. Aluminum and copper
   E. Lead and tin

55. Three different substances A, B, and oxygen, are mixed. A two-step reaction occurs. Which substance is the catalyst?

   Step 1 $A + O_2 \rightarrow AO_2$
   Step 2 $B + AO_2 \rightarrow A + BO_2$

   A. A
   B. B
   C. O_2
   D. AO_2
   E. BO_2

56. Aqueous hydrochloric acid forms a precipitate with which of the following reagents?

   A. Barium nitrate
   B. Iron(II) sulfate
   C. Mercury(I) nitrate
   D. Sodium nitrite
   E. Zinc sulfate

57. Which statement about ionization energy is true?

   A. It is represented by the equation $X + e^- \rightarrow X^- +$ energy.

   B. It decreases as atomic number increases in a period of the Periodic Table.

   C. It increases as atomic number increases in a group of the Periodic Table.

   D. It is the energy needed to remove the most loosely bound electron from an atom its ground state.

   E. It is the energy released as an electron is added to a gaseous nonmetallic atom in its ground state.

58. What is the molarity of hydrogen ions in a 0.001 M solution of NaOH?

   A. 0 M
   B. 1×10^{-14} M
   C. 1×10^{-11} M
   D. 1×10^{-7} M
   E. 1×10^{-3} M

GO ON TO THE NEXT PAGE. ➥

59. In which molecule does the central atom use sp^2 hybrid atomic orbitals in forming bonds?

    A. CS_2
    B. Cl_2O
    C. SO_2
    D. NH_3
    E. H_2S

60. How many moles of Al_2O_3 are formed when a mixture of 0.36 moles Al and 0.36 moles O_2 is ignited?

    A. 0.12
    B. 0.18
    C. 0.28
    D. 0.46
    E. 0.72

61. Observations in his laboratory that some α-particles directed at a gold foil are scattered backwards at angles greater than 90° permitted Ernest Rutherford to conclude that:

    A. all atoms are electrically neutral.

    B. electrons have a very small mass.

    C. negatively charged electrons are a part of all matter.

    D. the positively charged part of atoms moves at extremely high velocity.

    E. the positively charged part of atoms occupies an extremely small fraction of the volume of an atom.

62. What is the value of the reaction quotient Q for the cell at 25°C?

    Ni (s) | Ni^{2+} (0.10 M) || Cl^- (0.40 M) | Cl_2 (g) (0.50 atm) | Pt

    A. 0.080
    B. 0.032
    C. 0.016
    D. 0.13
    E. 0.31

63. Which intermolecular force is most responsible for the unusually high boiling point of water?

    A. Hydrogen bonds
    B. Ion-dipole interactions
    C. London dispersion forces
    D. Polar covalent bonds
    E. Van der Waals forces

64. The concentration of a saturated solution of a peptide is 0.00100 M at 25°C. What is the solution osmotic pressure in mmHg?

    A. 0.0245
    B. 0.760
    C. 18.6
    D. 24.5
    E. 156

65. An aqueous solution of $BaCl_2$ is added to each of 4 test tubes containing an aqueous solution of $AgNO_3$, Na_2CO_3, $(NH_4)_2SO_4$, and $Pb(NO_3)_2$. In which of the test tubes will precipitates form?

    I. $AgNO_3$

    II. Na_2CO_3

    III. $(NH_4)_2SO_4$

    IV. $Pb(NO_3)_2$

    A. I and IV only
    B. I and III only
    C. II, III, and IV only
    D. I, III, and IV only
    E. I, II, III, and IV

66. Addition of heat to the reaction causes the equilibrium to shift to the products. What implication about reaction is most probable?

    $CaCO_3$ (s) → CaO (s) + CO_2 (g)

    A. The reaction is endothermic.
    B. The reaction is exothermic.
    C. The reaction is bimolecular.
    D. The reaction is spontaneous.
    E. The reaction is nonspontaneous.

GO ON TO THE NEXT PAGE. ➡

67. If 60. mL of a 0.50 M solution is of HCl is diluted with 90. mL of water, what is the new concentration?

    A. 0.20 M
    B. 0.30 M
    C. 0.33 M
    D. 0.40 M
    E. 0.75 M

68. Which is a structural isomer of $CH_3CH_2CH_2CH_2OH$?

    A. $CH_3CH_2OCH_3$
    B. $CH_3CH_2CH_2OH$
    C. $CH_3CH_2CH_2CH_3$
    D. $CH_3CH_2CH_2CH_2Cl$
    E. $CH_3CH(OH)CH_2CH_3$

69. The rate constant, k, for the decomposition of N_2O_5 is 6.2×10^{-4} at 45°C and 2.1×10^{-3} at 55°C. What is the activation energy, E_a, for this reaction in kJ / mol?

    A. 46
    B. 110
    C. 2,500
    D. 25,000
    E. 110,000

70. What is the most common method for preparing small quantities of oxygen gas in the lab?

    A. The passing of steam over hot coke
    B. The reaction of strong acids with metals
    C. The condensation and distillation of air
    D. The reaction of metallic oxides with water
    E. The pyrolization of specific inorganic chlorate compounds

71. Which has a dipole moment of zero?

    A. CH_4
    B. CH_3Cl
    C. H_2O
    D. NH_3
    E. HF

72. Where does the reaction take place in a voltaic cell?

    $$2\,H_2O \rightarrow O_2 + 4\,H^+ + 4\,e^-$$

    A. Anode
    B. Cathode
    C. Salt bridge
    D. Voltmeter
    E. Wire

73. Fluorine-18 decays by positron emission. What is the product nuclide of this decay?

    A. Fluorine-17
    B. Neon-18
    C. Nitrogen-14
    D. Nitrogen-18
    E. Oxygen-18

74. Which is the correct electron configuration for an element with 24 electrons?

    A. $1s^2 2s^2 2p^6 3s^2 3p^6 3d^6$
    B. $1s^2 2s^2 2p^6 3s^2 3p^6 4s^2 3d^4$
    C. $1s^2 2s^2 2p^6 3s^2 3p^6 4s^1 4p^5$
    D. $1s^2 2s^2 2p^6 3s^2 3p^6 4s^1 3d^5$
    E. $1s^2 2s^2 3s^2 2p^6 3p^6 4s^2 3d^4$

75. The kinetics of the reaction $A + B \rightarrow C$ was investigated. When the initial concentration of A is doubled and the concentration of B is held constant, the rate doubles. When the initial concentration of B is doubled and the concentration of A is held constant, the rate increases four-fold. What is the rate law for this reaction?

    A. rate = k[A][B]
    B. rate = k[A]²[B]³
    C. rate = k[A][B]²
    D. rate = k[A]³[B]
    E. rate = k[A]²[B]

GO ON TO THE NEXT PAGE. ➡

SECTION II—PART A

Time—40 minutes

<u>Directions</u>: Write your answers in the space provided. Choose between question 2 and question 3.

You may use a calculator for this section. You may use the Periodic Table, the "Equations and Constants" sheets, and the table of standard reduction potentials provided on pages 469–473 of this book.

1. A saturated solution of magnesium hydroxide, $Mg(OH)_2$, has a magnesium ion concentration of 1.65×10^{-4} M at 25°C.

   (A) What is the value of the solubility product constant, K_{sp}, at 25°C?

   (B) What is the molar solubility of $Mg(OH)_2$ in 0.10 M $Mg(NO_3)_2$ solution at 25°C?

   (C) To 350. mL of a 0.150 M $Mg(NO_3)_2$ solution, 150. mL of a 0.500 M NaOH solution is added. What are the $[Mg^{2+}]$ and $[OH^-]$ in the resulting solution at 25°C?

   (D) If the temperature were raised to 50°C, what effect will this change have on solubility product constant, K_{sp}? Explain briefly.

GO ON TO THE NEXT PAGE. ➡

Answer EITHER question 2 (below) or question 3 (next page).

2. The data is obtained for the initial rate of the reaction:

$$2 NO (g) + O_2 (g) \rightarrow 2 NO_2 (g)$$

Data	[NO]	[O$_2$]	Initial rate; mol / L • s
Run 1	0.20 M	0.20 M	4.64×10^{-8}
Run 2	0.20	0.40	9.28×10^{-8}
Run 3	0.40	0.40	3.71×10^{-7}
Run 4	0.10	0.10	5.80×10^{-9}

(A) What is the rate law equation for the reaction?

(B) What is the value of the specific rate law constant, k, including units?

(C) This mechanism is suggested:

1. $NO (g) + O_2 (g) \rightleftharpoons NO_3 (g)$ *Equilibrium*
2. $NO_3(g) + NO (g) \rightarrow 2NO_2 (g)$ *Slow*

Show how the mechanism leads to the observed rate equation.

(D) The reaction rate is increased by a factor of times 5 when the temperature is increased from 1,400 K to 1500 K. What is the energy of activation for the reaction?

GO ON TO THE NEXT PAGE. ➡

3. Methane, CH_4, and chlorine, Cl_2, react to form carbon tetrachloride, CCl_4, and hydrogen chloride, HCl, at 25°C. The heat of the reaction, $\Delta H°$, is −429.8 kJ mol^{-1} of CH_4, reacted. The equation is:

$$CH_4 \text{ (g)} + 2Cl_2 \text{ (g)} \rightarrow CCl_4 \text{ (}\ell\text{)} + 4HCl \text{ (g)}$$

Substance	Heat of Formation, $\Delta H_f°$ kJ mol	Absolute Entropy, $S°$ J mol K
$C_{graphite}$ (s)	0	5.740
CH_4 (g)	−74.86	186.2
CCl_4 (ℓ)	?	216.4
Cl_2 (g)	0	223.0
HCl (g)	−92.31	186.8

(A) What is the standard heat of formation, $\Delta H_f°$, for carbon tetrachloride at 25°C?

(B) What is the standard entropy change, $\Delta S°$, for the reaction at 25°C?

(C) Theoretically, carbon tetrachloride is formed by the reaction:

$$C_{graphite} \text{ (s)} + 2Cl_2 \text{ (g)} \rightarrow CCl_4 \text{ (}\ell\text{)}$$

Calculate the standard free energy of formation, $\Delta G_f°$, for carbon tetrachloride, CCl_4.

(D) Calculate the value of the equilibrium constant, K, for the reaction in Part (C) at 25°C.

GO ON TO THE NEXT PAGE. ➡

SECTION II—PART B

Time—50 minutes

<u>Directions</u>: Write your answers in the space provided. Choose bewtween question 7 and 8.

You may NOT use a calculator for this section. You may use the Periodic Table, the "Equations and Constants" sheet, and the table of standard reduction potentials provided on pages 469–473 of this book.

4. Write the formulas to show the reactants and products for any FIVE of the laboratory situations described below. A reaction occurs in all cases. All solutions are aqueous unless otherwise indicated. Represent substances in solutions as ions if the substances are extensively ionized. Omit formulas for any ions or molecules that are unchanged by the reaction. You need not balance the equation.

(A) Sulfur trioxide gas is bubbled through a solution of sodium hydroxide.

(B) Solid zinc nitrate is treated with excess sodium hydroxide solution.

(C) A solution containing tin(II) ion is added to acidified potassium dichromate.

(D) Sodium dichromate is added to an acidified solution of sodium iodide.

GO ON TO THE NEXT PAGE. ➡

(E) Boron trifluoride is added to gaseous trimethylamine.

(F) Gaseous silane (SiH_4) is burned in excess oxygen.

(G) Chlorine gas is bubbled through a solution of potassium bromide.

(H) A suspension of copper(II) hydroxide in water is treated with an excess of ammonia.

5. A laboratory experiment is performed to verify the Laws of Multiple Proportions and Definite Proportions.

(A) Define:
   (i) The Law of Definite Composition

   (ii) The Law of Multiple Proportions

(B) (i) Fill in the table.

1.	Mass of stannic fluoride	1.9468 grams
2.	Mass of stannous fluoride	1.5669 grams
3.	Mass of tin	1.1869 grams
4.	Mass of fluorine in stannic fluoride	
5.	Mass of fluorine in stannous fluoride	
6.	Ratio of the fluorine mass in the two compounds	

(C) This weight and ratio data shows evidence of the Law of:
   A. Conservation of Matter
   B. Constant Composition
   C. Definite Composition
   D. Multiple Proportions

(D) Consider the data and complete the table for three compounds of iron and phosphorus.

		% Iron	% Phosphorus	$\dfrac{\text{g Iron}}{\text{g Phosphorus}}$	Formula
1.	Iron Phosphide	64.30	35.70		
2.	Iron Phosphide	78.28	21.71		
3.	Iron Phosphide	84.40	15.60		

(E) These percent data show evidence of the Law of:
   A. Conservation of Matter
   B. Constant Composition
   C. Definite Composition
   D. Multiple Proportions

GO ON TO THE NEXT PAGE. ➡

KAPLAN

6. Use principles of atomic structure and/or chemical bonding to explain the following. In each part your answer should include references to both substances.

(A) The atomic radius of Rb is larger than that of K.

(B) The boiling point of F_2 is lower than the boiling point of I_2.

(C) The carbon-to-carbon bond energy in CHCH is greater than it is in CH_3CH_3.

(D) The 2nd ionization energy of Na is greater than the 2nd ionization energy of Mg.

GO ON TO THE NEXT PAGE.

Answer EITHER question 7 (below) or question 8 (next page). Only one of these questions will be graded. If you start both questions, clearly indicate the question you do not want graded.

7. The questions refer to four 100 mL samples of aqueous solutions at 25°C in stoppered flasks.

   Flask 1: 0.10 M $CaCl_2$

   Flask 2: 0.10 M KNO_2

   Flask 3: 0.10 M CH_3COOH

   Flask 4: 0.10 M CH_3COCH_3

   (A) Which solution has the lowest electrical conductivity? Explain.

   (B) Which solution has the lowest freezing point? Explain.

   (C) Above which solution is the pressure of water vapor greatest? Explain.

   (D) Which solution has the highest pH? Explain.

GO ON TO THE NEXT PAGE. ➡

8. An electrochemical cell is assembled using the components shown in the representation.

$$\text{Cd (s) } | \text{ Cd}^{2+} \text{ (aq) (1.00) } || \text{ Ag}^+ \text{ (aq) (1.00 M) } | \text{ Ag (s)}$$

The salt bridge is filled with KNO_3 (aq).

(A) Identify the cathode of the cell and write the half-reaction that occurs there.

(B) Write the net ionic equation for the overall reaction that occurs as the cell operates and calculate the value of the standard cell potential, $E°$.

(C) How can the concentration of Cd^{2+} (aq) and the concentration of Ag^+ (aq) be changed in order to increase the cell potential, $E°$? Explain your answer.

(D) Will the value of the equilibrium constant, K, for this cell reaction be $K < 1$, $K = 1$, or $K > 1$? Explain how you have come to this conclusion.

STOP

Practice Test I Answers and Explanations

SECTION I ANSWERS AND EXPLANATIONS

1. A	16. E	31. D	46. C	61. E
2. A	17. D	32. C	47. D	62. B
3. E	18. E	33. C	48. C	63. A
4. C	19. C	34. E	49. A	64. C
5. E	20. D	35. D	50. B	65. E
6. B	21. E	36. A	51. D	66. A
7. D	22. C	37. B	52. E	67. A
8. A	23. C	38. D	53. E	68. E
9. D	24. C	39. E	54. A	69. B
10. D	25. B	40. D	55. A	70. E
11. E	26. E	41. D	56. C	71. A
12. E	27. A	42. D	57. D	72. A
13. B	28. B	43. A	58. C	73. E
14. A	29. E	44. A	59. C	74. D
15. D	30. B	45. C	60. B	75. C

1. A

The correct answer is: −9,230 kJ. The molar mass of sulfur is 32.06 g/mol. There are 31.19 moles in 1.00 kg of solid sulfur. Since 296 kJ of energy is released for each mole of sulfur reacted, a total of 9,230 kJ of energy is released. Since energy is released, the enthalpy change is negative.

2. A

The 250.0 mL sample has a mass of 197.3 g. The temperature change is 22.0°C. Using the equation $q = mc\Delta T$, the quantity of heat absorbed is calculated. Since this reaction is occurring at constant pressure, the heat absorbed equals the change in internal energy of the ethanol.

3. E

This is what you get when you divide the speed of light by the wavelength in meters:

$$v = \frac{c}{\lambda} = \frac{(3.00 \times 10^8 \text{ m / s})}{500 \text{ nm}} \times 1 \times 10^9 \frac{\text{nm}}{\text{m}} = 1 \times 10^{14} \text{ s}^{-1}$$

4. C

The high-energy end of the spectrum begins with very energetic and super high-frequency gamma rays that have more energy per photon than the more familiar X-rays.

5. E

This gets at the heart of quantum mechanics and its advancement from classical physics. It can't be known where an electron is and be known where it is going at the same time. It can only be given a quantum number address of where it is most likely to be found.

6. B

The "dumbbell" or hourglass shape is the closest description available for each of the three p orbitals.

7. D

The two orbitals beyond the nearest noble gas and in the outermost unfilled shell of an atom are the valence orbitals.

8. A

There are 6 valence electrons from the sulfur atom, 6 from each of four oxygen atoms, and two additional electrons due to the −2 charge.

9. D

Bond order is directly related to bond energy; as one increases, so does the other. Bond order is also inversely related to bond length; as one increases, the other decreases.

10. D

Geometric isomers have the same formula and the same bonds, but different arrangement of the atoms in space. Because of the different arrangement of the atoms (and thus electrons), geometric isomers have different physical properties such as melting and boiling points, color, and solubility.

11. E

This is an ideal gas law question in which you must account for the pressure due to a formed gas. Moles of CO_2 = 0.7500 g sucrose / 342 g sucrose $\cdot$ mol sucrose^{-1} $\times$ 12 mol CO_2 / mol sucrose = 0.0263.

$$P = \frac{nRT}{V} = \frac{(0.0263 \times 0.08206 \times 298.15)}{3.40\ L} = 0.190\ atm = 144\ mmHg.$$

12. E

The SO_3 has 24 bonding and nonbonding electrons, and the Lewis diagram shows a central S atom single-bonded to two O's and double-bonded to the third O. The number of nonbonding electrons is 24 − 8 = 16, and they are arranged so each of the oxygen atoms has an octet and S with its four bonds also has an octet. The three bonds are arranged in a planar triangle, which is a symmetrical shape and a nonpolar molecule, despite have polar bonds.

13. B

A Lewis electron diagram must be worked out to form octets for each of the atoms in the formula. The molecule C_2Cl_4 has the structure $CCl_2=CCl_2$. It consists of four single bonds that are sigma bonds, and one double bond. The first bond of the double bond is a σ, and second is a π bond. The answer is not carbon disulfide because CS_2 has two double bonds and hence two π bonds.

14. A

Lewis electron diagram shows that the molecule CS_2 has the structure S=C=S.

15. D

Lewis electron diagram shows that the molecule HCN has a single bond between the H and the C, and then a triple bond between the C and the N.

16. E

Ionic bonding generally occurs in substances consisting of two elements with one being a metal and the other a nonmetal.

17. D

The HCN is the only molecule in the group with polar bonds and a nonsymmetrical shape.

18. E

Ionic compounds have very high melting and boiling points and are generally solids at room temperature.

19. C

The diatomic elemental bromine has the same atoms bonding, resulting in an exactly zero electronegativity difference and a nonpolar bond.

20. D

Roman Numeral I is incorrect. To be an isotope, it must be the same element. This means it must have the same, not different atomic number. II is correct. Two isotopes of an element differ by their number of neutrons. III is a correct statement as well. The weighted average (by abundance) or the naturally occurring stable isotopes are added to compute the atomic weight.

21. E

The outer electronic structure s^2p^4 is characteristic of the Group 16 (VIA) elements, the oxygen family.

22. C

Oxygen has the molecular orbital structure $KK(\sigma_{2s})^2 (\sigma^*_{2s})^2 (\sigma_{2s})^2 (\sigma_{2p})^2 (\pi_{2p})^4 (\pi^*_{2px})^1 (\pi^*_{2py})^1$.

The bond order is: $\dfrac{[(\text{bonding electrons} - \text{anti-bonding electrons})]}{2} = \dfrac{[8-4]}{2} = 2$. With 2 unpaired electrons in the outermost orbitals, the bond order will be paramagnetic.

23. C

The molarity times the volume gives the moles of solute. If these moles are multiplied by the molar mass, the result will be grams of solute. Grams = $0.200 \text{ L} \times 0.100 \text{ mol} / \text{L} \times 294 \text{ g} / \text{mol} = 5.88$ g.

24. C

Raoult's Law: $P_2 = x_2 P_2^\circ$. Compute the vapor pressure using P_2° as the vapor pressure of the water and x_2 as the mole fraction of the *water*. First the mole fraction is computed from the molality of the sucrose:

$$X_1 = \frac{\text{moles of solute}}{\text{moles of solution}} = \frac{23.8 \text{ mol}}{23.8 + \dfrac{1,000 \text{ H}_2\text{O}}{18 \text{ g/ mol}}} = 0.300 \text{ and } X_2 = 1 - X_1 = 0.700$$

$P_2 = x_2 P_2^\circ = 0.700 \times (23.76 \text{ Torr}) = 16.6 \text{ Torr}$

25. B

This question tests your familiarity with laboratory procedures or classroom demonstrations. To produce chlorine gas, a mixture of hydrochloric acid and a $KMnO_4$ catalyst is heated. Chlorine is a highly irritating gas and must be handled carefully with the outside venting of a hood.

26. E

The kinetic molecular theory predicts that gas molecules take up very little space compared to the distance between the molecules. When pressure is applied to a gas, theory predicts the molecules move closer to each other, but that there is still plenty of distance between molecules.

27. A

A molecule with "a molar mass of many thousands" is poorly soluble in many solvents, may react with the solvent, and is highly sensitive to the application and removal of heat. Let the solvent move through the semipermeable membrane used in osmosis to give the measurement of molar mass.

28. B

$$K_{sp} = [Ba^{2+}][SO_4{}^{2-}] = 1.1 \times 10^{-10}$$

The solutions are each diluted to half their initial molarity when mixed.

$$Q = [Ba^{2+}][SO_4{}^{2-}] = [0.025][0.050] \cdot 1.3 \times 10^{-3}$$

Since $Q > K_{sp}$, the $BaSO_4$ precipitates, consuming much of the initial Ba^{2+} and $SO_4{}^{2-}$.

29. E

At equilibrium, $\Delta G = 0$ and $\Delta G = 0 = \Delta H - T\Delta S$. Rearrangement gives us $\Delta H = T\Delta S$.

30. B

^{85}Rb has an atomic number of 37, and therefore has $(85 - 37) = 48$ neutrons. Strontium, atomic number 38, will have a mass number of 86 if it has 48 neutrons, and this matches the answer. Kr has to have the mass number of 84 to hold 48 neutrons, and Rb-87 will have 50 neutrons.

31. D

This is what the titration curve looks like when a strong acid is titrated with a strong base. The only strong base listed among the answers is sodium hydroxide, NaOH.

32. C

A Brønsted–Lowry acid is a proton donor and a Brønsted–Lowry base is a proton acceptor. Answer choice (C) is the correct answer because H_2O can accept a proton, act as a base, and become H_3O^+. It can also donate a proton, act as an acid, and become OH^-.

33. C

In this question, search for an ionic compound that would consist of one metal and one nonmetal. Answer choice (C) fits this description and is an ionic substance.

34. E

The molar mass of $MgSO_4$ is 120. g/mol, and the molar mass of 7 H_2O is 126. g/mol. So the total molar mass of the hydrate is 246 g/mol. The percent water $= \left(\dfrac{126}{246 \times 100} \right) = 51.2\%$.

35. D

You can manipulate the ideal gas law equation ($PV = nRT$) to figure this problem out. Looking at answer choice (D): since n and R are constants, just look at $PV = T$. Rearrange the equations to get: $\dfrac{P}{T} = \dfrac{1}{V}$. When $\dfrac{P}{T}$ increases, $\dfrac{1}{V}$ should decrease. The graph shows the opposite relationship.

36. A

Formal charge = [(Group number) − (# of lone pair electrons) − (1/2 (# of bonding electrons)]

The structure of $COCl_2$ is a central carbon single-bonded to two chlorines and double-bonded to one oxygen.

Carbon's formal charge = [4 − 0 − 1/2 (8)] = 0
Chorine's formal charge = [7 − 6 − 1/2 (2)] = 0
Oxygen's formal charge = [6 − 4 − 1/2 (4)] = 0

37. B

The band of stability is an area on an isotope plot that contains all the "stable" nuclei. The axes for the plot are neutrons versus protons. As the atomic number for an atom rises, the ratio of neutrons to protons tends to increase. When an isotope falls to the left, or sits above the band of stability on the plot, it tends to be a beta emitter. When an isotope falls to the right or sits below the plot, it tends to be a positron emitter. All isotopes with atomic number greater than 83 are radioactive.

38. D

Because aluminum has a lower standard potential it will be oxidized, and the chlorine will be reduced.

Oxidation: Al (s) → Al^{3+} (aq) + 3e⁻ $E° = +1.66$ V
Reduction: Cl_2 (g) + 2e⁻ → 2Cl⁻ (aq) $E° = +1.36$ V

Voltage is the push on electrons and is not affected by the number of electrons. Do *not* multiply the voltages by 2 and 3, as the electrons are balanced. The voltage of a cell is calculated by adding the standard reduction potential of the oxidation reaction to the standard reduction potential of the reduction reaction. $E°$(cell) = 1.66 + (1.36) = +3.02 V.

39. E

Graphite is composed of layers of sheets made up of these carbon atoms. The individual carbon atoms of the graphite layer are held together by covalent bonds, and the sheets or layers are held together by dispersion forces, such as London forces. Because of these slippery layers, graphite is a lubricant.

40. D

This question tests an understanding of colligative properties and, more specifically, how to calculate the change in boiling point of a solution. The formula to use is: $\Delta T = K_f m$.

$$m = \left(\frac{\text{moles of solute)}}{\text{(kg of solvent}}\right) = \frac{1.00 \text{ mol}}{0.500 \text{ kg H}_2\text{O}} = 2.00 \text{ m}$$

$$\Delta T = K_f m = 1.86°\text{C / m (2.00 m)} = 3.72°\text{C}$$

Remember, the ΔT is a freezing point *depression*. New $FP = 0$ °C $- 3.72°\text{C} = -3.72°\text{C}$

41. D

There are five peripheral atoms around the central arsenic atom, so there must be five hybrid orbitals. To form five hybrid orbitals, five atomic orbitals must be combined. The sp^3d hybrid orbital is a combination of one s orbital, three p orbitals, and one d orbital.

42. D

The correct answer is: 147 mL. You solved correctly for the unknown volume, V_2, in expressions involving the ideal gas law: $P_1 V_1 = nRT_1$ and $P_2 V_2 = nRT_2$.

Rearrange the equations to solve for nR, and then they are set equal to each another: $\dfrac{P_1 V_1}{T_1} = nR = \dfrac{P_2 V_2}{T_2}$. Then solve for V_2: $V_2 = V_1 \times \dfrac{P_1}{P_2} \times \dfrac{T_1}{T_2}$.

43. A

Use $P = \dfrac{nRT}{V}$ to find the total pressure of the CO_2 and H_2O. Dalton's law of partial pressures is then applied. It's a classic lab technique to collect gases by bubbling them through water. The pressure is adjusted by subtracting the vapor pressure of the water at the collecting temperature to get the pressure of the "dry" gas.

44. A

Graham's Law predicts that the molar mass, μ, gets larger at a constant temperature, and the molecular velocity, v, decreases.

45. C

The square root of the molar mass ratio of nitrogen to helium, $\sqrt{\dfrac{N_2}{\text{He}}} = \sqrt{\dfrac{28}{4}} = 2.65$ is proportional to the ratio of the time it takes the gases to effuse.

46. C

Adding a product (i.e., NH_3 (g)) shifts the equilibrium towards the reactants, NO_2 (g) and H_2O (g), so their partial pressures increase. The partial pressure of O_2 (g) decreases because it reacts with the additional NH_3 (g).

47. D

A change in the pressure does not affect this equilibrium because there is the same number of moles on either side of the reaction. For all the other reactions, there are different numbers of moles on the two sides of the reaction, so a change in pressure shifts the equilibrium.

48. C

When 4.17% of the 0.30 mol of acetic acid dissociates to release hydrogen ions in the solution, there is 0.0125 mol of H^+ (aq) in the solution. This gives a pH of 1.60.

49. A

The correct answer is: activation energy. When the Arrhenius equation is algebraically arranged in the form of a straight line, it has the formula $\ln(k) = \left(\dfrac{-E_a}{R}\right)\left(\dfrac{1}{T}\right) + \ln(A)$. The slope of the line is $\dfrac{-E_a}{R}$, and is used to determine the activation energy, E_a.

50. B

The first law of thermodynamics states that the total energy of a system and its surroundings is constant. The first law of thermodynamics doesn't say anything about either the direction or the entropy of a process.

51. D

Solubility product is an equilibrium constant, K_{sp}, and insoluble salts produce very few product ions in solution, making their K_{sp} have values less than 1. The K_{sp} will change in value when the temperature is changed, and for salts such as $PbBr_2$ the units of K_{sp} is mol^3 / L^3.

52. E

The problem is solved using the Henderson-Hasselbalch equation:

$$pH = pK_a + \log \frac{[A^-]}{[HA]} = pK_a + \log \frac{[base]}{[acid]}$$

$$pH = pK_a + \log \frac{[acetate]}{[acetic\ acid]} = -\log(1.8 \times 10^{-5}) + \log\frac{4.00}{1.00}$$

53. E

Only neon is an elemental gas in its standard state, the requirements for the heat of formation. Bromine is a liquid; carbon is a solid; CO is a compound; and nitrogen in its standard state is diatomic.

54. A

By definition, bronze is an alloy made by combining copper and tin. It is much more expensive than brass, where the relatively less expensive zinc is used instead of expensive tin.

55. A

A catalyst is both consumed and regenerated in a chemical reaction. The overall reaction is: $B + O_2 \rightarrow BO_2$. In Step 1, A, the catalyst, reacts with oxygen, but is regenerated as a product in Step 2. AO_2 is an intermediate, B and O_2 are consumed, and BO_2 is a product.

56. C

The halides of Ag^+, Pb^{2+}, Hg_2^{2+}, and Cu^+ are insoluble, except that $PbCl_2$ is slightly soluble and HgI_2 is insoluble.

57. D

By definition, ionization energy is the energy needed to remove the most loosely bound electron from a gaseous atom in its ground state.

58. C

$K_w = [H^+][OH^-] = 1 \times 10^{-14}$. $[H^+] = 1 \times 10^{-14} / 1 \times 10^{-3} = 1 \times 10^{-11}$ M

59. C

This question tests the knowledge of hybridization orbitals.

Hybrid orbital	Geometry
sp	linear
sp^2	trigonal planar
sp^3	tetrahedral
sp^3d	trigonal bipyramidal
sp^3d^2	octahedral

The required structure has a trigonal planar geometry. The Lewis structure of SO_2 consists of two single bonds to the two oxygens, and a lone pair of electrons in a trigonal planar geometry. CS_2 is linear, Cl_2O and H_2S are bent, and NH_3 is trigonal pyramidal.

60. B

It is essential to write a balanced equation for this problem.

$4 Al + 3 O_2 \rightarrow 2 Al_2O_3$

Given equal moles of Al and O_2, inspection of the coefficients of Al and O_2 indicates the Al is the limiting reagent. Four moles of Al produce 2 moles of Al_2O_3 making the mole ratio is 1/2. Moles of Al_2O_3 = 1/2 moles of Al = $1/2 \times 0.36 = 0.18$ moles.

61. E

The α particle experiment permitted Rutherford to conclude all of the positive charge is concentrated in $1 / 100,000$ of the volume of the atom. Although answers A–C are correct statements, massive α particles have no trouble passing through a field of comparatively mass-less electrons.

62. B

The reaction is $Ni\ (s) + Cl_2\ (g) \rightarrow Ni^{2+} + 2\ Cl^-\ (aq)$

$$Q = \frac{[Ni^{2+}][Cl^-]^2}{[Cl_2]} = \frac{[0.10][0.40]^2}{[0.50]} = 0.032$$

63. A

Many of the unique physical characteristics of water are due to the presence of hydrogen bonds. When in doubt, remember this fact when asked why water has a given characteristic.

64. C

Osmotic pressure $= MRT = (0.00100)(0.0821)(298) = 0.0245$ atm $= 18.6$ mmHg

65. E

$BaCl_2$ reacts with all four, producing in I $AgCl$, in II $BaCO_3$, in III $BaSO_4$, and in IV $PbCl_2$.

The appropriate solubility rules are:

(1) Chlorides are generally soluble except for those of Ag^+, Pb^{2+}, Hg_2^{2+}, and Cu^+.

(2) Sulfates are generally soluble except for those of Pb^{2+}, Ba^{2+}, and Sr^{2+}.

(3) Carbonates are generally insoluble.

66. A

If adding heat shifts the equilibrium to the products, according to Le Châtelier's principle, the reaction must be endothermic because the heat is acting as a reactant.

67. A

The moles of HCl remain the same so the molarity $\times$ volume before and after the addition of water is the same. $M_f V_f = M_i V_i$. The final molarity $= 60 \times \dfrac{0.50}{150} = 0.20$ M

68. E

By definition, a structural isomer of a compound is alternate arrangement of the atoms with the same molecular formula. The initial compound has the formula $C_4H_{10}O$. Only (E) has the same molecular formula. Others are short carbons or have added elements not in the original.

69. B

Activation energy, E^*, is calculated from rate constant data at two temperatures by using the

Arrhenius equation: In $\dfrac{k_2}{k_1} = -\dfrac{E_a}{R}\left[\dfrac{1}{T_2} - \dfrac{1}{T_1}\right]$. Substituting the data gives:

In $\dfrac{2.1 \times 10^{-3}}{6.2 \times 10^{-4}} = \dfrac{E_a}{8.314 \text{ J/mol}}\left[\dfrac{1}{328} - \dfrac{1}{318}\right]$ and $E_a = 1.1 \times 10^5$ J/mol = 110 kJ/mo

70. E

The heating of chlorates such as potassium chlorate in the presence of an MnO_2 catalyst generates only oxygen as a gas, leaving behind KCl solid. $2\ KClO_3\ (s) \rightarrow 2\ KCl\ (s) + 3\ O_2\ (g)$.

71. A

If the dipole moment is equal to zero, the compound is completely nonpolar. Answer choice (A), CH_4, is a tetrahedral molecule in which the individual dipole moments.

72. A

In this half-reaction the electrons are being lost by the oxygen which represents an oxidation half-reaction. Oxidation reactions always take place at the anode (think "An Ox and a Red Cat").

73. E

Positron emission transforms a proton to a neutron in the nucleus and a positron that is emitted, leading to the formation of an atom with the next lower atomic number, oxygen.

74. D

The superscripts must add up to 24 in the correct answer, and the sublevels must be filled in the proper order. But this question is a bit tricky because the element with the atomic number 24 is chromium and this element is an exception to the rules of electron configuration. Normally, the $4s$ sublevel fills completely before the $3d$ sublevel receives any electrons. This would make answer choice (E) look pretty good. However, answer choice (D) is correct because in the case of chromium, an electron is distributed to each group within the $3d$ sublevel before filling the $4s$ sublevel to make its valence configuration $4s^1 3d^5$.

75. C

Based on the information in the question, if A is doubled with B held constant, the rate doubles. This means that A is a first order reactant. If B is doubled with A held constant, the rate quadruples. This means that B is a second order reactant. As a result, the rate law for this reaction would be: $k[A][B]^2$.

SECTION II—PART A ANSWERS AND EXPLANATIONS

1. A saturated solution of magnesium hydroxide, $Mg(OH)_2$, has a magnesium ion concentration of 1.65×10^{-4} M at 25°C.

   (A) What is the value of the solubility product constant, K_{sp}, at 25°C?

   ANSWER:

   $$[Mg^{2+}] = 1.65 \times 10^{-4} \text{ M}$$
   $$[OH^-] = 2 \times [Mg^{2+}] = 3.30 \times 10^{-4} \text{ M}$$
   $$K_{sp} = [Mg^{2+}][OH^-]^2 = [1.65 \times 10^{-4}][3.30 \times 10^{-4}]^2$$
   $$K_{sp} = 1.80 \times 10^{-11}$$
   $$\boldsymbol{K_{sp} = 1.80 \times 10^{-11}}$$

   (B) What is the molar solubility of $Mg(OH)_2$ in 0.10 M $Mg(NO_3)_2$ solution at 25°C.

   ANSWER:
   Let x equal the molar solubility of $Mg(OH)_2$.

	$Mg(OH)_2$ (s) $\rightleftharpoons$	Mg^{2+} (aq) +	$2OH^-$ (aq)
Start	0	0.10 M	0 M
Δ	$-x$	$+x$	$+2x$
Finish	Some $-x$	0.10	$2x$

   Note: x is small compared to 0.10, and is ignored.

   $$K_{sp} = [0.10][2x]^2 = 1.80 \times 10^{-11}$$

   $$x = \text{molar solubility} = 6.71 \times 10^{-6} \text{ M}$$

   $$\boldsymbol{x = \text{molar solubility} = 6.71 \times 10^{-6} \text{ M}}$$

   (C) To 350. mL of a 0.150 M $Mg(NO_3)_2$ solution, 150. mL of a 0.500 M NaOH solution is added. What are the $[Mg^{2+}]$ and $[OH^-]$ in the resulting solution at 25°C?

   ANSWER:

   $$[Mg^{2+}] = \frac{350. \text{ mL} \times 0.150 \text{ mol L}^{-1}}{500. \text{ mL}} = 0.105 \text{ M}$$

   $$[OH^-] = \frac{150. \text{ mL} \times 0.500 \text{ mol L}^{-1}}{500. \text{ mL}} = 0.150 \text{ M}$$

   Let x = amount of OH^- formed as a new equilibrium is established ("bounce back").

PRACTICE TEST I ANSWERS AND EXPLANATIONS

	$Mg(OH)_2(s)$ $\rightleftharpoons$	$Mg^{2+}(aq)$ +	$2OH^-(aq)$
Start	0	0.105 M	0.150 M
Δ_1	+0.075	−0.075	−0.150
Precipitate	Some	0.030	0
Δ_2	$-x/2$	$+x/2$	$+x$
Finish	Some $- x/2$	0.030	x

Note: $x/2$ is small compared to 0.030, and is ignored.

$$K_{sp} = [Mg^{2+}][OH^-]^2 = [0.030][x]^2 = 1.80 \times 10^{-11}$$
$$[Mg^{2+}] = 0.030 \text{ M}$$
$$[OH^-] = x = 2.45 \times 10^{-5} \text{ M}$$

(D) If the temperature were raised to 50°C, what effect will this change have on the solubility product constant, K_{sp}? Explain briefly.

ANSWER:
The solubility product constant will increase because the solubility of solids in liquids generally increases with increasing temperature.

2. The data is obtained for the initial rate of the reaction:

$$2 \text{ NO (g)} + O_2 \text{ (g)} \rightarrow 2 \text{ NO}_2(g)$$

Data	[NO]	[O_2]	Initial rate; mol / L • s
Run 1	0.20 M	0.20 M	4.64×10^{-8}
Run 2	0.20	0.40	9.28×10^{-8}
Run 3	0.40	0.40	3.71×10^{-7}
Run 4	0.10	0.10	5.80×10^{-9}

(A) What is the rate law equation for the reaction?

ANSWER:

$$\frac{\text{Run 2}}{\text{Run 1}} = \frac{9.28 \times 10^{-8}}{4.64 \times 10^{-8}} = \left[\frac{[0.20]}{[0.20]}\right]^m \times \left[\frac{[0.40]}{[0.20]}\right]^n$$

$$2 = [2]^n \text{ and } n = 1$$

$$\frac{\text{Run 3}}{\text{Run 2}} = \frac{3.71 \times 10^{-7}}{9.28 \times 10^{-8}} = \left[\frac{[0.40]}{[0.20]}\right]^m \times \left[\frac{[0.40]}{[0.40]}\right]^n$$

$$4 = [2]^m \text{ and } m = 2$$
$$\textbf{Rate} = \textbf{k[NO]}^2\textbf{[O}_2\textbf{]}$$

(B) What is the value of the specific rate law constant, k, including units?

ANSWER:
From Run 4 (or any convenient run):
Rate = 5.80×10^{-9} mol / L • s = k [0.20 M]^2 [0.10 M]
k = 5.8×10^{-6} 1/ M^2•s

(C) This mechanism is suggested:

1. $NO\ (g) + O_2\ (g) \rightleftharpoons NO_3\ (g)$ *Equilibrium*
2. $NO_3\ (g) + NO\ (g) \rightarrow 2NO_2\ (g)$ *Slow*

Show how the mechanism leads to the observed rate equation.

ANSWER:

$$\text{Rate} = k_2[NO_3][NO]$$

$$K_1 = \frac{[NO_3]}{[NO][O_2]}$$

$$[NO_3] = K_1\ [NO][O_2]$$

$$\textbf{Rate} = \boldsymbol{K_1 k_2 [NO]^2 [O_2] = k[NO]^2 [O_2]}$$

(D) The reaction rate is increased by a factor of times 5 when the temperature is increased from 1400 K to 1500 K. What is the energy of activation for the reaction?

ANSWER:

$$\ln \frac{k_2}{k_1} = -\frac{E_a}{R}\left[\frac{1}{T_2} - \frac{1}{T_1}\right]$$

$$\ln \frac{5}{1} = -\frac{E_a}{8.315 \text{ J/mol•K}}\left[\frac{1}{1{,}500} - \frac{1}{1{,}400}\right]$$

$$E_a = 280{,}000 \text{ J} = \textbf{280 kJ}$$

3. Methane, CH_4, and chlorine, Cl_2, react to form carbon tetrachloride, CCl_4, and hydrogen chloride, HCl, at 25°C. The heat of the reaction, $\Delta H°$, is –429.8 kJ / mol of CH_4, reacted. The equation is:

$$CH_4\ (g) + 2Cl_2\ (g) \rightarrow CCl_4\ (\ell) + 4HCl\ (g)$$

Substance	Heat of Formation, ΔH_f° kJ / mol	Absolute Entropy, S° J / mol $\cdot$ K
$C_{graphite}$ (s)	0	5.740
CH_4 (g)	−74.86	186.2
CCl_4 (ℓ)	?	216.4
Cl_2 (g)	0	223.0
HCl (g)	−92.31	186.8

(A) What is the standard heat of formation, ΔH_f°, for carbon tetrachloride at 25°C?

ANSWER:
$\Delta H^\circ = \Sigma \Delta H_f^\circ \text{ (Products)} - \Sigma \Delta H_f^\circ \text{ (Reactants)}$

$-429.8 \text{ kJ} = (\Delta H_f^\circ + (4 \times -92.31 \text{ kJ})) - (-74.86 + 0)$

$\Delta H_f^\circ = -135.4 \text{ kJ / mol}$

(B) What is the standard entropy change, ΔS°, for the reaction at 25°C?

ANSWER:
$\Delta S^\circ = \Sigma S_f^\circ \text{ (Products)} - \Sigma S_f^\circ \text{ (Reactants)}$

$\Delta S^\circ = (216.4 + (4 \times 186.8)) - (186.2 + (2 \times 223.0))$

$\Delta S^\circ = +331.4 \text{ J / mol} \cdot \text{K}$

(C) Theoretically, carbon tetrachloride is formed by the reaction:

$$C_{graphite} \text{ (s)} + 2Cl_2 \text{ (g)} \rightarrow CCl_4 \text{ (ℓ)}$$

Calculate the standard free energy of formation, ΔG_f°, for carbon tetrachloride, CCl_4.

ANSWER:
$$\Delta G^\circ = \Delta H^\circ - \frac{T\Delta S^\circ}{1000}$$

$$\Delta H^\circ = \Delta H_f^\circ(CCl_4) = -135.4 \text{ kJ / mol (from part (A))}$$
$$\Delta S^\circ = \Sigma S_f^\circ \text{ (Products)} - \Sigma S_f^\circ \text{ (Reactants)}$$
$$\Delta S^\circ = ((216.4) - (5.74 + (2 \times 223.0)))$$
$$\Delta S^\circ = -235.3 \text{ J / mol} \cdot \text{K}$$

$$\Delta G^\circ = -135.4 - \frac{((298 \text{ K})(-235.3 \text{ J/mol} \cdot \text{K}))}{1000}$$

$$\Delta G_f^\circ = -65.3 \text{ kJ / mol}$$

The $\Delta G°$ of the reaction is the $\Delta G_f°$ of CCl_4. Both of the reactants are elements with a $\Delta G_f°$ of zero.

(D) Calculate the value of the equilibrium constant, K, for the reaction in Part (C) at 25°C.

ANSWER:

$$\Delta G° = -RT \ln K$$

$$\ln K = -\frac{-65,300 \text{ J/mol}}{(8.314 \text{ J/mol·K})(298 \text{ K})} = 26.3$$

$$K = e^{26.3} \text{ (“inverse” } \ln 26.3) = 3 \times 10^{11}$$

SECTION II—PART B ANSWERS AND EXPLANATIONS

4. (A) Sulfur trioxide gas is bubbled through a solution of sodium hydroxide.

   ANSWER:
   $SO_3 + OH^- \rightarrow HSO_4^-$

   (B) Solid zinc nitrate is treated with excess sodium hydroxide solution.

   ANSWER:
   $Zn(NO_3)_2 + OH^- \rightarrow Zn(OH)_4^{2-} + NO_3^-$

   (C) A solution containing tin(II) ion is added to acidified potassium dichromate.

   ANSWER:
   $Sn^{2+} + H^+ + Cr_2O_7^{2-} \rightarrow Sn^{4+} + Cr^{3+} + H_2O$

   (D) Sodium dichromate is added to an acidified solution of sodium iodide.

   ANSWER:
   $Cr_2O_7^{2-} + H^+ + I^- \rightarrow Cr^{3+} + I_2 + H_2O$

   (E) Boron trifluoride is added to gaseous trimethylamine.

   ANSWER:
   $BF_3 + (CH_3)_3N \rightarrow F_3BN(CH_3)_3$

   (F) Gaseous silane (SiH_4) is burned in excess oxygen.

   ANSWER:
   $SiH_4 + O_2 \rightarrow SiO_2 + H_2O$

(G) Chlorine gas is bubbled through a solution of potassium bromide.

ANSWER:
$Cl_2 + Br^- \rightarrow Cl^- + Br_2$

(H) A suspension of copper(II) hydroxide in water is treated with an excess of ammonia.

ANSWER:
$Cu(OH)_2 \text{ (s)} + NH_3 \text{ (aq)} \rightarrow Cu(NH_3)_4^{2+} + OH^-$

5. A laboratory experiment is performed to verify the Laws of Multiple Proportions and Definite Proportions.

(A) Define:
(i) The Law of Definite Composition

ANSWER:
This law states that the relative masses of the elements in a given compound are fixed. The Law of Definite Composition is the same as the Law of Constant Composition.

(ii) The Law of Multiple Proportions

ANSWER:
This law states that when two elements, A and B, form two compounds, the relative amounts of B that combine with a fixed amount of A are in a ration of small integers. Example: In H_2O and H_2O_2 there are 8 g and 16 g, respectively, of oxygen per gram of hydrogen.

(B) (i) Fill in the table.

1.	Mass of stannic fluoride	1.9468 g
2.	Mass of stannous fluoride	1.5669 g
3.	Mass of tin	1.1869 g
4.	Mass of fluorine in stannic fluoride	**0.7599 g**
5.	Mass of fluorine in stannous fluoride	**0.3800 g**
6.	Ratio of the fluorine mass in the two compounds	**2.000**

(C) This weight and ratio data shows evidence of the Law of:
A. Conservation of Matter
B. Constant Composition
C. Definite Composition
D. Multiple Proportions

ANSWER: D

(D) Consider the data and complete the table for three compounds of iron and phosphorus.

		% Iron	% Phosphorus	$\dfrac{\text{g Iron}}{\text{g Phosphorus}}$	Empirical Formula
1.	Iron Phosphide	64.30	35.70	1.833	FeP
2.	Iron Phosphide	78.28	21.71	3.606	Fe_2P
3.	Iron Phosphide	84.40	15.60	5.410	Fe_3P

(E) These percent data show evidence of the Law of:
A. Conservation of Matter
B. Constant Composition
C. Definite Composition
D. Multiple Proportions

ANSWER: D

6. Use principles of atomic structure and/or chemical bonding to explain each statement. In each part, the answer will include references to both substances.

(A) The atomic radius of Rb is larger than that of K.

ANSWER:
K has 4 shells and Rb has 5 shells. Although there are more protons in the nucleus of Rb, it has more shielding of the valence electrons from the pull of the nuclear protons by the intervening shells, called the "shielding effect."

(B) The boiling point of F_2 is lower than the boiling point of I_2.

ANSWER:
Both F_2 and I_2 are nonpolar, and the only intermolecular attractive forces are London dispersion forces. London dispersion forces are momentary, instantaneous dipoles caused by an unequal electron distribution around the molecule. Since I_2 has more electrons than F_2, the valence electrons in I_2 are more probable to be in an uneven distribution. The more valence electrons a molecule has, the greater is the dispersion forces and the higher is the boiling point.

(C) The carbon-to-carbon bond energy in CHCH is greater than it is in CH_3CH_3.

ANSWER:
CHCH (acetylene) has a triple bond between the two carbon atoms, where CH_3CH_3 (ethene) has a single bond between the carbon atoms. More energy is required to break the triple bond in acetylene than to break the single bond in ethene; so the carbon-to-carbon bond energy in acetylene is greater.

(D) The 2nd ionization energy of Na is greater than the 2nd ionization energy of Mg.

ANSWER:
The second electron removed from a sodium atom comes from a stable electron structure similar to neon, which is in the 2nd energy level. The second electron removed from a magnesium atom comes from an unstable electron structure similar to sodium, which is in the 3rd energy level. The electrons in the 2nd level are closer to the nucleus, so the attraction is much greater than for electrons in the 3rd level.

7. The questions refer to four 100 mL samples of aqueous solutions at 25°C in stoppered flasks.

Flask 1: 0.10 M $CaCl_2$

Flask 2: 0.10 M KNO_2

Flask 3: 0.10 M CH_3COCH_3

Flask 4: 0.10 M CH_3COOH

(A) Which solution has the lowest electrical conductivity? Explain.

ANSWER:
Flask 4: 0.10 M CH_3COCH_3

Acetone is a nonelectrolyte, and does not break up or dissociate into ions in an aqueous solution.

(B) Which solution has the lowest freezing point? Explain.

ANSWER:
Flask 1: 0.10 M $CaCl_2$

The freezing-point depression is proportional to the concentration of solute particles. Since all of the solutes are at the same concentration, but the van't Hoff factor (i) is 3 for $CaCl_2$, 2 for KNO_2, 1+ for CH_3COOH, and 1 for CH_3COCH_3.

(C) Above which solution is the pressure of water vapor greatest? Explain.

ANSWER:
Flask 4: 0.10 M CH_3COCH_3

Solute particles in solution will lower the vapor pressure. This vapor pressure lowering of water is directly proportional to the concentration of solute particles in solution, and the highest vapor pressure will be the solution with the *least* number of solute particles. CH_3COCH_3 is the only nonelectrolyte, and has a van't Hoff factor of 1, resulting in the *fewest* solute molecules.

(D) Which solution has the highest pH? Explain.

ANSWER:
Flask 2: 0.10 M KNO_2

The NO_2^- ion liberated by the dissociation of KNO_2 in aqueous solution is a weak base. It is the only solution of the four with pH > 7.

8. An electrochemical cell is assembled using the components shown in the representation.

$$Cd\ (s)\ |\ Cd^{2+}\ (aq)\ (1.00)\ ||\ Ag^+\ (aq)\ (1.00\ M)\ |\ Ag\ (s)$$

The salt bridge is filled with KNO_3 (aq).

(A) Identify the cathode of the cell and write the half-reaction that occurs there.

ANSWER:
The cathode is the electrode in the half-cell where reduction takes place. The Ag^+ (aq) with a reduction potential of +0.80 V is more easily reduced than the Cd^{2+} (aq) with a reduction potential of −1.21 V. Hint: The convention for writing cell representations is to put the anode on the left and the cathode on the right. Beware, however, of "tricky" questions where the cells are reversed—use the half-cell potentials to check. The more active metal is the anode, the less active is the cathode. Cd is more active than Ag. So **Ag** is the correct answer.

(B) Write the net ionic equation for the overall reaction that occurs as the cell operates and calculate the value of the standard cell potential, $E°$.

ANSWER:

Cathode: Ag^+ (aq) + e^- → Ag (s) $\qquad\qquad\qquad E° = +0.80\ V$

Anode: Cd (s) → Cd^{2+} (aq) + 2 e^- $\qquad\qquad E° = +1.21\ V$

Cd (s) + 2 Ag^+ (aq) → Cd^{2+} (aq) + 2 Ag (s) $\qquad E° = +2.01\ V$

(C) How can the concentration of Cd^{2+} (aq) and the concentration of Ag^+ (aq) be changed in order to increase the cell potential, E_{cell}? Explain your answer.

ANSWER:
The [Cd^{2+} (aq)] should be decreased, and the [Ag^+ (aq)] should be increased in order to increase the $E°$ of the cell.

In the Nernst Equation, $E_{cell} = E° - \dfrac{RT}{nF}$ ln Q, the value of $Q = \dfrac{[\text{Products}]}{[\text{Reactants}]}$.

Decreasing the [Cd^{2+}] and increasing the [Ag^+] makes the value of $Q < 1$. The ln Q

is then a negative number, which makes the "ln" term additive to $E°$, making E_{cell}

larger.

(D) Will the value of the equilibrium constant, K for this cell reaction be $K < 1$, $K = 1$, or $K > 1$. Explain how you have come to this conclusion.

ANSWER:
The cell potential for this reaction is positive (+2.01 V), which shows the cell reaction is spontaneous. The ΔG can be calculated from the voltage and when the reaction is spontaneous, the ΔG is a negative number. The equation relating ΔG with the equilibrium constant, K, is $\Delta G° = -RT$ ln K. With a negative ΔG, the ln K term is positive leading to an equilibrium constant which is greater than 1 ($K > 1$).

Practice Test Two

On the following pages you'll find a complete practice examination in a format similar to that found on the AP Chemistry exam.

Section I consists of 75 multiple-choice questions and constitutes 45% of your final grade. Only a "bare bones" Periodic Table may be used; you may *not* use a calculator for this section. Use the answer sheet provided to record your multiple-choice answers. The time allotted for this section is 90 minutes.

Section II consists of eight free-response questions, each of which will have multiple parts. Section II makes up 55% of your total score. In addition to the Periodic Table, you may use the table of standard reduction potentials and the list of AP Chemistry equations and constants provided at the beginning of this section. Use the space provided to record your free-response answers. (On the actual AP Chemistry exam, you will be asked to record your answers to Section II on lined pages following each question.)

Be sure to pay careful attention to the instructions for Section II:

- You are allotted 40 minutes to complete Part A. You may use a calculator on this part only.

- There are three questions in Part A. You must answer Question 1, and you will be asked to choose between Question 2 and Question 3.

- On Part B, calculators are *not* permitted. You are allotted 50 minutes for this part.

- There are five questions in Part B. You must answer Questions 5 and 6, and choose between Questions 7 and 8. Question 4 has explicit instructions to write five of eight net reactions.

At the end of the practice test you will find an answer key, along with a discussion of the answers to each multiple-choice and free-response question. Scoring information is located on page 564.

Good luck!

PRACTICE TEST TWO ANSWER SHEET

1 Ⓐ Ⓑ Ⓒ Ⓓ Ⓔ   20 Ⓐ Ⓑ Ⓒ Ⓓ Ⓔ   39 Ⓐ Ⓑ Ⓒ Ⓓ Ⓔ   58 Ⓐ Ⓑ Ⓒ Ⓓ Ⓔ

2 Ⓐ Ⓑ Ⓒ Ⓓ Ⓔ   21 Ⓐ Ⓑ Ⓒ Ⓓ Ⓔ   40 Ⓐ Ⓑ Ⓒ Ⓓ Ⓔ   59 Ⓐ Ⓑ Ⓒ Ⓓ Ⓔ

3 Ⓐ Ⓑ Ⓒ Ⓓ Ⓔ   22 Ⓐ Ⓑ Ⓒ Ⓓ Ⓔ   41 Ⓐ Ⓑ Ⓒ Ⓓ Ⓔ   60 Ⓐ Ⓑ Ⓒ Ⓓ Ⓔ

4 Ⓐ Ⓑ Ⓒ Ⓓ Ⓔ   23 Ⓐ Ⓑ Ⓒ Ⓓ Ⓔ   42 Ⓐ Ⓑ Ⓒ Ⓓ Ⓔ   61 Ⓐ Ⓑ Ⓒ Ⓓ Ⓔ

5 Ⓐ Ⓑ Ⓒ Ⓓ Ⓔ   24 Ⓐ Ⓑ Ⓒ Ⓓ Ⓔ   43 Ⓐ Ⓑ Ⓒ Ⓓ Ⓔ   62 Ⓐ Ⓑ Ⓒ Ⓓ Ⓔ

6 Ⓐ Ⓑ Ⓒ Ⓓ Ⓔ   25 Ⓐ Ⓑ Ⓒ Ⓓ Ⓔ   44 Ⓐ Ⓑ Ⓒ Ⓓ Ⓔ   63 Ⓐ Ⓑ Ⓒ Ⓓ Ⓔ

7 Ⓐ Ⓑ Ⓒ Ⓓ Ⓔ   26 Ⓐ Ⓑ Ⓒ Ⓓ Ⓔ   45 Ⓐ Ⓑ Ⓒ Ⓓ Ⓔ   64 Ⓐ Ⓑ Ⓒ Ⓓ Ⓔ

8 Ⓐ Ⓑ Ⓒ Ⓓ Ⓔ   27 Ⓐ Ⓑ Ⓒ Ⓓ Ⓔ   46 Ⓐ Ⓑ Ⓒ Ⓓ Ⓔ   65 Ⓐ Ⓑ Ⓒ Ⓓ Ⓔ

9 Ⓐ Ⓑ Ⓒ Ⓓ Ⓔ   28 Ⓐ Ⓑ Ⓒ Ⓓ Ⓔ   47 Ⓐ Ⓑ Ⓒ Ⓓ Ⓔ   66 Ⓐ Ⓑ Ⓒ Ⓓ Ⓔ

10 Ⓐ Ⓑ Ⓒ Ⓓ Ⓔ   29 Ⓐ Ⓑ Ⓒ Ⓓ Ⓔ   48 Ⓐ Ⓑ Ⓒ Ⓓ Ⓔ   67 Ⓐ Ⓑ Ⓒ Ⓓ Ⓔ

11 Ⓐ Ⓑ Ⓒ Ⓓ Ⓔ   30 Ⓐ Ⓑ Ⓒ Ⓓ Ⓔ   49 Ⓐ Ⓑ Ⓒ Ⓓ Ⓔ   68 Ⓐ Ⓑ Ⓒ Ⓓ Ⓔ

12 Ⓐ Ⓑ Ⓒ Ⓓ Ⓔ   31 Ⓐ Ⓑ Ⓒ Ⓓ Ⓔ   50 Ⓐ Ⓑ Ⓒ Ⓓ Ⓔ   69 Ⓐ Ⓑ Ⓒ Ⓓ Ⓔ

13 Ⓐ Ⓑ Ⓒ Ⓓ Ⓔ   32 Ⓐ Ⓑ Ⓒ Ⓓ Ⓔ   51 Ⓐ Ⓑ Ⓒ Ⓓ Ⓔ   70 Ⓐ Ⓑ Ⓒ Ⓓ Ⓔ

14 Ⓐ Ⓑ Ⓒ Ⓓ Ⓔ   33 Ⓐ Ⓑ Ⓒ Ⓓ Ⓔ   52 Ⓐ Ⓑ Ⓒ Ⓓ Ⓔ   71 Ⓐ Ⓑ Ⓒ Ⓓ Ⓔ

15 Ⓐ Ⓑ Ⓒ Ⓓ Ⓔ   34 Ⓐ Ⓑ Ⓒ Ⓓ Ⓔ   53 Ⓐ Ⓑ Ⓒ Ⓓ Ⓔ   72 Ⓐ Ⓑ Ⓒ Ⓓ Ⓔ

16 Ⓐ Ⓑ Ⓒ Ⓓ Ⓔ   35 Ⓐ Ⓑ Ⓒ Ⓓ Ⓔ   54 Ⓐ Ⓑ Ⓒ Ⓓ Ⓔ   73 Ⓐ Ⓑ Ⓒ Ⓓ Ⓔ

17 Ⓐ Ⓑ Ⓒ Ⓓ Ⓔ   36 Ⓐ Ⓑ Ⓒ Ⓓ Ⓔ   55 Ⓐ Ⓑ Ⓒ Ⓓ Ⓔ   74 Ⓐ Ⓑ Ⓒ Ⓓ Ⓔ

18 Ⓐ Ⓑ Ⓒ Ⓓ Ⓔ   37 Ⓐ Ⓑ Ⓒ Ⓓ Ⓔ   56 Ⓐ Ⓑ Ⓒ Ⓓ Ⓔ   75 Ⓐ Ⓑ Ⓒ Ⓓ Ⓔ

19 Ⓐ Ⓑ Ⓒ Ⓓ Ⓔ   38 Ⓐ Ⓑ Ⓒ Ⓓ Ⓔ   57 Ⓐ Ⓑ Ⓒ Ⓓ Ⓔ

SECTION I

Time—90 minutes

Directions: Select the best answer choice and fill in the corresponding oval on the answer sheet. **You may NOT use a calculator for this section. You may use the Periodic Table provided on page 469 of this book.**

1. Which of the following statements about nuclear stability is correct?

    A. Heavier, more stable nuclei have somewhat larger numbers of protons than neutrons.

    B. A stable nucleus cannot undergo a nuclear reaction, even with the addition of external energy.

    C. Unstable nuclei do not spontaneously change to stable nuclei.

    D. Lighter nuclei tend to have equal numbers of protons and neutrons.

    E. Heavier nuclei have significantly more neutrons than protons.

2. Rank the following in order of increasing acidity: propionic acid ($K_a = 1.3 \times 10^{-5}$), benzoic acid ($K_a = 6.3 \times 10^{-5}$), hypobromous acid ($K_a = 2.6 \times 10^{-9}$).

    A. Benzoic acid < propionic acid < hypobromous acid

    B. Hypobromous acid < propionic acid < benzoic acid

    C. Propionic acid < benzoic acid < hypobromous acid

    D. Hypobromous acid < benzoic acid < propionic acid

    E. Benzoic acid < hypobromous acid < propionic acid

3. Which of the following compounds contain(s) no covalent bonds?

    $$KCl \quad PH_3 \quad O_2 \quad B_2H_6 \quad H_2SO_4$$

    A. KCl, PH_3, and B_2H_6 only

    B. KCl and H_2SO_4 only

    C. PH_3, O_2, and B_2H_6 only

    D. KCl only

    E. KCl and B_2H_6 only

4. Which of the following nuclear reactions is *incorrect*?

    A. $^{14}_{7}N + ^{4}_{2}He \rightarrow ^{17}_{8}O + ^{1}_{1}H$

    B. $^{9}_{4}Be + ^{4}_{2}He \rightarrow ^{12}_{6}C + ^{1}_{0}n$

    C. $^{30}_{15}P + ^{-30}_{-14}Si \rightarrow ^{0}_{-1}\beta$

    D. $^{3}_{1}H + ^{3}_{2}He \rightarrow ^{1}_{-1}\beta$

    E. None of the above

5. What is the hydrogen ion concentration in a solution of 0.00200 M potassium hydroxide?

    A. $[H^+] = \dfrac{K_w}{0.00200}$

    B. $[H^+] = K_w(0.00200)$

    C. $[H^+] = \dfrac{0.00200}{K_w}$

    D. $[H^+] = -\log\dfrac{K_w}{0.00200}$

    E. $[H^+] = -\log\dfrac{0.00200}{K_w}$

GO ON TO THE NEXT PAGE. ➡

6. What is the correct coefficient on the Fe^{3+} ion when the following reaction is balanced?

$$Fe^{2+} + ClO_3^- + H^+ \rightarrow Fe^{3+} + Cl^- + H_2O$$

   A.   2
   B.   3
   C.   5
   D.   6
   E.   7

7. The van der Waals equation includes 2 terms, "a" and "b," that are not present in the ideal gas law. Which of the following statements is true about these terms?

   A.   "a" corrects for attractive forces between gas particles and "b" corrects for the volume of the container.

   B.   "a" corrects for the external pressure of the gas and "b" corrects for the internal pressure of the gas.

   C.   "a" corrects for interactions between gas particles and "b" corrects for the volume of the gas particles.

   D.   "a" corrects for the volume of the gas particles and "b" corrects for the repulsive forces between gas particles.

   E.   None of the above

8. Which of the following molecular structures is not possible?

   A. OF_2

   B. SF_2

   C. OF_4

   D. SF_4

   E. O_2F_2

9. Which of the following is *not* a postulate of the kinetic molecular theory?

   A.   Gas molecules travel in random, straight paths.

   B.   The energy of a gas molecule is determined by quantum mechanics.

   C.   The collisions between gas molecules are elastic.

   D.   The absolute temperature of a substance is equal to the average kinetic energy of its particles.

   E.   The particles of a sample of gas have no volume.

10. Which parameter of a chemical reaction will change with a catalyst?

   A.   Free energy change
   B.   Entropy change
   C.   Equilibrium constant
   D.   Rate constant
   E.   Enthalpy change

11. A plot of $1/[NO_2]$ versus time for the decomposition of NO_2 was found to be linear. This means that the reaction:

   A.   is zero order with respect to NO_2.

   B.   is first order with respect to NO_2.

   C.   is second order reaction with respect to NO_2.

   D.   is third order reaction with respect to NO_2.

   E.   Order cannot be determined from the information given.

GO ON TO THE NEXT PAGE. ➡

12. Which of the following is not a characteristic of ionic substances?

    A. High melting point
    B. Fragility
    C. Crystalline (in the solid form)
    D. Deforms when struck
    E. Well-defined three-dimensional structure

13. Which of the following behaves most like an ideal gas?

    A. H_2

    B. He

    C. O_2

    D. CO_2

    E. Ne

14. The change in enthalpy of a system is equal to Q, the heat flow between the system and the surroundings, under which conditions?

    A. Constant volume
    B. Constant pressure
    C. Constant temperature
    D. Absence of pressure-volume work
    E. None of the above

15. Which statement about metals is *incorrect*?

    A. Metals exhibit higher electronegativities than non-metals.

    B. Metals are reducing agents.

    C. Metals form basic hydroxides.

    D. Metals exhibit low ionization potentials.

    E. Metals generally have one to five electrons in their outermost shell.

16. 40 L of an ideal gas at 25°C and 750 mmHg is allowed to expand to 50 L and the pressure is increased to 765 mmHg. What is the final temperature of the gas?

    A. $\dfrac{(298)(750)(50)}{(40)(765)}$

    B. $\dfrac{(298)(765)(50)}{(40)(750)}$

    C. $\dfrac{(298)(750)(40)}{(50)(765)}$

    D. $\dfrac{(750)(40)}{(50)(298)(765)}$

    E. $\dfrac{(298)(765)(40)}{(50)(750)}$

17. Rutherford's scattering experiments demonstrated:

    A. the existence of X-rays.
    B. the existence of α-particles.
    C. the nature of blackbody radiation.
    D. the mass-to-charge ration of the electron.
    E. the nuclear model of the atom.

18. Elements not found in nature, synthesized in nuclear reactions and involving the completion of the $5f$ atomic orbitals are known as:

    A. lanthanides.
    B. halogens.
    C. actinides.
    D. transition metals.
    E. rare gases.

GO ON TO THE NEXT PAGE. ➡

19. Which of the following statements is (are) true about the half-reactions shown below?

    I. As written, the standard potential for the overall reaction is –1.56 V.

    II. As written, the overall reaction is spontaneous.

    III. The sign of the potential of the overall reaction indicates whether or not the reaction is spontaneous.

    $$Ag \rightarrow Ag^+ + e^- \quad E° = -0.80 \text{ V}$$

    $$Zn^{2+} + 2\,e^- \rightarrow Zn \; E° = -0.76 \text{ V}$$

    A. I only
    B. II only
    C. III only
    D. I and III only
    E. II and III only

20. K_p for the reaction below is 1.36 at 499 K. Which of the following equations can be used to calculate K_c for this reaction?

    $$PCl_3 \text{ (g)} + Cl_2 \text{ (g)} \rightarrow PCl_5 \text{ (g)}$$

    A. $K_c = \dfrac{[(0.0821)(499)]^{-1}}{1.36}$

    B. $K_c = \dfrac{1.36(0.0821)}{499}$

    C. $K_c = \dfrac{1.36}{[(0.0821)(499)]^{-1}}$

    D. $K_c = 1.36 \cdot [(0.0821)(499)]^{-1}$

    E. $K_c = \dfrac{1.36(499)}{0.0821}$

$$\begin{array}{cc} \text{I} & \text{II} \\ \downarrow & \downarrow \\ \end{array}$$
$$\text{H—N—N—N}$$

21. The Third Law of Thermodynamics states that at absolute zero temperature, all perfect crystals have:

    A. the same crystal lattice.
    B. the same lattice energy.
    C. the same enthalpy.
    D. the same free energy.
    E. the same entropy.

22. Which of the following has the smallest ionic radius?

    A. Li^+
    B. Na^+
    C. K^+
    D. Rb^+
    E. Cs^+

23. The van der Waals equation for non-ideal gases differs from the ideal gas law in that it accounts for:

    I. the mass of each particle of gas.
    II. the volume of each particle of gas.
    III. the attractive forces between particles of gas.

    A. I only
    B. II only
    C. III only
    D. I and III only
    E. II and III only

24. What is the molarity of a sulfuric acid solution if 50.0 mL completely neutralizes 1.00 L of a 0.10 M potassium hydroxide solution?

    A. 1.0 M
    B. 0.10 M
    C. 2.0 M
    D. 0.20 M
    E. 10.0 M

GO ON TO THE NEXT PAGE. ➡

25. Which of the following is an intensive property of a system?

   A. Pressure
   B. Mass
   C. Enthalpy
   D. Volume
   E. None of the above

26. Which of the following salts produces the most basic aqueous solution?

   A. $Al(CN)_3$

   B. $KC_2H_3O_2$

   C. $FeCl_3$

   D. KCl

   E. $Pb(C_2H_3O_2)_2$

27. How many molecules are present in 0.20 g of hydrogen gas?

   A. $\dfrac{0.20}{1.008} \cdot 6.02 \times 10^{23}$

   B. $0.20 \cdot 1.008$

   C. $0.20 \cdot 2.016$

   D. $\dfrac{0.20}{2.016} \cdot 6.02 \times 10^{23}$

   E. $\dfrac{0.20}{6.02 \times 10^{23}} \cdot 2.016$

28. The least accurate of the volumetric measuring devices is a:

   A. pipet.
   B. buret.
   C. volumetric flask.
   D. graduated cylinder.
   E. beaker.

29. Which of the following describes the equilibrium constant for a spontaneous reaction?

   A. $K = 0$
   B. $K < 0$
   C. $K = 1$
   D. $K < 1$
   E. $K > 1$

30. Which of the following oxides is amphoteric?

   A. Na_2O

   B. ZnO

   C. MgO

   D. Cl_2O_7

   E. P_2O_5

31. It takes 250.0 J to raise the temperature of a 50.0 g sample of a metal by 10.0°C. What is the specific heat capacity of this metal?

   A. 5.00×10^{-4} J/g•K

   B. 1.25×10^{5} J/g•K

   C. 0.500 J/g•K

   D. 1.25×10^{2} kJ/g•K

   E. 50.0 J/g•K

32. What is the molecular geometry of IF_5?

   A. Tetrahedral
   B. Trigonal bipyramidal
   C. Square pyramidal
   D. Octahedral
   E. Seesaw

GO ON TO THE NEXT PAGE.

33. How much barium nitrate is required to prepare 250.0 mL of a 0.100 M solution? (The molar mass of barium nitrate is 199.344.)

A. $\dfrac{(250.0)(199.344)}{(1000)(0.100)}$

B. $\dfrac{(250.0)(0.100)}{(1000)(199.344)}$

C. $\dfrac{(250.0)(0.100)(199.344)}{(1000)}$

D. $\dfrac{(250.0)}{(1000)(199.344)(0.100)}$

E. None of the above

34. What is the oxidation number of chlorine in ClO_4^-?

A. +1
B. +3
C. +5
D. +7
E. +8

35. A neutral atom has the ground-state electron configuration $1s^2 2s^2 2p^6 3s^1$. It will gain or lose electrons to form an ion of charge:

A. −2.
B. −1.
C. +1.
D. +2.
E. +3.

36. The emission of an alpha particle from $^{226}_{88}Ra$ will yield:

A. $^{223}_{86}Rn$.

B. $^{222}_{86}Rn$.

C. $^{223}_{87}Fr$.

D. $^{222}_{87}Fr$.

E. $^{222}_{88}Ra$.

37. The Rydberg equation was a very useful result of Bohr's model of the hydrogen atom. The Rydberg equation gives:

A. the velocity of electrons as they move through spectral lines.

B. the rate of absorption of hydrogen atoms in the ultraviolet region.

C. the rate of emission of hydrogen atoms in the Lyman series.

D. the rate of emission of heated hydrogen atoms.

E. the frequencies of the series of lines in the hydrogen spectrum.

38. Which of the characteristics below is *not* necessary for a reaction to be used in a titration?

A. The reaction can have no side reactions.

B. The equilibrium constant of the reaction must be very large.

C. The reaction should proceed according to a definite chemical equation.

D. The reaction should proceed very slowly so that the endpoint is readily observable.

E. A method should be available to indicate when to stop the titration.

39. Which of the following molecules or ions is linear?

A. H_2O

B. ClO_2^-

C. NO_2^-

D. NO_2

E. NO_2^+

GO ON TO THE NEXT PAGE. ➡

40. Which of the following statements about equilibrium is correct?

   I. Equilibrium is reached when $\Delta G = 0$.

   II. $\Delta G° = -RT\ln K_{eq}$

   III. At equilibrium, $\Delta G°$ is dependent on pressure.

   A. I only
   B. II only
   C. III only
   D. I and II only
   E. I, II, and III

41. The triple point pressure of water is 4.58 mmHg and the triple point temperature is 273.16 K. From this we can conclude that:

   A. steam cannot exist at temperatures below 273.16 K.

   B. the vapor pressure of ice is 4.58 mmHg for temperatures below 273.16 K.

   C. the vapor pressure of water is less than 4.58 mmHg in most cases.

   D. liquid water cannot exist at pressures below 4.58 mmHg.

   E. ice cannot exist at pressures below 4.58 mmHg.

42. The molecular geometry of the ammonium ion, NH_4^+, is:

   A. trigonal planar.
   B. trigonal pyramidal.
   C. square planar.
   D. tetrahedral.
   E. octahedral.

43. Which of the following statements is (are) true about an oxygen atom in the ground state?

   I. Electrons in the $1s$ atomic orbital may be described using the quantum numbers (1, 0, 0, +1/2) and (1, 0, 0, -1/2).

   II. Electrons in the $2s$ atomic orbital may be described using the quantum numbers (2, 1, 1, +1/2) and (2, 1, 1, -1/2).

   III. The fourth quantum number, m_s, describes the ways that an electron may be aligned with a magnetic field.

   A. I only
   B. I and II only
   C. I and III only
   D. II and III only
   E. I, II, and III

44. LeChâtelier's principle states that:

   A. equilibrium is only reached under certain conditions of temperature and pressure.

   B. when stress is applied to a system at equilibrium, the reaction shifts in the direction that minimizes the stress.

   C. increasing the temperature while decreasing the pressure increases the equilibrium constant.

   D. neither temperature nor pressure has a major effect on equilibrium.

   E. equilibrium is eventually obtained, regardless of reaction conditions.

GO ON TO THE NEXT PAGE. ➡

45. Which of the following statements about carbon-containing compounds is (are) true?

    I. Carbon monoxide is produced when carbon is burned with insufficient oxygen.

    II. Carbon dioxide may undergo sublimation.

    III. Carbonic acid is a diprotic acid.

    A. I only
    B. II only
    C. III only
    D. I and III only
    E. I, II, and III

46. An ionic bond is formed between two ions. Which of the following has no effect on the strength of the bond?

    I. Doubling the charge on both ions
    II. Doubling the temperature
    III. Doubling the radii of both ions

    A. I only
    B. II only
    C. III only
    D. Both I and II
    E. Both II and III

47. Which of the alkali metals is most electronegative?

    A. Li
    B. Na
    C. K
    D. Rb
    E. Cs

48. How many carbon atoms are there in 27.3 g of trichloroacetic acid? (The molar mass of trichloroacetic acid is 163.5.)

    A. $\dfrac{(27.3)(2)(6.02 \times 10^{23})}{163.5}$

    B. $\dfrac{(27.3)(2)}{163.5}$

    C. $\dfrac{(27.3)(163.5)}{(2)(6.02 \times 10^{23})}$

    D. $\dfrac{(27.3)(6.02 \times 10^{23})(163.5)}{(2)(12.01)}$

    E. $\dfrac{(27.3)(6.02 \times 10^{23})}{163.5}$

49. The primary weakness of the Bohr model of the atom is that:

    A. it only works for the hydrogen atom.

    B. it treats the electron as a wave rather than a particle.

    C. it doesn't consider the role of the neutron.

    D. it neglects the radiation emitted by accelerating charged particles.

    E. it only allows for certain energy levels.

50. Which of the following statements is not true for the reaction shown below?

    $$Fe^{3+} + 1\ e^{-} \rightarrow Fe^{2+}$$

    A. Fe^{3+} is being reduced.

    B. The oxidation state of Fe has changed.

    C. Fe^{3+} is the oxidizing agent in this reaction.

    D. The reaction is similar to the reaction between magnesium metal and hydrogen gas.

    E. Both Fe^{3+} and Fe^{2+} are anions.

GO ON TO THE NEXT PAGE. ➡

51. How much heat is released when the temperature of 100 g of water decreases from 25°C to 5°C? (The specific heat of water is 4.18 J/g · K.)

    A. $(100)(4.18)(25-5)$

    B. $(100)(4.18)(5-25)$

    C. $\dfrac{2,000}{4.18}$

    D. $\dfrac{-2,000}{4.18}$

    E. $\dfrac{-20(4.18)}{100}$

52. Increasing the temperature of a reaction increases:

    A. the reaction order.

    B. the activation energy.

    C. the number of collisions in the correct orientation.

    D. the kinetic energy of the molecules.

    E. None of the above.

53. Glass is an example of an amorphous solid which can be characterized as:

    A. a malleable solid.
    B. crystal-like in structure.
    C. a good conductor.
    D. a molecular solid.
    E. a very viscous fluid.

54. The K_a of the ammonium ion is 5.6×10^{-10} at 25°C. What is the approximate pH of a 1.0 M ammonium chloride solution?

    A. 9
    B. 7
    C. 5
    D. 3
    E. 1

55. Which of the following pairs of elements does not have approximately the same electronegativity?

    A. C and S
    B. Co and Ni
    C. B and Al
    D. U and Pu
    E. Fe and Ni

56. Which of the following salts will produce a basic solution when dissolved in water?

    A. NH_4Cl

    B. $NaCl$

    C. $NaNO_3$

    D. Na_2SO_4

    E. Na_2CO_3

57. A pH 7 buffer solution contains H_2CO_3 and $NaHCO_3$. What must the ratio of $[NaHCO_3]/[H_2CO_3]$ be in order to maintain the solution at pH 7? The K_a of H_2CO_3 is 4.3×10^{-7}.

    A. 43
    B. 4.3
    C. 0.43
    D. 86
    E. 1.29

58. For which of the following compounds is hydrogen bonding an important component of the intermolecular forces?

    A. CH_3Cl

    B. CH_3OCH_3

    C. CH_3NH_2

    D. CH_3CH_3Cl

    E. None of the above

GO ON TO THE NEXT PAGE. ➡

59. The rate data for the reaction $A + B \rightarrow C$ is shown below.

[A]	[B]	rate
1.0	1.0	0.01
1.0	2.0	0.02
3.0	1.0	0.09

The reaction is:

A. first order in both A and B.
B. second order in both A and B.
C. first order in A and second order in B.
D. second order in A and first order in B.
E. second order in A and zero order in B.

60. During a redox reaction, the oxidizing agent:

A. gains electrons.
B. is oxidized.
C. has an increase in oxidation state.
D. is hydrolyzed.
E. loses electrons.

61. Which of the following statements about semiconductors is (are) true?

I. A p-type semiconductor is formed when silicon or germanium is doped with a Group III element.

II. Doping decreases the conductivity of a silicon or germanium crystal.

III. An n-type semiconductor is formed when silicon or germanium is doped with an element that produces non-bonded electrons.

A. I only
B. II only
C. III only
D. I and II only
E. I and III only

62. What is the frequency of light of wavelength 3×10^{-3} cm?

A. 1×10^{13}
B. 2.2×10^{-31}
C. 1×10^{7}
D. 9×10^{7}
E. 2.64×10^{-36}

63. Which of the following substances is a Lewis acid?

A. CCl_4
B. BF_3
C. I_2
D. NaH
E. $(CH_3)_3N$

64. Which of the following compounds has the shortest carbon-halogen bond?

A. CH_3F
B. CH_3Cl
C. CH_3Br
D. CH_3I
E. They are all equal.

65. All of the following statements about entropy change are true EXCEPT:

A. It is a measure of the energy dispersal.
B. The natural tendency is for it to increase.
C. It is not a state function under all conditions.
D. It can be defined both thermodynamically and statistically.
E. Its calculation is only possible for processes involving no temperature change.

GO ON TO THE NEXT PAGE. ➡

66. Which of the following is not a method for separating mixtures?

A. Filtration
B. Distillation
C. Selective precipitation
D. Absorption chromatography
E. Solvation

67. An atom containing two electrons which possess the following sets of quantum numbers $(3, 1, 1, -\frac{1}{2})$ and $(3, 1, 1, -\frac{1}{2})$ may not exist based on:

A. Pauli exclusion principle.
B. Lewis's law.
C. Hund's rule.
D. Heisenberg uncertainty principle.
E. Bohr model.

68. The natural logarithm of the rate constant of a reaction is:

A. directly proportional to temperature.
B. inversely proportional to temperature.
C. not affected by changes in temperature.
D. only affected by the activation energy.
E. independent of the activation energy and the temperature.

69. Which of the following compounds does not contain a covalent bond?

A. PH_3

B. $GeCl_4$

C. H_2S

D. CsF

E. CH_3Cl

70. In the most stable resonance form of the molecule whose skeleton structure is shown above, the bond orders of bonds I and II are:

A. I = 2 and II = 1
B. I = 2 and II = 20
C. I > 2 and II < 2
D. I > 2 and II > 2
E. I < 2 and II < 2

Match the descriptions in questions 71–73 with the choices given below. A choice may be used once, more than once, or not at all.

(A) BF_3

(B) C_2H_2

(C) $CHCl_3$

(D) XeF_4

(E) NO_2

71. A molecule that has an unpaired electron within its structure.

72. A molecule whose shape is square planar.

73. A molecule that contains 2 pi (π) bonds.

74. A weak acid has a K_a of 1.0×10^{-10}. What is the $[H_3O^+]$ ion concentration in a 0.01 M solution of this acid?

    A. 1.0×10^{-12} M

    B. 1.0×10^{-10} M

    C. 1.0×10^{-8} M

    D. 1.0×10^{-6} M

    E. 1.0×10^{-5} M

75. All of the following statements are correct EXCEPT:

    A. In all spontaneous processes, $\Delta S_{universe} > 0$.

    B. The entropy of a perfect crystal is taken to be 0 at 298 K.

    C. During freezing, the entropy of the system decreases.

    D. A spontaneous process is accompanied by a negative free energy change.

    E. The free energy change of a system at equilibrium is 0.

GO ON TO THE NEXT PAGE. ➡

SECTION II—PART A

Time—40 minutes

<u>Directions</u>: Write your answers in the space provided. Choose between question 2 and question 3.

You may use a calculator for this section. You may use the Periodic Table, the "Equations and Constants" sheets, and the table of standard reduction potentials provided on pages 469–473 of this book.

1. At 25°C, the K_{sp} for $CaCO_3$ is 3.36×10^{-9} and the K_{sp} for CaF_2 is 3.45×10^{-11}.

   (A) Write the K_{sp} expressions for $CaCO_3$ and CaF_2.

   (B) A concentrated aqueous solution of $Ca(NO_3)_2$ is added slowly to 1.00 liter of a well-stirred aqueous solution containing 0.100 mole of CO_3^{2-} and 0.00200 mole of F^- at 25°C. (Assume that the volume of the solution is not affected by the addition of $Ca(NO_3)_2$.)

      (1) Explain why $CaCO_3$ will precipitate first even though its K_{sp} is larger than the K_{sp} of CaF_2. Use calculations to support your explanation.

      (2) What is the concentration of the calcium ion when the $CaCO_3$ first precipitates?

      (3) As more $Ca(NO_3)_2$ is added to the mixture, CaF_2 begins to precipitate. At that stage, what percent of the CO_3^{2-} remains in solution?

GO ON TO THE NEXT PAGE. ➡

Answer EITHER question 2 (below) or question 3 (next page).

2. Body fat is a complex mixture with the approximate formula $C_{56}H_{108}O_6$. What volume of oxygen at 25°C and 1.00 atm would be required to completely "burn off" 5.00 pounds of fat?
(Note: 1 pound = 453.6 g.)

GO ON TO THE NEXT PAGE. ➡

3. A 0.200 g sample of hemoglobin is dissolved in enough water to produce 10.00 mL of solution. The osmotic pressure of the solution is 5.5 torr at 25°C.

(A) What is the molar mass of hemoglobin?

(B) If hemoglobin contains 0.33% iron by mass, how many iron atoms are there in one molecule of hemoglobin?

GO ON TO THE NEXT PAGE. ➡

SECTION II—PART B

Time—50 minutes

<u>Directions</u>: Write your answers in the space provided. Choose between question 7 and question 8.

You may NOT use a calculator for this section. You may use the Periodic Table, the "Equations and Constants" sheets, and the table of standard reduction potentials provided on pages 469–473 of this book.

4. Write the formulas to show the reactants and products for any FIVE of the laboratory situations described below. A reaction occurs in all cases. All solutions are aqueous unless otherwise indicated. Represent substances in solutions as ions if the substances are extensively ionized. Omit formulas for any ions or molecules that are unchanged by the reaction. You need not balance the equation.

(A) A solution of copper(II) nitrate is added to a solution of sulfuric acid.

(B) Carbon dioxide gas is bubbled through distilled water.

(C) Solid lithium hydride is added to water.

(D) A strip of aluminum metal is dropped into a solution of 6 M hydrochloric acid.

GO ON TO THE NEXT PAGE. ➡

(E) A solution of iron(III) chloride is added to a 1 M ammonia solution.

(F) Carbon dioxide gas is bubbled through a solution of calcium hydroxide.

(G) Hydrogen peroxide solution is added to a solution of sulfurous acid.

(H) A solution of lead(II) nitrate is added to a solution of potassium iodide.

GO ON TO THE NEXT PAGE. ➡

5. A 0.200 M solution of ammonia has a pH of 11.3. This ammonia solution is used in a titration in order to determine the concentration of an unknown nitric acid solution.

   (A) Describe the experimental procedure you would use to collect the necessary data for determination of the nitric acid concentration.

   (B) Sketch and carefully label the titration curve you would obtain in this experiment and explain how it could be used to determine the concentration of the unknown nitric acid solution.

   (C) Which of the acid-base indicators listed below would be best suited for a visual titration of nitric acid by ammonia, and why?

Indicator:	Color change:	pH region:
methyl orange	red → yellow	3.1–3.4
methyl red	red → yellow	4.2–6.3
bromothymol blue	yellow → blue	6.2–7.6
phenolphthalein	colorless → pink	8.3–10.0

GO ON TO THE NEXT PAGE. ➡

6. The phase diagram of a substance is shown below.

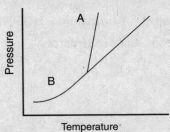

(A) Label each region of the diagram with the phase that is present.

(B) Sketch the heating curve that is expected if heat is added to the sample at a constant pressure, starting at point B.

(C) Starting at point A, describe what happens if the pressure is lowered at constant temperature.

Answer EITHER question 7 (below) or question 8 (next page).

7. Two resonance structures can be drawn for formamide, $HCONH_2$.

   (A) Draw a Lewis structure for each resonance structure.

   (B) Predict the bond angles about the carbon and nitrogen atoms for both resonance structures.

   (C) How could the experimental measurement of the H-N-H bond angle be used to determine which resonance structure is more important?

GO ON TO THE NEXT PAGE. ➡

8. The kinetic molecular theory describes the behavior of gas particles at the molecular level.

   (A) Charles's law states that the volume of a gas is directly proportional to the absolute temperature when pressure and moles of gas are constant. How can the kinetic molecular theory be used to explain Charles's law?

   (B) Dalton's law of partial pressures states that the total pressure of a mixture of gases is the sum of the partial pressures of all components in the mixture. How can the kinetic molecular theory be used to explain Dalton's law of partial pressures?

STOP

Practice Test II Answers and Explanations

SECTION I ANSWERS AND EXPLANATIONS

1.	D	16.	B	31.	C	46.	B	61.	E
2.	B	17.	E	32.	C	47.	A	62.	A
3.	D	18.	C	33.	C	48.	A	63.	B
4.	D	19.	C	34.	D	49.	A	64.	A
5.	A	20.	C	35.	C	50.	E	65.	C
6.	D	21.	E	36.	B	51.	B	66.	E
7.	C	22.	A	37.	E	52.	D	67.	A
8.	C	23.	E	38.	D	53.	E	68.	B
9.	B	24.	A	39.	E	54.	C	69.	D
10.	D	25.	A	40.	D	55.	C	70.	B
11.	C	26.	B	41.	D	56.	E	71.	E
12.	D	27.	D	42.	D	57.	B	72.	D
13.	B	28.	E	43.	C	58.	C	73.	B
14.	B	29.	E	44.	B	59.	D	74.	D
15.	A	30.	B	45.	E	60.	A	75.	B

1. D

Answer (A) is not true as they do not tend to have more protons than neutrons. Answer (B) is incorrect: plenty of stable nuclei undergo reactions with the addition of external energy. Answer (C) is incorrect: over time unstable nuclei eventually spontaneously change to stable nuclei, which is the goal of a radioactive unstable nucleus. Answer (E) is incorrect because a nucleus can be heavier simply because it has a lot of protons and neutrons. It does not need to have more neutrons than protons to be heavy. Answer (D) is correct: the lighter a nucleus is, the more similar the number of protons and neutrons.

2. B

As K_a increases, the strength of an acid increases. Rank the K_a's given from smallest to greatest and that will point you to the correct answer.

3. D

You want to look for a metal/nonmetal pairing here; i.e., ionic bonds. KCl is the only ionic bond in the bunch. All the others contain covalent bonds.

4. D

Answer choice (D) is wrong: the atomic numbers and mass numbers on the left do not add up to the same as the atomic and mass numbers on the right.

5. A

$$[OH^-][H^+] = K_w$$
$$[OH^-] = 0.00200$$

$$\frac{K_w}{0.00200} = [H^+]$$

6. D

Balancing redox reactions can be confusing. You need to break this original equation into two half-reactions:

Oxidation Reaction: $Fe^{2+} \rightarrow Fe^{3+} + e^-$

Reduction Reaction: $ClO_3^- \rightarrow Cl^-$

The reduction reaction is incomplete, and in order to balance it you must add oxygen to the right side of the equation. The way you do this is by adding three H_2O molecules. But now there are six hydrogen molecules on the right and none on the left. You counter this by adding six protons to the left side, giving you:

Reduction Reaction: $6H^+ + ClO_3^- \rightarrow Cl^- + 3H_2O$

PRACTICE TEST II ANSWERS AND EXPLANATIONS

Lastly, you need to balance the charge on both sides. There is a total charge of +5 on the left and −1 on the right. This means you need to add six electrons to the left side to give the balanced equation of:

Reduction Reaction: $6e^- + 6H^+ + ClO_3^- \rightarrow Cl^- + 3H_2O$

When completing the balancing of the reaction, you need to multiply the oxidation reaction by six so that the electrons will cancel out, giving you a final balanced reaction of:

$$6\ Fe^{2+} + 6H^+ + ClO_3^- \rightarrow Cl^- + 3H_2O + 6\ Fe^{3+}$$

7. C

The van der Waals equation corrects for volume and the effect of intermolecular forces. Answer choice (C) sums this up nicely.

8. C

The molecule oxygen tetrafluoride, OF_4, has 34 electrons that need to be placed around the molecule. Oxygen is the central atom and is single bonded to each of the four fluorine atoms. The maximum possible number of electrons that can be drawn with this structure is 32. This structure is not possible.

9. B

The kinetic molecular theory consists of the following postulates:

- The molecules of a gas have volumes that are negligible compared to the volume of the gas itself. [Answer choice (E)]

- The molecules of a gas are in a continuous, rapid, and random state of motion. [Answer choice (A)]

- The gas temperature determines the average kinetic energy of the gas molecules. [Answer choice (D)]

- Gas molecules collide in perfectly elastic collisions and with the walls of their container, and do not lose energy in the process. [Answer choice (C)]

10. D

A catalyst increases the rate of reaction by lowering the activation energy. This will result in a change in the rate constant.

11. C

When the graph of $\frac{1}{[A]}$ versus time produces a straight line, then the reaction is a second order reaction with the following rate reaction: rate $= k[A]^2$.

12. D

The only answer choice on this list that does not accurately describe ionic substances is (D): that they deform when they are struck. The other four answer choices accurately described ionic compounds.

13. B

Answer choice (B), helium, is thought of as a "near" ideal gas. It exhibits the relationships expected of an ideal gas between pressure and temperature and volume and temperature.

14. B

This is a good fact to know for the AP chemistry exam. $\Delta H = Q$ when pressure is held constant.

15. A

Metals are located on the left of the Periodic Table, nonmetals on the right. Electronegativity increases as you move from left to right on the table. It can therefore be concluded that answer choice (A) is incorrect because metals exhibit *lower* electronegativities than nonmetals.

16. B

This question involves ideal gas law equations:

$$\frac{P_1V_1}{T_1} = \frac{P_2V_2}{T_2}$$

$$\frac{(750)(40)}{(298)} = \frac{(765)(50)}{T_2}$$

$$T_2 = \frac{(765)(50)(298)}{(750)(40)}$$

17. E

This question tests your understanding of Rutherford's scattering experiments and their demonstration of the nuclear model of the atom.

18. C

Looking at the Periodic Table, the elements not found in nature made in nuclear reactions that involve the 5*f* orbitals are known as the actinides.

19. C

Roman numeral I is incorrect. It does not properly account for the fact that to balance the equation you need to multiply the top equation by two. This allows you to eliminate answer choices (A) and (D). Roman numeral II is incorrect: a negative overall E° represents a nonspontaneous reaction. This allows you to eliminate answers (B) and (E). This leaves you with just answer choice (C)—and Roman numeral III is indeed a correct statement.

20. C

$K_p = K_c (RT)^{\Delta n}$. Rearrange the equation to solve for K_c: $K_c = (K_p) \div ((RT)^{\Delta n})$. R is the molar gas constant of 0.0821. T is the temperature in Kelvin of 499K. Δn is calculated by adding the moles on the right and subtracting the moles on the left, which would be –1. Now just plug the numbers in to the equation for K_c:

$$K_c = \frac{1.36}{[(0.0821)(499)]^{-1}}$$

21. E

The third law of thermodynamics says that the entropy at zero Kelvin (absolute zero) for a perfectly arranged crystalline material is zero.

22. A

The ionic radius increase from right to left across the table and from top to bottom. Because Li is above Na, K, Rb, and Cs in Group 1 of the Periodic Table, it must have the smallest ionic radius.

23. E

This problem again tests your knowledge of the van der Waals equation that accounts for molecular volume and intermolecular force. Roman numerals II and III are correct.

24. A

This question tests your understanding of acid and base equivalents. H_2SO_4 has 2 equivalents per mole and KOH has 1. You must first calculate how many equivalents of OH^- there must be by calculating how many moles of KOH there were. 1 L of 0.10M KOH would be 0.10 moles of KOH, which would be 0.10 equivalents. To completely neutralize this base you need 0.10 equivalents of H^+. Since H_2SO_4 has 2 equivalents per mole, this would be only 0.05 moles of H_2SO_4, which would make it $\frac{0.050 \text{ moles}}{0.05 \text{L}} = 1.0\text{M}$.

25. A

An intensive property is one that independent of the quantity of substance present. The only example of such is answer choice (A), pressure.

26. B

When trying to determine which of these salts will produce the most basic aqueous solution, remember that you will be adding the salt into water. Answer choice (B) is correct because when water reacts with $KC_2H_3O_2$, it will form KOH, a strong base, and $HC_2H_3O_2$, a weak acid. Thus, it will produce a solution that is more basic than acidic. Answer choice (A) will produce $Al(OH)_3$, a weak base, and HCN, a weak acid. Answer choice (C) will produce $Fe(OH)_3$, a weak base, and HCl, a strong acid. Answer choice (D) will produce KOH, a strong base, and HCl, a strong acid. Answer choice (E) will produce $Pb(OH)_2$, a weak base, and $HC_2H_3O_2$, a weak acid.

27. D

For this problem you must first calculate how many moles there are in 0.2 g of hydrogen gas (H_2). Then you must convert this to molecules using Avogadro's number.

28. E

Just think about this one logically. Of the five devices listed, the beaker has the least accuracy simply because its units of measurement provide the largest room for error. Using a beaker, you can not measure out a volume as precisely as you might be able to with any of the other four listed in this question.

29. E

For a spontaneous reaction, the equilibrium constant, K, is known to be greater than 1.

30. B

An amphoteric substance is one that can be both an acid and a base. ZnO is the only one on the list that fits this description.

31. C

This problem involves the equation $q = c_p m \Delta T$.

$$250 \text{ J} = 50(c)(10)$$
$$250 = 500c$$
$$c = 0.5 \text{ J/g} \cdot \text{K}$$

32. C

A molecule that has 5 bonded pairs and 1 nonbonded pair of electrons has square-pyramidal geometry.

33. C

First you need to calculate how many moles of barium nitrate you would need to prepare 250 mL of a 0.100 M solution. Then you need to calculate how many grams would be in that many moles. Answer choice (C) shows the proper calculation.

$$\text{Molarity} = \frac{\text{moles solute}}{\text{liters solution}}$$

$$0.100 = \frac{\text{moles solute}}{0.250 \text{ liters solution}} = .025 \text{ moles solute}$$

$$0.025 \text{ moles solute} \times \frac{199.34 \text{ grams}}{1 \text{ mole}} = 4.98 \text{ grams}$$

34. D

Oxygen's oxidation number is always −2 except with peroxides. The four oxygen atoms give a combined −8, and in order for the entire compound to have a −1 oxidation number, the chloride must be +7.

35. C

Since we are told that it is a neutral atom, we can add together the superscripts to determine the atomic number of the element. There are 11 electrons in the neutral atom, which means we are looking at sodium. This atom will lose one electron to form a cation with a +1 charge.

36. B

An alpha particle is a helium atom: ^4_2He. The decay reaction will look as follows:

$$^{226}_{88}\text{Ra} \rightarrow \, ^4_2\text{He} + \, ^{222}_{86}\text{Rn}$$

You must make sure that the atomic numbers and mass number totals on the left and right are equal when you are finished.

37. E

The Rydberg equation is used to predict wavelength and can give the frequency of a series of lines in the hydrogen spectrum.

38. D

The reaction does not necessarily need to proceed slowly so that the endpoint is readily observable. Quick reactions can result in successful titration procedures. The other four answer choices are indeed necessary for a reaction to be used in a titration.

39. E

NO_2^+ has 16 total electrons. The compound consists of a central nitrogen double bonded to two oxygen atoms in a linear configuration.

40. D

Roman numeral I is a true statement: $\Delta G = 0$ represents a system in equilibrium. This allows you to eliminate answer choices (B) and (C). Roman numeral II contains a correct equation and Roman numeral III is an incorrect statement. Therefore answer choice (D) is correct I and II only.

41. D

A phase diagram plots temperature on the x-axis versus pressure on the y-axis.

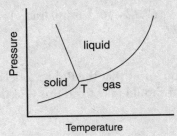

As you can see by looking at the chart, the pressure at the triple point, T, is the lowest pressure at which the liquid phase exists. All other points for the liquid section have a higher pressure than the triple point pressure. Answer (D) is correct. By looking at the chart you can prove to yourself why answers (A), (B), (C), and (E) are incorrect.

42. D

NH_4^+ has 4 bonded pairs and 0 nonbonded electron pairs, giving it a tetrahedral geometry.

43. C

Roman numeral I is true which allows you to eliminate answer choice (D). Roman numeral II is incorrect, and Roman numeral III is correct, leaving you with answer choice (C).

44. B

Answer choice (B) is an excellent definition of LeChâtelier's principle. You need to be familiar with this principle and how to apply it for the AP Chemistry exam.

45. E

All three of these statements about carbon-containing compounds are true.

46. B

The charge and the size of the radius of the ions play an important role in measuring the strength of a bond. The different sized radii and the changing charge will affect the strength with which the two ions pull towards each other. Changing temperature does not affect the strength of the bond.

47. A

Electronegativity increases as you go up the Periodic Table and as you move from left to right. Lithium is higher up the table than Na, K, Rb, and Cs, and is therefore the most electronegative of those choices.

48. A

The formula for acetic acid is CH_3COOH. This means that there are 2 moles of carbon in each mole of the trichloroacetic acid. To solve this problem you need to first calculate how many moles of trichloroacetic acid there are in 27.3 grams. You can then use Avogadro's number to figure out how many carbon atoms there are.

$$2\left(\frac{27.3 \text{ g}}{163.5 \text{ g/mol}}\right) = \text{moles of C}$$

$$2\left(\frac{27.3 \text{ g}}{163.5 \text{ g/mol}}\right)(6.02 \times 10^{23} \text{ atoms/mole}) = \text{atoms of C}$$

49. A

This question makes sure that you know that the Bohr model of the atom only works for the hydrogen atom. That is what you should take from this question.

50. E

The question wants to know which answer choice is NOT true. Answer choice (A) is true, because Fe^{3+} is adding an electron, it is being reduced. Answer choice (B) is true because the oxidation state is changing from +3 to +2. Answer choice (C) is true because the species being reduced is also known as the oxidizing agent. Answer choice (D) is true and answer choice (E) is false: Fe^{3+} and Fe^{2+} are cations.

51. B

This question uses the formula $Q = mc\Delta T = (100)(4.18)(5 - 25)$

52. D

Increasing the temperature of a reaction will definitely increase the kinetic energy of the molecules. Kinetic energy is the energy of motion and it is directly proportional to temperature. As temperature rises, so does the kinetic energy of a system.

53. E

Some argue that glass is a liquid and not a solid. How could this be? How could glass be solid to the touch if it is a liquid? An important characteristic of liquids that comes into play here is viscosity, which is resistance to flow. Usually, viscosity will increase when a liquid is cooled and when a liquid is cooled to a temperature lower than its freezing point, it changes into the solid form. But this is not always the case. Supercooling can occur, which allows the substance to remain in the liquid phase even though it is below its freezing point. As this is happening, if the viscosity grows to be large enough, the substance may never solidify and instead may form what is known as an amorphous solid. The molecules of an amorphous solid have a chaotic arrangement compared to most normal solids, but they are cohesive enough to give substances such as glass the appearance and some characteristics of a solid. Because of this, glass is often thought of in terms of being a very viscous fluid.

54. C

$$NH_4^+ + H_2O \rightleftharpoons H_3O^+ + NH_3$$

$$K_a = \frac{[H_3O^+][NH_3]}{[NH_4^+]}$$

Since the molar ratio of H_3O^+: NH_3Cl^- is 1:1, you can write the K_a expression as:

$$K_a = \frac{[X]^2}{[NH_4^+]}$$

You can then estimate the concentration of H_3O^+ by rearranging the equation and solving.

$$[NH_4^+]K_a = ([X])^2$$

$$\sqrt{[NH_4^+]K_a} = [X] = \sqrt{[1.0M](5.6 \times 10^{-10})} = 2.4 \times 10^{-5} = [H_3O^+]$$

$$pH = -\log[H_3O^+] = -\log[2.4 \times 10^{-5}] = 4.62 \approx 5.00$$

55. C

In general the trend is for electronegativity to increase as you go from left to right across the table and from bottom to top. The reason (C) is the best answer to this question is because it contains the only pair of elements that are from different "sections" of the table. Boron is a metalloid and aluminum is a metal, causing the electronegativities to be further apart than the other pairs.

56. E

Sodium carbonate, Na_2CO_3, is a basic compound that produces a basic solution when dissolved in water.

57. B

This question involves the Henderson-Hasselbalch equation:

$pH = pKa + \log\frac{[A^-]}{[HA]}$ We can set the pH = 7, and in fact it is easier to solve this problem with the pH in the log form: $-\log[1 \times 10^{-7}] = -\log[4.3 \times 10^{-7}] + \log\frac{[A^-]}{[HA]}$

Rearranging the above equation leaves you with:

$$\log[4.3] = +\log\frac{[A^-]}{[HA]}$$
$$4.3 = \frac{[A^-]}{[HA]}$$

58. C

Hydrogen bonding is common among alcohols, amines, and carboxylic acids. Answer choice (C), CH_3NH_2, is the only compound of that type among the answer choices.

59. D

If B is doubled with A remaining constant, the rate doubles. This means that B is a first order reactant. If A is tripled with B constant, the rate increases by nine-fold. This means that A is a second order reactant.

60. A

An oxidizing agent causes the other element involved in the reaction to be oxidized. This means that the oxidizing agent gets reduced. Remember that if an atom loses electrons it is oxidized, if it gains electrons it is reduced.

61. E

A p-type semiconductor is formed by adding Group 13 atoms to silicon, and that an n-type semiconductor is formed by adding Group 15 atoms to silicon. This adds an extra electron to the crystal lattice, which moves freely throughout the lattice, i.e., a nonbonded electron. Therefore Roman numerals I and III are true statements. Roman numeral II is false because doping *increases* conductivity. Therefore choice (E) is the correct answer.

62. A

This question uses the equation $\lambda v = c$. λ = wavelength; v = frequency; c = speed of light = 3 $\times 10^8$ m/s.

$$(3 \times 10^{-5}\text{m}) \, v = 3 \times 10^8 \text{ m/s}$$
$$v = 1 \times 10^{13}$$

63. B

A Lewis acid is known to be an electron acceptor. The only answer among the five choices here that has room to add electrons is BF_3.

64. A

CH_3F has the shortest carbon-halogen bond because fluorine is the most active of the halogens and pulls the electron more strongly than any of the other halogens.

65. C

All of the answer choices are true except for (C)—entropy *is* a state function under all conditions.

66. E

Solvation is not a process used to separate mixtures. In fact it is the process of binding a solute to one or more molecules of a solvent.

67. A

The Pauli exclusion principle states that two electrons of an atom cannot have the exact same quantum numbers.

68. B

Based on the equation $\ln K = \ln A - \dfrac{Ea}{RT}$; you can tell that the natural logarithm of the rate constant would be inversely proportional to the temperature.

69. D

In this problem you want to look for nonmetal/metal pairs. Answer choice (D) is the only ionic compound in the bunch.

70. B

The structure of HN_3 is most likely:

$$\overset{}{H} - \overset{+}{N} \equiv \overset{+}{N} - \overset{2-}{\underset{..}{\overset{..}{N}}}:$$

When two atoms are joined by a triple bond, the bond order is 3. When there is a single bond between two atoms, the bond order is 1. The bond order for the first bond in this structure is 1, and the bond order for the second bond in this structure is 3.

71. E

NO_2 has 17 electrons total and thus will have an unpaired electron when drawn.

72. D

Xenon tends to form compounds that disobey the octet rule and can bond up to 6 halogens. XeF_4 has 4 bonded pairs and 2 nonbonded pairs and thus has a square planer configuration.

73. B

C_2H_2 is an alkyne—it has a triple bond. It has two carbons triple bonded to each other, each bonding to an atom of hydrogen on the other side. The triple bond is composed of one sigma bond and two pi bonds.

74. D

When trying to solve this problem, try to use a good example, like this one:

$$HA + H_2O \rightleftharpoons H_3O^+ + A^-$$

$$K_a = \frac{[H_3O^+][A^-]}{[HA]}$$

$$1.0 \times 10^{-10} = \frac{[X][X]}{0.01}$$

$$X^2 = 1.0 \times 10^{-12}$$

$$X^2 = [H_3O^+] = 1.0 \times 10^{-6}$$

75. B

This answer choice goes against the third law of thermodynamics, which states that the entropy of a perfect crystal is taken to be 0 at 0K, not 298K.

SECTION II—PART A ANSWERS AND EXPLANATIONS

1. At 25°C, the K_{sp} for $CaCO_3$ is 3.36×10^{-9} and the K_{sp} for CaF_2 is 3.45×10^{-11}.

(A) Write the K_{sp} expressions for $CaCO_3$ and CaF_2.

ANSWER:

$$K_{sp} (CaCO_3) = [Ca^{2+}][CO_3^{2-}]$$

$$K_{sp} (CaF_2) = [Ca^{2+}][F^-]^2$$

(B) A concentrated aqueous solution of $Ca(NO_3)_2$ is added slowly to 1.00 liter of a well-stirred aqueous solution containing 0.100 mole of CO_3^{2-} and 0.00200 mole of F^- at 25°C. (Assume that the volume of the solution is not affected by the addition of $Ca(NO_3)_2$.)

(1) Explain why $CaCO_3$ will precipitate first even though its K_{sp} is larger than the K_{sp} of CaF_2. Use calculations to support your explanation.

ANSWER:
The order of precipitation is determined by molar solubility: the compound with the smaller molar solubility will precipitate first.

$K_{sp} = 3.36 \times 10^{-9}$	$CaCO_3(s) \rightleftharpoons Ca^{2+}(aq) + CO_3^{2-}(aq)$	
Species:	$[Ca^{2+}]$	$[CO_3^{2-}]$
Initial concentration:	0.00	0.00
Change in concentration:	$+x$	$+x$
Equilibrium concentration:	x	x

$$K_{sp} = [Ca^{2+}][CO_3^{2-}]$$
$$3.36 \times 10^{-9} = x^2$$
$$x = \text{molar solubility of } CaCO_3 = \mathbf{5.80 \times 10^{-5} \, M}$$

$K_{sp} = 3.45 \times 10^{-11}$	$CaF_2(s) \rightleftharpoons Ca^{2+}(aq) + 2F^-(aq)$	
Species:	$[Ca^{2+}]$	$[F^-]$
Initial concentration:	0.00	0.00
Change in concentration:	$+x$	$+2x$
Equilibrium concentration:	x	$2x$

$$K_{sp} = [Ca^{2+}][F^-]^2$$
$$3.45 \times 10^{-11} = (x)(2x)^2 = 4x^3$$
$$x = \text{molar solubility of } CaF_2 = \textbf{2.05} \times \textbf{10}^{-4} \textbf{ M}$$

Since the molar solubility of $CaCO_3$ is smaller than the molar solubility of CaF_2, the $CaCO_3$ will precipitate first.

(2) What is the concentration of the calcium ion when the CaCO3 first precipitates?

ANSWER:

$$K_{sp} = [Ca^{2+}][CO_3^{2-}]$$
$$3.36 \times 10^{-9} = [Ca^{2+}] \cdot (0.100 \text{ M})$$
$$[Ca^{2+}] = \textbf{3.36} \times \textbf{10}^{-8} \textbf{ M}$$

(3) As more $Ca(NO_3)_2$ is added to the mixture, CaF_2 begins to precipitate. At that stage, what percent of the CO_3^{2-} remains in solution?

ANSWER:
First we calculate the $[Ca^{2+}]$ when CaF_2 begins to precipitate. Then we calculate the $[CO_3^{2-}]$ at that point. We determine the percent of $[CO_3^{2-}]$ remaining, by dividing this number by the *initial* $[CO_3^{2-}]$ and multiplying the result by 100.

$$K_{sp} = [Ca^{2+}][F^-]^2$$
$$3.45 \times 10^{-11} = [Ca^{2+}] \cdot (0.0020\text{M})^2$$
$$[Ca^{2+}] = 8.63 \times 10^{-6} \text{ M}$$

$$K_{sp} = [Ca^{2+}][CO_3^{2-}]$$
$$3.36 \times 10^{-9} = (8.63 \times 10^{-6} \text{ M})[CO_3^{2-}]$$
$$[CO_3^{2-}] = 3.89 \times 10^{-4} \text{ M}$$

$$\text{Percent of } [CO_3^{2-}] \text{ remaining} = \frac{3.89 \times 10^{-4} \text{ M}}{0.100 \text{ M}} \times 100 = \textbf{0.389\%}$$

PRACTICE TEST II ANSWERS AND EXPLANATIONS

2. Body fat is a complex mixture with the approximate formula $C_{56}H_{108}O_6$. What volume of oxygen at 25°C and 1.00 atm would be required to completely "burn off" 5.00 pounds of fat? (Note: 1 pound = 453.6 g.)

ANSWER:
(1) Write balanced combustion reaction:
$$C_{56}H_{108}O_6 + 80\ O_2 \rightarrow 56\ CO_2 + 54\ H_2O$$

(2) Calculate molar mass of fat:
$$56(12.011) + 108(1.0079) + 6(16.00) = 877.47\ g/mol$$

(3) Convert 5 lbs fat to moles:
$$5\ lbs(453.6\ g/lb)(1\ mol/877.47\ g) = 2.58\ mol\ fat$$

(4) Use balanced equation to calculate moles O_2:
$$2.58\ mol\ fat(80\ mol\ O_2/mol\ fat) = 207\ mol\ O_2$$

(5) Use ideal gas law to calculate L O_2:
$$V = nRT/P = (207\ mol)(0.0821\ L \cdot atm/mol \cdot K)(298K)/1.00\ atm = 5060\ L$$

3. A 0.200 g sample of hemoglobin is dissolved in enough water to produce 10.00 mL of solution. The osmotic pressure of the solution is 5.5 torr at 25°C.

(A) What is the molar mass of hemoglobin?

ANSWER:
Osmotic pressure equation: $\pi = MRT = \dfrac{nRT}{V}$

$$\frac{5.5\ torr}{(760\ torr/atm)} = \frac{n(0.0821\ L \cdot atm/mol \cdot K)(298\ K)}{10.00 \times 10^{-3}\ L}$$

$$n = 2.96 \times 10^{-6}\ mol\ hemoglobin$$

$$molar\ mass = \frac{0.200\ g\ hemoglobin}{2.96 \times 10^{-6}\ mol\ hemoglobin} = 6.76 \times 10^4\ g/mol$$

(B) If hemoglobin contains 0.33% iron by mass, how many iron atoms are there in one molecule of hemoglobin?

ANSWER:
$$100\ g\ hemoglobin = 0.33\ g\ Fe;\ \frac{0.33\ g\ Fe}{55.85\ g/mol} = 5.91 \times 10^{-3}\ mol\ Fe$$

$$100\ g\ hemoglobin \cdot \left(\frac{1\ mol}{6.76 \times 10^4\ g/mol}\right) = 1.48 \times 10^{-3}\ mol\ hemoglobin$$

$$\frac{mol\ Fe}{mol\ hemoglobin} = \frac{5.91 \times 10^{-3}\ mol\ Fe}{1.48 \times 10^{-3}\ mol\ hemoglobin} = 4$$

SECTION II—PART B ANSWERS AND EXPLANATIONS

4. Write the formulas to show the reactants and products for any FIVE of the laboratory situations described below. A reaction occurs in all cases. All solutions are aqueous unless otherwise indicated. Represent substances in solutions as ions if the substances are extensively ionized. Omit formulas for any ions or molecules that are unchanged by the reaction. You need not balance the equation.

(A) A solution of copper(II) nitrate is added to a solution of sulfuric acid.

ANSWER:
No reaction: $CuSO_4$ is soluble so copper(II) remains in solution as the blue coordination complex $Cu(H_2O)_6^{2+}$

(B) Carbon dioxide gas is bubbled through distilled water.

ANSWER:
$CO_2 + H_2O \rightarrow H_2CO_3$

(C) Solid lithium hydride is added to water.

ANSWER:
$LiH(s) + H_2O \rightarrow Li^+ + OH^- + H_2(g)$

(D) A strip of aluminum metal is dropped into a solution of 6 M hydrochloric acid.

ANSWER:
$Al(s) + H^+ \rightarrow Al^{3+} + H_2(g)$

(E) A solution of iron(III) chloride is added to a 1 M ammonia solution.

ANSWER:
$Fe^{3+} + OH^- \rightarrow Fe(OH)_3(s)$

(F) Carbon dioxide gas is bubbled through a solution of calcium hydroxide.

ANSWER:
$Ca^{2+} + CO_3^{2-} \rightarrow CaCO_3(s)$ **OR** $Ca^{2+} + HCO_3^{2-} \rightarrow Ca(HCO_3)_2(s)$

(G) Hydrogen peroxide solution is added to a solution of sulfurous acid.

ANSWER:
$H_2O_2 + SO_3^{2-} \rightarrow H_2O + SO_4^{2-}$

(H) A solution of lead(II) nitrate is added to a solution of potassium iodide.

ANSWER:
$Pb^{2+} + I^- \rightarrow PbI_2(s)$

5. A 0.200 M solution of ammonia has a pH of 11.3. This ammonia solution is used in a titration in order to determine the concentration of an unknown nitric acid solution.

(A) Describe the experimental procedure you would use to collect the necessary data for determination of the nitric acid concentration.

ANSWER:
(1) Ammonia of known concentration is placed in a flask.

(2) Unknown nitric acid solution is placed in a buret.

(3) pH electrode is placed in the flask.

(4) Initial pH reading is taken.

(5) Small volume (1–2 mL) of nitric acid added from buret.

(6) Flask swirled to mix.

(7) pH reading taken and recorded.

(8) Repeat addition of nitric acid until pH begins to drop sharply, then switch to smaller increments of nitric acid (a few tenths of a mL–several drops).

(9) When the pH no longer changes drastically with addition of nitric acid, switch back to larger increments and continue until the curve is essentially flat.

(B) Sketch and carefully label the titration curve you would obtain in this experiment and explain how it could be used to determine the concentration of the unknown nitric acid solution.

ANSWER:
• x-axis = mL nitric acid added
• y-axis = pH
• Initial pH (0 mL added) = 11.3
• Buffer region (small, negative slope) between pH 11 and about pH 7–8
• Equivalence point between pH 5-6 (students should recognize NH_4^+ as a weak acid)
• Final pH between 1 and 2 (depends on concentration of nitric acid)

Finding $[HNO_3]$:
• Determine mL base added at equivalence point
• Convert mL base to mol base using its concentration
• Convert mol base to mol acid using stoichiometry (here, 1:1)
• Convert mL acid used to L acid used (divide by 1,000)
• Convert mol acid to molarity by dividing mol acid by L acid used in titration

(C) Which of the acid-base indicators listed below would be best suited for a visual titration of nitric acid by ammonia, and why?

Indicator	Color change	pH region
methyl orange	red → yellow	3.1–3.4
methyl red	red → yellow	4.2–6.3
bromothymol blue	yellow → blue	6.2–7.6
phenolphthalein	colorless → pink	8.3–10.0

ANSWER:

Methyl red. At the equivalence point, the main acid in the solution is the ammonium ion. This is a weak acid so pH will be slightly less than 7. The range for methyl orange is too low and the range for bromothymol blue is too high.

6. The phase diagram of a substance is shown below.

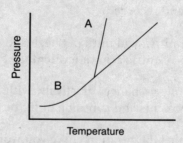

(A) Label each region of the diagram with the phase that is present.

ANSWER:

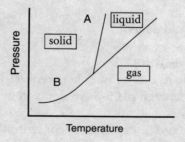

(B) Sketch the heating curve that is expected if heat is added to the sample at a constant pressure, starting at point B.

ANSWER:

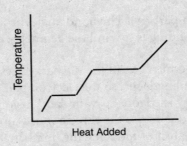

Important points:
- Substance goes from solid to liquid to gas
- Upward slope = heating the solid
- Short flat line = melting
- Upward slope = heating the liquid
- Longer flat line = vaporization
- Upward slope = heating the steam

(C) Starting at point A, describe what happens if the pressure is lowered at constant temperature.

ANSWER:
First, nothing happens, then:
- Solid turns to liquid as phase boundary is passed
- Substance stays liquid for a while
- Pressure drops further: liquid becomes a gas as second phase boundary is passed

7. Two resonance structures can be drawn for formamide, $HCONH_2$.

(A) Draw a Lewis structure for each resonance structure.

ANSWER:

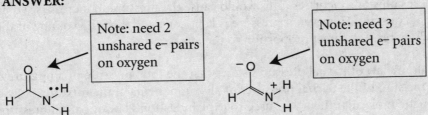

(B) Predict the bond angles about the carbon and nitrogen atoms for both resonance structures.

ANSWER:
See Lewis structures in part (A). The bond angles around carbon are 120° in both structures since the bonding is sp^2 and the geometry is trigonal planar.

The bond angle around nitrogen in the uncharged resonance structure on the left is about 108°. The bonding is sp^3, so the electron pair geometry is tetrahedral and the molecular geometry is trigonal pyramidal. The lone pair takes up more space than a bonded atom, so it causes the three bonded atoms to move slightly closer together. This has the effect of reducing the bond angle by a degree or two. The bond angle around nitrogen in the structure on the right is approximately 120°, since the bonding is sp^2 and the geometry is trigonal planar.

(C) How could the experimental measurement of the H-N-H bond angle be used to determine which resonance structure is more important?

The actual experimental measurement will be somewhere between 120° and 108°. The closer it is to 120°, the more important the charged resonance structure (the one on the right in part A). The closer it is to 108°, the more important the resonance structure on the left in part A (where the nitrogen is sp^3 hybridized).

8. The kinetic molecular theory describes the behavior of gas particles at the molecular level.

(A) Charles's law states that the volume of a gas is directly proportional to the absolute temperature when pressure and moles of gas are constant. How can the kinetic molecular theory be used to explain Charles's law?

ANSWER:
When temperature increases, the average kinetic energy and therefore the average velocity of gas particles increases. The kinetic molecular theory explains that gas pressure is due to elastic collisions between molecules and between molecules and the walls of the container. An increase in particle velocity means that the force of these collisions increases. This would cause the pressure to increase if volume were kept constant. However, if pressure is kept constant, then the volume of the gas will increase in direct proportion to the temperature.

(B) Dalton's law of partial pressures states that the total pressure of a mixture of gases is the sum of the partial pressures of all components in the mixture. How can the kinetic molecular theory be used to explain Dalton's law of partial pressures?

ANSWER:
The kinetic molecular theory treats all gases as identical, volumeless, noninteracting particles. Therefore, the identity of the gas is irrelevant. The pressure of a gas results from the collisions between particles or between particles and the walls of the container. The particles could all be one type of atom or molecule, or they could different. At a given temperature, all the gas particles have the same average kinetic energy so they will exert the same force, regardless of their chemical identity. The total force on the container is therefore the sum of the pressures exerted by all the gases in the container, which is exactly what Dalton's law states.

SCORING YOUR PRACTICE TESTS

Don't take your practice test scores too literally. They are intended to give you an approximate idea of your performance. There is no way to determine precisely how well you have scored for the following reasons.

* Practice test conditions do not precisely mirror real test conditions.
* While the multiple-choice questions are scored by computer, the free-response questions are graded manually by faculty consultants. New scoring criteria are established for every test administration.
* Various statistical factors and formulas are taken into account on the real test.
* For each AP grade, the composite score range changes from year to year (and from subject to subject).

Section I: Multiple-Choice

This section accounts for 45 percent of your final grade.

$$\underline{\hspace{2cm}} - (¼ \times \underline{\hspace{2cm}}) \times 1 = \underline{\hspace{2cm}} \text{ (maximum of 75 points)}$$

Number correct   Number wrong   Multiple-Choice Score
(out of 75)   (Round to the nearest
whole number.)

Section II: Free-Response

This section accounts for 55 percent of your final grade. Of course, it will be almost impossible for you to accurately score your own essays. In general, assign yourself one to two points for every component of the question you got right, i.e., a correct formula, and two to three points for a difficult equation or explanation.

Question 1 $\underline{\hspace{2cm}} \times 2 = \underline{\hspace{2cm}}$ Question 5 $\underline{\hspace{2cm}} \times 1.7 = \underline{\hspace{2cm}}$
(max of 9 points)   (max of 8 points)

Question 2 or 3 $\underline{\hspace{2cm}} \times 2 = \underline{\hspace{2cm}}$ Question 6 $\underline{\hspace{2cm}} \times 1.7 = \underline{\hspace{2cm}}$
(max of 9 points)   (max of 8 points)

Question 4 $\underline{\hspace{2cm}} \times 1.9 = \underline{\hspace{2cm}}$ Question 7 or 8 $\underline{\hspace{2cm}} \times 1.7 = \underline{\hspace{2cm}}$
(max of 15 points)   (max of 8 points)

Free-Response Score = $\underline{\hspace{2cm}}$

Composite Score

Multiple-Choice Score $\underline{\hspace{2cm}}$ + Free-Response Score $\underline{\hspace{2cm}}$ = $\underline{\hspace{2cm}}$
(Composite Score)

Conversion Chart (approximate)	
Composite Score Range	**AP Grade**
100–160	5
82–99	4
57–81	3
35–56	2
0–34	1

KAPLAN

APPENDIXES

Answering the Reaction Question

As we indicated in chapter 1, Question 4 consists of eight word-descriptions of reactions. You are must answer any five by writing the *net reaction* for the description. Reactions need not be balanced and phases need not be included. You must not include any molecules or ions that are *unchanged* by the reaction.

Each of the five reactions is worth three points, awarded as follows:

- One point is awarded for listing the reactants correctly. If a *spectator* ion is included on the *reactant* side, that point is lost.

- Two points are awarded for identifying the products correctly.

- If an equation is written in its molecular form when it should be ionic, only 1 point will be awarded for the reaction.

- All ion charges must be correct.

One of the first things you need to do is familiarize yourself thoroughly with the list of common ions given in Appendix B. Then you need to memorize the solubility rules given in chapter 4.

Word-reactions can be presented in an almost endless variety of ways, and it is an impossible task to try and memorize each and every specific example. The secret to success on this question is to be able to place a reaction in its proper category. Sometimes, these categories overlap: For example, a reaction could be considered both a synthesis and a combustion reaction. A number of categories follow, along with examples and solutions.

Redox Reactions

Synthesis

Magnesium powder is burned in an excess of pure nitrogen gas.

$$Mg + N_2 \rightarrow Mg_3N_2$$

Phosphorus is burned in an excess of pure oxygen.

$$P + O_2 \rightarrow P_4O_{10}$$

Decomposition

An electric current is passed through molten sodium bromide.

$$Na^+ + Br^- \rightarrow Na + Br_2$$

Potassium chlorate is heated in the presence of manganese dioxide, a catalyst.

$$KClO_3 \rightarrow KCl + O_2$$

Single Replacement

Solid copper is added to a solution of silver nitrate.

$$Cu + Ag^+ \rightarrow Cu^{2+} + Ag$$

Bromine is added to a solution of sodium iodide.

$$Br_2 + I^- \rightarrow Br^- + I_2$$

Sodium metal is added to water.

$$Na + H_2O \rightarrow Na^+ + OH^- + H_2$$

Combustion

Carbon monoxide is burned in an excess of pure oxygen.

$$CO + O_2 \rightarrow CO_2$$

Redox with Commonly Used Oxidizing and Reducing Agents

At least one of the reactions given in Question 4 of the AP test will fall in this category. The following table lists a number of commonly used agents and the products they form:

Oxidizing Agents		Reducing Agents	
Agent	**Product**	**Agent**	**Product**
MnO_4^- (acidic soln.)	Mn^{2+}	Free Halogens (e.g., Cl_2 in dil. base)	ClO^-
MnO_4^- (basic soln.)	MnO_2	Free Halogens (e.g., Br_2 in con. base)	BrO_2^-
MnO_2 (acidic soln.)	Mn^{2+}	NO_2^-	NO_3^-
HNO_3 (con.)	NO_2	SO_3^{2-} or SO_2	SO_4^{2-}
HNO_3 (dil.)	NO		
Metallic ions (e.g., Sn^{4+})	Metallic ions (e.g., Sn^{2+})		
Free Halogens (e.g., Cl_2)	Halide ions (e.g., Cl^-)		
Peroxides (e.g., Na_2O_2)	Hydroxides (e.g., NaOH)		
	$HClO_4$		
$Cr2O_2^{2-}$ (acidic soln.)			

Hydrogen peroxide is added to an acidified aqueous solution of potassium iodide.

$$H_2O_2 + H^+ + I^- \rightarrow H_2O + I_2$$

An aqueous solution of tin (II) chloride is added to an acidified solution of potassium dichromate.

$$Sn^{2+} + H^+ + Cr_2O_7^{2-} \rightarrow Sn^{4+} + Cr^{3+} + H_2O$$

Powdered iron is added to a solution of iron (III) nitrate.

$$Fe + Fe^{3+} \rightarrow Fe^{2+}$$

Nonredox Reactions

Decomposition

Calcium carbonate is heated strongly.

$$CaCO_3 \rightarrow CaO + CO_2$$

Precipitation

Aqueous solutions of barium nitrate and sodium sulfate are mixed.

$$Ba^{2+} + SO_4{}^{2-} \rightarrow BaSO_4$$

Aqueous solutions of silver acetate and calcium bromide are mixed.

$$Ag^+ + Br^- \rightarrow AgBr$$

An aqueous solution of barium hydroxide is added to a dilute solution of sulfuric acid.

$$Ba^{2+} + OH^- + H^+ + SO_4{}^{2-} \rightarrow BaSO_4 + H_2O$$

Acid-Base Reactions

Aqueous solutions of potassium hydroxide and dilute nitric acid are mixed.

$$H^+ + OH^- \rightarrow H_2O \text{ (neutralization)}$$

Equimolar amounts of sodium dihydrogen phosphate and sodium hydroxide are mixed.

$$H_2PO_4{}^- + OH^- \rightarrow HPO_4{}^{2-} + H_2O \text{ (conjugate acid-base reaction)}$$

Solid sodium oxide is added to water.

$$Na_2O + H_2O \rightarrow NaOH \text{ (base anhydride)}$$

Gaseous carbon dioxide is bubbled into water.

$$CO_2 + H_2O \rightarrow H_2CO_3 \text{ (acid anhydride producing a weak acid)}$$

Solid sodium acetate is added to water.

$$C_2H_3O_2{}^- + H_2O \rightarrow OH^- + HC_2H_3O_2 \text{ (hydrolysis)}$$

Ammonia gas is reacted with boron trifluoride.

$$NH_3 + BF_3 \rightarrow NH_3BF_3 \text{ (Lewis acid-base reaction)}$$

Formation of Complex Ions

Aqueous ammonia is added to solid silver chloride.

$$NH_3 + AgCl \rightarrow [Ag(NH_3)_2]^+ + Cl^-$$

A solution of iron (II) nitrate is mixed with a solution of sodium cyanide.

$$Fe^{2+} + CN^- \rightarrow [Fe(CN)_6]^{4-}$$

Organic Reactions

Substitution

Gaseous methane is reacted with liquid bromine.

$$CH_4 + Br_2 \rightarrow CH_3Br + HBr$$

Chlorine gas is reacted with benzene.

$$Cl_2 + C_6H_6 \rightarrow C_6H_5Cl + HCl$$

Addition

Ethene gas is bubbled into aqueous bromine.

$$C_2H_4 + Br_2 \rightarrow C_2H_4Br_2$$

Esterification

Methanol and acetic acid are reacted in the presence of sulfuric acid.

$$CH_3OH + HC_2H_3O_2 \rightarrow H_2O + CH_3O_2CCH_3$$

Combustion

Ethanol is burned in an excess of pure oxygen.

$$C_2H_5OH + O_2 \rightarrow CO_2 + H_2O$$

PRACTICE, PRACTICE, PRACTICE

What do you do now? Work on as many reactions as you can, preferably from actual AP exams. Remember that the first step is to ask: What kind of reaction is this? In addition, there is a book published by Flinn Scientific entitled *The Ultimate Chemical Equations Handbook* (Student Edition) [www.flinnsci.com] if you want to saturate yourself with equations.

Practice Questions

Here are eight more reactions for you to try. But don't panic! During the AP examination, you only need to answer five of the eight items.

(A) A strip of copper is immersed in dilute nitric acid

(B) Potassium permanganate solution is added to an acidic solution of hydrogen peroxide

(C) Concentrated hydrochloric acid is added to solid manganese (II) sulfide

(D) Excess chlorine gas is passed over hot iron filings

(E) Water is added to a sample of solid magnesium nitride

(F) Excess sulfur dioxide gas is bubbled through a dilute solution of potassium hydroxide

(G) Excess potassium cyanide is added to a suspension of silver bromide

(H) Solutions of potassium phosphate and zinc nitrate are mixed

And the answers are:

(A) $Cu + H^+ + NO_3^- \rightarrow Cu^{2+} + NO + H_2O$ (redox)

(B) $MnO_4^- + H_2O_2 \rightarrow Mn^{2+} + O_2 + H_2O$ (redox)

(C) $H^+ + MnS \rightarrow H_2S + Mn^{2+}$ (acid-base; formation of a gas)

(D) $Fe + Cl_2 \rightarrow FeCl_3$ (redox synthesis)

(E) $Mg_3N_2 + H_2O \rightarrow Mg(OH)_2 + NH_3$ (acid-base)

(F) $SO_2 + OH^- \rightarrow HSO_3^-$ (acid anhydride plus base)

(G) $AgCl + KCN \rightarrow [Ag(CN)_2]^- + Cl^-$ (formation of complex ion)

(H) $Zn^{2+} + PO_4^{3-} \rightarrow Zn_3(PO_4)_2$ (precipitation)

Reference Tables

SELECTED PHYSICAL CONSTANTS

Name	Symbol	Value
Avogadro constant	N_A	6.022×10^{23} mol^{-1}
Boltzmann constant	k	1.381×10^{-23} J K^{-1}
Elementary charge	e	1.602×10^{-19} C
Faraday constant	$F (N_A e)$	9.649×10^5 C mol^{-1}
Speed of light	c	2.998×10^8 m s^{-1}
Planck constant	h	6.626×10^{-34} J s
Molar gas constant	R	8.315 J mol^{-1} K^{-1} 0.08206 dm^3 atm mol^{-1} K^{-1}
Nernst constant at 25°C	RT/F	0.02569 V (J C^{-1})
Atomic mass unit	u	1.661×10^{-27} kg
Proton mass	m_p	1.673×10^{-27} kg 1.007 u
Neutron mass	m_n	1.675×10^{-27} kg 1.009 u
Electron mass	m_e	9.109×10^{-31} kg 0.0005486 u
Standard state pressure	P	100.0 kPa (exact)
Standard atmosphere (atm)	P	1 atm = 101.3 kPa = 760.0 mmHg = 760.0 torr
Triple point of water	T_{tr}	273.16 K (exact)
Standard temperature	T	273.15 K (0°C)
Thermodynamic reference temperature	T	298.15 K (25°C)
Standard temperature and pressure	STP	273.15 K and 1 atm
Molar volume of ideal gas at STP	V_m	22.41 dm^3 mol^{-1}

ALPHABETICAL LIST OF THE ELEMENTS

Element	Symbol	Atomic Number	Atomic Mass*	Element	Symbol	Atomic Number	Atomic Mass*
Actinium	Ac	89	(227)	Mercury	Hg	80	200.6
Aluminum	Al	13	26.98	Molybdenum	Mo	42	95.94
Americium	Am	95	(243)	Neodymium	Nd	60	144.2
Antimony	Sb	51	121.8	Neon	Ne	10	20.18
Argon	Ar	18	39.95	Neptunium	Np	93	(237)
Arsenic	As	33	74.92	Nickel	Ni	28	58.69
Astatine	At	85	(210)	Nielsbohrium	Ns	107	(262)
Barium	Ba	56	137.3	Niobium	Nb	41	92.91
Berkelium	Bk	97	(247)	Nitrogen	N	7	14.01
Beryllium	Be	4	9.012	Nobelium	No	102	(253)
Bismuth	Bi	83	209.0	Osmium	Os	76	190.2
Boron	B	5	10.81	Oxygen	O	8	16.00
Bromine	Br	35	79.90	Palladium	Pd	46	106.4
Cadmium	Cd	48	11 2.4	Phosphorus	P	15	30.97
Calcium	Ca	20	40.08	Platinum	Pt	78	195.1
Californium	Cf	98	(249)	Plutonium	Pu	94	(242)
Carbon	C	6	12.01	Polonium	Po	84	(2 10)
Cerium	Ce	58	140.1	Potassium	K	19	39.10
Cesium	Cs	55	132.9	Praseodymium	Pr	59	140.9
Chlorine	C1	1 7	35.45	Promethium	Pm	61	(147)
Chromium	Cr	24	52.00	Protactinium	Pa	91	(231)
Cobalt	Co	27	58.93	Radium	Ra	88	(226)
Copper	Cu	29	63.55	Radon	Rb	37	85.47
Erbium	Er	68	167.3	Ruthenium	Ru	44	101.1
Europium	Eu	63	152.0	Rutherfordium	Rf	104	(257)
Fermium	Fm	100	(253)	Samarium	Sm	62	150.4
Fluorine	F	9	19.00	Scandium	Sc	21	44.96
Francium	Fr	87	(223)	Seaborgium	Sg	106	(263)
Gadolinium	Gd	64	157.3	Selenium	Se	34	78.96
Gallium	Ga	31	69.72	Silicon	Si	14	28.09
Germanium	Ge	32	72.59	Silver	Ag	47	107.9
Gold	Au	79	197.0	Sodium	Na	11	22.99
Hafnium	Hf	72	178.5	Strontium	Sr	38	87.62
Hahnium,	Ha	105	(260)	Sulfur	S	16	32.07
Hassium	Hs	108	(265)	Tantalum	Ta	73	180.9
Helium	He	2	4.003	Technetium	Tc	43	(99)
Holmium	Ho	67	164.9	Tellurium	Te	52	127.6
Hydroge n	H	1	1.008	Terbium	Tb	65	158.9
Indium	In	49	114.8	Thallium	Tl	81	204.4
Iodine	I	53	126.9	Thorium	Th	90	232.0
Iridium	Ir	77	192.2	Thulium	Tm	69	168.9
Iron	Fe	26	55.85	Tin	Sn	50	118.7
Krypton	Kr	36	83.80	Titanium	Ti	22	47.88
Lanthanum	La	57	138.9	Tungsten	W	74	183.9
Lawrencium	Lr	103	(257)	Uranium	U	92	238.0
Lead	Pb	82	207.2	Vanadium	V	23	50.94
Lithium	Li	3	6.941	Xenon	Xe	54	131.3
Lutetium	Lu	71	175.0	Ytterbium	Yb	70	l73.0
Magnesium	Mg	12	24.31	Yttrium	Y	39	88.91
Manganese	Mn	25	54.94	Zinc	Zn	30	65.39
Meitnerium	Mt	109	(266)	Zirconium	Zr	40	91.22
Mendelevium	Md	101	(256)				

* Atomic masses are measured relative to carbon-12, which has an assigned mass of 12.000. . . u. Numbers in parentheses are the mass numbers of the longest-lived isotope of the element.

VAPOR PRESSURE OF WATER

t_c /°C	VP /mmHg	VP /kPa	t_c /°C	VP /mmHg	VP /kPa
0	4.585	0.6113	26	25.22	3.363
1	4.929	0.6572	27	26.75	3.567
2	5.296	0.7061	28	28.37	3.782
3	5.686	0.7581	29	30.06	4.008
4	6.102	0.8136	30	31.84	4.246
5	6.545	0.8726	40	55.37	7.381
6	7.016	0.9354	50	92.59	12.34
7	7.516	1.002	60	149.5	19.93
8	8.048	1.073	70	233.8	31.18
9	8.612	1.148	80	355.3	47.37
10	9.212	1.228	90	525.9	70.12
11	9.848	1.313	100	760.0	101.3
12	10.52	1.403	105	906.0	120.8
13	11.24	1.498	110	1,074	143.2
14	11.99	1.599	115	1,268	169.0
15	12.79	1.706	120	1,489	198.5
16	13.64	1.819	130	2,025	270.0
17	14.54	1.938	140	2,709	361.2
18	15.48	2.064	150	3,568	475.7
19	16.48	2.198	160	4,633	617.7
20	17.54	2.339	170	5,937	791.5
21	18.66	2.488	180	7,515	1,002
22	19.84	2.645	190	9,407	1,254
23	21.08	2.810	200	11,650	1,554
24	22.39	2.985	210	14,300	1,906
25	23.77	3.169	220	17,390	2,318

IONIZATION CONSTANTS OF WEAK ACIDS AND BASES

(298.15 K and 100 kPa)

$$K_w = [H_3O^+] \cdot [OH^-] = 1.0 \times 10^{-14}$$

Monoprotic Acids (Acids with one ionizable hydrogen atom)

Formula	Name	K_a
HIO_3	Iodic acid	1.6×10^{-1}
HNO_2	Nitrous acid	7.2×10^{-4}
HF	Hydrofluoric acid	6.6×10^{-4}
$HCHO_2$	Formic acid	1.8×10^{-4}
$HC_3H_5O_3$	Lactic acid	1.4×10^{-4}
$HC_7H_5O_2$	Benzoic acid	6.3×10^{-5}
$HC_4H_7O_2$	Butanoic acid	1.5×10^{-5}
HN_3	Hydrazoic acid	1.9×10^{-5}
$HC_2H_3O_2$	Acetic acid	1.8×10^{-5}
$HC_3H_5O_2$	Propanoic acid	1.3×10^{-5}
$HOCl$	Hypochlorous acid	2.9×10^{-8}
HCN	Hydrocyanic acid	6.2×10^{-10}
HC_6H_5O	Phenol	1.0×10^{-10}
H_2O_2	Hydrogen peroxide	2.2×10^{-12}

Polyprotic Acids (Acids with more than one ionizable hydrogen atom)

Formula	Name	K_{a1}	K_{a2}	K_{a3}
H_2SO_4	Sulfuric acid	Large	1.1×10^{-2}	
H_2CrO_4	Chromic acid	5.0	1.5×10^{-6}	
$H_2C_2O_4$	Oxalic acid	5.4×10^{-2}	5.3×10^{-5}	
H_3PO_3	Phosphorous acid	3.7×10^{-2}	2.1×10^{-7}	
H_2SO_3	Sulfurous acid	1.3×10^{-2}	6.2×10^{-8}	
H_2SeO_3	Selenous acid	2.3×10^{-3}	5.4×10^{-9}	
$H_2C_3H_2O_4$	Malonic acid	1.5×10^{-3}	2.0×10^{-6}	
$H_2C_8H_4O_4$	Phthalic acid	1.1×10^{-3}	3.9×10^{-6}	
$H_2C_4H_4O_6$	Tartaric acid	9.2×10^{-4}	4.3×10^{-5}	
H_2CO_3	Carbonic acid	4.4×10^{-7}	4.7×10^{-11}	
H_3PO_4	Phosphoric acid	7.1×10^{-3}	6.3×10^{-8}	4.2×10^{-13}
H_3AsO_4	Arsenic acid	6.0×10^{-3}	1.0×10^{-7}	3.2×10^{-12}
$H_3C_6H_5O_7$	Citric acid	7.4×10^{-4}	1.7×10^{-5}	4.0×10^{-7}

Bases

Formula	Name	K_b
$(CH_3)_2NH$	Dimethylamine	6.9×10^{-4}
CH_3NH_2	Methylamine	4.2×10^{-4}
$CH_3CH_2NH_2$	Ethylamine	4.3×10^{-4}
$(CH_3)_3N$	Trimethylamine	6.3×10^{-5}
NH_3	Ammonia	1.8×10^{-5}
N_2H_4	Hydrazine	8.5×10^{-7}
C_5H_5N	Pyridine	1.5×10^{-9}
$C_6H_5NH_2$	Aniline	7.4×10^{-10}

SOLUBILITY PRODUCT CONSTANTS AT 298.15 K

Formula	K_{sp}	Formula	K_{sp}
Bromides		**Iodides**	
AgBr	5.35×10^{-13}	AgI	8.52×10^{-17}
CuBr	6.27×10^{-9}	CuI	1.27×10^{-12}
$PbBr_2$	6.60×10^{-6}	PbI_2	9.80×10^{-28}
Carbonates		**Phosphates**	
Ag_2CO_3	8.46×10^{-12}	Ag_3PO_4	8.89×10^{-17}
$BaCO_3$	2.58×10^{-9}	$Ca_3(PO_4)_2$	2.07×10^{-33}
$CaCO_3$	3.36×10^{-9}	$Cd_3(PO_4)_2$	2.53×10^{-33}
$MgCO_3$	6.82×10^{-6}	$Cu_3(PO_4)_2$	1.40×10^{-37}
$SrCO_3$	5.60×10^{-10}	$Mg_3(PO_4)_2$	1.04×10^{-24}
$ZnCO_3$	8×10^{-28}	$Ni_3(PO_4)_2$	4.74×10^{-32}
Chlorides		**Sulfates**	
AgCl	1.77×10^{-10}	Ag_2SO_4	1.20×10^{-5}
CuCl	1.72×10^{-7}	$BaSO_4$	1.08×10^{-10}
$PbCl_2$	1.70×10^{-5}	$CaSO_4$	4.93×10^{-5}
Chromates		$PbSO_4$	2.53×10^{-8}
Ag_2CrO_4	1.12×10^{-12}	**Sulfides**	
Fluorides		CdS	1.40×10^{-29}
BaF_2	1.84×10^{-6}	CuS	1.27×10^{-36}
CaF_2	3.45×10^{-11}	HgS	1.55×10^{-52}
MgF_2	5.16×10^{-11}	MnS	4.65×10^{-14}
PbF_2	3.30×10^{-8}	PbS	9.05×10^{-29}
Hydroxides		ZnS	2.93×10^{-25}
$Ca(OH)_2$	5.02×10^{-6}	**Sulfites**	
$Cd(OH)_2$	7.20×10^{-15}	Ag_2SO_3	1.50×10^{-14}
$Fe(OH)_2$	4.87×10^{-17}	$BaSO_3$	5.00×10^{-10}
$Fe(OH)_3$	2.79×10^{-39}		
$Mg(OH)_2$	5.61×10^{-12}		
$Ni(OH)_2$	5.48×10^{-16}		
$Zn(OH)_2$	3.00×10^{-17}		

STANDARD ELECTRODE (REDUCTION) POTENTIALS AT 298.15K

Half-reaction	$E°/V$	Half-reaction	$E°/V$
$F_2 + 2e^- \rightarrow 2F^-$	2.87	$O_2 + 2H_2O + 4e^- \rightarrow 4OH^-$	0.40
$Ag^{2+} + e^- \rightarrow Ag^+$	1.99	$Cu^{2+} + 2e^- \rightarrow Cu$	0.34
$Co^{3+} + e^- \rightarrow Co^{2+}$	1.82	$Hg_2Cl_2 + 2e^- \rightarrow 2Hg + 2Cl^-$	0.34
$H_2O_2 + 2H^+ + 2e^- \rightarrow 2H_2O$	1.78	$AgCl + e^- \rightarrow Ag + Cl^-$	0.22
$Ce^{4+} + e^- \rightarrow Ce^{3+}$	1.70	$SO_4^{2-} + 4H^+ + 2e^- \rightarrow H_2SO_3 + H_2O$	0.20
$PbO_2 + 4H^+ + SO_4^{2-} + 2e^- \rightarrow PbSO_4 + 2H_2O$	1.69	$Cu^{2+} + e^- \rightarrow Cu^+$	0.16
$MnO_4^- + 4H^+ + 3e^- \rightarrow MnO_2 + 2H_2O$	1.68	$2H^+ + 2e^- \rightarrow H_2$	**0.00**
$2e^- + 2H^+ + IO_4^- \rightarrow IO_3^- + H_2O$	1.60	$Fe^{3+} + 3e^- \rightarrow Fe$	−0.036
$MnO_4^- + 8H^+ + 5e^- \rightarrow Mn^{2+} + 4H_2O$	1.51	$Pb^{2+} + 2e^- \rightarrow Pb$	−0.13
$Au^{3+} + 3e^- \rightarrow Au$	1.50	$Sn^{2+} + 2e^- \rightarrow Sn$	−0.14
$PbO_2 + 4H^+ + 2e^- \rightarrow Pb^{2+} + 2H_2O$	1.46	$Ni^{2+} + 2e^- \rightarrow Ni$	−0.23
$Cl_2 + 2e^- \rightarrow 2Cl^-$	1.36	$PbSO_4 + 2e^- \rightarrow Pb + SO_4^{2-}$	−0.35
$Cr_2O_7^{2-} + 14H^+ + 6e^- \rightarrow 2Cr^{3+} + 7H_2O$	1.33	$Cd^{2+} + 2e^- \rightarrow Cd$	−0.40
$O_2 + 4H^+ + 4e^- \rightarrow 2H_2O$	1.23	$Fe^{2+} + 2e^- \rightarrow Fe$	−0.44
$MnO_2 + 4H^+ + 2e^- \rightarrow Mn^{2+} + 2H_2O$	1.21	$Cr^{3+} + e^- \rightarrow Cr^{2+}$	−0.50
$IO_3^- + 6H^+ + 5e^- \rightarrow \frac{1}{2}I_2 + 3H_2O$	1.20	$Cr^{3+} + 3e^- \rightarrow Cr$	−0.73
$Br_2 + 2e^- \rightarrow 2Br^-$	1.09	$Zn^{2+} + 2e^- \rightarrow Zn$	−0.76
$VO_2^+ + 2H^+ + e^- \rightarrow VO^{2+} + H_2O$	1.00	$2H_2O + 2e^- \rightarrow H_2 + 2OH^-$	−0.83
$AuCl_4^- + 3e^- \rightarrow Au + 4Cl^-$	0.99	$Mn^{2+} + 2e^- \rightarrow Mn$	−1.18
$NO_3^- + 4H^+ + 3e^- \rightarrow NO + 2H_2O$	0.96	$Al^{3+} + 3e^- \rightarrow Al$	−1.66
$ClO_2 + e^- \rightarrow ClO_2^-$	0.95	$H_2 + 2e^- \rightarrow 2H^-$	−2.23
$2Hg^{2+} + 2e^- \rightarrow Hg_2^{2+}$	0.91	$Mg^{2+} + 2e^- \rightarrow Mg$	−2.37
$Ag^+ + e^- \rightarrow Ag$	0.80	$La^{3+} + 3e^- \rightarrow La$	−2.37
$Hg_2^{2+} + 2e^- \rightarrow 2Hg$	0.80	$Na^+ + e^- \rightarrow Na$	−2.71
$Fe^{3+} + e^- \rightarrow Fe^{2+}$	0.77	$Ca^{2+} + 2e^- \rightarrow Ca$	−2.76
$O_2 + 2H^+ + 2e^- \rightarrow H_2O_2$	0.68	$Ba^{2+} + 2e^- \rightarrow Ba$	−2.90
$MnO_4^- + e^- \rightarrow MnO_4^{2-}$	0.56	$K^+ + e^- \rightarrow K$	−2.92
$I_2 + 2e^- \rightarrow 2I^-$	0.54	$Li^+ + e^- \rightarrow Li$	−3.05
$Cu^+ + e^- \rightarrow Cu$	0.52		

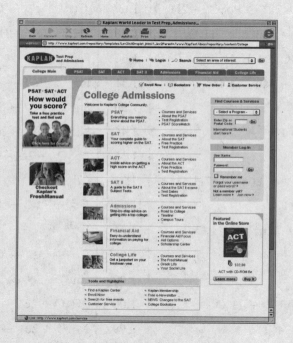